STUDENT'S SOLUTIONS MANUAL

GEX PUBLISHING SERVICES

STATISTICS: INFORMED DECISIONS USING DATA

FIFTH EDITION

Michael Sullivan, III

Joliet Junior College

PEARSON

Boston Columbus Indianapolis New York San Francisco

Amsterdam Cape Town Dubai London Madrid Milan Munich Paris Montreal Toronto

Delhi Mexico City São Paulo Sydney Hong Kong Seoul Singapore Taipei Tokyo

Copyright © 2017, 2013, 2009 by Pearson Education, Inc.
Publishing as Pearson, 501 Boylston Street, Boston, MA 02116.

ISBN-13: 978-0-13-413540-3
ISBN-10: 0-13-413540-7

4 17

www.pearsonhighered.com

Table of Contents

Preface

This solutions manual accompanies *Statistics: Informed Decisions Using Data, 5e* by Michael Sullivan, III. The Instructor's Solutions Manual contains detailed solutions to all exercises in the text as well as the Consumer Reports® projects and the case studies. The Student's Solutions Manual contains detailed solutions to all odd exercises in the text and all solutions to chapter reviews and tests. A concerted effort has been made to make this manual as user-friendly and error free as possible.

Chapter 1
Data Collection

Section 1.1

1. Statistics is the science of collecting, organizing, summarizing, and analyzing information in order to draw conclusions and answer questions. In addition, statistics is about providing a measure of confidence in any conclusions.

3. Individual

5. Statistic; Parameter

7. 18% is a parameter because it describes a population (all of the governors).

9. 32% is a statistic because it describes a sample (the high school students surveyed).

11. 0.366 is a parameter because it describes a population (all of Ty Cobb's at-bats).

13. 23% is a statistic because it describes a sample (the 6076 adults studied).

15. Qualitative

17. Quantitative

19. Quantitative

21. Qualitative

23. Discrete

25. Continuous

27. Continuous

29. Discrete

31. Nominal

33. Ratio

35. Ordinal

37. Ratio

39. The population consists of all teenagers 13 to 17 years old who live in the United States. The sample consists of the 1028 teenagers 13 to 17 years old who were contacted by the Gallup Organization.

41. The population consists of all of the soybean plants in this farmer's crop. The sample consists of the 100 soybean plants that were selected by the farmer.

43. The population consists of all women 27 to 44 years of age with hypertension. The sample consists of the 7373 women 27 to 44 years of age with hypertension who were included in the study.

45. Individuals: Alabama, Colorado, Indiana, North Carolina, Wisconsin.
Variables: Minimum age for driver's license (unrestricted); mandatory belt use seating positions, maximum allowable speed limit (rural interstate) in 2011.
Data for minimum age for driver's license: 17, 17, 18, 16, 18;
Data for mandatory belt use seating positions: front, front, all, all, all;
Data for maximum allowable speed limit (rural interstate) 2011: 70, 75, 70, 70, 65 (mph.)
The variable *minimum age for driver's license* is continuous; the variable *mandatory belt use seating positions* is qualitative; the variable *maximum allowable speed limit (rural interstate) 2011* is continuous (although only discrete values are typically chosen for speed limits.)

47. (a) The research objective is to determine if adolescents who smoke have a lower IQ than nonsmokers.

    (b) The population is all adolescents aged 18–21. The sample consisted of 20,211 18-year-old Israeli military recruits.

    (c) Descriptive statistics: The average IQ of the smokers was 94, and the average IQ of nonsmokers was 101.

    (d) The conclusion is that individuals with a lower IQ are more likely to choose to smoke.

49. (a) The research objective is to determine the proportion of adult Americans who believe the federal government wastes 51 cents or more of every dollar.

(b) The population is all adult Americans aged 18 years or older.

(c) The sample is the 1017 American adults aged 18 years or older that were surveyed.

(d) Descriptive statistics: Of the 1017 individuals surveyed, 35% indicated that 51 cents or more is wasted.

(e) From this study, one can infer that many Americans believe the federal government wastes much of the money collected in taxes.

51. *Jersey number* is nominal (the numbers generally indicate a type of position played). However, if the researcher feels that lower caliber players received higher numbers, then *jersey number* would be ordinal since players could be ranked by their number.

53. (a) The research question is to determine if the season of birth affects mood later in life.

(b) The sample consisted of the 400 people the researchers studied.

(c) The season in which you were born (winter, spring, summer, or fall) is a qualitative variable.

(d) According to the article, individuals born in the summer are characterized by rapid, frequent swings between sad and cheerful moods, while those born in the winter are less likely to be irritable.

(e) The conclusion was that the season at birth plays a role in one's temperament.

55. The values of a discrete random variable result from counting. The values of a continuous random variable result from a measurement.

57. We say data vary, because when we draw a random sample from a population, we do not know which individuals will be included. If we were to take another random sample, we would have different individuals and therefore different data. This variability affects the results of a statistical analysis because the results would differ if a study is repeated.

59. Age could be considered a discrete random variable. A random variable can be discrete by allowing, for example, only whole numbers to be recorded.

Section 1.2

1. The response variable is the variable of interest in a research study. An explanatory variable is a variable that affects (or explains) the value of the response variable. In research, we want to see how changes in the value of the explanatory variable affect the value of the response variable.

3. Confounding exists in a study when the effects of two or more explanatory variables are not separated. So any relation that appears to exist between a certain explanatory variable and the response variable may be due to some other variable or variables not accounted for in the study. A lurking variable is a variable not accounted for in a study, but one that affects the value of the response variable. A confounding variable is an explanatory variable that was considered in a study whose effect cannot be distinguished from a second explanatory variable in the study.

5. Cross-sectional studies collect information at a specific point in time (or over a very short period of time). Case-control studies are retrospective (they look back in time). Also, individuals that have a certain characteristic (such as cancer) in a case-control study are matched with those that do not have the characteristic. Case-control studies are typically superior to cross-sectional studies. They are relatively inexpensive, provide individual level data, and give longitudinal information not available in a cross-sectional study.

7. There is a perceived benefit to obtaining a flu shot, so there are ethical issues in intentionally denying certain seniors access to the treatment.

9. This is an observational study because the researchers merely observed existing data. There was no attempt by the researchers to manipulate or influence the variable(s) of interest.

11. This is an experiment because the explanatory variable (teaching method) was intentionally varied to see how it affected the response variable (score on proficiency test).

13. This is an observational study because the survey only observed preference of Coke or Pepsi. No attempt was made to manipulate or influence the variable of interest.

15. This is an experiment because the explanatory variable (carpal tunnel treatment regimen) was intentionally manipulated in order to observe potential effects on the response variable (level of pain).

17. **(a)** This is a cohort study because the researchers observed a group of people over a period of time.

    (b) The response variable is whether the individual has heart disease or not. The explanatory variable is whether the individual is happy or not.

    (c) There may be confounding due to lurking variables. For example, happy people may be more likely to exercise, which could affect whether they will have heart disease or not.

19. **(a)** This is an observational study because the researchers simply administered a questionnaire to obtain their data. No attempt was made to manipulate or influence the variable(s) of interest. This is a cross-sectional study because the researchers are observing participants at a single point in time.

    (b) The response variable is body mass index. The explanatory variable is whether a TV is in the bedroom or not.

    (c) Answers will vary. Some lurking variables might be the amount of exercise per week and eating habits. Both of these variables can affect the body mass index of an individual.

    (d) The researchers attempted to avoid confounding due to other variables by taking into account such variables as "socioeconomic status."

    (e) No. Since this was an observational study, we can only say that a television in the bedroom is associated with a higher body mass index.

21. **(a)** This is a cross-sectional study because information was collected at a specific point in time (or over a very short period of time).

    (b) The explanatory variable is delivery scenario (caseload midwifery, standard hospital care, or private obstetric care).

    (c) The two response variables are (1) cost of delivery, which is quantitative, and (2) type of delivery (vaginal or not), which is quantitative.

23. Answers will vary. This is a prospective, cohort observational study. The response variable is whether the worker had cancer or not, and the explanatory variable is the amount of electromagnetic field exposure. Some possible lurking variables include eating habits, exercise habits, and other health-related variables such as smoking habits. Genetics (family history) could also be a lurking variable. This was an observational study, and not an experiment, so the study only concludes that high electromagnetic field exposure is associated with higher cancer rates. The author reminds us that this is an observational study, so there is no direct control over the variables that may affect cancer rates. He also points out that while we should not simply dismiss such reports, we should consider the results in conjunction with results from future studies. The author concludes by mentioning known ways (based on extensive study) of reducing cancer risks that can currently be done in our lives.

Section 1.3

1. The frame is a list of all the individuals in the population.

3. Sampling without replacement means that no individual may be selected more than once as a member of the sample.

5. Answers will vary. We will use one-digit labels and assign the labels across each row (i.e. *Pride and Prejudice* – 0, *The Sun Also Rises* – 1, and so on). In Table I of Appendix A, starting at row 5, column 11, and proceeding downward, we obtain the following labels: 8, 4, 3
In this case, the 3 books in the sample would be *As I Lay Dying*, *A Tale of Two Cities*, and *Crime and Punishment*. Different labeling order, different starting points in Table I in Appendix A, or use of technology will likely yield different samples.

7. (a) {616, 630}, {616, 631}, {616, 632},
{616, 645}, {616, 649}, {616, 650},
{630, 631}, {630, 632}, {630, 645},
{630, 649}, {630, 650}, {631, 632},
{631, 645}, {631, 649}, {631, 650},
{632, 645}, {632, 649}, {632, 650},
{645, 649}, {645, 650}, {649, 650}

(b) There is a 1 in 21 chance that the pair of
courses will be EPR 630 and EPR 645.

9. (a) Starting at row 5, column 22, using two-
digit numbers, and proceeding
downward, we obtain the following
values: 83, 94, 67, 84, 38, 22, 96, 24, 36,
36, 58, 34,.... We must disregard 94 and
96 because there are only 87 faculty
members in the population. We must
also disregard the second 36 because we
are sampling without replacement. Thus,
the 9 faculty members included in the
sample are those numbered 83, 67, 84,
38, 22, 24, 36, 58, and 34.

(b) Answers will vary depending on the type
of technology used. If using a TI-84
Plus, the sample will be: 4, 20, 52, 5, 24,
87, 67, 86, and 39.

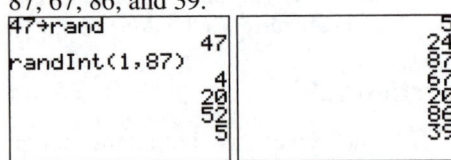

Note: We must disregard the second 20
because we are sampling without
replacement.

11. (a) Answers will vary depending on the
technology used (including a table of
random digits). Using a TI-84 Plus
graphing calculator with a seed of 17 and
the labels provided, our sample would be
North Dakota, Nevada, Tennessee,
Wisconsin, Minnesota, Maine, New
Hampshire, Florida, Missouri, and
Mississippi.

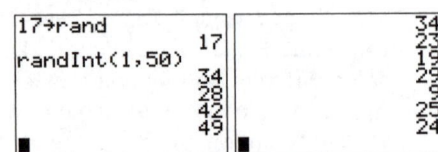

(b) Repeating part (a) with a seed of 18, our
sample would be Michigan,
Massachusetts, Arizona, Minnesota,
Maine, Nebraska, Georgia, Iowa, Rhode
Island, Indiana.

13. (a) The list provided by the administration
serves as the frame. Number each student
in the list of registered students, from 1 to
19,935. Generate 25 random numbers,
without repetition, between 1 and 19,935
using a random number generator or
table. Select the 25 students with these
numbers.

(b) Answers will vary.

15. Answers will vary. Members should be
numbered 1–32, though other numbering
schemes are possible (e.g. 0–31). Using a
table of random digits or a random-number
generator, four different numbers (labels)
should be selected. The names corresponding
to these numbers form the sample.

Section 1.4

1. Stratified random sampling may be
appropriate if the population of interest can be
divided into groups (or strata) that are
homogeneous and nonoverlapping.

3. Convenience samples are typically selected in
a nonrandom manner. This means the results
are not likely to represent the population.
Convenience samples may also be self-
selected, which will frequently result in small
portions of the population being
overrepresented.

5. Stratified sample

7. False. In many cases, other sampling
techniques may provide equivalent or more
information about the population with less
"cost" than simple random sampling.

9. True. Because the individuals in a
convenience sample are not selected using
chance, it is likely that the sample is not
representative of the population.

11. Systematic sampling. The quality-control
manager is sampling every 8th chip, starting
with the 3rd chip.

13. Cluster sampling. The airline surveys all
passengers on selected flights (clusters).

15. Simple random sampling. Each known user of
the product has the same chance of being
included in the sample.

17. Cluster sampling. The farmer samples all
trees within the selected subsections (clusters).

19. Convenience sampling. The research firm is relying on voluntary response to obtain the sample data.

21. Stratified sampling. Shawn takes a sample of measurements during each of the four time intervals (strata).

23. The numbers corresponding to the 20 clients selected are 16, $16+25=41$, $41+25=66$, $66+25=91$, $91+25=116$, 141, 166, 191, 216, 241, 266, 291, 316, 341, 366, 391, 416, 441, 466, 491.

25. Answers will vary. To obtain the sample, number the Democrats 1 to 16 and obtain a simple random sample of size 2. Then number the Republicans 1 to 16 and obtain a simple random sample of size 2. Be sure to use a different starting point in Table I or a different seed for each stratum.

    For example, using a TI-84 Plus graphing calculator with a seed of 38 for the Democrats and 40 for the Republicans, the numbers selected would be 6, 9 for the Democrats and 14, 4 for the Republicans. If we had numbered the individuals down each column, the sample would consist of Haydra, Motola, Thompson, and Engler.

    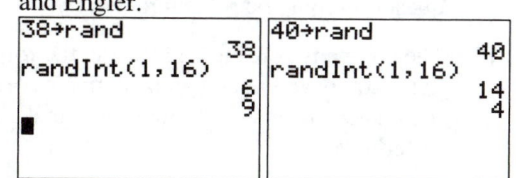

27. **(a)** $\dfrac{N}{n} = \dfrac{4502}{50} = 90.04 \to 90$; Thus, $k = 90$.

    (b) Randomly select a number between 1 and 90. Suppose that we select 15. Then the individuals to be surveyed will be the 15th, 105th, 195th, 285th, and so on up to the 4425th employee on the company list.

29. Simple Random Sample:
    Number the students from 1 to 1280. Use a table of random digits or a random-number generator to randomly select 128 students to survey.

    Stratified Sample:
    Since class sizes are similar, we would want to randomly select $\dfrac{128}{32} = 4$ students from each class to be included in the sample.

Cluster Sample:
Since classes are similar in size and makeup, we would want to randomly select $\dfrac{128}{32} = 4$ classes and include all the students from those classes in the sample.

31. Answers will vary. One design would be a stratified random sample, with two strata being commuters and noncommuters, as these two groups each might be fairly homogeneous in their reactions to the proposal.

33. Answers will vary. One design would be a cluster sample, with the clusters being city blocks. Randomly select city blocks and survey every household in the selected blocks.

35. Answers will vary. Since the company already has a list (frame) of 6600 individuals with high cholesterol, a simple random sample would be an appropriate design.

37. **(a)** For a political poll, a good frame would be all registered voters who have voted in the past few elections since they are more likely to vote in upcoming elections.

    (b) Because each individual from the frame has the same chance of being selected, there is a possibility that one group may be over- or underrepresented.

    (c) By using a stratified sample, the strategist can obtain a simple random sample within each strata (political party) so that the number of individuals in the sample is proportionate to the number of individuals in the population.

39. Answers will vary.

Section 1.5

1. A closed question is one in which the respondent must choose from a list of prescribed responses. An open question is one in which the respondent is free to choose his or her own response. Closed questions are easier to analyze, but limit the responses. Open questions allow respondents to state exactly how they feel, but are harder to analyze due to the variety of answers and possible misinterpretation of answers.

3. Bias means that the results of the sample are not representative of the population. There are three types of bias: sampling bias, response bias, and nonresponse bias. Sampling bias is due to the use of a sample to describe a population. This includes bias due to convenience sampling. Response bias involves intentional or unintentional misinformation. This would include lying to a surveyor or entering responses incorrectly. Nonresponse bias results when individuals choose not to respond to questions or are unable to be reached. A census can suffer from response bias and nonresponse bias, but would not suffer from sampling bias.

5. (a) Sampling bias. The survey suffers from undercoverage because the first 60 customers are likely not representative of the entire customer population.

   (b) Since a complete frame is not possible, systematic random sampling could be used to make the sample more representative of the customer population.

7. (a) Response bias. The survey suffers from response bias because the question is poorly worded.

   (b) The survey should inform the respondent of the current penalty for selling a gun illegally and the question should be worded as "Do you approve or disapprove of harsher penalties for individuals who sell guns illegally?" The order of "approve" and "disapprove" should be switched from one individual to the next.

9. (a) Nonresponse bias. Assuming the survey is written in English, non-English speaking homes will be unable to read the survey. This is likely the reason for the very low response rate.

   (b) The survey can be improved by using face-to-face or phone interviews, particularly if the interviewers are multi-lingual.

11. (a) The survey suffers from sampling bias due to undercoverage and interviewer error. The readers of the magazine may not be representative of all Australian women, and advertisements and images in the magazine could affect the women's view of themselves.

    (b) A well-designed sampling plan not in a magazine, such as a cluster sample, could make the sample more representative of the population.

13. (a) Response bias due to a poorly worded question

    (b) The question should be reworded in a more neutral manner. One possible phrasing might be "Do you believe that a marriage can be maintained after an extramarital relation?"

15. (a) Response bias. Students are unlikely to give honest answers if their teacher is administering the survey.

    (b) An impartial party should administer the survey in order to increase the rate of truthful responses.

17. No. The survey still suffers from sampling bias due to undercoverage, nonresponse bias, and potentially response bias.

19. It is very likely that the order of these two questions will affect the survey results. To alleviate the response bias, either question B could be asked first, or the order of the two questions could be rotated randomly.

21. The company is using a reward in the form of the $5.00 payment and an incentive by telling the reader that his or her input will make a difference.

23. For random digit dialing, the frame is anyone with a phone (whose number is not on a do-not-call registry). Even those with unlisted numbers can still be reached through this method.
    Any household without a phone, households on the do-not-call registry, and homeless individuals are excluded. This could result in sampling bias due to undercoverage if the excluded individuals differ in some way than those included in the frame.

25. It is extremely likely, particularly if households on the do-not-call registry have a trait that is not part of those households that are not on the registry.

27. Some nonsampling errors presented in the article as leading to incorrect exit polls were poorly trained interviewers, interviewer bias, and over representation of female voters.

29. – 31. Answers will vary.

33. The *Literary Digest* made an incorrect prediction due to sampling bias (an incorrect frame led to undercoverage) and nonresponse bias (due to the low response rate).

35. (a) Answers will vary. Stratified sampling by political affiliation (Democrat, Republican, etc.) could be used to ensure that all affiliations are represented. One question that could be asked is whether or not the person plans to vote in the next election. This would help determine which registered voters are likely to vote.

(b) Answers will vary. Possible explanations are that presidential election cycles get more news coverage or perhaps people are more interested in voting when they can vote for a president as well as a senator. During non-presidential cycles it is very informative to poll likely registered voters.

(c) Answers will vary. A higher percentage of Democrats in polls versus turnout will lead to overstating the predicted Democrat percentage of Democratic votes.

37. Nonresponse can be addressed by conducting callbacks or offering rewards.

39. Conducting a presurvey with open questions allows the researchers to use the most popular answers as choices on closed-question surveys.

41. Provided the survey was conducted properly and randomly, a high response rate will provide more representative results. When a survey has a low response rate, only those who are most willing to participate give responses. Their answers may not be representative of the whole population.

43. There is more than one type of CD. This can be interpreted as a medium used to store music or information electronically: a compact disk. It could also be understood as a special type of savings account: a certificate of deposit. The question can be improved by asking, "Do you own any certificates of deposit, which are a special type of savings account at a bank?"

Section 1.6

1. (a) An experimental unit is a person, object, or some other well-defined item upon which a treatment is applied.

(b) A treatment is a condition applied to an experimental unit. It can be any combination of the levels of the explanatory variables.

(c) A response variable is a quantitative or qualitative variable that measures a response of interest to the experimenter.

(d) A factor is a variable whose effect on the response variable is of interest to the experimenter. Factors are also called explanatory variables.

(e) A placebo is an innocuous treatment, such as a sugar pill, administered to a subject in a manner indistinguishable from an actual treatment.

(f) Confounding occurs when the effect of two explanatory variables on a response variable cannot be distinguished.

3. In a single-blind experiment, subjects do not know which treatment they are receiving. In a double-blind experiment, neither the subject nor the researcher(s) in contact with the subjects knows which treatment is received.

5. Blocking

7. (a) The research objective of the study was to determine the association between number of times one chews food and food consumption.

(b) The response variable is food consumption; quantitative.

(c) The explanatory variable is chew level (100%, 150%, 200%); qualitative.

(d) The experimental units are the 45 individuals aged 18 to 45 who participated in the study.

(e) Control is used by determining a baseline number of chews before swallowing; same type of food is used in the baseline as in the experiment; same time of day (lunch); age (18 to 45).

(f) Randomization reduces the effect of the order in which the treatments are administered. For example, perhaps the first time through the subjects are more diligent about their chewing than the last time through the study.

9. (a) The response variable is the achievement test scores.

(b) Answers may vary. Some factors are teaching methods, grade level, intelligence, school district, and teacher.
Fixed: grade level, school district, teacher
Set at predetermined levels: teaching method

(c) The treatments are the new teaching method and the traditional method. There are 2 levels of treatment.

(d) The factors that are not controlled are dealt with by random assignment into the two treatment groups.

(e) Group 2, using the traditional teaching method, serves as the control group.

(f) This experiment has a completely randomized design.

(g) The subjects are the 500 first-grade students from District 203 recruited for the study.

(h)

11. (a) This experiment has a matched-pairs design.

(b) The response variable is the level of whiteness.

(c) The explanatory variable or factor is the whitening method. The treatments are Crest Whitestrips Premium in addition to brushing and flossing, and just brushing and flossing alone.

(d) Answers will vary. One other possible factor is diet. Certain foods and tobacco products are more likely to stain teeth. This could impact the level of whiteness.

(e) Answers will vary. One possibility is that using twins helps control for genetic factors such as weak teeth that may affect the results of the study.

13. (a) This experiment has a completely randomized design.

(b) The population being studied is adults with insomnia.

(c) The response variable is the terminal wake time after sleep onset (WASO).

(d) The explanatory variable or factor is the type of intervention. The treatments are cognitive behavioral therapy (CBT), muscle relaxation training (RT), and the placebo.

(e) The experimental units are the 75 adults with insomnia.

(f)

15. **(a)** This experiment has a completely randomized design.

(b) The population being studied is adults over 60 years old and in good health.

(c) The response variable is the standardized test of learning and memory.

(d) The factor set to predetermined levels (explanatory variable) is the drug. The treatments are 40 milligrams of ginkgo 3 times per day and the matching placebo.

(e) The experimental units are the 98 men and 132 women over 60 years old and in good health.

(f) The control group is the placebo group.

(g)

17. **(a)** This experiment has a matched-pairs design.

(b) The response variable is the distance the yardstick falls.

(c) The explanatory variable or factor is hand dominance. The treatment is dominant versus non-dominant hand.

(d) The experimental units are the 15 students.

(e) Professor Neil used a coin flip to eliminate bias due to starting on the dominant or non-dominant hand first on each trial.

(f)

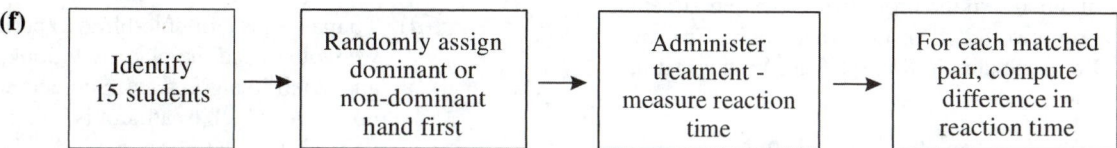

19. (a) This experiment has a randomized block design.

(b) The response variable is the score on the recall exam.

(c) The explanatory variable or factor is the type of advertising. The treatments are print, radio, and television.

(d) Level of education is the variable that serves as the block.

(e)

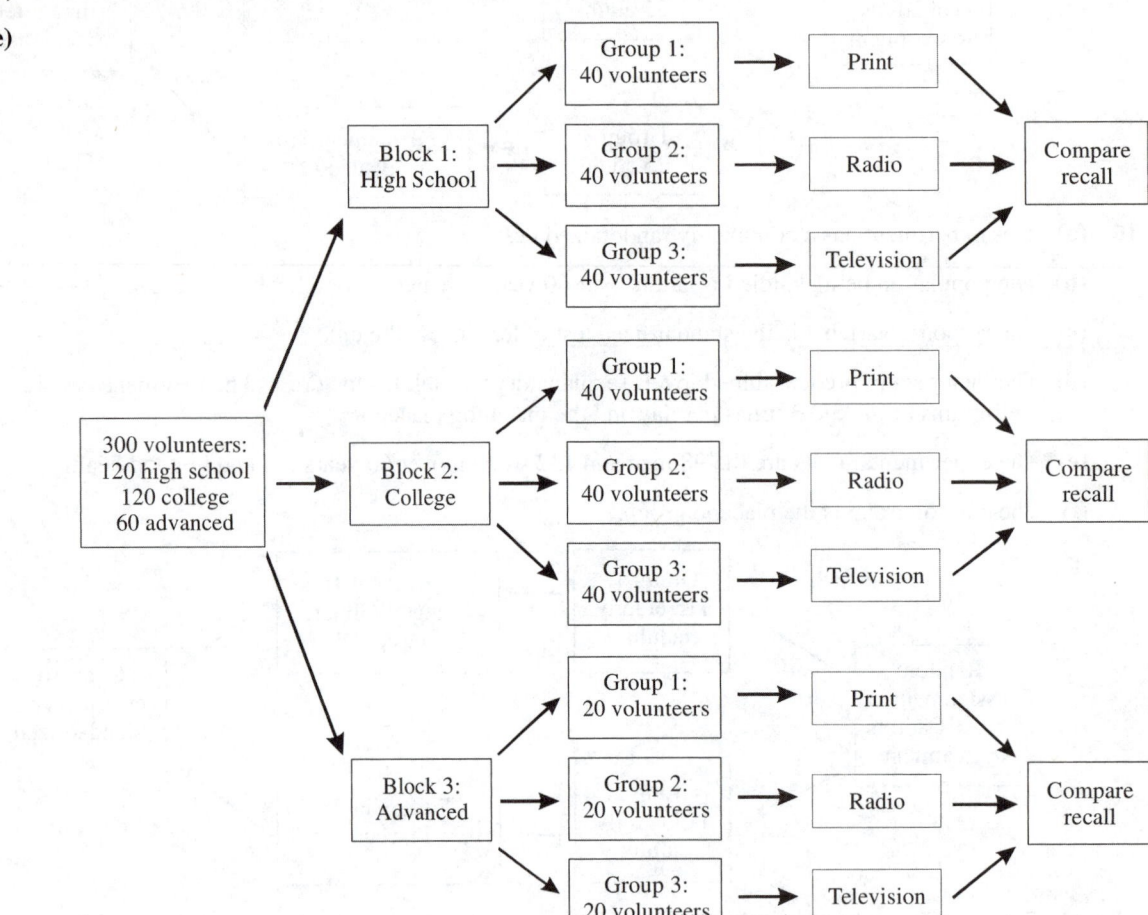

21. Answers will vary. Using a TI-84 Plus graphing calculator with a seed of 195, we would pick the volunteers numbered 8, 19, 10, 12, 13, 6, 17, 1, 4, and 7 to go into the experimental group. The rest would go into the control group. If the volunteers were numbered in the order listed, the experimental group would consist of Ann, Kevin, Christina, Eddie, Shannon, Randy, Tom, Wanda, Kim, and Colleen.

23. (a) This is an observational study because there is no intent to manipulate an explanatory variable or factor. The explanatory variable or factor is whether the individual is a green tea drinker or not, which is qualitative.

(b) Some lurking variables include diet, exercise, genetics, age, gender, and socioeconomic status.

(c) The experiment is a completely randomized design.

(d) To make this a double-blind experiment, we would need the placebo to look, taste, and smell like green tea. Subjects would not know which treatment is being delivered. In addition, the individuals administering the treatment and measuring the changes in LDL cholesterol would not know the treatment either.

(e) The factor that is manipulated is the tea, which is set at three levels; qualitative.

(f) Answers will vary. Other factors you might want to control in this experiment include age, exercise, and diet of the participants.

(g) Randomization could be used by numbering the subjects from 1 to 120. Randomly select 40 subjects and assign them to the placebo group. Then randomly select 40 from the remaining 80 subjects and assign to the one cup of green tea group. The remaining subjects will be assigned to the two cups of green tea group. By randomly assigning the subjects to the treatments, the expectation is that uncontrolled variables (such as genetic history, diet, exercise, etc.) are neutralized (even out).

(h) Exercise is a confounding variable because any change in the LDL cholesterol cannot be attributed to the tea. It may be the exercise that caused the change in LDL cholesterol.

---

25. Answers will vary. A completely randomized design is probably best.

27. Answers will vary. A randomized block design blocked by type of car is likely best.

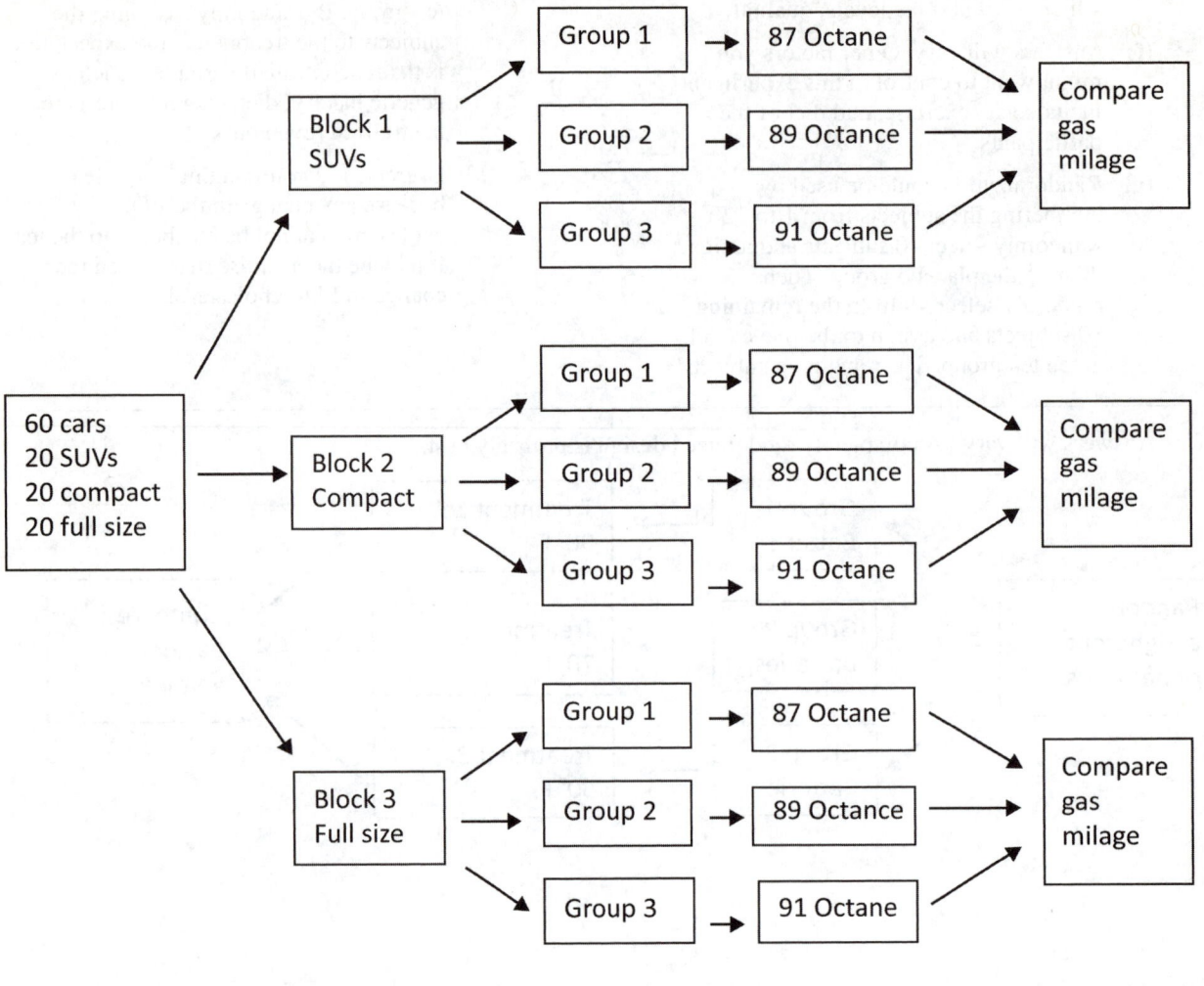

---

29. (a) The response variable is blood pressure.

(b) Three factors that have been identified are daily consumption of salt, daily consumption of fruits and vegetables, and the body's ability to process salt.

(c) The daily consumption of salt and the daily consumption of fruits and vegetables can be controlled. The body's ability to process salt cannot be controlled. To deal with variability of the body's ability to process salt, randomize experimental units to each treatment group.

(d) Answers will vary. Three levels of treatment might be a good choice – one level below the recommended daily allowance, one equal to the recommended daily allowance, and one above the recommended daily allowance.

31. Answers will vary.

33. Answers will vary. Control groups are needed in a designed experiment to serve as a baseline against which other treatments can be compared.

35. In a randomized block design, experimental units are divided into homogeneous groups called *blocks* before being randomly assigned to a treatment within each block. In a stratified random sample, the population is subdivided into homogeneous groups called *strata* before a simple random sample is drawn from each strata. The purpose of blocking is to remove any variability in the response variable that may be attributable to the block.

Chapter 1 Review Exercises

1. Statistics is the science of collecting, organizing, summarizing, and analyzing information in order to draw conclusions.

2. The population is the group of individuals that is to be studied.

3. A sample is a subset of the population.

4. An observational study uses data obtained by studying individuals in a sample without trying to manipulate or influence the variable(s) of interest. Observational studies are often called *ex post facto* studies because the value of the response variable has already been determined.

5. In a designed experiment, a treatment is applied to the individuals in a sample in order to isolate the effects of the treatment on the response variable.

6. The three major types of observational studies are (1) cross-sectional studies, (2) case-control studies, and (3) cohort studies.

   Cross-sectional studies collect data at a specific point in time or over a short period of time. Cohort studies are prospective and collect data over a period of time, sometimes over a long period of time. Case-controlled studies are retrospective, looking back in time to collect data either from historical records or from recollection by subjects in the study. Individuals possessing a certain characteristic are matched with those that do not.

7. The process of statistics refers to the approach used to collect, organize, analyze, and interpret data. The steps are to
   (1) identify the research objective,
   (2) collect the data needed to answer the research question,
   (3) describe the data, and
   (4) perform inference.

8. The three types of bias are sampling bias, nonresponse bias, and response bias. Sampling bias occurs when the techniques used to select individuals to be in the sample favor one part of the population over another. Bias in sampling is reduced when a random process is to select the sample. Nonresponse bias occurs when the individuals selected to be in the sample that do not respond to the survey have different opinions from those that do respond. This can be minimized by using callbacks and follow-up visits to increase the response rate.

Response bias occurs when the answers on a survey do not reflect the true feelings of the respondent. This can be minimized by using trained interviewers, using carefully worded questions, and rotating question and answer selections.

9. Nonsampling errors are errors that result from undercoverage, nonresponse bias, response bias, and data-entry errors. These errors can occur even in a census. Sampling errors are errors that result from the use of a sample to estimate information about a population. These include random error and errors due to poor sampling plans, and result because samples contain incomplete information regarding a population.

10. The following are steps in conducting an experiment:

    (1) *Identify the problem to be solved.*
        Give direction and indicates the variables of interest (referred to as the claim).

    (2) *Determine the factors that affect the response variable.*
        List all variables that may affect the response, both controllable and uncontrollable.

    (3) *Determine the number of experimental units.*
        Determine the sample size. Use as many as time and money allow.

    (4) *Determine the level of each factor.*
        Factors can be controlled by fixing their level (e.g. only using men) or setting them at predetermined levels (e.g. different dosages of a new medicine). For factors that cannot be controlled, random assignment of units to treatments helps average out the effects of the uncontrolled factor over all treatments.

    (5) *Conduct the experiment.*
        Carry out the experiment using an equal number of units for each treatment. Collect and organize the data produced.

    (6) *Test the claim.*
        Analyze the collected data and draw conclusions.

11. "Number of new automobiles sold at a dealership on a given day" is quantitative because its values are numerical measures on which addition and subtraction can be performed with meaningful results. The variable is discrete because its values result from a count.

12. "Weight in carats of an uncut diamond" is quantitative because its values are numerical measures on which addition and subtraction can be performed with meaningful results. The variable is continuous because its values result from a measurement rather than a count.

13. "Brand name of a pair of running shoes" is qualitative because its values serve only to classify individuals based on a certain characteristic.

14. 73% is a statistic because it describes a sample (the 1011 people age 50 or older who were surveyed).

15. 70% is a parameter because it describes a population (all the passes completed by Cardale Jones in the 2015 Championship Game).

16. Birth year has the *interval* level of measurement since differences between values have meaning, but it lacks a true zero.

17. Marital status has the *nominal* level of measurement since its values merely categorize individuals based on a certain characteristic.

18. Stock rating has the *ordinal* level of measurement because its values can be placed in rank order, but differences between values have no meaning.

19. Number of siblings has the *ratio* level of measurement because differences between values have meaning and there is a true zero.

20. This is an observational study because no attempt was made to influence the variable of interest. Sexual innuendos and curse words were merely observed.

21. This is an experiment because the researcher intentionally imposed treatments (experimental drug vs. placebo) on individuals in a controlled setting.

22. This was a cohort study because participants were identified to be included in the study and then followed over a period of time with data being collected at regular intervals (every 2 years).

23. This is convenience sampling since the pollster simply asked the first 50 individuals she encountered.

24. This is a cluster sample since the ISP included all the households in the 15 randomly selected city blocks.

25. This is a stratified sample since individuals were randomly selected from each of the three grades.

26. This is a systematic sample since every 40th tractor trailer was tested using a random start with the 12th tractor trailer.

27. (a) Sampling bias; undercoverage or nonrepresentative sample due to a poor sampling frame. Cluster sampling or stratified sampling are better alternatives.

    (b) Response bias due to interviewer error. A multilingual interviewer could reduce the bias.

    (c) Data-entry error due to the incorrect entries. Entries should be checked by a second reader.

28. Answers will vary. Using a TI-84 Plus graphing calculator with a seed of 1990, and numbering the individuals from 1 to 21, we would select individuals numbered 14, 6, 10, 17, and 11. If we numbered the businesses down each column, the businesses selected would be Jiffy Lube, Nancy's Flowers, Norm's Jewelry, Risky Business Security, and Solus, Maria, DDS.

29. Answers will vary. The first step is to select a random starting point among the first 9 bolts produced. Using row 9, column 17 from Table I in Appendix A, he will sample the 3rd bolt produced, then every 9th bolt after that until a sample size of 32 is obtained. In this case, he would sample bolts 3, 12, 21, 30, and so on, until bolt 282.

30. Answers will vary. The goggles could be numbered 00 to 99, then a table of random digits could be used to select the numbers of the goggles to be inspected. Starting with row 12, column 1 of Table 1 in Appendix A and reading down, the selected labels would be 55, 96, 38, 85, 10, 67, 23, 39, 45, 57, 82, 90, and 76.

31. **(a)** To determine the ability of chewing gum to remove stains from teeth

 (b) This is an experimental design because the teeth were separated into groups that were assigned different treatments.

 (c) Completely randomized design

 (d) Percentage of stain removed

 (e) Type of stain remover (gum or saliva); Qualitative

 (f) The 64 stained bovine incisors

 (g) The chewing simulator could impact the percentage of the stain removed.

 (h) Gum A and B remove significantly more stain.

32. **(a)** Matched-pairs

 (b) Reaction time; Quantitative

 (c) Alcohol consumption

 (d) Food consumption; caffeine intake

 (e) Weight, gender, etc.

 (f) To act as a placebo to control for the psychosomatic effects of alcohol

 (g) Alcohol delays the reaction time significantly in seniors for low levels of alcohol consumption; healthy seniors that are not regular drinkers.

---

33. **(a)** This experiment has a randomized block design.

 (b) The response variable is the exam grade.

 (c) The factor "Notecard use" is set at predetermined levels. The treatments are "with notecard" and "without notecard."

 (d) The experimental units are the instructor's statistics students.

 (e)

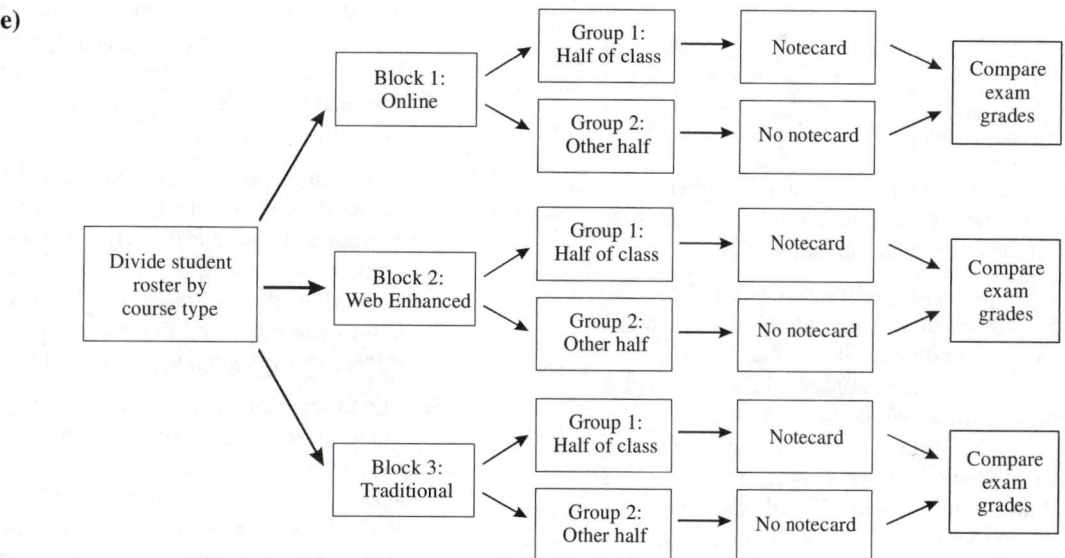

34. Answers will vary. Since there are ten digits (0 – 9), we will let a 0 or 1 indicate that (a) is to be the correct answer, 2 or 3 indicate that (b) is to be the correct answer, and so on. Beginning with row 1, column 8 of Table 1 in Appendix A, and reading downward, we obtain the following:

2, 6, 1, 4, 1, 4, 2, 9, 4, 3, 9, 0, 6, 4, 4, 8, 6, 5, 8, 5

Therefore, the sequence of correct answers would be:

b, d, a, c, a, c, b, e, c, b, e, a, d, c, c, e, d, c, e, c

35. (a) Answers will vary. One possible diagram is shown below.

(b) Answers will vary. One possible diagram is shown below.

36. A matched-pairs design is an experimental design where experimental units are matched up so they are related in some way.

In a completely randomized design, the experimental units are randomly assigned to one of the treatments. The value of the response variable is compared for each treatment. In a matched-pairs design, experimental units are matched up on the basis of some common characteristic (such as husband-wife or twins). The differences between the matched units are analyzed.

37. Answers will vary.

38. Answers will vary.

39. Randomization is meant to even out the effect of those variables that are not controlled for in a designed experiment. Answers to the randomization question may vary; however, each experimental unit must be randomly assigned. For example, a researcher might randomly select 25 experimental units from the 100 units and assign them to treatment #1. Then the researcher could randomly select 25 from the remaining 75 units and assign them to treatment #2, and so on.

Chapter 1 Test

1. Collect information, organize and summarize the information, analyze the information to draw conclusions, provide a measure of confidence in the conclusions drawn from the information collected.

2. The process of statistics refers to the approach used to collect, organize, analyze, and interpret data. The steps are to
(1) identify the research objective,
(2) collect the data needed to answer the research question,
(3) describe the data, and
(4) perform inference.

3. The time to complete the 500-meter race in speed skating is quantitative because its values are numerical measurements on which addition and subtraction have meaningful results. The variable is continuous because its values result from a measurement rather than a count. The variable is at the *ratio* level of measurement because differences between values have meaning and there is a true zero.

4. Video game rating is qualitative because its values classify games based on certain characteristics but arithmetic operations have no meaningful results. The variable is at the *ordinal* level of measurement because its values can be placed in rank order, but differences between values have no meaning.

5. The number of surface imperfections is quantitative because its values are numerical measurements on which addition and subtraction have meaningful results. The variable is discrete because its values result from a count. The variable is at the *ratio* level of measurement because differences between values have meaning and there is a true zero.

6. This is an experiment because the researcher intentionally imposed treatments (brand-name battery versus plain-label battery) on individuals (cameras) in a controlled setting. The response variable is the battery life.

7. This is an observational study because no attempt was made to influence the variable of interest. Fan opinions about the asterisk were merely observed. The response variable is whether or not an asterisk should be placed on Barry Bonds' 756th homerun ball.

8. A *cross-sectional study* collects data at a specific point in time or over a short period of time; a *cohort study* collects data over a period of time, sometimes over a long period of time (prospective); a *case-controlled study* is retrospective, looking back in time to collect data.

9. An experiment involves the researcher actively imposing treatments on experimental units in order to observe any difference between the treatments in terms of effect on the response variable. In an observational study, the researcher observes the individuals in the study without attempting to influence the response variable in any way. Only an experiment will allow a researcher to establish causality.

10. A control group is necessary for a baseline comparison. This accounts for the placebo effect that says that some individuals will respond to any treatment. Comparing other treatments to the control group allows the researcher to identify which, if any, of the other treatments are superior to the current treatment (or no treatment at all). Blinding is important to eliminate bias due to the individual or experimenter knowing which treatment is being applied.

11. The steps in conducting an experiment are to (1) identify the problem to be solved, (2) determine the factors that affect the response variable, (3) determine the number of experimental units, (4) determine the level of each factor, (5) conduct the experiment, and (6) test the claim.

12. Answers will vary. The franchise locations could be numbered 01 to 15 going across. Starting at row 7, column 14 of Table I in Appendix, and working downward, the selected numbers would be 08, 11, 03, and 02. The corresponding locations would be Ballwin, Chesterfield, Fenton, and O'Fallon.

13. Answers will vary. Using the available lists, obtain a simple random sample from each stratum and combine the results to form the stratified sample. Start at different points in Table I or use different seeds in a random number generator. Using a TI-84 Plus graphing calculator with a seed of 14 for Democrats, 28 for Republicans, and 42 for Independents, the selected numbers would be Democrats: 3946, 8856, 1398, 5130, 5531, 1703, 1090, and 6369
Republicans: 7271, 8014, 2575, 1150, 1888, 3138, and 2008
Independents: 945, 2855, and 1401

14. Answers will vary. Number the blocks from 1 to 2500 and obtain a simple random sample of size 10. The blocks corresponding to these numbers represent the blocks analyzed. All trees in the selected blocks are included in the sample. Using a TI-84 Plus graphing calculator with a seed of 12, the selected blocks would be numbered 2367, 678, 1761, 1577, 601, 48, 2402, 1158, 1317, and 440.

15. Answers will vary. $\frac{600}{14} \approx 42.86$, so we let $k = 42$. Select a random number between 1 and 42 that represents the first slot machine inspected. Using a TI-84 Plus graphing calculator with a seed of 132, we select machine 18 as the first machine inspected. Starting with machine 18, every 42nd machine thereafter would also be inspected (60, 102, 144, 186, …, 564).

16. In a completely randomized design, the experimental units are randomly assigned to one of the treatments. The value of the response variable is compared for each treatment. In a randomized block design, the experimental units are first divided according to some common characteristic (such as gender). Then each experimental unit within each block is randomly assigned to one treatment. Within each block, the value of the response variable is compared for each treatment, but not between blocks. By blocking, we prevent the effect of the blocked variable from confounding with the treatment.

17. (a) Sampling bias due to voluntary response

    (b) Nonresponse bias due to the low response rate

    (c) Response bias due to poorly worded questions.

    (d) Sampling bias due to poor sampling plan (undercoverage)

18. (a) This experiment has a matched-pairs design.

(b) The subjects are the 159 social drinkers who participated in the study.

(c) Treatments are the types of beer glasses (straight glass or curved glass).

(d) The response variable is the time to complete the drink; quantitative.

(e) The type of glass used in the first week is randomly determined. This is to neutralize the effect of drinking out of a specific glass first.

(f)

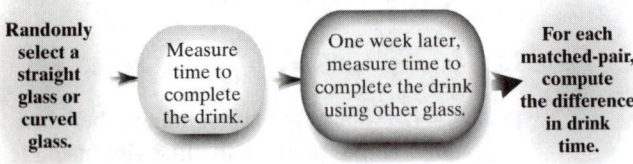

19. (a) This experiment has a completely randomized design.

(b) The factor set to predetermined levels is the topical cream concentration. The treatments are 0.5% cream, 1.0% cream, and a placebo (0% cream).

(c) The study is double-blind if neither the subjects, nor the person administering the treatments, are aware of which topical cream is being applied.

(d) The control group is the placebo (0% topical cream).

(e) The experimental units are the 225 patients with skin irritations.

(f)

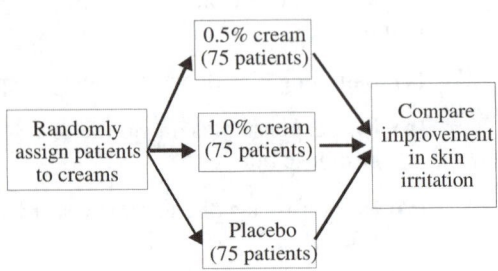

20. (a) This was a cohort study because participants were identified to be included in the study and then followed over a long period of time with data being collected at regular intervals (every 4 years).

(b) The response variable is bone mineral density. The explanatory variable is weekly cola consumption.

(c) The response variable is quantitative because its values are numerical measures on which addition and subtraction can be performed with meaningful results.

(d) The researchers observed values of variables that could potentially impact bone mineral density (besides cola consumption), so their effect could be isolated from the variable of interest.

(e) Answers will vary. Some possible lurking variables that should be accounted for are smoking status, alcohol consumption, physical activity, and calcium intake (form and quantity).

(f) The study concluded that women who consumed at least one cola per day (on average) had a bone mineral density that was significantly lower at the femoral neck than those who consumed less than one cola per day. The study cannot claim that increased cola consumption *causes* lower bone mineral density because it is only an observational study. The researchers can only say that increased cola consumption is *associated* with lower bone mineral density for women.

21. A confounding variable is an explanatory variable that cannot be separated from another explanatory variable. A lurking variable is an explanatory variable that was not considered in the study but affects the response variable in the study.

Case Study: Chrysalises for Cash

Reports will vary. The reports should include the following components:

Step 1: *Identify the problem to be solved.* The entrepreneur wants to determine if there are differences in the quality and emergence time of broods of the black swallowtail butterfly depending on the following factors: (a) early brood season versus late brood season; (b) carrot plants versus parsley plants; and (c) liquid fertilizer versus solid fertilizer.

Step 2: *Determine the explanatory variables that affect the response variable.* Some explanatory variables that may affect the quality and emergence time of broods are the brood season, the type of plant on which the chrysalis grows, fertilizer used for plants, soil mixture, weather, and the level of sun exposure.

Step 3: *Determine the number of experimental units.* In this experiment, a sample of 40 caterpillars/butterflies will be used.

Step 4: *Determine the level of the explanatory variables:*

- **Brood season** – We wish to determine the differences in the number of deformed butterflies and in the emergence times depending on whether the brood is from the early season or the late season. We use a total of 20 caterpillars/butterflies from the early brood season and 20 caterpillars/butterflies from the late brood season.

- **Type of plant** – We wish to determine the differences in the number of deformed butterflies and in the emergence times depending on the type of plant on which the caterpillars are placed. A total of 20 caterpillars are placed on carrot plants and 20 are placed on parsley plants.

- **Fertilizer** – We wish to determine the differences in the number of deformed butterflies and in the emergence times depending on the type of fertilizer used on the plants. A total of 20 chrysalises grow on plants that are fed liquid fertilizer and 20 grow on plants that are fed solid fertilizer.

- **Soil mixture** – We control the effects of soil by growing all plants in the same mixture.

- **Weather** – We cannot control the weather, but the weather will be the same for each chrysalis grown within the same season. For chrysalises grown in different seasons, we expect the weather might be different and thus part of the reason for potential differences between seasons. Also, we can control the amount of watering that is done.

- **Sunlight exposure** – We cannot control this variable, but the sunlight exposure will be the same for each chrysalis grown within the same season. For chrysalises grown in different seasons, we expect the sunlight exposure might be different and thus part of the reason for potential differences between seasons.

Step 5: *Conduct the experiment.*

(a) We fill eight identical pots with equal amounts of the same soil mixture. We use four of the pots for the early brood season and four of the pots for the late brood season.

For the early brood season, two of the pots grow carrot plants and two grow parsley plants. One carrot plant is fertilized with a liquid fertilizer, one carrot plant is fertilized with a solid fertilizer, one parsley plant is fertilized with the liquid fertilizer, and one parsley plant is fertilized with the solid fertilizer. We place five black swallowtail caterpillars of similar age into each of the four pots.

Similarly, for the late brood season, two of the pots grow carrot plants and two grow parsley plants. One carrot plant is fertilized with a liquid fertilizer, one carrot plant is fertilized with a solid fertilizer, one parsley plant is fertilized with the liquid fertilizer, and one parsley plant is fertilized with the solid fertilizer. We place five black swallowtail caterpillars of similar age into each of the four pots.

(b) We determine the number of deformed butterflies and in the emergence times for the caterpillars/butterflies from each pot.

Step 6: *Test the claim.* We determine whether any differences exist depending on season, plant type, and fertilizer type.

Conclusions:

Early versus late brood season: From the data presented, more deformed butterflies occur in the late season than in the early season. Five deformed butterflies occurred in the late season, while only one occurred in the early season. Also, the emergence time seems to be longer in the early season than in the late season. In the early season, all but one of the 20 emergence times were between 6 and 8 days. In the late season, all 20 of the emergence times were between 2 and 5 days.

Parsley versus carrot plants: From the data presented, the plant type does not seem to affect the number of deformed butterflies that occur. Altogether, three deformed butterflies occur from parsley plants and three deformed butterflies occur from the carrot plants. Likewise, the plant type does not seem to affect the emergence times of the butterflies.

Liquid versus solid fertilizer: From the data presented, the type of fertilizer seems to affect the number of deformed butterflies that occur. Five deformed butterflies occurred when the solid fertilizer was used, while only one occurred when the liquid fertilizer was used. The type of fertilizer does not seem to affect emergence times.

Chapter 2
Organizing and Summarizing Data

Section 2.1

1. Raw data are the data as originally collected, before they have been organized or coded.

3. The relative frequencies should add to 1, although rounding may cause the answers to vary slightly.

5. (a) The largest segment in the pie chart is for "Washing your hands" so the most commonly used approach to beat the flu bug is washing your hands. 61% of respondents selected this as their primary method for beating the flu.

(b) The smallest segment in the pie chart is for "Drinking Orange Juice" so the least used method is drinking orange juice. 2% of respondents selected this as their primary method for beating the flu.

(c) 25% of respondents felt that flu shots were the best way to beat the flu.

7. (a) The highest bar corresponds to the position OF (outfield), so OF is the position with the most MVPs.

(b) The bar for first base (1B) reaches the line for 15. Thus, there were 15 MVPs who played first base.

(c) The bar for outfield (OF) is 30 on the vertical axis. The bar for first base (1B) reaches 15. Since $30 - 15 = 15$, there were 15 more MVPs who played outfield than first base.

(d) Each of the three outfield positions should be reported as MVPs, rather than treating the three positions as one position.

9. (a) 69% of the respondents believe divorce is morally acceptable.

(b) 23% believe divorce is morally wrong. So, 240 million * 0.23 = 55.2 million adult Americans believe divorce is morally wrong.

(c) This statement is inferential, since it is a generalization based on the observed data.

11. (a) The proportion of 18–34 year old respondents who are more likely to buy when made in America is 0.42. For 34–44 year olds, the proportion is 0.61.

(b) The 55+ age group has the greatest proportion of respondents who are more likely to buy when made in America.

(c) The 18–34 age group has a majority of respondents who are less likely to buy when made in America.

(d) As age increases, so does the likelihood that a respondent will be more likely to buy a product that is made in America.

13. (a) Total students surveyed = 125 + 324 + 552 + 1257 + 2518 = 4776
Relative frequency of "Never"
= 125 / 4776 ≈ 0.0262, and so on.

Response	Relative Frequency
Never	0.0262
Rarely	0.0678
Sometimes	0.1156
Most of the time	0.2632
Always	0.5272

(b) 52.72%

(c) $0.0262 + 0.0678 = 0.0940$ or 9.40%

(d)

"How Often Do You Wear Your Seat Belt?"

(e)

(f)

(g) The statement is inferential since it is inferring something about the entire population based on the results of a sample survey.

15. (a) Total adults surveyed = 377 + 192 + 132 + 81 + 243 = 1025
Relative frequency of "More than 1 hour a day" = $377 / 1025 \approx 0.3678$, and so on.

Response	Relative Frequency
More than 1 hr a day	0.3678
Up to 1 hr a day	0.1873
A few times a week	0.1288
A few times a month or less	0.0790
Never	0.2371

(b) 0.2371 (about 24%)

(c)

(d)

(e)

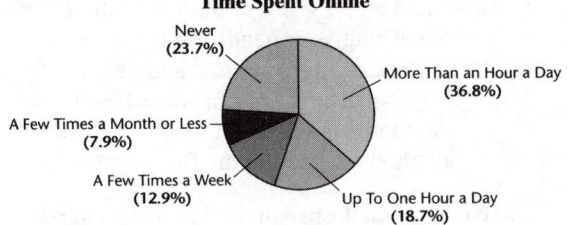

(f) The statement provides an estimate, but no level of confidence is given.

17. (a) Total adults = 1936
Relative frequency for "none" is:
$173/1936 = 0.09$, and so on.

Number of Texts	Rel. Freq. (Adults)
None	0.089
1 to 10	0.505
11 to 20	0.129
21 to 50	0.129
51 to 100	0.069
101+	0.079

(b) Total teens = 627
Relative frequency for "none" is:
$13/627 = 0.021$, and so on.

Number of Texts	Rel. Freq. (Teens)
None	0.021
1 to 10	0.220
11 to 20	0.110
21 to 50	0.180
51 to 100	0.180
101+	0.289

(c)

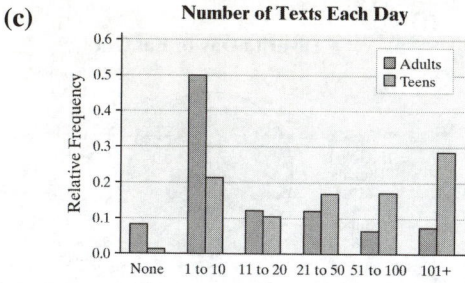

Number of Texts Each Day

(d) Answers will vary. Adults are much more likely to send fewer texts per day, while teens are much more likely to do more texting.

19. (a) Total males = 99; Relative frequency for "Professional Athlete" is 40/99 = 0.404, and so on.

Total number of females = 100; Relative frequency for "Professional Athlete" is 18/100 = 0.18, and so on.

Dream Job	Men	Women
Professional Athlete	0.404	0.180
Actor/Actress	0.263	0.370
President of the United States	0.131	0.130
Rock Star	0.131	0.130
Not Sure	0.071	0.190

(b)

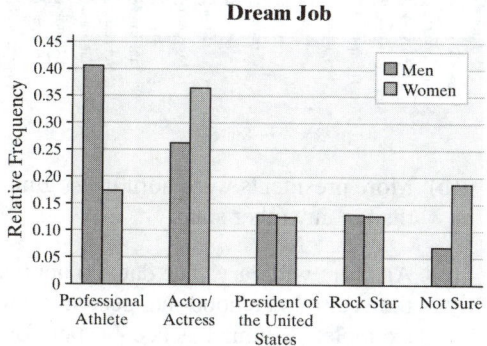

Dream Job

(c) Answers will vary. Males are much more likely to want to be a professional athlete. Women are more likely to aspire to a career in acting than men. Men's desire to become athletes may be influenced by the prominence of male sporting figures in popular culture. Women may aspire to careers in acting due to the perceived glamour of famous female actresses.

21. (a), (b)
Total number of Trading Days = 30; relative frequency for Down is 15/30 = 0.5, and so on.

Winner	Freq.	Rel. Freq.
Down	15	0.500
No Change	2	0.067
Up	13	0.433

(c)

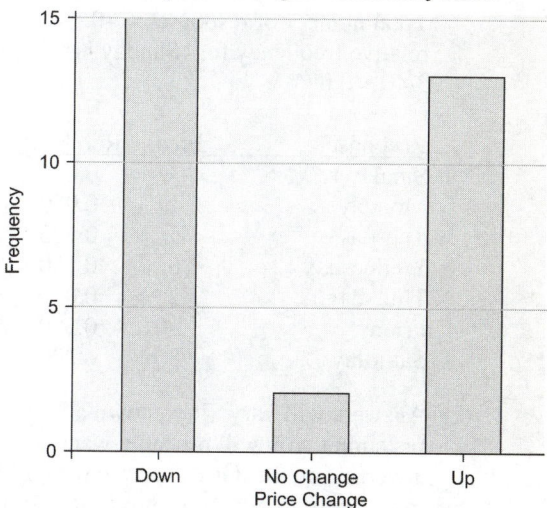

Daily Price Change of Walt Disney Stock

(d)

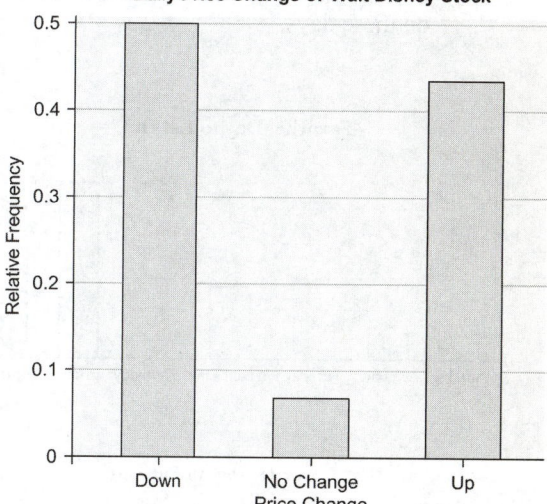

Daily Price Change of Walt Disney Stock

(e)

Daily Price Change in Walt Disney Stock

23. (a), (b)

Total number of responses = 40; relative frequency for "Sunday" is 3/40 = 0.075.

Response	Freq.	Rel. Freq.
Sunday	3	0.075
Monday	2	0.050
Tuesday	5	0.125
Wednesday	6	0.150
Thursday	2	0.050
Friday	14	0.350
Saturday	8	0.200

(c) Answers will vary. If you own a restaurant, you will probably want to advertize on the days when people will be most likely to order takeout: Friday. You might consider avoiding placing an ad on Monday and Thursday, since the readers are least likely to choose to order takeout on these days.

(d)

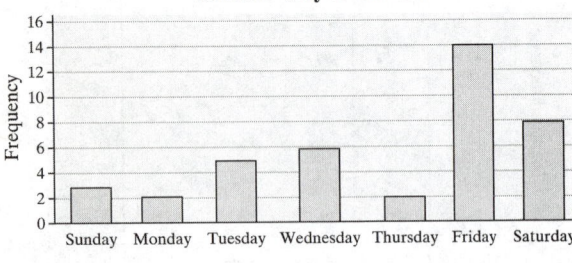

Favorite Day to Eat Out

(e)

Favorite Day to Eat Out

(f)

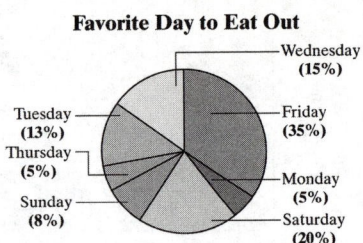

Favorite Day to Eat Out

25. (a)

State	AR	CA	CT	GA	HI	IL
Freq.	1	1	1	1	1	1

State	IA	KY	MA	MO	NE
Freq.	1	1	4	1	1

State	NH	NJ	NY	NC	OH
Freq.	1	2	4	2	7

State	PA	SC	TX	VT	VA
Freq.	1	1	2	2	8

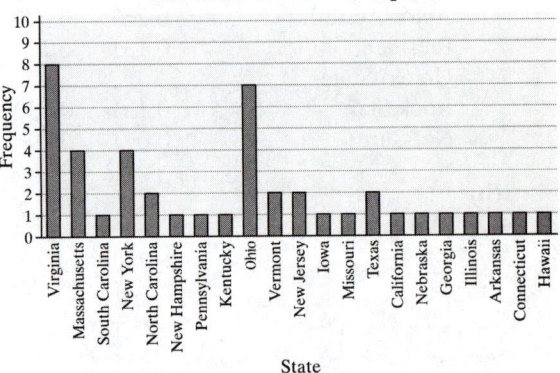

The U.S. Presidents' Birthplaces

(b) More presidents were born in Virginia than in any other state.

(c) Answers will vary. The data do not take the year of statehood into account. For example, Virginia has been a state for roughly 62 years more than California. The population of the United States was more concentrated in the east in the early years, so it was more likely that the president would be from that part of the country.

27. Answers will vary.

29. (a) The researcher wants to determine if online homework improves student learning over traditional pencil-and-paper homework.

(b) This study is an experiment because the researcher is actively imposing treatments (the homework style) on subjects.

(c) Answers will vary. Some examples are same teacher, same semester, and same course.

(d) Assigning different homework methods to entire classes could confound the results because there may be differences between the classes. The instructor may give more instruction to one class than the other. The instructor is not blinded, so he or she may treat one group differently from the other.

(e) *Number of students*: quantitative, discrete
Average age: quantitative, continuous
Average exam score: quantitative, continuous
Type of homework: qualitative
College experience: qualitative

(f) Letter grade is a qualitative variable at the ordinal level of measurement.
Answers will vary. It is possible that ordering the data from A to F is better because it might give more "weight" to the higher grade and the researcher wants to show that a higher percent of students passed using the online homework.

(g) The graph being displayed is a side-by-side relative frequency bar graph.

(h) Yes; the "whole" is the set of students who received a grade for the course for each homework method.

(i) The table shows that the two groups with no prior college experience had roughly the same average exam grade. From the bar graph, we see that the students using online homework had a lower percent for As, but had a higher percent who passed with a C or better.

31. Answers will vary. If the goal is to illustrate the levels of importance, then arranging the bars in a bar chart in decreasing order makes sense. Sometimes it is useful to arrange the categorical data in a bar chart in alphabetical order. A pie chart does not readily allow for arranging the data in order.

33. No, the percentages do not sum to 100%.

Section 2.2

1. classes

3. class width

5. True

7. False. The distribution shape shown is skewed right.

9. (a) The value with the highest frequency is 8.

(b) The value with the lowest frequency is 2.

(c) The value of 7 was observed 15 times.

(d) The value of 5 was observed 11 times and the value of 4 was observed 7 times. Therefore, the value of 5 was observed 4 more times than the value of 4 (e.g. $11 - 7 = 4$).

(e) $\dfrac{15}{100} = 0.15$ or 15% of the time a 7 was observed.

(f) The distribution is approximately bell-shaped.

11. (a) Total frequency = 2 + 3 + 13 + 42 + 58 + 40 + 31 + 8 + 2 + 1 = 200

(b) 10 (e.g. $70 - 60 = 10$)

(c)

IQ Score (class)	Frequency
60–69	2
70–79	3
80–89	13
90–99	42
100–109	58
110–119	40
120–129	31
130–139	8
140–149	2
150–159	1

(d) The class "100 – 109" has the highest frequency.

(e) The class "150 – 159" has the lowest frequency

(f) $\dfrac{8+2+1}{200} = 0.055 = 5.5\%$

(g) No, there were no IQs above 159.

13. (a) Likely skewed right. Most household incomes will be to the left (perhaps in the $50,000 to $150,000 range), with fewer higher incomes to the right (in the millions).

(b) Likely bell-shaped. Most scores will occur near the middle range, with scores tapering off equally in both directions.

(c) Likely skewed right. Most households will have, say, 1 to 4 occupants, with fewer households having a higher number of occupants.

(d) Likely skewed left. Most Alzheimer's patients will fall in older-aged categories, with fewer patients being younger.

15. (a) Total number of households =
$16 + 18 + 12 + 3 + 1 = 50$
Relative frequency of 0 children = 16/50
= 0.32, and so on.

Number of Children Under Five	Relative Frequency
0	0.32
1	0.36
2	0.24
3	0.06
4	0.02

(b) $\dfrac{12}{50} = 0.24$ or 24% of households have two children under the age of 5.

(c) $\dfrac{18+12}{50} = \dfrac{30}{50} = 0.6$ or 60% of households have one or two children under the age of 5.

17. From the legend, 1|0 represents 10, so the original data set is 10, 11, 14, 21, 24, 24, 27, 29, 33, 35, 35, 35, 37, 37, 38, 40, 40, 41, 42, 46, 46, 48, 49, 49, 53, 53, 55, 58, 61, 62.

19. From the legend, 1|2 represents 1.2, so the original data set is 1.2, 1.4, 1.6, 2.1, 2.4, 2.7, 2.7, 2.9, 3.3, 3.3, 3.3, 3.5, 3.7, 3.7, 3.8, 4.0, 4.1, 4.1, 4.3, 4.6, 4.6, 4.8, 4.8, 4.9, 5.3, 5.4, 5.5, 5.8, 6.2, 6.4.

21. (a) There are six classes.

(b) Lower class limits: 10, 14, 18, 22, 26, 30
Upper class limits: 13.9, 17.9, 21.9, 25.9, 29.9, 33.9

(c) The class width can be found by subtracting consecutive lower class limits. For example, $14 - 10 = 4$. Therefore, the class width is 4 (players).

23. (a) Total frequency =
$4 + 7 + 17 + 91 + 282 + 206 = 607$
Relative frequency for 10–13.9 is
$4/607 = 0.0066$, and so on.

Speed (Km/hr)	Relative Frequency
10–13.9	0.0066
14–17.9	0.0115
18–21.9	0.0280
22–25.9	0.1499
26–29.9	0.4646
30–33.9	0.3394

(b)

(c)

The percentage of players who had a top speed between 30 and 33.9 km/h is 33.94%. The percent of players who had a top speed less than 13.9 km/h is 0.66%.

25. (a) The data are discrete. The possible values for the number of color televisions in a household are countable.

(b), (c)
The relative frequency for 0 color televisions is 1/40 = 0.025, and so on.

Number of Color TVs	Frequency	Relative Frequency
0	1	0.025
1	14	0.350
2	14	0.350
3	8	0.200
4	2	0.050
5	1	0.025

(d) The relative frequency is 0.2, so 20% of the households surveyed had 3 color televisions.

(e) $0.05 + 0.025 = 0.075$
7.5% of the households in the survey had 4 or more color televisions.

(f)

Televisions in Household

(g)
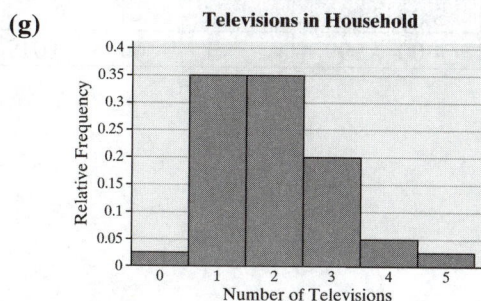
Televisions in Household

(h) The distribution is skewed right.

27. (a), (b) Relative frequency of a Gini Index of 20–24.9 = 5/136 = 0.037, and so on.

Gini Index	Freq.	Rel. Freq.
20–24.9	5	0.037
25–29.9	16	0.118
30–34.9	28	0.206
35–39.9	27	0.199
40–44.9	20	0.147
45–49.9	17	0.125
50–54.9	13	0.096
55–59.9	5	0.037
60–64.9	5	0.037

(c)
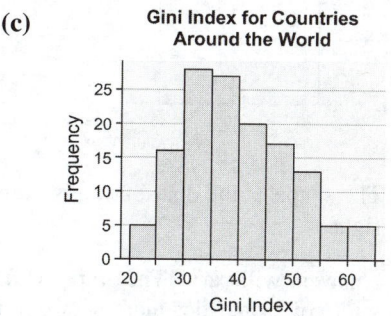
Gini Index for Countries Around the World

(d)
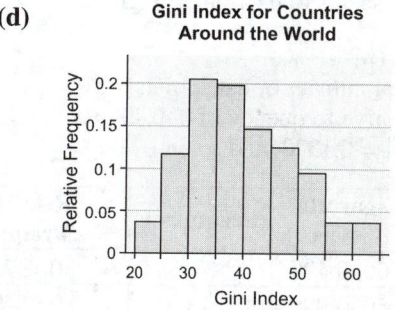
Gini Index for Countries Around the World

(e) The shape of the distribution is skewed right.

(f) Relative frequency of a Gini Index of 20–29.9 = 21/136 = 0.154, and so on.

Gini Index	Freq.	Rel. Freq.
20–29.9	21	0.154
30–39.9	55	0.404
40–49.9	37	0.272
50–59.9	18	0.132
60–69.9	5	0.037

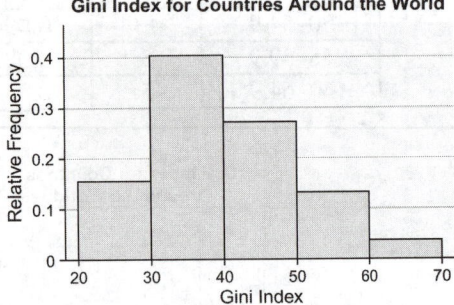

The shape of the distribution is skewed right.

(g) Answers will vary. The graph with a class width of 5 provides more detail, so it seems to be a superior graph.

29. **(a), (b)**
Total number of data points = 51
Relative frequency of 0–0.499 is
7/51 = 0.1373, and so on.

Cigarette Tax	Frequency	Relative Frequency
0.00–0.499	7	0.1373
0.50–0.999	13	0.2549
1.00–1.499	7	0.1373
1.50–1.999	8	0.1569
2.00–2.499	5	0.0980
2.50–2.999	5	0.0980
3.00–3.499	3	0.0588
3.50–3.999	2	0.0392
4.00–4.499	1	0.0196

(c)

(d)

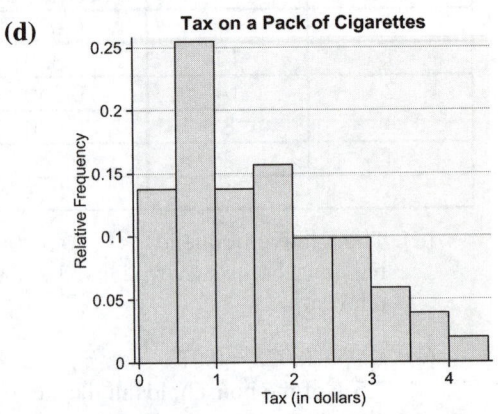

(e) The distribution appears to be right skewed.

(f) Relative frequency of 0–0.999 is:
20/51 = 0.3922, and so on.

Cigarette Tax	Frequency	Relative Frequency
0.00–0.999	20	0.3922
1.00–1.999	15	0.2941
2.00–2.999	10	0.1961
3.00–3.999	5	0.0980
4.00–4.999	1	0.0196

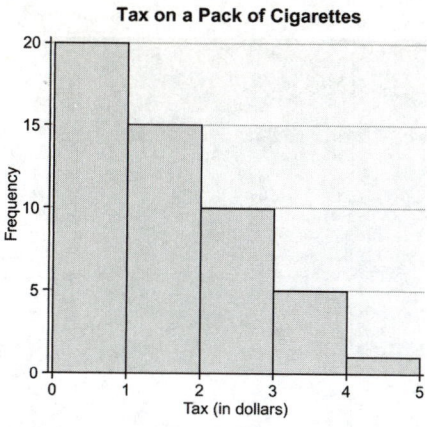

Tax on a Pack of Cigarettes

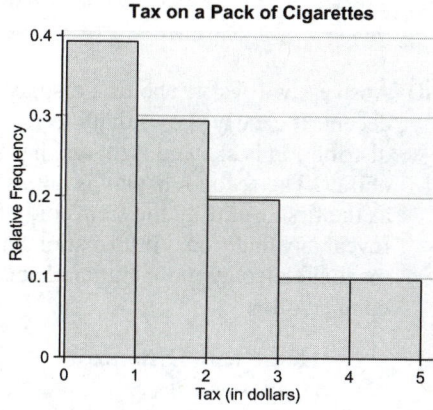

Tax on a Pack of Cigarettes

The distribution is right skewed.

(g) Answers will vary. The first distribution gives a more detailed pattern and does a nice job summarizing the data.

31. Answers will vary. One possibility follows.

(a) Choose a lower class limit of first class of 0 with a class width of 200.

(b), (c)
Relative frequency for 0–199 is 4/51 = 0.0784, and so on.

Violent Crime Rate	Frequency	Relative Frequency
0–199.9	4	0.0784
200–399.9	26	0.5098
400–599.9	17	0.3333
600–799.9	3	0.0588
800–999.9	0	0.0000
1000–1199.9	0	0.0000
1200–1399.9	1	0.0196

(d)

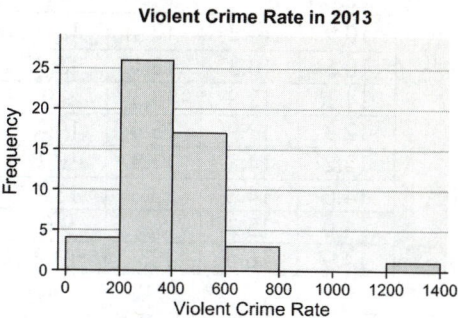

Violent Crime Rate in 2013

(e)

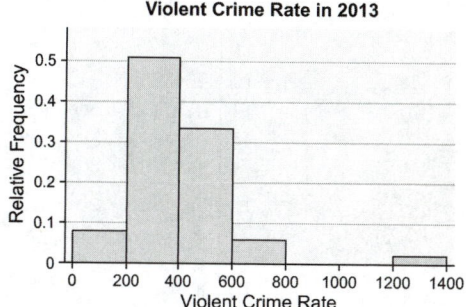

Violent Crime Rate in 2013

(f) The distribution is skewed right.

33. (a)   President Ages at Inauguration

```
4 | 23
4 | 6677899
5 | 0011112244444
5 | 555566677778
6 | 0111244
6 | 589
```
Legend: 4 | 2 represents 42 years.

(b) The distribution appears to be roughly symmetric and bell-shaped.

35. (a)   Fat in McDonald's Breakfast

```
0 | 39
1 | 1266
2 | 1224577
3 | 0012267
4 | 6
5 | 159
```
Legend: 5 | 1 represents 51 grams of fat.

(b) The distribution appears to be roughly symmetric and bell-shaped.

37. (a) Five Year Rate of Return Rounded to the nearest tenth:

10.9	14.2	12.4	13.6	13.0
10.5	10.3	13.1	15.7	14.9
14.1	12.8	13.3	9.9	15.6
12.3	13.9	13.4	19.4	13.4
12.2	14.8	11.9	10.1	13.6
14.6	14.8	13.5	13.9	13.2
14.0	15.2	8.3	9.0	8.7
14.9	16.0	13.7	13.9	12.8

(b) Five Year Rate of Return

```
 8 | 37
 9 | 09
10 | 1359
11 | 9
12 | 23488
13 | 0123445667999
14 | 01268899
15 | 267
16 | 0
17 |
18 |
19 | 4
```

Legend: 8|3 represents 8.3%

(c) The distribution is bell-shaped.

39. (a) Violent crime rates rounded to the nearest tens:

450	1240	350	260	410	560	320
600	490	220	450	350	320	280
430	380	500	270	240	640	200
470	240	120	260	300	410	
420	210	480	610	470	210	
310	410	410	190	250	140	
280	350	450	290	350	190	
550	260	230	560	250	300	

(b) Violent Crime Rates by State, 2013

```
 1 | 2499
 2 | 0112344556667889
 3 | 0012255558
 4 | 1111235557789
 5 | 0566
 6 | 014
 7 |
 8 |
 9 |
10 |
11 |
12 | 4
```

Legend: 1|2 represents 120 violent crimes per 100,000 population

(c) Violent Crime Rates by State, 2013

```
 1 | 24
 1 | 99
 2 | 0112344
 2 | 556667889
 3 | 00122
 3 | 55558
 4 | 111123
 4 | 5557789
 5 | 0
 5 | 566
 6 | 014
 6 |
 7 |
 7 |
 8 |
 8 |
 9 |
 9 |
10 |
10 |
11 |
11 |
12 | 4
```

Legend: 1|2 represents 120 violent crimes per 100,000 population

(d) Answers will vary. The first display is decent. It clearly shows that the distribution is skewed right and has an outlier. The second display is not as good as the first. Splitting the stems did not reveal any additional information and has made the display more cluttered and cumbersome.

41. (a)   **Home Run Distances**

```
McGwire           Bonds
              32 | 0 0
              33 |
         10   34 | 7
         00   35 | 0
       9000   36 | 0 0 0 1 5
      70000   37 | 0 0 5 5 5 5
   85500000   38 | 0 0 0 0 0 5
      80000   39 | 0 0 1 4 6
        900   40 | 0 0 0 0 4 5
      00000   41 | 0 0 0 0 0 0 0 0 0 0 1 5 5 6 7 7
    5300000   42 | 0 0 0 0 0 0 0 0 9
    0000000   43 | 0 0 0 0 0 5 5 6
        000   44 | 0 0 0 0 2
     820000   45 | 0 4
        100   46 |
       8000   47 |
          0   48 | 8
              49 |
          0   50 |
         00   51 |
          7   52 |
              53 |
              54 |
          0   55 |
```

Legend: 0 | 34 | 7 represents 340 feet for McGwire and 347 feet for Bonds.

(b) Answers will vary. For both players, the distances of home runs mainly fall from 360 to 450 feet. McGwire has quite a few extremely long distances.

43. Answers will vary. It is disconcerting that some schools have a negative ROI.

45.

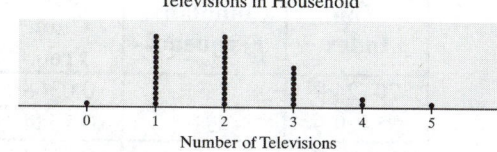

47. Because the data are quantitative, either a stem-and-leaf plot or a histogram would be appropriate. There were 20 people who spent less than 30 seconds, 7 people spent at least 30 seconds but less than 60 seconds, etc. One possible histogram is:

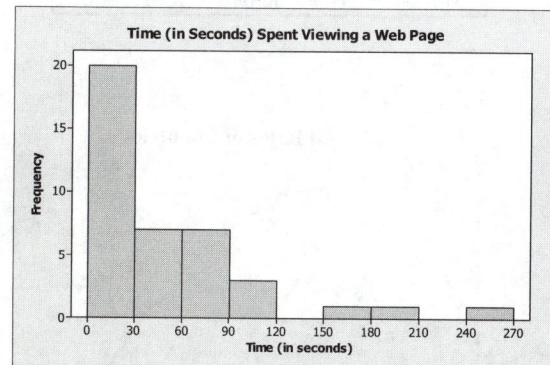

The data appear to be skewed right with a gap and one potential outlier. It seems as if the majority of surfers spent less than one minute viewing the page, while a few surfers spent several minutes viewing the page.

49. Classes should not overlap to avoid any confusion as to which class an observation belongs to.

51. There is no such thing as the correct choice for a class width, however some choices are better than others. For example, if the class width is

too small, the histogram will show many gaps between the bars. If the class width is too large, the histogram may not provide enough detail.

53. Answers will vary. The exercise illustrates the fact that there is no such thing as the "correct" histogram. However, some histograms are better than others and class width can affect the shape of a graph.

Section 2.3

1. An ogive is a line graph of cumulative frequencies or cumulative relative frequencies against upper class limits.

3. True

5. **(a)** 0.5; The class width is the difference between successive class midpoints. (e.g. $1.25 - 0.75 = 0.5$). There are 6 classes represented in the graph. The 8 plotted points are for the 6 class midpoints plus two additional points to connect the graph to the horizontal axis on the ends.

(b) The midpoint of the first class is 1.25. The lower limit of the first class is 1.00 and the upper limit is 1.49.

(c) The midpoint of the last class is 3.75. The lower limit of the last class is 3.50 and the upper limit is 3.99.

(d) The lower and upper limits of the class with 25 students are 2.50 and 2.99, respectively.

(e) The lower and upper limits of the class with the fewest students are 1.00 and 1.49, respectively.

7. **(a)** 5; The class width is the difference between successive class midpoints. (e.g. $24.9 - 19.9 = 5$).

(b) From the graph, it appears that approximately 10% of all four-year universities have a graduation rate below 34.9%.

(c) From the graph, it appears that approximately 60% of all four-year universities have a graduation rate below 59.9%.

(d) From the graph, it appears that approximately 5% of all four-year universities have a graduation rate above 90%.

9. (a) From the graph, it appears the unemployment rate in 2011 was about 9%.

(b) The highest unemployment rate was about 9.8%. This occurred in 2010.

(c) The highest inflation rate was about 4.3%. This occurred in 2008.

(d) The unemployment rate and inflation rate were closest in 2001. The unemployment rate and inflation rate were furthest in 2009.

(e) The misery index for 1999 was approximately 4.2 + 1.8 = 6. The misery index for 2014 was approximately 6.5 + 1.5 = 8. According to the misery index, the year 2014 was more "miserable" than the year 1999.

(f) Since 2010, the misery index has been declining due to the decreases in unemployment each year.

11. (a), (b)

Speed (km/h)	Cumulative Frequency	Cumulative Relative Frequency
10–13.9	4	0.0066
14–17.9	11	0.0181
18–21.9	28	0.0461
22–25.9	119	0.1960
26–29.9	401	0.6606
30–33.9	607	1.0000

(c)

(d)

13. (a), (b)
Cumulative frequency of a Gini Index of 25–29.9 = 5 + 16 = 21, and so on. The relative cumulative frequency of 25–29.9 is 21/136 = −.1544, and so on.

Gini Index	Cumulative Frequency	Rel. Cum. Freq.
20–24.9	5	0.0368
25–29.9	21	0.1544
30–34.9	49	0.3603
35–39.9	76	0.5588
40–44.9	96	0.7059
45–49.9	113	0.8309
50–54.9	126	0.9265
55–59.9	131	0.9632
60–64.9	136	1.0000

(c)

(d)

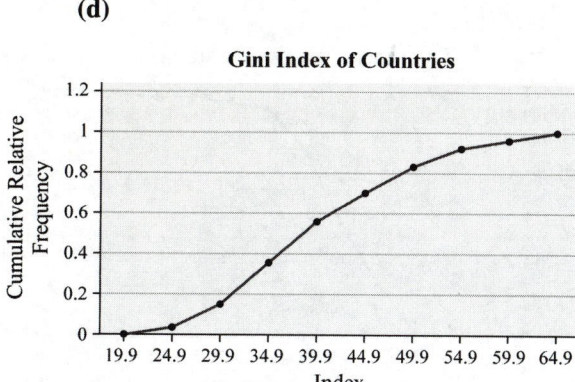

Gini Index of Countries

Answers will vary. The time-series plot shows that debt as a percent of gross domestic product remained relatively stable around 60% from 1996 to 2007 and then began to increase steadily from 2008 to 2014.

19. (a)

Return (Percent)	Consumer Defense	Industrial
−20 to −10.01	0.0435	0.0000
−10 to −0.01	0.0217	0.0235
0 to 9.99	0.0543	0.1882
10 to 19.99	0.5217	0.2471
20 to 29.99	0.2391	0.3059
30 to 39.99	0.0543	0.1647
40 to 49.99	0.0326	0.0235
50 to 59.99	0.0109	0.0235
60 to 69.99	0.0000	0.0000
70 to 79.99	0.0217	0.0118
80 to 89.99	0.0000	0.0118

15. (a)

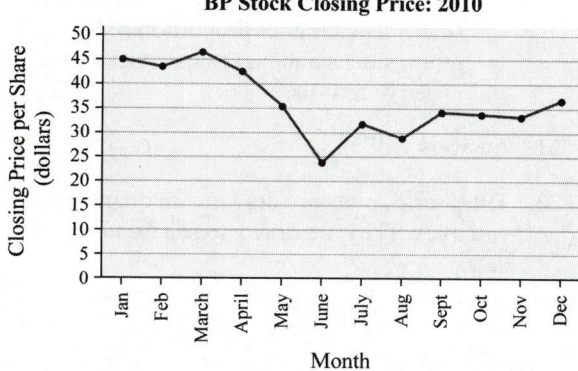

BP Stock Closing Price: 2010

(b) The value of the BP stock at the end of May 2010 was 35.72 and was only 24.02 at the end of June 2010. The percentage change in the BP stock price from May to June 2010 was
$(24.02-35.72)/35.72 = -0.328$, which is a decrease of 32.8%.

(b)

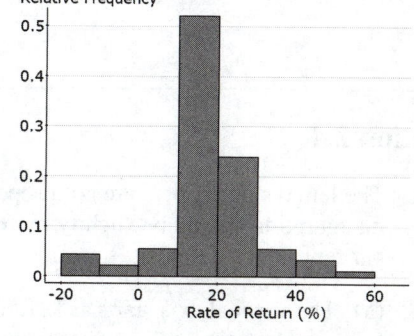

Three-Year Rate of Return in Consumer Defensive S

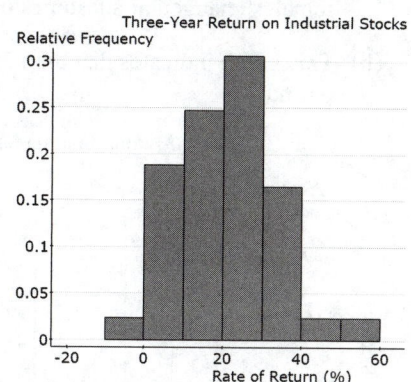

Three-Year Return on Industrial Stocks

17. (a)

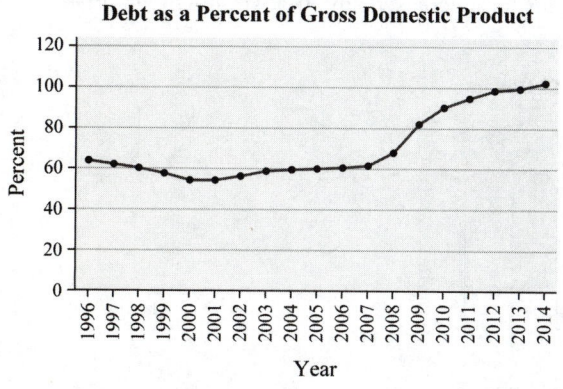

Debt as a Percent of Gross Domestic Product

(c)

(d)

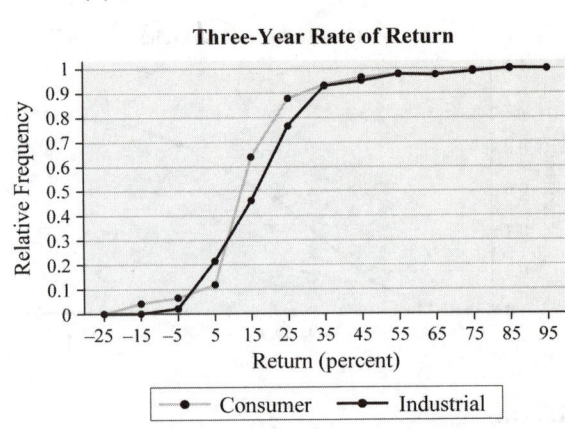

(e) Answers will vary. However, industrial stocks had a higher proportion of high-return stocks and had fewer stocks with negative returns.

21. Answers will vary.

23. Time-series plots are drawn with quantitative variables. They are drawn to see trends in the data.

Section 2.4

1. The lengths of the bars are not proportional. For example, the bar representing the cost of Clinton's inauguration should be slightly more than 9 times as long as the one for Carter's cost, and twice as long as the bar representing Reagan's cost.

3. **(a)** The vertical axis starts at $21,500 instead of $0. This tends to indicate that the median earnings for females changed at a faster rate than actually occurred.

(b) This graph indicates that the median earnings for females has decreased slightly over the given time period.

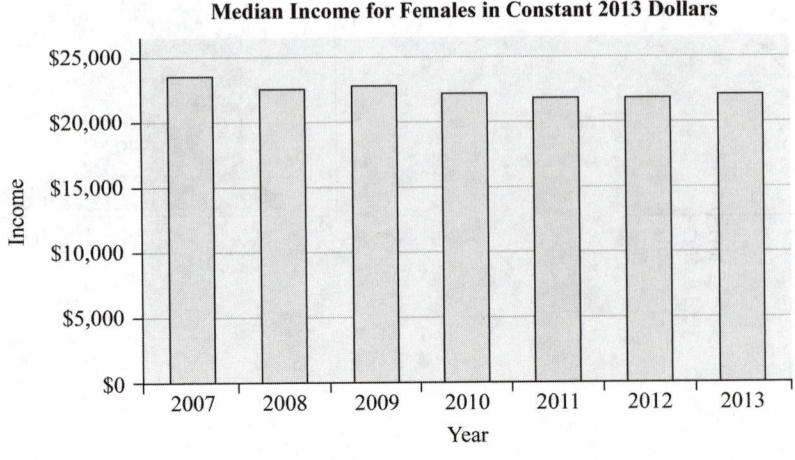

5. The bar for 12p–6p covers twice as many hours as the other bars. By combining two 3-hour periods, this bar looks larger compared to the others, making afternoon hours look more dangerous. If this bar were split into two periods, the graph may give a different impression. For example, the graph may show that daylight hours are safer.

7. Answers will vary. This graph is misleading because it does not take into account the size of the population of each state. For example, Vermont is going to pay less in total taxes than California simply because its population is so much lower. There are many variables that should be considered on per capita basis. For example, this graph would be less misleading if it was drawn to represent taxes paid per capita (per person).

9. (a) The graphic is misleading because the bars are not proportional. The bar for housing should be a little more than twice the length of the bar for transportation, but it is not.

(b) The graphic could be improved by adjusting the bars so that their lengths are proportional.

11. (a) Answers will vary. Here is a time-series plot that a politician might use to support the position that health care is increasing.

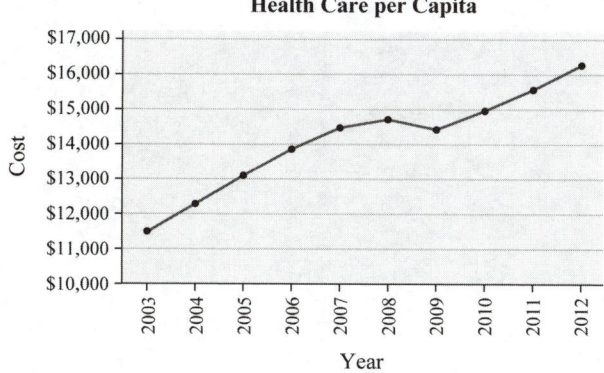

Health Care per Capita

(b) Answers will vary. Here is a time-series plot that the health care industry might use to refute the opinion of the politician.

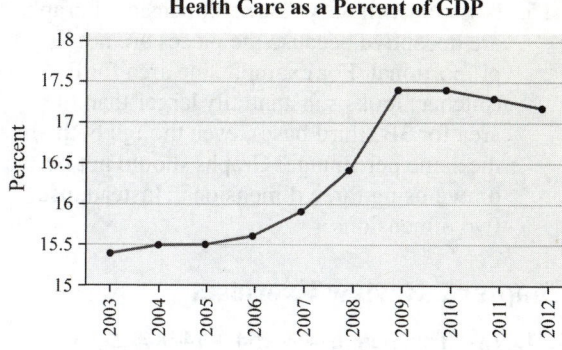

Health Care as a Percent of GDP

(c) Answers will vary. Changing the scale on the graph will affect the message. The message is also affected by using the variable "Health Care per Capita" rather than "Health Care as a Percent of GDP."

13. (a) A graph that is not misleading will use a vertical scale starting at 0% and bars of equal width. One example:

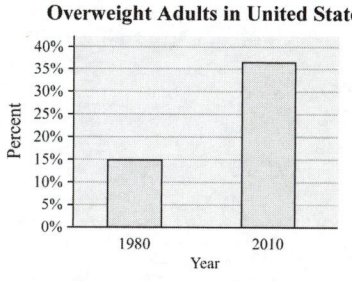

Overweight Adults in United States

(b)

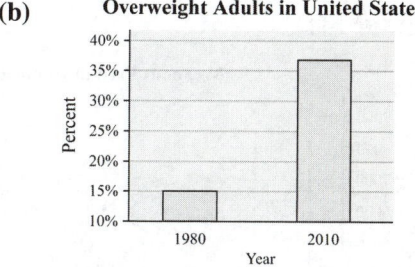

Overweight Adults in United States

This graphic is misleading because the vertical scale starts at 10% instead of 0% without indicating a gap. This might cause the reader to think that the proportion of overweight adults in the United States is increasing more quickly than they really are.

15. Answers will vary. Three-dimensional graphs are deceptive because the pieces are not proportional. For example, the area for P (pitcher) looks substantially larger than the area for 3B (third base), even though both are the same percentage. Graphs should not be drawn using three dimensions. Instead, use two dimensions.

Chapter 2 Review Exercises

1. **(a)** There are $614 + 154 + 1448 = 2216$ participants.

   (b) The relative frequency of the respondents indicating that it makes no difference is

$$\frac{1448}{2216} \approx 0.653$$

(c) A Pareto chart is a bar chart where the bars are in descending order.

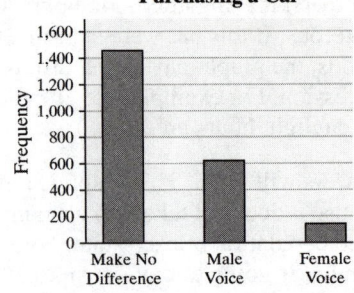

Convincing Voice in Purchasing a Car

(d) Answers will vary.

2. **(a)** Total homicides $= 8438 + 1486 + 685 + 1621 = 12230$
Relative frequency for firearms is $8438/12230 = 0.6899$, and so on.

Type of Weapon	Relative Frequency
Firearms	0.6899
Knives or cutting instruments	0.1215
Personal weapons	0.0560
Other weapon	0.1325

(b) The relative frequency is 0.6899, so 68.99% of the homicides were committed using a firearm.

(c)

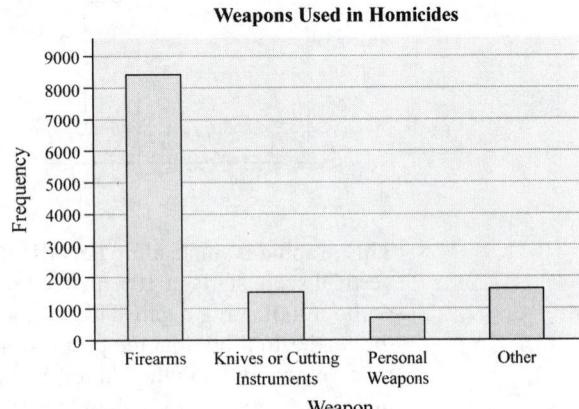

Weapons Used in Homicides

(d)

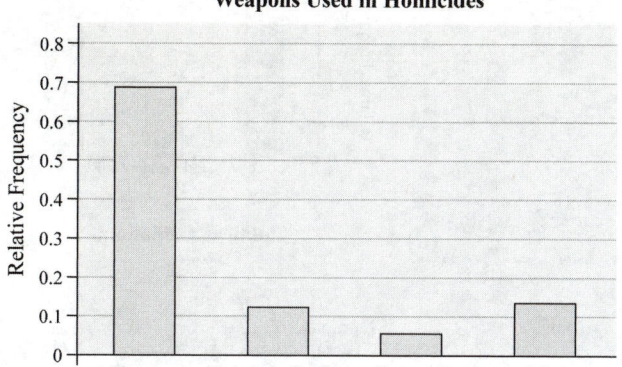

Weapons Used in Homicides

(e)

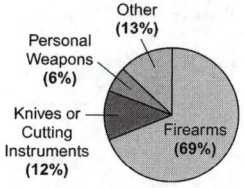

Weapons Used in Homicides

(d) The distribution is roughly symmetric and bell-shaped.

3. (a), (b), and **(c)**

Total births (in thousands) = 3 + 275 + 902 + 1128 + 1044 + 487 + 110 + 7 + 1 = 3957

Relative frequency for 10–14 year old mothers = $3/3957 \approx 0.0008$, and so on.

Cumulative frequency for 15–19 year old mothers = 3 + 275 = 278, and so on.

Cumulative relative frequency for 15–19 year old mothers = $278/3957 \approx 0.0703$, and so on.

Age of Mother	Rel. Freq.	Cumul. Freq.	Cumul. Rel. Freq.
10 – 14	0.0008	3	0.0008
15 – 19	0.0695	278	0.0703
20 – 24	0.2280	1180	0.2982
25 – 29	0.2851	2308	0.5833
30 – 34	0.2638	3352	0.8471
35 – 39	0.1231	3839	0.9702
40 – 44	0.0278	3949	0.9980
45 – 49	0.0018	3956	0.9997
50 – 54	0.0003	3957	1.0000

(g)

(h)

(i)

(j) From the relative frequency table, the relative frequency of 20–24 is 0.2280, so the percentage is 22.80%.

(k) $\dfrac{1044 + 487 + 110 + 7 + 1}{3957} = \dfrac{1649}{3957} \approx 0.4167$

41.67% of live births were to mothers aged 30 years or older.

4. **(a), (b)**

Affiliation	Frequency	Relative Frequency
Democrat	46	0.46
Independent	16	0.16
Republican	38	0.38

(c)

(d)

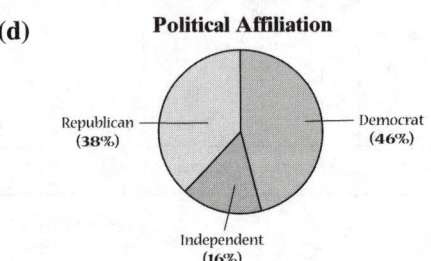

(e) Democrat appears to be the most common affiliation in Naperville.

5. **(a), (b), (c),** and **(d)**

Family Size	Freq.	Rel. Freq.	Cumul. Freq.	Cumul. Rel. Freq.
0	7	0.1167	7	0.1167
1	7	0.1167	14	0.2333
2	18	0.3000	32	0.5333
3	20	0.3333	52	0.8667
4	7	0.1167	59	0.9833
5	1	0.0167	60	1.0000

(e) The distribution is more or less symmetric.

(f)

(g) From the relative frequency table, the relative frequency of two children is 0.3000, so 30% of the couples have two children.

(h) From the frequency table, the relative frequency of at least two children (i.e. two or more) is
0.3000 + 0.3333 + 0.1167 + 0.0167 = 0.7667 or 76.67%. So, 76.67% of the couples have at least two children.

(i)

6. (a), (b)

Homeownership Rate	Frequency	Relative Frequency
45–49.9	1	0.0196
50–54.9	3	0.0588
55–59.9	1	0.0196
60–64.9	12	0.2353
65–69.9	17	0.3333
70–74.9	15	0.2941
75–75.9	2	0.0392

(c)

(d)

(e) The distribution is skewed left.

(f)

Homeownership Rate	Frequency	Relative Frequency
40–49.9	1	0.0196
50–59.9	4	0.0784
60–69.9	29	0.5686
70–79.9	17	0.3333

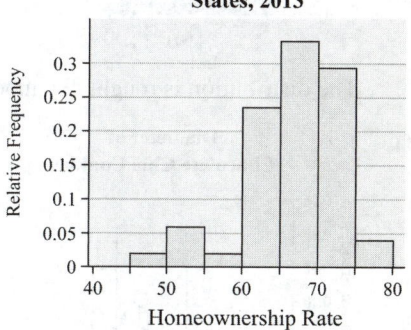

The distribution is skewed left.

Answers will vary. Both class widths give a good overall picture of the distribution. The first class width provides a little more detail to the graph, but not necessarily enough to be worth the trouble. An intermediate value, say a width of 8, might be a reasonable compromise.

7. (a), (b), (c), and **(d)**

Answers will vary. Using 2.2000 as the lower class limit of the first class and 0.0200 as the class width, we obtain the following.

Class	Freq.	Rel. Freq.	Cumul. Freq.	Cumul. Rel.Freq.
2.2000 – 2.2199	2	0.0588	2	0.0588
2.2200 – 2.2399	3	0.0882	5	0.1471
2.2400 – 2.2599	5	0.1471	10	0.2941
2.2600 – 2.2799	6	0.1765	16	0.4706
2.2800 – 2.2999	4	0.1176	20	0.5882
2.3000 – 2.3199	7	0.2059	27	0.7941
2.3200 – 2.3399	5	0.1471	32	0.9412
2.3400 – 2.3599	1	0.0294	33	0.9706
2.3600 – 2.3799	1	0.0294	34	1

(e)

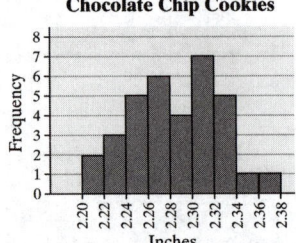

Diameter of Chocolate Chip Cookies

The distribution is roughly symmetric.

(f)

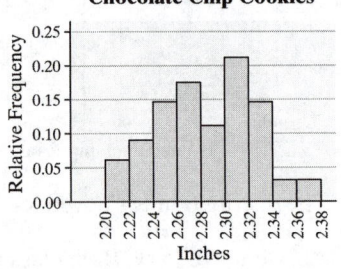

Diameter of Chocolate Chip Cookies

8.

Hours Spent Online

```
12 |
13 | 467
14 | 05578
15 | 1236
16 | 456
17 | 113449
18 | 066889
19 | 2
20 | 168
21 | 119
22 | 29
23 | 48
24 | 4
25 | 7
26 |
```

Legend: 13 | 4 = average 13.4 hours per week.

The distribution is slightly skewed right.

9. (a) Grade inflation seems to be happening in colleges. GPAs have increased every time period for all schools.

(b) Answers may vary. GPAs increased about 5.6% for public schools. GPAs increased about 6.8% for private schools. Private schools have higher grade inflation because the GPAs are higher and they are increasing faster.

(c) The graph is misleading because it starts at 2.6 on the vertical axis.

10. (a) Answers will vary. The adjusted gross income share of the top 1% of earners shows steady increases overall, with a few minor exceptions. The adjusted gross income share of the bottom 50% of earners shows steady decreases overall, with a few minor exceptions.

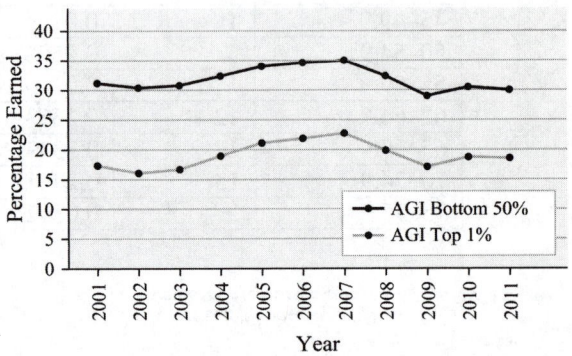

Percentage of Adjusted Gross Incomes (AGI) Earned

(b) Answers will vary. The income tax share of the top 1% of earners shows steady increases overall, with few exceptions, including a notable decrease from 2007 to 2008. The income tax share of the bottom 50% of earners shows steady decreases over time.

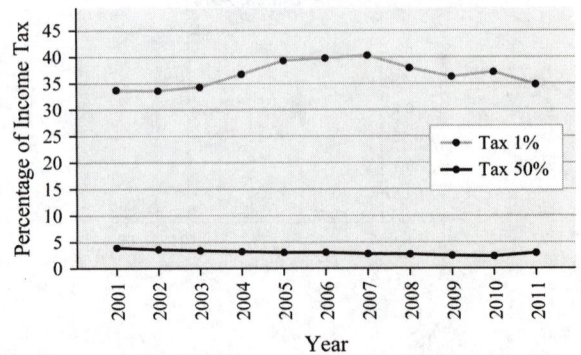

Income Tax Share of Earners

11. (a) Graphs will vary. One way to mislead would be to start the vertical scale at a value other than 0. For example, starting the vertical scale at $30,000 might make the reader believe that college graduates earn more than three times what a high school graduate earns (on average).

(b) A graph that does not mislead would use equal widths for the bars and would start the vertical scale at $0. Here is an example of a graph that is not misleading:

2013 Average Earnings

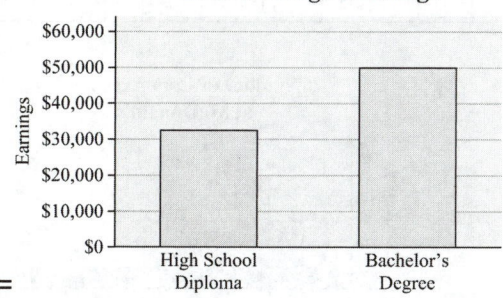

12. (a) Flats are preferred the most (40%) and extra-high heels are preferred the least (1%).

(b) The graph is misleading because the bar heights and areas for each category are not proportional.

Chapter 2 Test

1. (a) A 5 Star rating was the most popular rating with 1675 votes.

(b) $35 + 67 + 246 + 724 + 1675 = 2747$ postings were posted on Yelp for Hot Doug's restaurant.

(c) $1675 - 724 = 951$
There were 951 more 5 Star ratings than 4 Star ratings.

(d) There were 1675 5 Star ratings out of a total of 2747 ratings. $\dfrac{1675}{2747} \approx 0.6098$
Approximately 61% of all ratings were 5 Star ratings.

(e) No, it is not appropriate to describe the shape of the distribution as skewed right. The data represented by the graph are qualitative, so the bars in the graph could be placed in any order.

2. (a) There were 1005 responses. The relative frequency who indicated they preferred new tolls was $\dfrac{412}{1005} = 0.4100$, and so on.

Response	Freq.	Rel. Freq.
New Tolls	412	0.4100
Inc. Gas Tax	181	0.1801
No New Roads	412	0.4100

(b) The relative frequency is 0.1801, so the percentage of respondents who would like to see an increase in gas taxes is 18.01%.

(c)

(d)

(e)

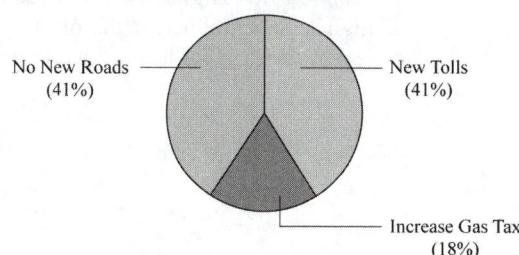

3. (a), (b)

Education	Freq.	Rel. Freq.
No high school diploma	9	0.18
High school graduate	16	0.32
Some college	9	0.18
Associate's degree	4	0.08
Bachelor's degree	8	0.16
Advanced degree	4	0.08

(c)

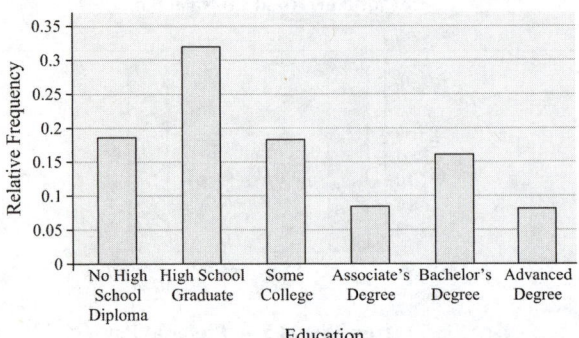

Educational Attainment of Commuters

(d) **Educational Attainment of Commuters**

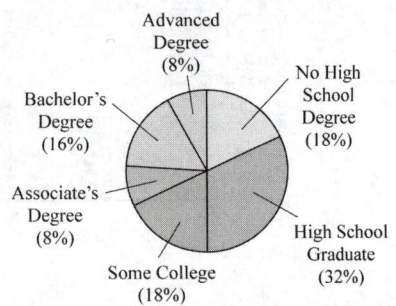

(e) The largest bar (and largest pie segment) corresponds to "High School Graduate," so high school graduate is the most common educational level of a commuter.

4. (a), (b), (c), and (d)

No. of Cars	Freq.	Rel. Freq.	Cumul. Freq.	Cumul. Rel. Freq.
1	5	0.10	5	0.10
2	7	0.14	12	0.24
3	12	0.24	24	0.48
4	6	0.12	30	0.60
5	8	0.16	38	0.76
6	5	0.10	43	0.86
7	2	0.04	45	0.90
8	4	0.08	49	0.98
9	1	0.02	50	1

(e)

Number of Cars Arriving at McDonald's

The distribution is skewed right.

(f)

Number of Cars Arriving at McDonald's

(g) The relative frequency of exactly 3 cars is 0.24. So, for 24% of the weeks, exactly three cars arrived between 11:50 am and 12:00 noon.

(h) The relative frequency of 3 or more cars
$$= 0.24 + 0.12 + 0.16 + 0.10$$
$$+ 0.04 + 0.08 + 0.02 = 0.76$$
So, for 76% of the weeks, three or more cars arrived between 11:50 am and 12:00 noon.

(i)

5. Answers may vary. One possibility follows:

(a), (b)

Using a lower class limit of the first class of 20 and a class width of 10:
Total number of data points = 40
Relative frequency of 20 – 29 = 1/40 = 0.025, and so on.

HDL Cholesterol	Frequency	Relative Frequency
20–29	1	0.025
30–39	6	0.150
40–49	10	0.250
50–59	14	0.350
60–69	6	0.150
70–79	3	0.075

(c)

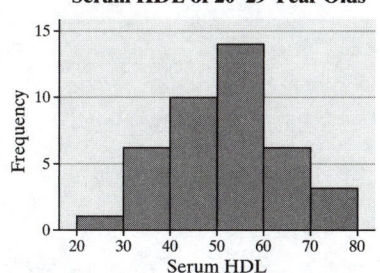

Serum HDL of 20–29 Year Olds

(d)

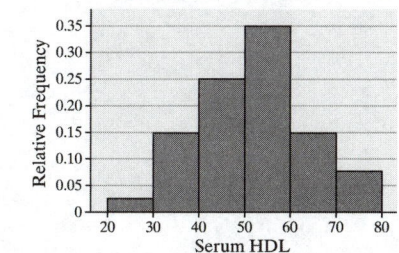

Serum HDL of 20–29 Year Olds

(e) The distribution appears to be roughly bell-shaped.

6. (a), (b)

Closing Price	Cumul. Freq.	Cumul. Rel. Freq.
50,000 – 99,999	4	0.08
100,000 – 149,999	17	0.34
150,000 – 199,999	36	0.72
200,000 – 249,999	43	0.86
250,000 – 299,999	46	0.92
300,000 – 349,999	48	0.96
350,000 – 399,999	49	0.98
400,000 – 449,999	49	0.98
450,000 – 499,999	50	1

(c) The cumulative relative frequency for the class $150,000–$199,000 is 0.72. Therefore, 72% of the homes sold for less than $200,000.

(d)

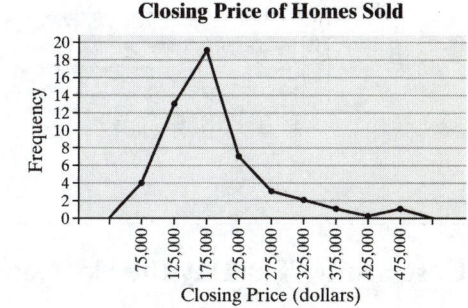

Closing Price of Homes Sold

The distribution is skewed right.

(e)

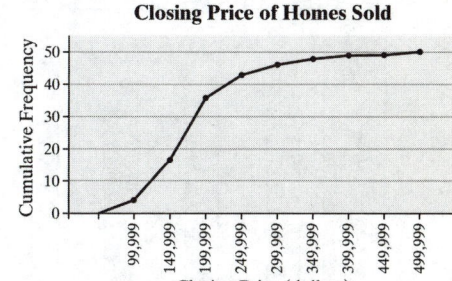

Closing Price of Homes Sold

(f)

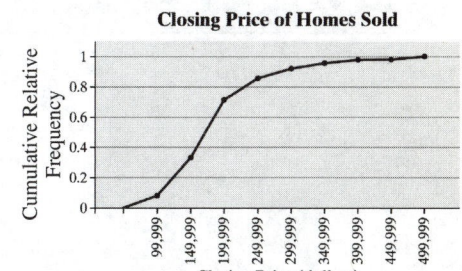

Closing Price of Homes Sold

7. The stem-and-leaf diagram below shows an approximately uniform distribution.

Time Spent on Homework

```
 4 | 0567
 5 | 26
 6 | 13
 7 | 01338
 8 | 59
 9 | 1369
10 | 3899
11 | 0018
12 | 556
```

Legend: 4 | 0 represents 40 minutes.

8. The curves in the figure below appear to follow the same trend. Birth rate increases as per capita income increases.

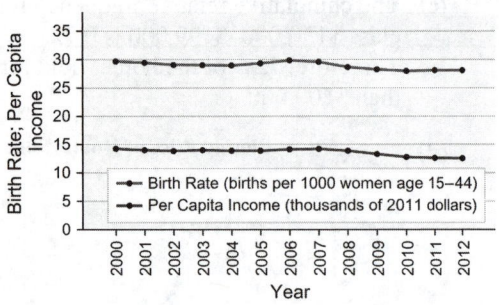

9. Answers may vary. It is difficult to interpret this graph because it is not clear whether the scale is represented by the height of the steps, the width of the steps, or by the graphics above the steps. The graphics are misleading because they must be increased in size both vertically and horizontally to avoid distorting the image. Thus, the resulting areas are not proportionally correct. The graph could be redrawn using bars whose widths are the same and whose heights are proportional based on the given percentages. The use of graphics should be avoided, or a standard size graphic representing a fixed value could be used and repeated as necessary to illustrate the given percentages.

Case Study: The Day the Sky Roared

1.

During the April 3-4, 1974 outbreak, about 44% of the tornadoes exceeded F-3 on the Fujita Wind Damage Scale. This was much greater than the 1% that typically occurs.

2. The histogram will vary depending on the class width.

3. Histograms may vary depending on class widths. For comparison purposes, the same class width was used for each histogram.

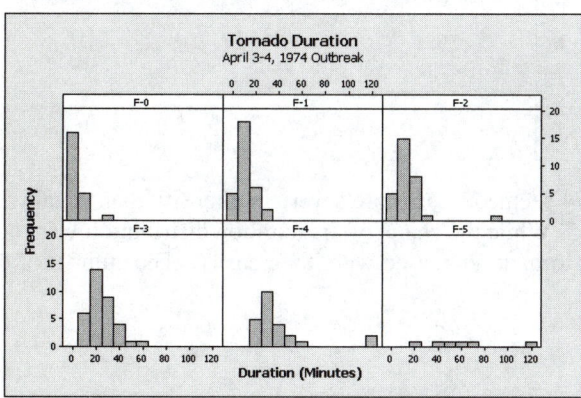

The distributions all appear to be skewed right, though the distribution for F-5 tornadoes is difficult to see due to the low sample size. There is an obvious shift in the distributions. As the strength of the tornado increases, the duration of the tornado increases.

4. There were 305 deaths during the outbreak. Of these, 259 were due to the more severe tornadoes.

$$\frac{259}{305} \approx 0.8492$$

Roughly 85% of the deaths during the outbreak were due to the more severe tornadoes. This may be a little high, but it is consistent since it is greater than 70%.

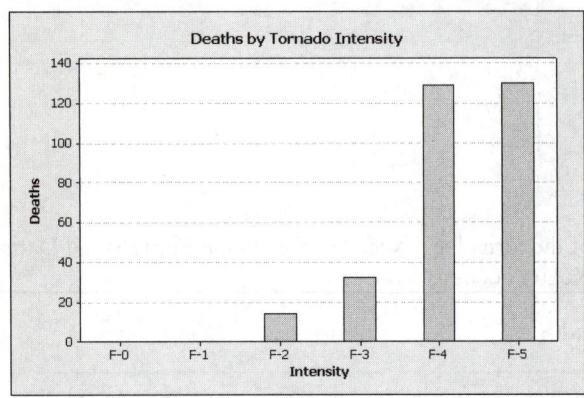

5. The provided data is not sufficient to determine whether or not tornadoes are more likely to strike rural areas. Some research at Texas A&M University indicates that tornadoes are more likely to occur in urban or suburban areas, possibly due to greater temperature differences. The data does indicate that the number of deaths decreases as the population of the community increases. The higher the population density, the greater the chance that a tornado is detected and reported early, thereby providing more time for residents to take shelter.

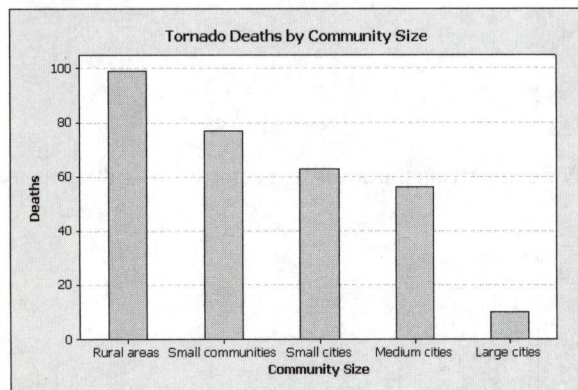

6. Answers will vary. The outbreak of April 3-4, 1974 seemed to be more severe in intensity than usual with 20% of the tornadoes being classified as F-4 or F-5. While the shape of the duration distribution was roughly the same for each intensity level, the duration of a tornado increased with its intensity. The number of deaths decreased as the community size increased.

Chapter 3
Numerically Summarizing Data

Section 3.1

1. A statistic is resistant if its value is not sensitive to extreme data values.

3. HUD uses the median because the data are skewed. Explanations will vary. One possibility is that the price of homes has a distribution that is skewed to the right, so the median is more representative of the typical price of a home.

5. $\dfrac{10,000+1}{2} = 5000.5$. The median is between the 5000th and the 5001st ordered values.

7. $\bar{x} = \dfrac{20+13+4+8+10}{5} = \dfrac{55}{5} = 11$

9. $\mu = \dfrac{3+6+10+12+14}{5} = \dfrac{45}{5} = 9$

11. $\mu = \dfrac{218,469,636}{82,566} = \$2,646$

The mean price per ticket was $2,646.

13. $\sum x_i = 34.0 + 33.2 + 37.0 + 29.4 + 23.6 + 25.9 = 183.1$

Mean $= \bar{x} = \dfrac{\sum x_i}{n} = \dfrac{183.1}{6} = 30.52$ mpg

Data in order: 23.6, 25.9, 29.4, 33.2, 34.0, 37.0

Median $= \dfrac{29.4+33.2}{2} = \dfrac{62.6}{2} = 31.3$ mpg

No data value occurs more than once, so there is no mode.

15. $\sum x_i = 3960 + 4090 + 3200 + 3100 + 2940 + 3830 + 4090 + 4040 + 3780 = 33,030$ psi

Mean $= \bar{x} = \dfrac{\sum x_i}{n} = \dfrac{33,030}{9} = 3670$ psi

Data in order: 2940, 3100, 3200, 3780, 3830, 3960, 4040, 4090, 4090

Median = the 5th ordered data value = 3830 psi

Mode = 4090 psi (because it is the only data value to occur twice)

17. (a) The histogram is skewed to the right, suggesting that the mean is greater than the median. That is, $\bar{x} > M$.

(b) The histogram is symmetric, suggesting that the mean is approximately equal to the median. That is, $\bar{x} = M$.

(c) The histogram is skewed to the left, suggesting that the mean is less than the median. That is, $\bar{x} < M$.

19. (a) Traditional Lecture:

$\sum x_i = 70.8 + 69.1 + 79.4 + 67.6$
$\quad + 85.3 + 78.2 + 56.2 + 81.3$
$\quad + 80.9 + 71.5 + 63.7 + 69.8$
$\quad + 59.8$
$\quad = 933.6$

Mean $= \bar{x} = \dfrac{\sum x_i}{n} = \dfrac{933.6}{13} \approx 71.82$

Data in order:
56.2, 59.8, 63.7, 67.6, 69.1, 69.8, 70.8, 71.5, 78.2, 79.4, 80.9, 81.3, 85.3
Median $= M = 70.8$

Flipped Classroom:

$\sum x_i = 76.4 + 71.6 + 63.4 + 72.4$
$\quad + 77.9 + 91.8 + 78.9 + 76.8$
$\quad + 82.1 + 70.2 + 91.5 + 77.8$
$\quad + 76.5$
$\quad = 1007.3$

Mean $= \bar{x} = \dfrac{\sum x_i}{n} = \dfrac{1007.3}{13} \approx 77.48$

Data in order:
63.4, 70.2, 71.6, 72.4, 76.4, 76.5, 76.8, 77.8, 77.9, 78.9, 82.1, 91.5, 91.8
Median $= M = 76.8$

(b) If the score of 59.8 had been incorrectly recorded as 598 in the traditional lecture, then the incorrect mean and median would be:

$\sum x_i = 70.8 + 69.1 + 79.4 + 67.6$
$\quad + 85.3 + 78.2 + 56.2 + 81.3$
$\quad + 80.9 + 71.5 + 63.7 + 69.8$
$\quad + 598$
$\quad = 1471.8$

$$\text{Mean} = \overline{x} = \frac{\sum x_i}{n} = \frac{1471.8}{13} \approx 113.22$$

Data in order:
56.2, 63.7, 67.6, 69.1, 69.8, 70.8, 71.5,
78.2, 79.4, 80.9, 81.3, 85.3, 598
$\text{Median} = M = 71.5$

The incorrect entry causes the mean to increase substantially while the median only slightly changes. This illustrates that the median is resistant while the mean is not resistant.

21. (a) $\sum x_i = 76 + 60 + 60 + 81 + 72 + 80 + 80$
$+ 68 + 73$

$= 650$

$$\mu = \frac{\sum x_i}{N} = \frac{650}{9} \approx 72.2 \text{ beats per minute}$$

(b) Samples and sample means will vary.

(c) Answers will vary.

23. The distribution is relatively symmetric as is evidenced by both the histogram and the fact that the mean and median are approximately equal. Therefore, the mean is the better measure of central tendency.

25. To create the histogram, we choose the lower class limit of the first class to be 0.78 and the class width to be 0.02. The resulting classes and frequencies follow:

Class	Freq.	Class	Freq.
0.78 − 0.79	1	0.88 − 0.89	10
0.80 − 0.81	1	0.90 − 0.91	9
0.82 − 0.83	4	0.92 − 0.93	4
0.84 − 0.85	8	0.94 − 0.95	2
0.86 − 0.87	11		

Weight of Plain M&Ms

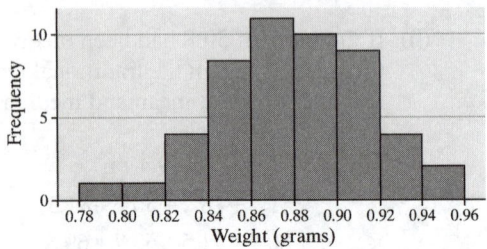

To find the mean, we add all of the data values and divide by the sample size: $\sum x_i = 43.73$;

$$\overline{x} = \frac{\sum x_i}{n} = \frac{43.73}{50} \approx 0.875 \text{ grams}$$

To find the median, we arrange the data in order. The median is the mean of the 25th and 26th data values:

$$M = \frac{0.87 + 0.88}{2} = 0.875 \text{ grams}$$

The mean is approximately equal to the median, suggesting that the distribution is symmetric. This is confirmed by the histogram (though is does appear to be slightly skewed left). The mean is the better measure of central tendency.

27. To create the histogram, we choose the lower class limit of the first class to be 0 and the class width to be 5. The resulting classes and frequencies follow:

Class	Freq.	Class	Freq.
0 − 4	3	25 − 29	4
5 − 9	1	30 − 34	5
10 − 14	0	35 − 39	3
15 − 19	4	40 − 44	1
20 − 24	4		

Hours Worked per Week

To find the mean, we add all of the data values and divide by the sample size: $\sum x_i = 550$;

$$\overline{x} = \frac{\sum x_i}{n} = \frac{550}{25} = 22 \text{ hours}$$

To find the median, we arrange the data in order. The median is the 13th data value:
$M = 25 \text{ hours}$

The mean is smaller than the median, suggesting that the distribution is skewed left. This is confirmed by the histogram. The median is the better measure of central tendency.

29. (a) The frequencies are:
Liberal = 10
Moderate = 12
Conservative = 8
The mode political view is moderate.

(b) Yes. Rotating the choices will help to avoid response bias that might be caused by the wording of the question.

31. Sample size of 5:
All data recorded correctly: $\bar{x} = 99.8$; $M = 100$
106 recorded at 160: $\bar{x} = 110.6$; $M = 100$
Sample size of 12:
All data recorded correctly: $\bar{x} \approx 100.4$; $M = 101$
106 recorded at 160: $\bar{x} \approx 104.9$; $M = 101$
Sample size of 30:
All data recorded correctly: $\bar{x} = 100.6$; $M = 99$
106 recorded at 160: $\bar{x} = 102.4$; $M = 99$
For each sample size, the mean becomes larger while the median remains the same. As the sample size increases, the impact of the incorrectly recorded data value on the mean decreases.

33. The sum of the nineteen readable scores is $19 \cdot 84 = 1596$. The sum of all twenty scores is $20 \cdot 82 = 1640$. Therefore, the unreadable score is $1640 - 1596 = 44$.

35. Answers will vary. $\bar{x} = 0.061$; $M = 0$
The histogram is skewed to the right, so we would expect to report the median as the measure of central tendency. However, a median of 0 does not tell the whole story of the fatal accidents, so it could be argued that both the mean and median should be reported. With this data, because the median is 0, at least half of all fatal accidents involved drivers with no alcohol in their system.

Blood Alcohol Level of Driver in Fatal Accidents

37. (a) Mean:
$$\sum x_i = 30 + 30 + 45 + 50 + 50 + 50 + 55 + 55$$
$$+ 60 + 75$$
$$= 500$$
$$\mu = \frac{\sum x_i}{N} = \frac{500}{10} = 50$$
The mean is $50,000.

Median: The ten data values are in order. The median is the mean of the 5th and 6th data values: $M = \dfrac{50 + 50}{2} = \dfrac{100}{2} = 50$.
The median is $50,000.

Mode: The mode is $50,000 (the most frequent salary).

(b) Add $2500 ($2.5 thousand) to each salary to form the 2nd: New data set: 32.5, 32.5, 47.5, 52.5, 52.5, 52.5, 57.5, 57.5, 62.5, 77.5

2nd Mean:
$$\sum x_i = 32.5 + 32.5 + 47.5 + 52.5 + 52.5$$
$$+ 52.5 + 57.5 + 57.5 + 62.5 + 77.5$$
$$= 525$$

$$\mu_{2nd} = \frac{\sum x_i}{N} = \frac{525}{10} = 52.5$$ The mean for the 2nd data set is $52,500.

2nd Median: The ten data values are in order. The median is the mean of the 5th and 6th data values:
$$M_{2nd} = \frac{52.5 + 52.5}{2} = \frac{105}{2} = 52.5.$$
The median for the 2nd data set is $52,500.

2nd Mode: The mode for the 2nd data set is $52,500 (the most frequent new salary).

All three measures of central tendency increased by $2500, which was the amount of the raises.

(c) Multiply each original data value by 1.05 to generate the 3rd data set: 31.5, 31.5, 47.25, 52.5, 52.5, 52.5, 57.75, 57.75, 63, 78.75

3rd Mean:
$$\sum x_i = 31.5 + 31.5 + 47.25 + 52.5 + 52.5$$
$$+ 52.5 + 57.75 + 57.75 + 63 + 78.75$$
$$= 525$$

$$\mu_{3rd} = \frac{\sum x_i}{N} = \frac{525}{10} = 52.5$$

The mean of the 3rd data set is $52,500.

3rd Median: The ten data values are in order. The median is the mean of the 5th and 6th data values:

$$M_{3rd} = \frac{52.5 + 52.5}{2} = \frac{105}{2} = 52.5.$$

The median of the 3rd data set is $52,500.

3rd Mode: The mode of the 3rd data set is $52,500 (the most frequent new salary).

All three measures of central tendency increased by 5%, which was the amount of the raises.

(d) Add $25 thousand to the largest data value to form the new data set: 30, 30, 45, 50, 50, 50, 55, 55, 60, 100

4th Mean:
$$\sum x_i = 30 + 30 + 45 + 50 + 50 + 50 + 55 + 55 + 60 + 100$$
$$= 525$$

$$\mu_{4th} = \frac{\sum x_i}{N} = \frac{525}{10} = 52.5$$

The mean of the 4th data set is $52,500.

4th Median: The ten data values are in order. The median is the mean of the 5th and 6th data values:

$$M_{4th} = \frac{50 + 50}{2} = \frac{100}{2} = 50.$$

The median of the 4th data set is $50,000.

4th Mode: The mode of the 4th data set is $50,000 (the most frequent salary).

The mean was increased by $2500, but the median and mode remained unchanged.

39. The largest value is 0.95 and the smallest is 0.79. After deleting these two values, we have:

$$\sum x_i = 41.99; \quad \overline{x} = \frac{\sum x_i}{n} = \frac{41.99}{48} \approx 0.875 \text{ grams}.$$

The mean after deleting these two data values is 0.875 grams. The trimmed mean is more resistant than the regular mean because the most extreme values are omitted before the mean is computed.

41. (a) The data are discrete.

(b) To construct the histogram, we first organize the data into a frequency table:

Number of Drinks	Frequency
0	23
1	17
2	4
3	3
4	2
5	1

Number of Days High School Students Consume Alcohol Each Week

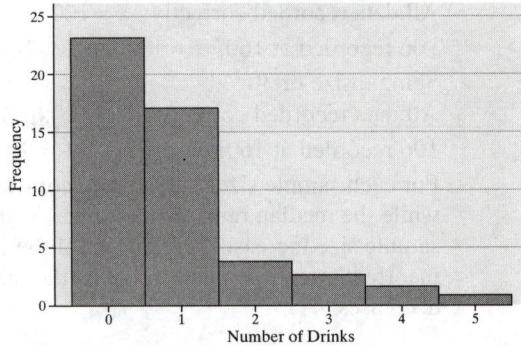

The distribution is skewed right.

(c) Since the distribution is skewed right, we would expect the mean to be greater than the median.

(d) To find the mean, we add all of the data values and divide by the sample size.

$$\sum x_i = 47; \quad \overline{x} = \frac{\sum x_i}{n} = \frac{47}{50} = 0.94$$

To find the median, we arrange the data in order. The median is the mean of the 25th and 26th data values. $M = \frac{1+1}{2} = 1$

This tells us that the mean can be less than the median in skewed-right data. Therefore, the rule *mean greater than median implies the data are skewed right* is not always true.

(e) The mode is 0 (the most frequent value).

(f) Yes, Carlos's survey likely suffers from sampling bias. It is difficult to get truthful responses to this type of question. Carlos would need to ensure that the identity of the respondents is anonymous.

43. The median is resistant because it is the "middle observation" and increasing the largest value or decreasing the smallest value does not affect the median. The mean is not resistant because it is a function of the sum of the data values. Changing the magnitude of one value changes the sum of the values, and thus affects the mean.

45. (a) The median describes the typical U.S. household's net worth better. Household net worth is expected to be skewed right, and the mean will be affected by the outliers.

    (b) Household net worth is expected to be skewed right.

    (c) There are a few households with very high net worth

47. The distribution is skewed right, so the median amount of money lost is less than the mean amount lost.

Section 3.2

1. zero

3. True

5. From Section 3.1, Exercise 7, we know that $\bar{x} = 11$.

x_i	$\bar{x}$	$x_i - \bar{x}$	$(x_i - \bar{x})^2$
20	11	$20 - 11 = 9$	$9^2 = 81$
13	11	$13 - 11 = 2$	$2^2 = 4$
4	11	$4 - 11 = -7$	$(-7)^2 = 49$
8	11	$8 - 11 = -3$	$(-3)^2 = 9$
10	11	$10 - 11 = -1$	$(-1)^2 = 1$
		$\sum(x_i - \bar{x}) = 0$	$\sum(x_i - \bar{x})^2 = 144$

$$s^2 = \frac{\sum(x_i - \bar{x})^2}{n-1} = \frac{144}{5-1} = \frac{144}{4} = 36$$

$$s = \sqrt{s^2} = \sqrt{\frac{\sum(x_i - \bar{x})^2}{n-1}} = \sqrt{\frac{144}{5-1}} = \sqrt{36} = 6$$

7. From Section 3.1, Exercise 9, we know that $\mu = 9$.

x_i	μ	$x_i - \mu$	$(x_i - \mu)^2$
3	9	$3 - 9 = -6$	$(-6)^2 = 36$
6	9	$6 - 9 = -3$	$(-3)^2 = 9$
10	9	$10 - 9 = 1$	$1^2 = 1$
12	9	$12 - 9 = 3$	$3^2 = 9$
14	9	$14 - 9 = 5$	$5^2 = 25$
		$\sum(x_i - \mu) = 0$	$\sum(x_i - \mu)^2 = 80$

$$\sigma^2 = \frac{\sum(x_i - \mu)^2}{N} = \frac{80}{5} = 16$$

$$\sigma = \sqrt{\sigma^2} = \sqrt{\frac{\sum(x_i - \mu)^2}{N}} = \sqrt{\frac{80}{5}} = \sqrt{16} = 4$$

9. $\bar{x} = \dfrac{6 + 52 + 13 + 49 + 35 + 25 + 31 + 29 + 31 + 29}{10}$

$$= \frac{300}{10} = 30$$

x_i	$\bar{x}$	$x_i - \bar{x}$	$(x_i - \bar{x})^2$
6	30	$6 - 30 = -24$	$(-24)^2 = 576$
52	30	$52 - 30 = 22$	$22^2 = 484$
13	30	$13 - 30 = -17$	$(-17)^2 = 289$
49	30	$49 - 30 = 19$	$19^2 = 361$
35	30	$35 - 30 = 5$	$5^2 = 25$
25	30	$25 - 30 = -5$	$(-5)^2 = 25$
31	30	$31 - 30 = 1$	$1^2 = 1$
29	30	$29 - 30 = -1$	$(-1)^2 = 1$
31	30	$31 - 30 = 1$	$1^2 = 1$
29	30	$29 - 30 = -1$	$(-1)^2 = 1$
		$\sum(x_i - \bar{x}) = 0$	$\sum(x_i - \bar{x})^2 = 1764$

$$s^2 = \frac{\sum(x_i - \bar{x})^2}{n-1} = \frac{1764}{10-1} = \frac{1764}{9} = 196$$

$$s = \sqrt{s^2} = \sqrt{\frac{\sum(x_i - \bar{x})^2}{N}} = \sqrt{\frac{1764}{10-1}} = \sqrt{196} = 14$$

11. Range = Largest Value − Smallest Value
= $37 - 23.6 = 13.4$ miles per gallon

To calculate the sample variance and the sample standard deviation, we use the computational formulas:

x_i	x_i^2
34.0	1156.00
33.2	1102.24
37.0	1369.00
29.4	864.36
23.6	556.96
25.9	670.81
$\sum x_i = 183.1$	$\sum x_i^2 = 5719.37$

$$s^2 = \frac{\sum x_i^2 - \frac{(\sum x_i)^2}{n}}{n-1} = \frac{5719.37 - \frac{(183.1)^2}{6}}{6-1}$$

$$\approx 26.3537 \text{ (mpg)}^2$$

$$s = \sqrt{s^2} = \sqrt{26.3537} \approx 5.134 \text{ mpg}$$

13. Range = Largest Value − Smallest Value
= $4090 - 2940 = 1150$ psi

To calculate the sample variance and the sample standard deviation, we use the computational formulas:

x_i	x_i^2
3960	15,681,600
4090	16,728,100
3200	10,240,000
3100	9,610,000
2940	8,643,600
3830	14,668,900
4090	16,728,100
4040	16,321,600
3780	14,288,400
$\sum x_i = 33,020$	$\sum x_i^2 = 122,910,300$

$$s^2 = \frac{\sum x_i^2 - \frac{(\sum x_i)^2}{n}}{n-1} = \frac{122,910,300 - \frac{(33,030)^2}{9}}{9-1}$$

$$\approx 211,275 \text{ (psi)}^2$$

$$s = \sqrt{\frac{122,910,300 - \frac{(33,030)^2}{9}}{9-1}} \approx 459.6 \text{ psi}$$

15. Histogram (b) depicts a higher standard deviation because the data is more dispersed, with data values ranging from 30 to 75. In histogram (a), the data values only range from 40 to 60.

17. (a) Traditional Classroom
Range = $85.3 - 56.2 = 29.1$

Flipped Classroom
Range = $91.8 - 63.4 = 28.4$

Using the range as the measure of dispersion, the traditional classroom has slightly more dispersion in exam scores.

(b) To calculate the sample standard deviation, we use the computational formula:

Traditional Classroom

x_i	x_i^2
70.8	5012.64
69.1	4774.81
79.4	6304.36
67.6	4569.76
85.3	7276.09
78.2	6115.24
56.2	3158.44
81.3	6609.69
80.9	6544.81
71.5	5112.25
63.7	4057.69
69.8	4872.04
59.8	3576.04
$\sum x_i = 933.6$	$\sum x_i^2 = 67,983.86$

$$s = \sqrt{\frac{\sum x_i^2 - \frac{(\sum x_i)^2}{n}}{n-1}} = \sqrt{\frac{67983.86 - \frac{(933.6)^2}{13}}{13-1}}$$

$$\approx 8.837$$

Flipped Classroom

x_i	x_i^2
76.4	5836.96
71.6	5126.56
63.4	4019.56
72.4	5241.76
77.9	6068.41
91.8	8427.24
78.9	6225.21
76.8	5898.24
82.1	6740.41
70.2	4928.04
91.5	8372.25
77.8	6052.84
76.5	5852.25
$\sum x_i = 1007.3$	$\sum x_i^2 = 78789.73$

$$s = \sqrt{\frac{\sum x_i^2 - \frac{\left(\sum x_i\right)^2}{n}}{n-1}} = \sqrt{\frac{78789.73 - \frac{(1007.3)^2}{13}}{13-1}}$$

$$\approx 7.850$$

Using the standard deviation as the measure of dispersion, the traditional classroom has slightly more dispersion in exam scores.

(c) If the score of 59.8 had been incorrectly recorded as 598 in the traditional lecture, then:

$$\sum x_i = 1471.8 \quad \sum x_i^2 = 422,011.82$$

$$s = \sqrt{\frac{\sum x_i^2 - \frac{\left(\sum x_i\right)^2}{n}}{n-1}} = \sqrt{\frac{422,011.82 - \frac{(1471.8)^2}{13}}{13-1}}$$

$$\approx 145.88$$

Range $= 598 - 56.2 = 541.8$

Both the range and the standard deviation are significantly larger because of the one extreme larger value. This demonstrates that the range and standard deviation are not resistant measures of dispersion.

19. (a) We use the computational formula:

x_i	x_i^2
76	5776
60	3600
60	3600
81	6561
72	5184
80	6400
80	6400
68	4624
73	5329
$\sum x_i = 650$	$\sum x_i^2 = 47,474$

$$\sigma = \sqrt{\frac{\sum x_i^2 - \frac{\left(\sum x_i\right)^2}{N}}{N}} = \sqrt{\frac{47,474 - \frac{(650)^2}{9}}{9}}$$

$$\approx 7.7 \text{ beats/minute}$$

(b) Samples and sample standard deviations will vary.

(c) Answers will vary.

21. (a) Ethan:

$$\sum x_i = 9 + 24 + 8 + 9 + 5 + 8 + 9 + 10 + 8 + 10$$
$$= 100$$
$$\mu = \frac{\sum x_i}{N} = \frac{100}{10} = 10 \text{ fish}$$

Range = Largest Value – Smallest Value
$$= 24 - 5 = 19 \text{ fish}$$

Drew:

$$\sum x_i = 15 + 2 + 3 + 18 + 20 + 1 + 17 + 2 + 19 + 3$$
$$= 100$$
$$\mu = \frac{\sum x_i}{N} = \frac{100}{10} = 10 \text{ fish}$$

Range = Largest Value – Smallest Value
$$= 20 - 1 = 19 \text{ fish}$$

Both fishermen have the same mean and range, so these values do not indicate any differences between their catches per day.

(b) From the dotplots, it appears that Ethan is a more consistent fisherman.

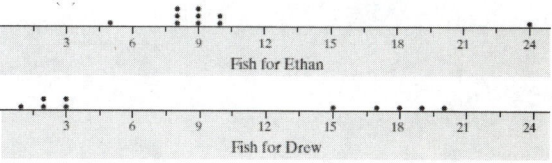

(c) Ethan:

$N = 10$; $\sum x_i = 100$

$\sum x_i^2 = 9^2 + 24^2 + 8^2 + 9^2 + 5^2 + 8^2 + 9^2$
$\qquad + 10^2 + 8^2 + 10^2$
$\qquad = 1236$

$\sigma = \sqrt{\dfrac{\sum x_i^2 - \dfrac{(\sum x_i)^2}{N}}{N}} = \sqrt{\dfrac{1236 - \dfrac{(100)^2}{10}}{10}}$

≈ 4.9 fish

Drew:

$N = 10$; $\sum x_i = 100$

$\sum x_i^2 = 15^2 + 2^2 + 3^2 + 18^2 + 20^2 + 1^2 + 17^2$
$\qquad + 2^2 + 19^2 + 3^2$
$\qquad = 1626$

$\sigma = \sqrt{\dfrac{\sum x_i^2 - \dfrac{(\sum x_i)^2}{N}}{N}} = \sqrt{\dfrac{1626 - \dfrac{(100)^2}{10}}{10}}$

≈ 7.9 fish

Yes, now there appears to be a difference in the two fishermen's records. Ethan had a more consistent fishing record, which is indicated by the smaller standard deviation.

(d) Answers will vary. One possibility follows: The range is limited as a measure of dispersion because it does not take all of the data values into account. It is obtained by using only the two most extreme data values. Since the standard deviation utilizes all of the data values, it provides a better overall representation of dispersion.

23. (a) We use the computational formula:

$\sum x_i = 43.73$; $\sum x_i^2 = 38.3083$; $n = 50$;

$s = \sqrt{\dfrac{\sum x_i^2 - \dfrac{(\sum x_i)^2}{n}}{n-1}} = \sqrt{\dfrac{38.3083 - \dfrac{(43.73)^2}{50}}{50-1}}$

≈ 0.036 g

(b) The histogram is approximately symmetric, so the Empirical Rule is applicable.

(c) Since 0.803 is two standard deviations below the mean [0.803 = 0.875 − 2(0.036)] and 0.943 is two standard deviations above the mean [0.947 = 0.875 + 2(0.036)], the Empirical Rule predicts that approximately 95% of the M&Ms will weigh between 0.803 and 0.947 grams.

(d) All except 2 of the M&Ms weigh between 0.803 and 0.947 grams. Thus, the actual percentage is 48/50 = 96%.

(e) Since 0.911 is one standard deviation above the mean [0.911 = 0.875 + 0.036], the Empirical Rule predicts that 13.5% + 2.35% + 0.15% = 16% of the M&Ms will weigh more than 0.911 grams.

(f) Six of the M&Ms weigh more than 0.911 grams. Thus, the actual percentage is 6/50 = 12%.

25. Car 1:

$\sum x_i = 3352$; $\sum x_i^2 = 755{,}712$; $n = 15$

Measures of Center:

$\bar{x} = \dfrac{\sum x_i}{n} = \dfrac{3352}{15} \approx 223.5$ miles

$M = 223$ miles (8th value in the ordered data)
Mode: 220, 223, and 233

Measures of Dispersion:
Range = Largest Value − Smallest Value
$\qquad = 271 - 178 = 93$ miles

$s^2 = \dfrac{\sum x_i^2 - \dfrac{(\sum x_i)^2}{n}}{n-1} = \dfrac{755{,}712 - \dfrac{(3352)^2}{15}}{15-1}$

≈ 475.1 (miles)^2

$s = \sqrt{s^2} = \sqrt{\dfrac{755{,}712 - \dfrac{(3352)^2}{15}}{15-1}} \approx 21.8$ miles

Car 2:

$\sum x_i = 3558$; $\sum x_i^2 = 877{,}654$; $n = 15$

Measures of Center:

$\bar{x} = \dfrac{\sum x_i}{n} = \dfrac{3558}{15} = 237.2$ miles

$M = 230$ miles (8th value in the ordered data)
Mode: 217 and 230

Measures of Dispersion:

Range = Largest Value − Smallest Value
= 326 − 160 = 166 miles

$$s^2 = \frac{\sum x_i^2 - \frac{(\sum x_i)^2}{n}}{n-1} = \frac{877{,}654 - \frac{(3558)^2}{15}}{15-1}$$

$$\approx 2406.9 \text{ (miles)}^2$$

$$s = \sqrt{s^2} = \sqrt{\frac{877{,}654 - \frac{(3558)^2}{15}}{15-1}} \approx 49.1 \text{ miles}$$

We expect that the distribution for Car 1 is symmetric since the mean and median are approximately equal. We expect that the distribution for Car 2 is skewed right slightly since the mean is larger than the median. Both distributions have similar measures of center, but Car 2 has more dispersion, which can be seen by its larger range, variance, and standard deviation. This means that the distance Car 1 can be driven on 10 gallons of gas is more consistent. Thus, Car 1 is probably the better car to buy.

27. (a)

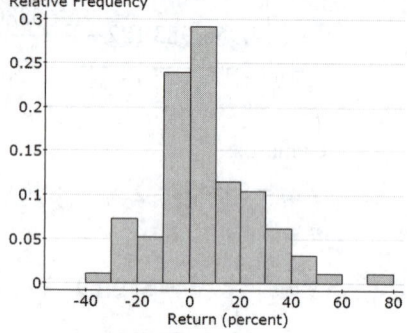

One Year Rate of Return on Consumer Cyclical Stocks

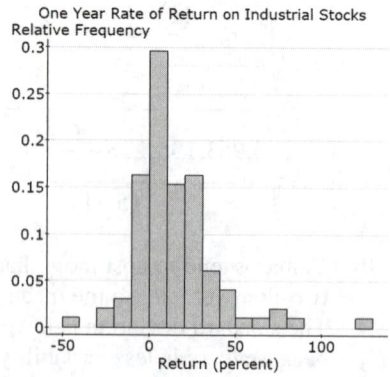

One Year Rate of Return on Industrial Stocks

Both distributions are skewed to the right. It appears that industrial stocks have a greater dispersion.

(b) Consumer Cyclical Stocks:
$\bar{x} = 6.595\%$; $M = 3.915\%$
Industrial Stocks:
$\bar{x} = 14.425\%$; $M = 9.595\%$

Industrial stocks have both a higher mean and median rate of return.

(c) Consumer Cyclical Stocks:
$s = 19.078\%$

Industrial Stocks:
$s = 23.851\%$

Industrial stocks are riskier since they have a larger standard deviation. The investor is paying for the higher return. For some investors, the higher returns is worth the cost.

29. (a) Since 70 is two standard deviations below the mean [70 = 100 − 2(15)] and 130 is two standard deviations above the mean [130 = 100 + 2(15)], the Empirical Rule predicts that approximately 95% of people have an IQ between 70 and 130.

(b) Since about 95% of people have an IQ score between 70 and 30, then approximately 5% of people have an IQ score either less than 70 or greater than 130.

(c) Approximately $5\% / 2 = 2.5\%$ of people have an IQ score greater than 130.

31. (a) Approximately 95% of the data will be within two standard deviations of the mean. Now, 325 − 2(30) = 265 and 325 + 2(30) = 385. Thus, about 95% of pairs of kidneys will be between 265 and 385 grams.

(b) Since 235 is three standard deviations below the mean [235 = 325 − 3(30)] and 415 is three standard deviations above the mean [415 = 325 + 3(30)], the Empirical Rule predicts that about 99.7% of pairs of kidneys weighs between 235 and 415 grams.

(c) Since about 99.7% of pairs of kidneys weighs between 235 and 415 grams, then about 0.3% of pairs of kidneys weighs either less than 235 or more than 415 grams.

(d) Since 295 is one standard deviation below the mean [295 = 325 – 30] and 385 is two standard deviations above the mean [385 = 325 + 2(30)], the Empirical Rule predicts that approximately 34% + 34% + 13.5% = 81.5% of pairs of kidneys weighs between 295 and 385 grams.

33. In Professor Alpha's class, only about 2.5% of the students will get an A, since 90% is 2 standard deviations above the mean. In Professor Omega's class, approximately 16% will get an A, since 90% in that class is only 1 standard deviation above the mean. Assuming you intend to earn an A, you will likely choose Professor Omega's class.
On the other hand, if a student only wants to get a passing grade, then Professor Alpha would be a better choice. Approximately 95% of Professor Alpha's class scores between 70% and 90%.

35. **(a)** By Chebyshev's inequality, at least

$$\left(1 - \frac{1}{k^2}\right) \cdot 100\% = \left(1 - \frac{1}{3^2}\right) \cdot 100\% \approx 88.9\%$$

of gasoline prices have prices within three standard deviations of the mean.

(b) By Chebyshev's inequality, at least

$$\left(1 - \frac{1}{k^2}\right) \cdot 100\% = \left(1 - \frac{1}{2.5^2}\right) \cdot 100\% = 84\%$$

of gasoline prices have prices within $k = 2.5$ standard deviations of the mean. Now, $3.06 - 2.5(0.06) = 2.91$ and $3.06 + 2.5(0.06) = 3.21$. Thus, the gasoline prices that are within 2.5 standard deviations of the mean are from $2.91 to $3.21.

(c) Since 2.94 is $k = 2$ standard deviations below the mean [2.94 = 3.06 – 2(0.06)] and 3.18 is $k = 2$ standard deviations above the mean [3.18 = 3.06 + 2(0.06)], Chebyshev's theorem predicts that at least

$$\left(1 - \frac{1}{k^2}\right) \cdot 100\% = \left(1 - \frac{1}{2^2}\right) \cdot 100\% = 75\%$$

of gas stations have prices between $2.94 and $3.18 per gallon.

37. When calculating the variability in team batting averages, we are finding the variability of means. When calculating the variability of all players, we are finding the variability of individuals. Since there is more variability among individuals than among means, the teams will have less variability.

39. Sample size of 5:
All data recorded correctly: $s \approx 5.3$
106 recorded incorrectly as 160: $s \approx 27.9$

Sample size of 12:
All data recorded correctly: $s \approx 14.7$
106 recorded incorrectly as 160: $s \approx 22.7$

Sample size of 30:
All data recorded correctly: $s \approx 15.9$
106 recorded incorrectly as 160: $s \approx 19.2$

As the sample size increases, the impact of the misrecorded observation on the standard deviation decreases.

41. **(a)** We use the computational formula:
Coupes

$$\sum x_i = 263880 \ ; \ \sum x_i^2 = 4,956,983,122 \ ;$$

$$n = 15 \ ; \ \overline{x} = \frac{263,880}{15} = \$17,592$$

$$s = \sqrt{\frac{\sum x_i^2 - \frac{(\sum x_i)^2}{n}}{n - 1}}$$

$$= \sqrt{\frac{4,956,983,122 - \frac{(263880)^2}{15}}{15 - 1}} \approx \$4741.96$$

Camaros:

$$\sum x_i = 295458 \ ; \ \sum x_i^2 = 5,963,595,276 \ ;$$

$$n = 15 \ ;$$

$$\overline{x} = \frac{295,458}{15} = \$19,697.20$$

$$s = \sqrt{\frac{\sum x_i^2 - \frac{(\sum x_i)^2}{n}}{n - 1}}$$

$$= \sqrt{\frac{5,963,595,276 - \frac{(295,458)^2}{15}}{15 - 1}} \approx \$3206.02$$

(b) Camaros tend to cost more than the typical two-door vehicle, so the mean is higher. The standard deviation is lower for Camaros because there is less variability in prices of one specific vehicle than for all two-door vehicles.

43. $\sum x_i = 183.1$;

$$\overline{x} = \frac{\sum x_i}{n} = \frac{183.1}{6} = 30.5167 \text{ mpg}$$

x_i	$x_i - \overline{x}$	$|x_i - \overline{x}|$
34.0	3.4833	3.4833
33.2	2.6833	2.6833
37.0	6.4833	6.4833
29.4	−1.1167	1.1167
23.6	−6.9167	6.9167
25.9	−4.6167	4.6167
$\sum(x_i - \overline{x}) = 0$		$\sum|x_i - \overline{x}| = 25.3$

$$\text{MAD} = \frac{\sum |x_i - \overline{x}|}{n} = \frac{25.3}{6} \approx 4.217 \text{ mpg}$$

This is somewhat less than the sample standard deviation of $s \approx 5.134$ mpg, which we found in Problem 11.

45. (a) Reading from the graph, the average annual return for a portfolio that is 10% foreign is 14.9%. The level of risk is 14.7%.

(b) To best minimize risk, 30% should be invested in foreign stocks. According to the graph, a 30% investment in foreign stocks has the smallest standard deviation (level of risk) at about 14.3%.

(c) Answers will vary. One possibility follows: The risk decreases because a portfolio including foreign stocks is more diversified.

(d) According to Chebyshev's theorem, at least 75% of returns are within $k = 2$ standard deviations of the mean. Thus, at least 75% of returns are between $\overline{x} - ks = 15.8 - 2(14.3) = -12.8\%$ and $\overline{x} + ks = 15.8 + 2(14.3) = 44.4\%$. By Chebyshev's theorem, at least 88.9% of returns are within $k = 3$ standard deviations of the mean. Thus, at least 88.9% of returns are between $\overline{x} - ks = 15.8 - 3(14.3) = -27.1\%$ and $\overline{x} + ks = 15.8 + 3(14.3) = 58.7\%$. An investor should not be surprised if he or she has a negative rate of return. Chebyshev's theorem indicates that a negative return is fairly common.

47. Answers will vary.

49. No. It is not appropriate to compare standard deviations for two units of measure. Assuming the same variable is being measured, each should be reported in the same unit of measure (such as converting the inches to centimeters).

51. None of the measures of center introduced in this section is resistant. The range, standard deviation, and the variance are all sensitive to extreme observations.

53. The range only uses two observations from a data set, whereas the standard deviation uses all the observations.

55. Standard deviation measures spread by determining the average distance from the mean for the observations.
Data set 1:

x_i	$\overline{x}$	$x_i - \overline{x}$	$(x_i - \overline{x})^2$
3	4	−1	1
4	4	0	0
5	4	1	1
		$\sum(x_i - \overline{x}) = 0$	$\sum(x_i - \overline{x})^2 = 2$

$$s = \sqrt{\frac{\sum(x_i - \overline{x})^2}{n-1}} = \sqrt{\frac{2}{3-1}} = 1$$

Data set 2:

x_i	$\overline{x}$	$x_i - \overline{x}$	$(x_i - \overline{x})^2$
0	4	−4	16
4	4	0	0
8	4	4	16
		$\sum(x_i - \overline{x}) = 0$	$\sum(x_i - \overline{x})^2 = 32$

$$s = \sqrt{\frac{\sum(x_i - \overline{x})^2}{n-1}} = \sqrt{\frac{32}{3-1}} = 4$$

Data set 2 has the higher standard deviation because the observations are further from the mean, on average.

57. 15, 15, 15, 15, 15, 15, 15, 15

59. Even though the mean wait time increased, patrons were happy because there was less variability in wait time (so fast pass decreases standard deviation wait times). This suggests that people get less upset at long lines than they do at not having their expectations (as to how long they will need to wait in line) satisfied.

Section 3.3

1. To find the mean, we use the formula $\mu = \dfrac{\sum x_i f_i}{\sum f_i}$. To find the standard deviation, we choose to use the

computational formula $\sigma = \sqrt{\sigma^2} = \sqrt{\dfrac{\sum x_i^2 f_i - \dfrac{\left(\sum x_i f_i\right)^2}{\sum f_i}}{\sum f_i}}$. We organize our computations of x_i, $\sum f_i$, $\sum x_i f_i$,

and $\sum x_i^2 f_i$ in the table that follows:

Class	Midpoint, x_i	Frequency, f_i	$x_i f_i$	x_i^2	$x_i^2 f_i$
0 – $19,999	$\dfrac{0+20,000}{2}=10,000$	344	3,440,000	100,000,000	34,400,000,000
20,000 – 39,999	$\dfrac{20,000+40,000}{2}=30,000$	98	2,940,000	900,000,000	88,200,000,000
40,000 – 59,999	50,000	52	2,600,000	2,500,000,000	130,000,000,000
60,000 – 79,999	70,000	19	1,330,000	4,900,000,000	93,100,000,000
80,000 – 99,999	90,000	13	1,170,000	8,100,000,000	105,300,000,000
100,000 – 119,999	110,000	6	660,000	12,100,000,000	72,600,000,000
120,000 – 139,999	130,000	2	260,000	16,900,000,000	33,800,000,000
		$\sum f_i = 534$	$\sum x_i f_i = 12,400,000$		$\sum x_i^2 f_i = 557,400,000,000$

With the table complete, we compute the population mean and population standard deviation:

$$\mu = \frac{\sum x_i f_i}{\sum f_i} = \frac{12,400,000}{534} \approx \$23,220.97$$

$$s = \sqrt{s^2} = \sqrt{\frac{\sum x_i^2 f_i - \dfrac{\left(\sum x_i f_i\right)^2}{\sum f_i}}{\sum f_i - 1}} = \sqrt{\frac{557,400,000,000 - \dfrac{(12,400,000)^2}{534}}{534 - 1}} \approx \$22,484.51$$

3. To find the mean, we use the formula $\bar{x} = \dfrac{\sum x_i f_i}{\sum f_i}$. To find the standard deviation, we choose to use the

computational formula $s = \sqrt{s^2} = \sqrt{\dfrac{\sum x_i^2 f_i - \dfrac{\left(\sum x_i f_i\right)^2}{\sum f_i}}{\left(\sum f_i\right)-1}}$. We organize our computations of x_i, $\sum f_i$, $\sum x_i f_i$,

and $\sum x_i^2 f_i$ in the table that follows:

Class	Midpoint, x_i	Frequency, f_i	$x_i f_i$	x_i^2	$x_i^2 f_i$
$61-64$	$\dfrac{61+65}{2}=63$	31	1,953	3,969	123,039
$65-67$	$\dfrac{65+68}{2}=66.5$	67	4,455.5	4,422.25	296,290.75
$68-69$	69	198	13,662	4,761	942,678
70	70	195	13,650	4,900	955,500
$71-72$	72	120	8,640	5,184	622,080
$73-76$	75	89	6,675	5,625	500,625
$77-80$	79	50	3,950	6,241	312,050
		$\sum f_i = 750$	$\sum x_i f_i = 52,985.5$		$\sum x_i^2 f_i = 3,752,262.75$

With the table complete, we compute the sample mean and sample standard deviation:

$$\bar{x} = \frac{\sum x_i f_i}{\sum f_i} = \frac{52,985.5}{750} \approx 70.6°F; \quad s = \sqrt{s^2} = \sqrt{\frac{\sum x_i^2 f_i - \dfrac{\left(\sum x_i f_i\right)^2}{\sum f_i}}{\left(\sum f_i\right)-1}} = \sqrt{\frac{3,752,262.75 - \dfrac{(52,985.5)^2}{750}}{750-1}} \approx 3.5°F$$

5. (a) To find the mean, we use the formula $\mu = \dfrac{\sum x_i f_i}{\sum f_i}$. To find the standard deviation, we choose to use the

computational formula $\sigma = \sqrt{\sigma^2} = \sqrt{\dfrac{\sum x_i^2 f_i - \dfrac{\left(\sum x_i f_i\right)^2}{\sum f_i}}{\sum f_i}}$. We organize our computations of x_i, $\sum f_i$,

$\sum x_i f_i$, and $\sum x_i^2 f_i$ in the table that follows:

Class	Midpoint, x_i	Frequency, f_i	$x_i f_i$	x_i^2	$x_i^2 f_i$
$15-19$	$\dfrac{15+20}{2}=17.5$	44	770	306.25	13,475
$20-24$	$\dfrac{20+25}{2}=22.5$	404	9,090	506.25	204,525
$25-29$	27.5	1,204	33,110	756.25	910,525
$30-34$	32.5	1,872	60,840	1,056.25	1,977,300
$35-39$	37.5	1,000	37,500	1,406.25	1,406,250
$40-44$	42.5	332	14,110	1,806.25	599,675
$45-49$	47.5	44	2090	2,256.25	99,275
$50-54$	52.5	19	997.5	2,756.25	52,368.75
		$\sum f_i = 4,919$	$\sum x_i f_i = 158,507.5$		$\sum x_i^2 f_i = 5,263,393.75$

With the table complete, we compute the population mean and population standard deviation:

$$\mu = \frac{\sum x_i f_i}{\sum f_i} = \frac{158,507.5}{4,919} \approx 32.2 \text{ years}$$

$$\sigma = \sqrt{\sigma^2} = \sqrt{\frac{\sum x_i^2 f_i - \dfrac{\left(\sum x_i f_i\right)^2}{\sum f_i}}{\sum f_i}} = \sqrt{\frac{5,263,393.75 - \dfrac{(158,507.5)^2}{4,919}}{4,919}} \approx 5.6 \text{ years}$$

(b)

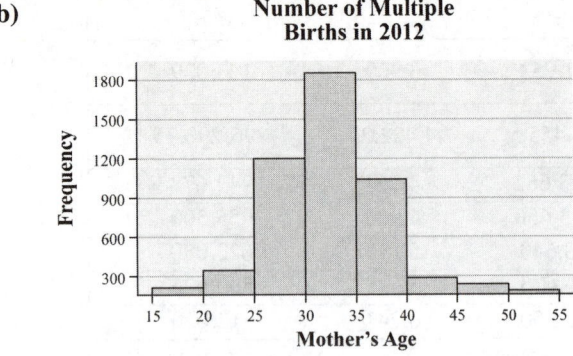

Number of Multiple Births in 2012

(c) By the Empirical Rule, 95% of the observations will be within 2 standard deviation of the mean. Now,
$\mu - 2\sigma = 32.2 - 2(5.6) = 21.0$ and
$\mu + 2\sigma = 32.2 + 2(5.6) = 43.4$, so 95% of the mothers of multiple births will be between 21.0 and 43.4 years of age.

---

7. We organize our computations of x_i, $\sum f_i$, $\sum x_i f_i$, and $\sum x_i^2 f_i$ in the table that follows, with frequencies listed in thousands:

Class	Midpoint, x_i	Frequency, f_i	$x_i f_i$	x_i^2	$x_i^2 f_i$
$0-0.499$	$\dfrac{0+0.5}{2} = 0.25$	7	1.75	0.0625	0.4375
$0.5-0.999$	0.75	13	9.75	0.5625	7.3125
$1-1.499$	1.25	7	8.75	1.5625	10.9375
$1.5-1.999$	1.75	8	14	3.0625	24.5
$2-2.499$	2.25	5	11.25	5.0625	25.3125
$2.5-2.999$	2.75	5	13.75	7.5625	37.8125
$3-3.499$	3.25	3	9.75	10.5625	31.6875
$3.5-3.999$	3.75	2	7.5	14.0625	28.125
$4-4.499$	4.25	1	4.25	18.0625	18.0625
		$\sum f_i = 51$	$\sum x_i f_i = 80.75$		$\sum x_i^2 f_i = 184.1875$

With the table complete, we compute the estimated population mean and standard deviation:

$$\mu = \frac{\sum x_i f_i}{\sum f_i} = \frac{80.75}{51} \approx \$1.5833 \; ; \; \sigma = \sqrt{\sigma^2} = \sqrt{\frac{\sum x_i^2 f_i - \dfrac{\left(\sum x_i f_i\right)^2}{\sum f_i}}{\left(\sum f_i\right)}} = \sqrt{\frac{184.1875 - \dfrac{(80.75)^2}{51}}{51}} \approx \$1.051$$

From the raw data, we find: $\sum x_i = 78.094$; $\sum x_i^2 = 170.84626$; $N = 51$

$$\mu = \frac{\sum x_i}{N} = \frac{78.094}{51} \approx \$1.5312 \; ; \; \sigma = \sqrt{\sigma^2} = \sqrt{\frac{\sum x_i^2 - \dfrac{\left(\sum x_i\right)^2}{N}}{N}} = \sqrt{\frac{170.846296 - \dfrac{(78.094)^2}{51}}{51}} \approx \$1.003$$

The approximations from the grouped data are good estimates of the actual results from the raw data.

9. GPA $= \bar{x}_w = \dfrac{\sum w_i x_i}{\sum w_i} = \dfrac{5(3) + 3(4) + 4(4) + 3(2)}{5+3+4+3} = \dfrac{49}{15} \approx 3.27$

11. Cost per pound $= \bar{x}_w = \dfrac{\sum w_i x_i}{\sum w_i} = \dfrac{4(\$3.50) + 3(\$2.75) + 2(\$2.25)}{4+3+2} = \dfrac{\$26.75}{9} \approx \$2.97 / \text{pound}$

13. (a) Male: We organize our computations of x_i, $\sum f_i$, $\sum x_i f_i$, and $\sum x_i^2 f_i$ in the following table, with frequencies listed in thousands.

Class	Midpoint, x_i	Frequency, f_i	$x_i f_i$	x_i^2	$x_i^2 f_i$
0 – 9	5	20,700	103,500	25	517,500
10 – 19	15	21,369	320,535	225	4,808,025
20 – 29	25	21,417	535,425	625	13,385,625
30 – 39	35	19,455	680,925	1,225	23,832,375
40 – 49	45	20,839	937,755	2,025	42,198,975
50 – 59	55	20,785	1,143,175	3,025	62,874,625
60 – 69	65	14,739	958,035	4,225	62,272,275
70 – 79	75	7,641	573,075	5,625	42,980,625
≥ 80	85	4,230	359,550	7,225	30,561,750
		$\sum f_i = 151{,}175$	$\sum x_i f_i = 5{,}611{,}975$		$\sum x_i^2 f_i = 283{,}431{,}775$

With the table complete, we compute the population mean and population standard deviation:

$$\mu = \frac{\sum x_i f_i}{\sum f_i} = \frac{5{,}611{,}975}{151{,}175} \approx 37.12 \text{ years}$$

$$\sigma = \sqrt{\sigma^2} = \sqrt{\frac{\sum x_i^2 f_i - \frac{\left(\sum x_i f_i\right)^2}{\sum f_i}}{\sum f_i}} = \sqrt{\frac{283{,}431{,}775 - \frac{(5{,}611{,}975)^2}{151{,}175}}{151{,}175}} \approx 22.29 \text{ years}$$

(b) Female: We organize our computations of x_i, $\sum f_i$, $\sum x_i f_i$, and $\sum x_i^2 f_i$ in the following table, with frequencies listed in thousands.

Class	Midpoint, x_i	Frequency, f_i	$x_i f_i$	x_i^2	$x_i^2 f_i$
0 – 9	5	19,826	99,130	25	495,650
10 – 19	15	20,475	307,125	225	4,606,875
20 – 29	25	21,355	533,875	625	13,346,875
30 – 39	35	20,011	700,385	1,225	24,513,475
40 – 49	45	21,532	968,940	2,025	43,602,300
50 – 59	55	22,058	1,213,190	3,025	66,725,450
60 – 69	65	16,362	1,063,530	4,225	69,129,450
70 – 79	75	9,474	710,550	5,625	53,291,250
≥ 80	85	6,561	557,685	7225	47,403,225
		$\sum f_i = 157{,}654$	$\sum x_i f_i = 6{,}154{,}410$		$\sum x_i^2 f_i = 323{,}114{,}550$

With the table complete, we compute the population mean and population standard deviation:

$$\mu = \frac{\sum x_i f_i}{\sum f_i} = \frac{6{,}154{,}410}{157{,}654} \approx 39.04 \text{ years}$$

$$\sigma = \sqrt{\sigma^2} = \sqrt{\frac{\sum x_i^2 f_i - \frac{\left(\sum x_i f_i\right)^2}{\sum f_i}}{\sum f_i}} = \sqrt{\frac{323{,}114{,}550 - \frac{(6{,}154{,}410)^2}{157{,}654}}{157{,}654}} \approx 22.93 \text{ years}$$

(c) Females have a higher mean age.

(d) Females also have more dispersion in age, which is indicated by the larger standard deviation.

15.

Class	f_i	CF
$0 - 499$	5	5
$500 - 999$	17	22
$1,000 - 1,499$	36	58
$1,500 - 1,999$	121	179
$2,000 - 2,499$	119	298
$2,500 - 2,999$	81	379
$3,000 - 3,499$	47	426
$3,500 - 3,999$	45	471
$4,000 - 4,499$	22	493
$4,500 - 4,999$	7	500

The distribution contains $n = 500$ data values. The position of the median is

$\dfrac{n+1}{2} = \dfrac{500+1}{2} = 250.5$, which is in the fifth

class, $2,000 - 2,499$. Then,

$$M = L + \dfrac{\dfrac{n}{2} - CF}{f} \cdot i$$

$$= 2,000 + \dfrac{\dfrac{500}{2} - 179}{119}(2,500 - 2,000)$$

$$\approx 2,298.3 \text{ ft}^2$$

17. From the table in Problem 1, the highest frequency is 344. So, the modal class is $0 - \$19,999$.

Section 3.4

1. z-score

3. quartiles

5. 34-week gestation:

$$z = \dfrac{x - \mu}{\sigma} = \dfrac{2,400 - 2,600}{660} \approx -0.30$$

40-week gestation:

$$z = \dfrac{x - \mu}{\sigma} = \dfrac{3,300 - 3,500}{470} \approx -0.43$$

The weight of the 34-week gestation baby is 0.30 standard deviation below the mean, while the weight of the 40-week gestation baby is 0.43 standard deviation below the mean. Thus, the 40-week gestation baby weighs less relative to the gestation period.

7. 75-inch man: $z = \dfrac{x - \mu}{\sigma} = \dfrac{75 - 69.6}{3.0} = 1.8$

70-inch woman: $z = \dfrac{x - \mu}{\sigma} = \dfrac{70 - 64.1}{3.8} \approx 1.55$

The height of the 75-inch man is 1.8 standard deviations above the mean, while the height of a 70-inch woman is 1.55 standard deviations above the mean. Thus, the 75-inch man is relatively taller than the 70-inch woman.

9. Clayton Kershaw:

$$z = \dfrac{x - \mu}{\sigma} = \dfrac{1.77 - 3.430}{0.721} \approx -2.30$$

Felix Hernandez:

$$z = \dfrac{x - \mu}{\sigma} = \dfrac{2.14 - 3.598}{0.762} \approx -1.91$$

Felix Hernandez's ERA was 1.91 standard deviations below the mean for his league, while Clayton Kershaw's ERA was 2.30 standard deviations below the mean for his league. Since lower ERAs are better, Kershaw had the better year relative to his peers.

11. 100-meter backstroke:

$$z = \dfrac{x - \mu}{\sigma} = \dfrac{45.3 - 48.62}{0.98} \approx -3.39$$

200-meter backstroke:

$$z = \dfrac{x - \mu}{\sigma} = \dfrac{99.32 - 106.58}{2.38} \approx -3.05$$

Ryan is 3.39 standard deviations below the mean in the 100-meter backstroke and is 3.05 standard deviations below the mean in the 200-meter backstroke. Thus, Ryan is better in the 100-meter backstroke.

13. $z = \dfrac{x - \mu}{\sigma}$

$$\dfrac{x - 200}{26} = 1.5$$

$$x - 200 = 1.5(26)$$

$$x - 200 = 39$$

$$x = 239$$

An applicant must make a minimum score of 239 to be accepted into the school.

15. (a) 15% of 3- to 5-month-old males have a head circumference that is 41.0 cm or less, and $(100 - 15)\% = 85\%$ of 3- to 5-month-old males have a head circumference that is greater than 41.0 cm.

(b) 90% of 2-year-old females have a waist circumference that is 52.7 cm or less, and $(100 - 90)\% = 10\%$ of 2-year-old females have a waist circumference that is more than 52.7 cm.

(c) The heights at each percentile decrease (except for the 40–49 age group) as the age increases. This implies that adult males are getting taller.

17. (a) 25% of the states have a violent crime rate that is 252.4 crimes per 100,000 population or less, and $(100 - 25)\% = 75\%$ of the states have a violent crime rate more than 252.4. 50% of the states have a violent crime rate that is 333.8 crimes per 100,000 population or less, while $(100 - 50)\% = 50\%$ of the states have a violent crime rate more than 333.8. 75% of the states have a violent crime rate that is 454.5 crimes per 100,000 population or less, and $(100 - 75)\% = 25\%$ of the states have a violent crime rate more than 454.5.

(b) $IQR = Q_3 - Q_1 = 454.5 - 252.4 = 202.1$ crimes per 100,000 population. This means that the middle 50% of all observations have a range of 241.9 crimes per 100,000 population.

(c) $LF = Q_1 - 1.5(IQR)$
$= 252.4 - 1.5(202.1) = -50.75$
$UF = Q_3 + 1.5(IQR)$
$= 454.5 + 1.5(202.1) = 757.65$
Since 1,459 is above the upper fence, the Washington, D.C. crime rate is an outlier.

(d) Skewed right. The difference between Q_1 and Q_2 (81) is quite a bit less than the difference between Q_2 and Q_3 (120.7), and the outlier is in the right tail of the distribution, which implies that the distribution is skewed right.

19. (a) An IQ of 100 corresponds to the 50th percentile. A person with an IQ of 100 has an IQ that is as high or higher than 50 percent of the population.

(b) An IQ of 120 corresponds to roughly the 90th percentile. A person with an IQ of 120 has an IQ that is as high or higher than 90 percent of the population.

(Answers will vary slightly, but they should be near the 90th percentile.)

(c) If an individual has an IQ in the 60th percentile, their score would be 105. A person with an IQ of 105 has an IQ that is as high or higher than 60 percent of the population. (Answers will vary slightly, but they should be near 105.)

21. (a) $z = \dfrac{x - \mu}{\sigma} \approx \dfrac{36.3 - 38.775}{3.416} \approx -0.72$

(b) By hand/TI-83 or 84/StatCrunch:
$Q_1 = \dfrac{36.3 + 37.4}{2} = 36.85$ mpg
$Q_2 = \dfrac{38.3 + 38.4}{2} = 38.35$ mpg
$Q_3 = \dfrac{40.6 + 41.4}{2} = 41.0$ mpg
Note: Results from MINITAB differ:
$Q_1 = 36.575$ mpg, $Q_2 = 38.35$ mpg, $Q_3 = 41.2$ mpg

(c) By hand/TI-83 or 84/StatCrunch:
$IQR = Q_3 - Q_1 = 41.0 - 36.85 = 4.15$ mpg.
This means that the middle 50% of observations have a range of 4.15 mpg.
Note: Results from MINITAB differ:
$IQR = 41.2 - 36.575 = 4.625$ mpg

(d) By hand/TI-83 or 84/StatCrunch:
$LF = Q_1 - 1.5(IQR)$
$= 36.85 - 1.5(4.15) = 30.625$ mpg
$UF = Q_3 + 1.5(IQR)$
$= 41.0 + 1.5(4.15) = 47.225$ mpg
Yes, 47.5 mpg is an outlier.
Note: Results from MINITAB differ:
$LF = Q_1 - 1.5(IQR)$
$= 36.575 - 1.5(4.625) = 29.6375$ mpg
$UF = Q_3 + 1.5(IQR)$
$= 41.2 + 1.5(4.625) = 48.1375$ mpg
There are no outliers using MINITAB's quartiles.

23. (a) There are $n = 45$ data values, and we put them in ascending order:

–0.18	–0.08	0.00	0.05	0.09
–0.18	–0.08	0.00	0.05	0.09
–0.17	–0.07	0.01	0.05	0.10
–0.15	–0.07	0.01	0.05	0.10
–0.14	–0.07	0.02	0.06	0.14
–0.10	–0.02	0.02	0.06	0.17
–0.10	–0.02	0.03	0.07	0.17
–0.10	–0.02	0.03	0.08	0.25
–0.10	–0.01	0.04	0.08	0.30

The second quartile (median) is the value that lies in the 23rd position, which is 0.02. So, $Q_2 = M = 0.02$.

The first quartile is the median of the bottom 23 data values, which is the value that lies in the 12th position. This value is –0.07, so $Q_1 = -0.07$.

The third quartile is the median of the top 23 data values, which is the value that lies in the 34th position. This value is 0.07, so $Q_3 = 0.07$.

Note: Results from MINITAB differ: $Q_1 = -0.075$, $Q_2 = 0.02$, and $Q_3 = 0.075$.

Interpretation: Using the by-hand results, 25% of the monthly returns are less than or equal to the first quartile, –0.07, and about 75% of the monthly returns are greater than –0.07; 50% of the monthly returns are less than or equal to the second quartile, 0.02, and about 50% of the monthly returns are greater than 0.02; about 75% of the monthly returns are less than or equal to the third quartile, 0.07, and about 25% of the monthly returns are greater than 0.07.

(b) $IQR = Q_3 - Q_1 = 0.07 - (-0.07) = 0.14$

$LF = Q_1 - 1.5(IQR)$

$\quad = -0.07 - 1.5(0.14) = -0.28$

$UF = Q_3 + 1.5(IQR)$

$\quad = 0.07 + 1.5(0.14) = 0.28$

The return 0.3 is an outlier because it is greater than the upper fence.

Note: If using MINITAB, the result will be:
$IQR = Q_3 - Q_1 = 0.075 - (-0.075) = 0.15$

$LF = Q_1 - 1.5(IQR)$

$\quad = -0.075 - 1.5(0.15) = -0.3$

$UF = Q_3 + 1.5(IQR)$

$\quad = 0.075 + 1.5(0.15) = 0.3$

There are no outliers using MINITAB's quartiles.

25. To find the upper fence, we must find the third quartile and the interquartile range. There are $n = 20$ data values, and we put them in ascending order:

345	429	461	471	505
346	437	466	480	515
358	442	466	489	516
372	442	470	490	549

The first quartile is the median of the bottom 10 data values, which is the mean of the data values that lie in the 5th and 6th positions. These values are 429 and 437, so $Q_1 = \dfrac{429 + 437}{2} = 433$ min.

The third quartile is the median of the top 10 data values, which is the mean of the data values that lie in the 15th and 16st positions. These values are 489 and 490, so

$Q_3 = \dfrac{489 + 490}{2} = 489.5$ min.

$IQR = 489.5 - 433 = 56.5$ min.
$UF = Q_3 + 1.5(IQR)$

$\quad = 489.5 + 1.5(56.5) = 574.25$ min.

The customer is contacted if more than 574 minutes are used.

Note: Results from MINITAB differ:
$Q_1 = 431$ minutes and $Q_3 = 489.8$ minutes
$IQR = 489.8 - 431 = 58.8$ minutes
$UF = 489.8 + 1.5(58.8) = 578$ min.

Using MINITAB, the customer is contacted if more than 578 minutes are used.

27. (a) To find outliers, we must find the first and third quartiles and the interquartile range. There are $n = 50$ data values, and we put them in ascending order:

0	0	188	347	547
0	0	203	367	567
0	67	244	375	579
0	82	262	389	628
0	83	281	403	635
0	95	289	454	650
0	100	300	476	671
0	149	310	479	719
0	159	316	521	736
0	181	331	527	12,777

The first quartile is the median of the bottom 25 data values, which is the value that lies in the 13th position. So, $Q_1 = \$67$.

The third quartile is the median of the top 25 data values, which is the value that lies in the 38th position. So, $Q_3 = \$479$.

$$IQR = 479 - 67 = \$412$$
$$LF = Q_1 - 1.5(IQR)$$
$$= 67 - 1.5(412) = -\$551$$
$$UF = Q_3 + 1.5(IQR)$$
$$= 479 + 1.5(412) = \$1,097$$

Note: Results from MINITAB differ:
$Q_1 = \$50$ and $Q_3 = \$490$
$IQR = 490 - 50 = \$440$ (Note that the MINITAB-calculated IQR is $439.)
$$LF = 50 - 1.5(440) = -\$610$$
$$UF = 490 + 1.5(440) = \$1,150$$

So, the only outlier is $12,777 because it is greater than the upper fence.

(b) To create the histogram, we choose the lower class limit of the first class to be 0 and the class width to be 100. The resulting classes and frequencies follow:

Class	Freq.	Class	Freq.
$0 - 99$	16	$500 - 599$	5
$100 - 199$	5	$600 - 699$	4
$200 - 299$	5	$700 - 799$	2
$300 - 3999$	8	⋮	
$400 - 499$	4	$12,700 - 12,799$	1

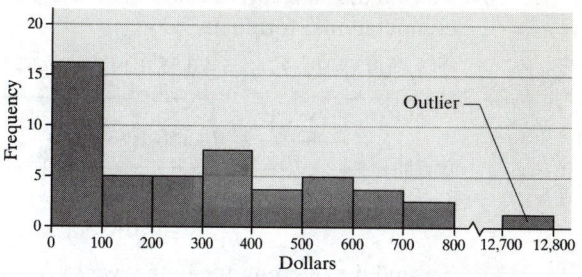

Students Weekly Income

(c) Answers will vary. One possibility is that a student may have provided his or her annual income instead of his or her weekly income.

29. From Problem 21 in Section 3.1 and Problem 19 in Section 3.2, we have $\mu \approx 72.2$ beats per minute and $\sigma \approx 7.7$ beats per minute.

Student	Pulse, x_i	z-score, z_i
P. Bernpah	76	$\dfrac{76 - 72.2}{7.7} = 0.49$
M. Brooks	60	$\dfrac{60 - 72.2}{7.7} = -1.58$
J. Honeycutt	60	$\dfrac{60 - 72.2}{7.7} = -1.58$
C. Jefferson	81	$\dfrac{81 - 72.2}{7.7} = 1.14$
C. Kurtenbach	72	$\dfrac{72 - 72.2}{7.7} = -0.03$
J. Laotka	80	$\dfrac{80 - 72.2}{7.7} = 1.01$
K. McCarthy	80	$\dfrac{80 - 72.2}{7.7} = 1.01$
T. Ohm	68	$\dfrac{68 - 72.2}{7.7} = -0.55$
K. Wojdyla	73	$\dfrac{73 - 72.2}{7.7} = 0.10$

The mean of the z-scores is 0.001 and the population standard deviation is 0.994. These are off slightly from the true mean of 0 and the true standard deviation of 1 because of rounding.

31. (a) To find the standard deviation, we use the computational formula:

$$\sum x_i = 9{,}049 \, ; \; \sum x_i^2 = 4{,}158{,}129 \, ; \; n = 20$$

$$s = \sqrt{\frac{\sum x_i^2 - \frac{(\sum x_i)^2}{n}}{n-1}} = \sqrt{\frac{4{,}158{,}129 - \frac{(9{,}049)^2}{20}}{20-1}}$$

$$\approx 58.0 \text{ minutes}$$

To find the interquartile range, we look back to the solution of Problem 25. There we found $Q_1 = 433$ minutes, $Q_3 = 489.5$ minutes, and $IQR = 489.5 - 433 = 56.5$ minutes.

Note: Results from MINITAB differ: $Q_1 = 431$ minutes, $Q_3 = 489.8$ minutes, and $IQR = 489.8 - 431 = 58.8$ minutes.

(b) Again, to find the standard deviation, we use the computational formula:

$$\sum x_i = 8{,}703 \, ; \; \sum x_i^2 = 4{,}038{,}413 \, ; \; n = 20$$

$$s = \sqrt{\frac{\sum x_i^2 - \frac{(\sum x_i)^2}{n}}{n-1}} = \sqrt{\frac{4{,}038{,}413 - \frac{(8{,}703)^2}{20}}{20-1}}$$

$$\approx 115.0 \text{ minutes}$$

To find the interquartile range, we first put the $n = 20$ data values in ascending order:

0	429	461	471	505
345	437	466	480	515
358	442	466	489	516
372	442	470	490	549

The first quartile is the median of the bottom 10 data values, which is the mean of the data values that lie in the 5th and 6th positions. These values are 429 and 437, so

$$Q_1 = \frac{429 + 437}{2} = 433 \text{ minutes.}$$

The third quartile is the median of the top 10 data values, which is the mean of the data values that lie in the 15th and 16th positions. These values are 489 and 490,

so $Q_3 = \dfrac{489 + 490}{2} = 489.5$ minutes.

So, $IQR = 489.5 - 433 = 56.5$ minutes.

Note: Results from MINITAB differ: $Q_1 = 431$ minutes, $Q_3 = 489.8$ minutes, and $IQR = 489.8 - 431 = 58.8$ minutes.

Changing the value 346 to 0 causes the standard deviation to nearly double in size, while the interquartile range does not change at all. This illustrates the property of resistance. The standard deviation is not resistant, but the interquartile range is resistant.

33. Since the percentile of a score is rounded to the nearest integer, it is possible for two different scores to have the same percentile.

35. No, an outlier should not always be removed. When an outlier is discovered it should be investigated to find the cause.

37. Answers will vary. Comparing z-scores allows a unitless comparison of the number of standard deviations an observation is from the mean.

39. Answers will vary. The first quartile is the 25th percentile, which means 25% of the observations are less than or equal to the value and 75% of the observations are above the value. The second quartile is the 50th percentile, which means 50% of the observations are less than or equal to the value and 50% of the observations are above the value. The third quartile is the 75th percentile, which means 75% of the observations are less than or equal to the value and 25% of the observations are above the value.

Section 3.5

1. The five-number summary consists of the minimum value in the data set, the first quartile, the median, the third quartile, and the maximum value in the data set.

3. (a) The median is to the left of the center of the box and the right line is substantially longer than the left line, so the distribution is skewed right.

(b) Reading the boxplot, the five-number summary is: 0, 1, 3, 6, 16.

5. (a) For the variable x: $M = 40$

(b) For the variable y: $Q_3 = 52$

(c) The variable y has more dispersion. This can be seen by the much broader range (span of the lines) and the much broader interquartile range (span of the box).

(d) The distribution of the variable x is symmetric. This can be seen because the median is near the center of the box and the horizontal lines are approximately the same in length.

(e) The distribution of the variable y is skewed right. This can be seen because the median is to the left of the center of the box and the right line is substantially longer than the left line.

7.

Statistics Exams Scores

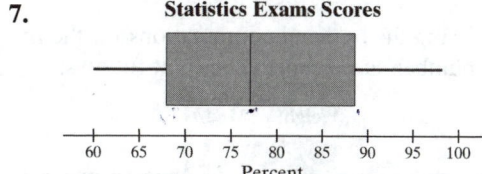

Percent

9. (a) Notice that the $n = 44$ data values are already arranged in order (moving down the columns). The smallest value (youngest president) in the data set is 42. The largest value (oldest president) in the data set is 69.

The second quartile (median) is the data value that lies in between the 22nd and 23nd positions. These values are 54 and 55, so

$$Q_2 = M = \frac{54+55}{2} = 54.5 .$$

The first quartile is the median of the bottom 22 data values, which is the data value that lies between the 11th and 12th positions. So, $Q_1 = \frac{50+51}{2} = 50.5 .$

The third quartile is the median of the top 22 data values, which is the data value that lies in the 33rd and 34th positions. So,

$$Q_3 = \frac{57+58}{2} = 57.5 .$$

So, the five-number summary is:
42, 50.5, 54.5, 57.5, 69

(b) $IQR = 57.5 - 50.5 = 7$

$LF = Q_1 - 1.5\left(IQR\right) = 50.5 - 1.5(7) = 40$

$UF = Q_3 + 1.5\left(IQR\right) = 57.5 + 1.5(7) = 68$

Thus, 69 is an outlier.

Age of Presidents at Inauguration

Age

(c) The median is near the center of the box and the horizontal lines are approximately the same in length, so the distribution is approximately symmetric, with an outlier.

11. (a) We arrange the $n = 30$ data values into ascending order.

16	20	22	25	29
17	20	23	25	30
18	21	23	25	32
19	21	24	25	33
19	21	24	26	33
20	22	25	29	35

The smallest data value is 16. The largest data value is 35.

The second quartile (median) is the mean of the data values that lie in the 15th and 16th positions. So, $Q_2 = M = 23.5 .$

The first quartile is the median of the bottom 15 data values, which is the mean of the data values that lie in the $\frac{15+1}{2} = 8^{th}$ position, which is 20. So, $Q_1 = 20 .$

The third quartile is the median of the top 15 data values, which is the data value that lies in the 23rd position, which is 26. So, $Q_3 = 26 .$

So, the five-number summary is:
16, 20, 23.5, 26, 35

$IQR = 26 - 20 = 6$

$LF = Q_1 - 1.5\left(IQR\right)$

$= 20 - 1.5(6) = 11$

$$UF = Q_3 + 1.5(IQR)$$
$$= 26 + 1.5(6) = 35$$

Thus, there are no outliers.

Age of Mother at Time of First Birth

15   20   25   30   35

Age of Mother

(b) The right line is a little longer than the left line. The distribution is slightly skewed right.

13. To find the five-number summary, we arrange the $n = 50$ data values in ascending order:

0.79	0.84	0.87	0.88	0.91
0.81	0.84	0.87	0.88	0.91
0.82	0.85	0.87	0.89	0.91
0.82	0.85	0.87	0.89	0.91
0.83	0.86	0.87	0.89	0.92
0.83	0.86	0.88	0.90	0.93
0.84	0.86	0.88	0.90	0.93
0.84	0.86	0.88	0.90	0.93
0.84	0.86	0.88	0.90	0.94
0.84	0.86	0.88	0.91	0.95

The smallest data value is 0.79 grams, and the largest data value is 0.95 grams.

The second quartile (median) is the mean of the data values that lie in the 25th and 26th positions, which are 0.87 and 0.88. So,

$$Q_2 = M = \frac{0.87 + 0.88}{2} = 0.875 \text{ grams.}$$

The first quartile is the median of the bottom 25 data values, which is the value that lies in the 13th position. So, $Q_1 = 0.85$ grams.

The third quartile is the median of the top 25 data values, which is value that lies in the 38th position. So, $Q_3 = 0.90$ grams.

So, the five-number summary is:
0.79, 0.85, 0.875, 0.90, 0.95
$IQR = 0.90 - 0.85 = 0.05$ grams

$$LF = Q_1 - 1.5(IQR)$$
$$= 0.85 - 1.5(0.05) = 0.775 \text{ grams}$$
$$UF = Q_3 + 1.5(IQR)$$
$$= 0.90 + 1.5(0.05) = 0.975 \text{ grams}$$

So, there are no outliers.

Note: Results from MINITAB differ:
0.79, 0.8475, 0.875, 0.90, 0.95
$IQR = 0.0525$, $LF = 0.76875$, $UF = 0.97875$

Using the by-hand computations for the five-number summary, the boxplot follows:

Weight of M&Ms

0.80   0.82   0.84   0.86   0.88   0.90   0.92   0.94   0.96

Weight (in grams)

Since the range of the data between the minimum value and the median is roughly the same as the range between the median and the maximum value, and because the range of the data between the first quartile and the median is the same as the range of the data between the median and third quartile, the distribution is symmetric.

15. (a) To find the five-number summary for each vitamin type, we arrange each data set in ascending order:

Centrum

2.15	2.57	2.80	3.12	3.85
2.15	2.60	2.95	3.25	3.92
2.23	2.63	3.02	3.30	4.00
2.25	2.67	3.02	3.35	4.02
2.30	2.73	3.03	3.53	4.17
2.38	2.73	3.07	3.63	4.33

Generic Brand

4.97	5.55	6.17	6.50	7.17
5.03	5.57	6.23	6.50	7.18
5.25	5.77	6.30	6.57	7.25
5.35	5.78	6.33	6.60	7.42
5.38	5.92	6.35	6.73	7.42
5.50	5.98	6.47	7.13	7.58

For Centrum, the smallest value is 2.15, and the largest value is 4.33. For the generic brand, the smallest value is 4.97, and the largest value is 7.58.

Since both sets of data contain $n = 30$ data points, the quartiles are in the same positions for both sets.

The second quartile (median) is the mean of the values that lie in the 15th and 16th positions. For Centrum, these values are both 3.02. So, $Q_2 = M = \dfrac{3.02 + 3.02}{2} = 3.02$.
For the generic brand, these values are 6.30 and 6.33, so $Q_2 = M = \dfrac{6.30 + 6.33}{2} = 6.315$.

The first quartile is the median of the bottom 15 data values. This is the value that lies in the 8th position. So, for Centrum, $Q_1 = 2.60$. For the generic brand, $Q_1 = 5.57$.

The third quartile is the median of the top 15 data values. This is the value that lies in the 23rd position. So, for Centrum, $Q_3 = 3.53$. For the generic brand, $Q_3 = 6.73$.

So, the five-number summaries are:
Centrum: 2.15, 2.60, 3.02, 3.53, 4.33
Generic: 4.97, 5.57, 6.315, 6.73, 7.58

The fences for Centrum are:
$$LF = Q_1 - 1.5(IQR)$$
$$= 2.60 - 1.5(3.53 - 2.60) = 1.205$$
$$UF = Q_3 + 1.5(IQR)$$
$$= 3.53 + 1.5(3.53 - 2.60) = 4.925$$

The fences for the generic brand are:
$$LF = Q_1 - 1.5(IQR)$$
$$= 5.57 - 1.5(6.73 - 5.57) = 3.83$$
$$UF = Q_3 + 1.5(IQR)$$
$$= 6.73 + 1.5(6.73 - 5.57) = 8.47$$

So, neither data set has any outliers.

Note: Results from MINITAB differ:
Centrum: 2.15, 2.593, 3.02, 3.555, 4.33
$\quad\quad LF = 1.15$ and $UF = 4.998$
Generic: 4.97, 5.565, 6.315, 6.83, 7.58
$\quad\quad LF = 3.6675$ and $UF = 8.7275$

Using the by-hand computations for the five-number summaries, the side-by-side boxplots follow:

(b) From the boxplots, we can see that the generic brand has both a larger range and a larger interquartile range. Therefore, the generic brand has more dispersion.

(c) From the boxplots, we can see that the Centrum vitamins dissolve in less time than the generic vitamins. That is, Centrum vitamins dissolve faster.

17. (a) For depth of earthquake:
$\mu = 19.483$ km ; $M = 7.26$ km ;
Range $= 534.05$ km ; $\sigma = 49.498$ km ;
$Q_1 = 2.99$ km; $Q_3 = 15.38$ km

For earthquake magnitudes:
$\mu = 1.6967$; $M = 1.38$;
Range $= 6.8$; $\sigma = 1.2061$;
$Q_1 = 0.9$; $Q_3 = 2.01$

For depth of earthquake, the mean is much larger than the median, so the distribution of depth is likely skewed to the right. For earthquake magnitude, the mean is larger than the median and the distance from Q_1 to M is less than the distance from M to Q_3, which suggest that the distribution of magnitude is skewed to the right.

(b)

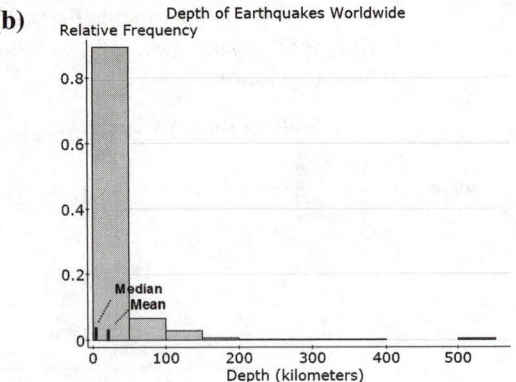

Depth of Earthquakes Worldwide

Both histograms appear to be skewed right.

(c)

There are a large number of outliers for both the depth and the magnitude of the earthquakes.

(d) For depth of earthquake:

$Q_1 = 2.99$ km; $Q_3 = 15.38$ km;

$$IQR = Q_3 - Q_1 = 15.38 - 2.99 = 12.39$$
$$LF = Q_1 - 1.5(IQR)$$
$$= 2.99 - 1.5(12.39) = -15.595$$
$$UF = Q_3 + 1.5(IQR)$$
$$= 15.38 + 1.5(12.39) = 33.965$$

For earthquake magnitudes:

$Q_1 = 0.9$; $Q_3 = 2.01$;

$$IQR = Q_3 - Q_1 = 2.01 - 0.9 = 1.11$$
$$LF = Q_1 - 1.5(IQR)$$
$$= 0.9 - 1.5(1.11) = -0.765$$
$$UF = Q_3 + 1.5(IQR)$$
$$= 2.01 + 1.5(1.11) = 3.675$$

19. (a) This is a completely randomized design. The researchers randomly assigned the subjects to the experimental group and the treatment group, then observed the outcome for each group.

(b) The experimental units are the 30 subjects who participated in the study.

(c) The response variable is the number of advantageous cards selected. This is a discrete quantitative variable.

(d) According to the article, factors that might impact the response variable are impulsivity, age, body mass index (BMI), and hunger. The factor that was manipulated was hunger, which had two levels (breakfast or no breakfast).

(e) In an attempt to create groups that were similar in terms of hunger, impulsivity, body mass index, and age, the subjects were randomly assigned to one of two treatment groups. In addition, the expectation is that randomization "evens out" the effect of other explanatory variables not considered. The researchers verified that the impulsivity, ages, and body mass index of each group were not significantly different.

(f) The statistics in this study are $\bar{x}_1 = 33.36$ and $\bar{x}_2 = 25.86$, which are the means for the subjects in group 1 and group 2, respectively.

(g) According to the article, the researchers concluded that hunger improves advantageous decision making.

21. Using the boxplot: If the median is left of the center in the box, and the right whisker is longer than the left whisker, the distribution is skewed right. If the median is in the center of the box, and the left and right whiskers are roughly the same length, the distribution is symmetric. If the median is in the right of the center of the box, and the left whisker is longer than the right whisker, the distribution is skewed left.

Using the quartiles: If the distance from the median to the first quartile is less than the distance from the median to the third quartile, or the distance from the median to the minimum value in the data set is less than the distance from the median to the maximum value in the data set, then the distribution is skewed right. If the distance from the median to the first quartile is the same as the distance from the median to the third quartile, or the distance from the median to the minimum is the same as the distance from the median to the maximum, the distribution is symmetric. If the distance from the median to the first quartile is more than the distance from the median to the third quartile, or the distance from the median to the minimum value in the data set is more than the distance from the median to the maximum value in the data set, the distribution is skewed left.

Chapter 3 Review Exercises

1. (a) $\sum x_i = 793.8 + 793.1 + 792.4 + 794.0 + 791.4$
$+ 792.4 + 791.7 + 792.3 + 789.6 + 794.4$
$= 7,925.1$

$$\text{Mean} = \bar{x} = \frac{\sum x_i}{n} = \frac{7,925.1}{10} = 792.51 \text{ m/sec}$$

Data in order:
789.6, 791.4, 791.7, 792.3, 792.4,
792.4, 793.1, 793.8, 794.0, 794.4

$$\text{Median} = \frac{792.4 + 792.4}{2} = 792.4 \text{ m/sec}$$

(b) Range = Largest Value – Smallest Value
$= 794.4 - 789.6 = 4.8$ m/sec

To calculate the sample variance and the sample standard deviation, we use the computational formulas:

x_i	x_i^2
793.8	630,118.44
793.1	629,007.61
792.4	627,897.76
794.0	630,436
791.4	626,313.96
792.4	627,897.76
791.7	626,788.89
792.3	627,739.29
789.6	623,468.16
794.4	631,071.36
$\sum x_i = 7925.1$	$\sum x_i^2 = 6,280,739.23$

$$s^2 = \frac{\sum x_i^2 - \frac{\left(\sum x_i\right)^2}{n}}{n-1}$$

$$= \frac{6,280,739.23 - \frac{(7,925.1)^2}{10}}{10-1}$$

$$\approx 2.03 \text{ (m/sec)}^2$$

$$s = \sqrt{\frac{6,280,739.23 - \frac{(7,925.1)^2}{10}}{10-1}}$$

$$\approx 1.42 \text{ m/sec}$$

2. (a) Add up the 9 data values: $\sum x = 91,610$

$$\bar{x} = \frac{\sum x}{n} = \frac{91,610}{9} \approx \$10,178.9$$

Data in order:
5500, 7200, 7889, 8998, 9980, 10995, 12999, 13999, 14050
The median is in the 5th position, so
$M = \$9,980$.

(b) Range = Largest Value − Smallest Value
$$= 14,050 - 5,500 = \$8,550$$

To find the interquartile range, we must find the first and third quartiles. The first quartile is the median of the bottom four data values, which is the mean of the values in the 2nd and 3rd positions. So,

$$Q_1 = \frac{7,200 + 7,889}{2} = 7,544.5 .$$

The third quartile is the median of the top four data values, which is the mean of the values in the 7th and 8th positions. So,

$$Q_3 = \frac{12,999 + 13,999}{2} = 13,499 .$$

Finally, the interquartile range is:
IQR = 13,499 − 7,544.5 = $5,954.5.

Note: Results from MINITAB differ: $Q_1 = \$7,545$ and $Q_3 = \$13,499$, so $IQR = \$5,954$. (Note that the MINITAB-calculated IQR is $5,955.)

To calculate the sample standard deviation, we use the computational formulas:

Data, x_i	x_i^2
14,050	197,402,500
13,999	195,972,001
12,999	168,974,001
10,995	120,890,025
9,980	99,600,400
8,998	80,964,004
7,889	62,236,321
7,200	51,840,000
5,550	30,250,000
$\sum x_i = 91,610$	$\sum x_i^2 = 1,008,129,252$

$$s = \sqrt{\frac{\sum x_i^2 - \frac{\left(\sum x_i\right)^2}{n}}{n-1}}$$

$$= \sqrt{\frac{1,008,129,252 - \frac{(91,610)^2}{9}}{9-1}} \approx \$3,074.9$$

(c) Add up the 9 data values: $\sum x = 118,610$

$$\bar{x} = \frac{\sum x}{n} = \frac{118,610}{9} \approx \$13,178.9$$

Data in order:
5500, 7200, 7889, 8998, 9980, 10995, 12999, 13999, 41050

The median is in the 5th position, so
$M = \$9,980$.

Range = 41,050 − 5,500 = $35,550

The first quartile is the mean of the values in the 2nd and 3rd positions. So,

$$Q_1 = \frac{7,200 + 7,889}{2} = 7,544.5 .$$

The third quartile is the mean of the values in the 7th and 8th positions. So,

$$Q_3 = \frac{12,999 + 13,999}{2} = 13,499 .$$

Finally, the interquartile range is:
IQR = 13,499 − 7,544.5 = $5,954.5
(or $5,954 if using MINITAB).

To calculate the sample standard deviation, we again use the computational formulas:

Data, x_i	x_i^2
41,050	1,685,102,500
13,999	195,972,001
12,999	168,974,001
10,995	120,890,025
9,980	99,600,400
8,998	80,964,004
7,889	62,236,321
7,200	51,840,000
5,550	30,250,000
$\sum x_i = 118,610$	$\sum x_i^2 = 2,495,829,252$

$$s = \sqrt{\frac{\sum x_i^2 - \frac{\left(\sum x_i\right)^2}{n}}{n-1}}$$

$$= \sqrt{\frac{2,495,829,252 - \frac{(118,610)^2}{9}}{9-1}}$$

$$\approx \$10,797.5$$

The mean, range, and standard deviation all changed considerably by the incorrectly entered data value. The median and interquartile range did not change. The median and interquartile range are resistant, while the mean, range, and standard deviation are not resistant.

3. (a) $\mu = \dfrac{\sum x}{N} = \dfrac{983}{17} \approx 57.8$ years

Data in order:
44, 46, 50, 51, 55, 56, 56, 56, 58, 59, 62, 62, 62, 64, 65, 68, 69

The median is the data value in the

$\dfrac{17+1}{2} = 9^{\text{th}}$ position. So, $M = 58$ years.

The data is bimodal: 56 years and 62 years. Both have frequencies of 3.

(b) Range = 69 − 44 = 25 years
To calculate the population standard deviation, we use the computational formula:

Data, x_i	x_i^2
44	1936
56	3136
51	2601
46	2116
59	3481
56	3136
58	3364
55	3025
65	4225
64	4096
68	4624
69	4761
56	3136
62	3844
62	3844
62	3844
50	2500
$\sum x_i = 983$	$\sum x_i^2 = 57{,}669$

$$\sigma = \sqrt{\dfrac{\sum x_i^2 - \dfrac{(\sum x_i)^2}{N}}{N}} = \sqrt{\dfrac{57{,}669 - \dfrac{(983)^2}{17}}{17}}$$
$$\approx 6.98 \text{ years}$$

(c) Answers will vary.

4. (a) To construct the histogram, we first organize the data into a frequency table:

Tickets Issued	Frequency
0	18
1	11
2	1

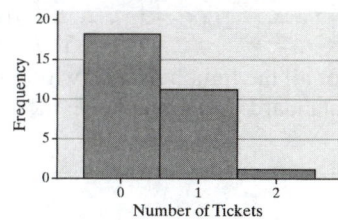

Number of Tickets Issued in a Month

The distribution is skewed right.

(b) Since the distribution is skewed right, we would expect the mean to be greater than the median.

(c) To find the mean, we add all of the data values and divide by the sample size.

$$\sum x_i = 13 \; ; \; \overline{x} = \dfrac{\sum x_i}{n} = \dfrac{13}{30} = 0.4$$

To find the median, we arrange the data in order. The median is the mean of the mean of the 15th and 16th data values.

$$M = \dfrac{0+0}{2} = 0$$

(d) The mode is 0 (the most frequent value).

5. (a) By the Empirical Rule, approximately 99.7% of the data will be within 3 standard deviations of the mean. Now, 600 − 3(53) = 441 and 600 + 3(53) = 759. Thus, about 99.7% of light bulbs have lifetimes between 441 and 759 hours.

(b) Since 494 is exactly 2 standard deviations below the mean [494 = 600 − 2(53)] and 706 is exactly 2 standard deviations above the mean [706 = 600 + 2(53)], the Empirical Rule predicts that about 95% of the light bulbs will have lifetimes between 494 and 706 hours.

(c) Since 547 is exactly 1 standard deviation below the mean [547 = 600 − 1(53)] and 706 is exactly 2 standard deviations above the mean [706 = 600 + 2(53)], the Empirical Rule predicts that about 34 + 47.5 = 81.5% of the light bulbs will have lifetimes between 547 and 706 hours.

(d) Since 441 hours is 3 standard deviations below the mean [441 = 600 − 3(53)], the Empirical Rule predicts that 0.15% of light bulbs will last less than 441 hours. Thus, the company should expect to replace about 0.15% of the light bulbs.

(e) By Chebyshev's theorem, at least
$$\left(1-\frac{1}{k^2}\right)\cdot 100\% = \left(1-\frac{1}{2.5^2}\right)\cdot 100\% = 84\%$$
of all the light bulbs are within $k = 2.5$ standard deviations of the mean.

(f) Since 494 is exactly $k = 2$ standard deviations below the mean [494 = 600 − 2(53)] and 706 is exactly 2 standard deviations above the mean [706 = 600 + 2(53)], Chebyshev's inequality indicates that at least
$$\left(1-\frac{1}{k^2}\right)\cdot 100\% = \left(1-\frac{1}{2^2}\right)\cdot 100\% = 75\%$$
of the light bulbs will have lifetimes between 494 and 706 hours.

---

6. (a) To find the mean, we use the formula $\bar{x}=\dfrac{\sum x_i f_i}{\sum f_i}$. To find the standard deviation in part (b), we will

choose to use the computational formula $s=\sqrt{s^2}=\sqrt{\dfrac{\sum x_i^2 f_i - \dfrac{\left(\sum x_i f_i\right)^2}{\sum f_i}}{\left(\sum f_i\right)-1}}$. We organize our

computations of x_i, $\sum f_i$, $\sum x_i f_i$, and $\sum x_i^2 f_i$ in the table that follows:

Class	Midpoint, x_i	Frequency, f_i	$x_i f_i$	x_i^2	$x_i^2 f_i$
0−9	$\dfrac{0+10}{2}=5$	125	625	25	3,125
10−19	$\dfrac{10+20}{2}=15$	271	4,065	225	60,975
20−29	25	186	4,650	625	116.250
30−39	35	121	4,235	1,225	148,225
40−49	45	54	2,430	2,025	109,350
50−59	55	62	3,410	3,025	187,550
60−69	65	43	2,795	4,225	181,675
70−79	75	20	1,500	5,625	112,500
80−89	85	13	1,105	7,225	93,925
		$\sum f_i = 895$	$\sum x_i f_i = 24{,}815$		$\sum x_i^2 f_i = 1{,}013{,}575$

With the table complete, we compute the sample mean:
$$\bar{x}=\frac{\sum x_i f_i}{\sum f_i}=\frac{24{,}815}{895}=27.7 \text{ minutes}$$

(b) $s=\sqrt{s^2}=\sqrt{\dfrac{\sum x_i^2 f_i - \dfrac{\left(\sum x_i f_i\right)^2}{\sum f_i}}{\left(\sum f_i\right)-1}}=\sqrt{\dfrac{1{,}013{,}575-\dfrac{(24{,}815)^2}{895}}{895-1}}\approx 19.1 \text{ minutes}$

7. $\text{GPA} = \bar{x}_w = \dfrac{\sum w_i x_i}{\sum w_i}$

$= \dfrac{5(4)+4(3)+3(4)+3(2)}{5+4+3+3} = \dfrac{50}{15} \approx 3.33$

8. Female: $z = \dfrac{x-\mu}{\sigma} = \dfrac{160-156.5}{51.2} \approx 0.07$

Male: $z = \dfrac{x-\mu}{\sigma} = \dfrac{185-183.4}{40.0} = 0.04$

The weight of the 160-pound female is 0.07 standard deviation above the mean, while the weight of the 185-pound male is 0.04 standard deviation above the mean. Thus, the 160-pound female is relatively heavier.

9. (a) The two-seam fastball

(b) The two-seam fastball

(c) The four-seam fastball

(d) There is an outlier at approximately 88 mph.

(e) Symmetric

(f) Skewed right

10. (a) Add up the 57 data values: $\sum x_i = 134,227$

$\mu = \dfrac{\sum x_i}{n} = \dfrac{134,227}{57} \approx 2,354.9$ words

To find the median and quartiles, we must arrange the data in order:

135	1,340	1,883	2,446	3,801
559	1,355	2,015	2,449	3,838
698	1,425	2,073	2,463	3,967
985	1,437	2,130	2,480	4,059
996	1,507	2,137	2,546	4,388
1,087	1,526	2,158	2,821	4,467
1,125	1,571	2,170	2,906	4,776
1,128	1,668	2,217	2,978	5,433
1,172	1,681	2,242	3,217	8,445
1,175	1,729	2,283	3,318	
1,209	1,802	2,308	3,319	
1,337	1,807	2,406	3,634	

The median is the data value that lies in the 29th position, which is M = 2,137 words.

(b) The second quartile is the median, $Q_2 = M = 2,137$ words.

The first quartile is the median of the bottom 29 data values, which is the data value that lies in the 15th position. Thus, $Q_1 = 1,425$ words.

The third quartile is the median of the top 29 data values, which is the data value that lies in the 43rd position. Thus, $Q_3 = 2,906$ words.

Note: Results from MINITAB differ: $Q_1 = 1,390$, $Q_2 = 2,137$, and $Q_3 = 2,942$.

We use the by-hand quartiles for the interpretations: 25% of the inaugural addresses had 1,425 words or less, 75% of the inaugural addresses had more than 1,425 words; 50% of the inaugural addresses had 2,137 words or less, 50% of the inaugural addresses had more than 2,137 words; 75% of the inaugural addresses had 2,906 words or less, 25% of the inaugural addresses had more than 2,906 words.

(c) The smallest value is 135 and the largest is 8,445. Combining these with the quartiles, we obtain the five-number summary: 135, 1,425, 2,137, 2,906, 8,445

Note: results from MINITAB differ: 135, 1,390, 2,137, 2,942, 8,445

(d) To calculate the sample standard deviation, we use the computational formulas:

We add up the 57 data values: $\sum x_i = 134,227$.

We square each of the 57 data values and add up the results: $\sum x_i^2 = 424,825,701$

$\sigma = \sqrt{\dfrac{\sum x_i^2 - \dfrac{(\sum x_i)^2}{N}}{N}}$

$= \sqrt{\dfrac{424,825,701 - \dfrac{(134,227)^2}{57}}{57}}$

$\approx 1,381.2$ words

The interquartile range is:
$IQR = Q_3 - Q_1$
$= 2,906 - 1,425 = 1,481$ words

Note: results from MINITAB differ:
$IQR = 2,942 - 1,390 = 1,552$ words

(e) $LF = Q_1 - 1.5(IQR)$

$\quad = 1,425 - 1.5(1,481) = -796.5$ words

$UF = Q_3 + 1.5(IQR)$

$\quad = 2,906 + 1.5(1,481) = 5127.5$ words

So, there are two outliers: 5,433 and 8,445.

Note: results from MINITAB differ:
$LF = 1,390 - 1.5(1,552) = -938$ words
$UF = 2,942 + 1.5(1,552) = 5,270$ words
Using MINITAB, there are two outliers:
5,433 and 8,445.

(f) We use the by-hand results to construct the boxplot.

Number of Words in an Inaugural Speech

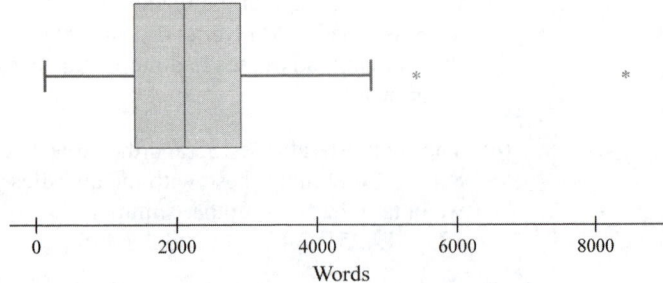

(g) The distribution is skewed right. The distance between the first and second quartile is 712. The distance between the second and third quartile is larger at 769. We can tell because the median is slightly left of the center in the box and the right whisker is longer than the left whisker (even without considering the outlier.)

(h) The median is the better measure of center because the distribution is skewed right and the outliers inflate the value of the mean.

(i) The interquartile range is the better measure of dispersion because the outliers inflate the value of the standard deviation.

11. This means that 85% of 19-year-old females have a height that is 67.1 inches or less, and $(100 - 85)\% = 15\%$ of 19-year-old females have a height that is more than 67.1 inches.

12. The median is used for three measures since it is likely the case that one of the three measures is extreme relative to the other two, thus substantially affecting the value of the mean. Since the median is resistant to extreme values, it is the better measure of central tendency.

Chapter 3 Test

1. (a) $\sum x_i = 48 + 88 + 57 + 109 + 111 + 93 + 71 + 63$

$\quad = 640$

$$\text{Mean} = \bar{x} = \frac{\sum x_i}{n} = \frac{640}{8} = 80 \text{ min}$$

(b) Data in order:
48, 57, 63, 71, 88, 93, 109, 111
The median is the mean of the values in the 4th and 5th positions:

$$\text{Median} = \frac{71 + 88}{2} = 79.5 \text{ min}$$

(c) $\sum x_i = 48 + 88 + 57 + 1009 + 111 + 93 + 71 + 63$

$\quad = 1540$

$$\text{Mean} = \bar{x} = \frac{\sum x_i}{n} = \frac{1540}{8} = 192.5 \text{ min}$$

Data in order:
48, 57, 63, 71, 88, 93, 111, 1009
The median is the mean of the values in the 4th and 5th positions:

$$\text{Median} = \frac{71 + 88}{2} = 79.5 \text{ min}$$

The mean was changed substantially by the incorrectly entered data value. The median did not change. The median is resistant, while the mean is not resistant.

2. The mode type of larceny is "From motor vehicles." It has the highest frequency.

3. Range = Largest Value − Smallest Value
$\quad = 111 - 48 = 63$ minutes

4. (a) To calculate the sample standard deviation, we use the computational formula:

x_i	x_i^2
48	2,304
88	7,744
57	3,249
109	11,881
111	12,321
93	8,649
71	5,041
63	3,969
$\sum x_i = 640$	$\sum x_i^2 = 55,158$

$$s = \sqrt{\frac{\sum x_i^2 - \frac{\left(\sum x_i\right)^2}{n}}{n-1}}$$

$$= \sqrt{\frac{55,158 - \frac{(640)^2}{8}}{8-1}} \approx 23.8 \, \text{min}$$

(b) To find the interquartile range, we must find the first and third quartiles. The first quartile is the median of the bottom four data values, which is the mean of the values in the 2nd and 3rd positions. So,

$$Q_1 = \frac{57+63}{2} = 60 \, \text{min.}$$

The third quartile is the median of the top four data values, which is the mean of the values in the 6th and 7th positions. So,

$$Q_3 = \frac{93+109}{2} = 101 \, \text{min.}$$

Finally, the interquartile range is:
IQR = 101 − 60 = 41 min.
Interpretation: The middle 50% of all the times students spent on the assignment have a range of 41 minutes.

(c) The interquartile range is resistant; the standard deviation is not resistant.

5. (a) By the Empirical Rule, approximately 99.7% of the data will be within 3 standard deviations of the mean. Now, 4302 − 3(340) = 3282 and 4302 + 3(340) = 5322. So, about 99.7% of toner cartridges will print between 3282 and 5322 pages.

(b) Since 3622 is 2 standard deviations below the mean [3622 = 4302 − 2(340)] and 4982 is 2 standard deviations above the mean [4982 = 4302 + 2(340)], the Empirical Rule predicts that about 95% of the toner cartridges will print between 3622 and 4982 pages.

(c) Since 3622 is 2 standard deviations below the mean [3622 = 4302 − 2(340)], the Empirical Rule predicts that 0.15 + 2.35 = 2.5% of the toner cartridges will last less than 3622 pages. So, the company can expect to replace about 2.5% of the toner cartridges.

(d) By Chebyshev's theorem, at least

$$\left(1 - \frac{1}{k^2}\right) \cdot 100\% = \left(1 - \frac{1}{1.5^2}\right) \cdot 100\% \approx 55.6\%$$

of all the toner cartridges are within $k = 1.5$ standard deviations of the mean.

(e) Since 3282 is $k = 3$ standard deviations below the mean [3282 = 4302 − 3(340)] and 5322 is 3 standard deviations above the mean [5322 = 4302 + 3(340)], by Chebyshev's inequality, at least

$$\left(1 - \frac{1}{k^2}\right) \cdot 100\% = \left(1 - \frac{1}{3^2}\right) \cdot 100\% \approx 88.9\%$$

of the toner cartridges will print between 3282 and 5322 pages.

6. (a) To find the mean, we use the formula $\bar{x} = \dfrac{\sum x_i f_i}{\sum f_i}$. To find the standard deviation in part (b), we will use

the computational formula $s = \sqrt{s^2} = \sqrt{\dfrac{\sum x_i^2 f_i - \dfrac{\left(\sum x_i f_i\right)^2}{\sum f_i}}{\left(\sum f_i\right) - 1}}$. We organize our computations of x_i, $\sum f_i$,

$\sum x_i f_i$, and $\sum x_i^2 f_i$ in the table that follows:

Class	Midpoint, x_i	Frequency, f_i	$x_i f_i$	x_i^2	$x_i^2 f_i$
$40-49$	$\dfrac{40+50}{2} = 45$	8	360	2,025	16,200
$50-59$	$\dfrac{50+60}{2} = 55$	44	2,420	3,025	133,100
$60-69$	65	23	1,495	4,225	97,175
$70-79$	75	6	450	5,625	33,750
$80-89$	85	107	9,095	7,225	773,075
$90-99$	95	11	1,045	9,025	99,275
$100-109$	105	1	105	11,025	11,025
		$\sum f_i = 200$	$\sum x_i f_i = 14,970$		$\sum x_i^2 f_i = 1,163,600$

With the table complete, we compute the sample mean: $\bar{x} = \dfrac{\sum x_i f_i}{\sum f_i} = \dfrac{14,970}{200} \approx 74.9$ minutes

(b) $s = \sqrt{s^2} = \sqrt{\dfrac{\sum x_i^2 f_i - \dfrac{\left(\sum x_i f_i\right)^2}{\sum f_i}}{\left(\sum f_i\right) - 1}} = \sqrt{\dfrac{1,163,600 - \dfrac{(14,970)^2}{200}}{200 - 1}} \approx 14.7$ minutes

---

7. Cost per pound $= \bar{x}_w = \dfrac{\sum w_i x_i}{\sum w_i x_i}$

$= \dfrac{2(\$2.70) + 1(\$1.30) + \frac{1}{2}(\$1.80)}{2 + 1 + \frac{1}{2}}$

$\approx \$2.17 / \text{lb}$

8. (a) Material A: $\sum x_i = 64.04$ and $n = 10$, so

$\bar{x}_A = \dfrac{64.04}{10} = 6.404$ million cycles.

Material B: $\sum x_i = 113.32$ and $n = 10$,

so $\bar{x}_B = \dfrac{113.32}{10} = 11.332$ million cycles.

(b) Notice that each set of $n = 10$ data values is already arranged in order. For each set, the median is the mean of the values in the 5th and 6th positions.

$M_A = \dfrac{5.69 + 5.88}{2} = 5.785$ million cycles

$M_B = \dfrac{8.20 + 9.65}{2} = 8.925$ million cycles

(c) To find the sample standard deviation, we choose to use the computational formula: Material A:

$\sum x_i = 64.04$ and $\sum x_i^2 = 472.177$, so

$s_A = \sqrt{\dfrac{\sum x_i^2 \sum x_i^2 - \dfrac{\left(\sum x_i\right)^2}{n}}{n - 1}}$

$= \sqrt{\dfrac{472.177 - \dfrac{(64.04)^2}{10}}{10 - 1}}$

≈ 2.626 million cycles

Material B:

$\sum x_i = 113.32$ and $\sum x_i^2 = 1{,}597.4002$, so

$$s_B = \sqrt{\dfrac{1597.4002 - \dfrac{(113.32)^2}{10}}{10-1}}$$

≈ 5.900 million cycles

Material B has more dispersed failure times because it has a much larger standard deviation.

(d) For each set, the first quartile is the data value in the 3rd position and the third quartile is the data value in the 8th position.

Material A:
3.17; 4.52; 5.785; 8.01; 11.92

Material B:
5.78; 6.84; 8.925; 14.71; 24.37

(e) Before drawing the side-by-side boxplots, we check each data set for outliers.

Fences for Material A:
$LF = Q_1 - 1.5(IQR)$
$\quad = 4.52 - 1.5(8.01 - 4.52)$
$\quad = -0.715$ million cycles
$UF = Q_3 + 1.5(IQR)$
$\quad = 8.01 + 1.5(8.01 - 4.52)$
$\quad = 13.245$ million cycles
Material A has no outliers.

Fences for Material B:
$LF = 6.84 - 1.5(14.71 - 6.84)$
$\quad = -4.965$ million cycles
$UF = 14.71 + 1.5(14.71 - 6.84)$
$\quad = 26.515$ million cycles
Material B has no outliers

Bearing Failures

Material A

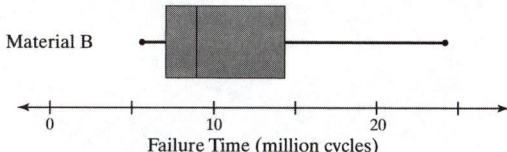

Material B

Failure Time (million cycles)

Annotated remarks will vary. Possible remarks: Material A generally has lower failure times. Material B has more dispersed failure times.

(f) In both boxplots, the median is to the left of the center of the box and the right line is substantially longer than the left line, so both distributions are skewed right. In terms of quartiles, the distance from the median to the first quartile is less than the distance from the median to the third quartile, so the distribution is skewed right.

9. Notice that the data set is already arranged in order. The second quartile (median) is the mean of the values in the 25th and 26th positions, which are both 5.60. So,

$Q_2 = \dfrac{5.60 + 5.60}{2} = 5.60$ grams.

The first quartile is the median of the bottom 25 data values, which is the value in the 13th position. So, $Q_1 = 5.58$ grams.

The third quartile is the median of the top 25 data values, which is the value in the 38th position. So, $Q_3 = 5.66$ grams.

$LF = Q_1 - 1.5(IQR)$
$\quad = 5.58 - 1.5(5.66 - 5.58) = 5.46$ grams
$UF = Q_3 + 1.5(IQR)$
$\quad = 5.66 + 1.5(5.66 - 5.58) = 5.78$ grams

So, the quarter that weighs 5.84 grams is an outlier.

Note: Results from MINITAB differ:
$Q_1 = 5.58$, $Q_2 = 5.60$, $Q_3 = 5.6625$,
$LF = 5.45625$, and $UF = 5.78625$.

For the interpretations, we use the by-hand quartiles: 25% of quarters have a weight that is 5.58 grams or less, 75% of the quarters have a weight more than 5.58 grams; 50% of the quarters have a weight that is 5.60 grams or less, 50% of the quarters have a weight more than 5.60 grams; 75% of the quarters have a weight that is 5.66 grams or less, 25% of the quarters have a weight that is more than 5.66 grams.

10. SAT: $z = \dfrac{x - \mu}{\sigma} = \dfrac{610 - 515}{114} \approx 0.83$

ACT: $z = \dfrac{x - \mu}{\sigma} = \dfrac{27 - 21.0}{5.1} \approx 1.18$

Armando's SAT score is 0.83 standard deviation above the mean, his ACT score is 1.18 standard deviations above the mean. So, Armando should report his ACT score since it is more standard deviations above the mean.

11. This means that 15% of 10-year-old males have a height that is 53.5 inches or less, and (100 − 15)% = 85% of 10-year-old males have a height that is more than 53.5 inches.

12. You should report the median. Income data will be skewed right, which means the median will be less than the mean.

13. (a) Report the mean since the distribution is symmetric.

    (b) Histogram I has more dispersion. The range of classes is larger.

14. Answers may vary

Case Study: Who Was "A Mourner"?

1. The table that follows gives the length of each word, line by line in the passage. A listing is also provided of the proper names, numbers, abbreviation, and titles that have been omitted from the data set.

   Word length:
   Line 1:   3, 7, 8, 3, 7, 3, 3, 6, 2, 3, 3, 2, 3, 4
   Line 2:   3, 8, 2, 3, 7, 4, 2, 10, 6, 3
   Line 3:   4, 9, 3, 7, 4, 2, 4, 2, 6, 4, 3, 7, 5, 2
   Line 4:   8, 2, 4, 4, 3, 3, 7, 2
   Line 5:   7, 3, 4, 2, 11, 2, 6, 5, 4, 8, 2, 3, 7
   Line 6:   2, 4, 6, 4, 3, 5, 6, 2, 3, 5, 5, 5, 5
   Line 7:   6, 5, 4, 8, 8, 2, 3, 8, 7, 2, 3
   Line 8:   6, 3, 6, 2, 3, 9, 3, 6, 4, 3, 3, 7
   Line 9:   3, 5, 2, 9, 3, 8, 8, 2, 6, 4, 3, 4, 5
   Line 10:   2, 3, 3, 4, 2, 7, 5, 6, 8, 4, 3, 7, 6
   Line 11:   6, 5, 2, 3, 6, 12
   Line 12:   5, 6, 2, 5, 2, 3, 1, 7, 6, 3, 5, 4, 4, 1
   Line 13:   6, 3

   Omission list:
   Line 2:   Richardson, 22d
   Line 4:   Frogg Lane, Liberty-Tree, Monday
   Line 7:   appear'd
   Line 11:   *Wolfe's Summit of human Glory*

2. Mean = 4.5; Median = 4; Mode = 3; Standard deviation ≈ 2.21; Sample variance ≈ 4.90; Range = 11; Minimum = 1; Maximum = 12; Sum = 649; Count = 143

Answers will vary. None of the provided authors match both the measures of central tendency and the measures of dispersion well. In other words, there is no clear cut choice for the author based on the information provided. Based on measures of central tendency, James Otis or Samuel Adams would appear to be the more likely candidates for A MOURNER. Based on measures of dispersion, Tom Sturdy seems the more likely choice. Still, the unknown author's mean word length differs considerably from that of Sturdy, and the unknown author's standard deviation differs considerably from those of Otis and Adams.

3. Comparing the two Adams summaries, both the measures of center and the measures of variability differ considerably for the two documents. For example, the means differ by 0.08 and the standard deviation differs by 0.19, not to mention the differences in word counts and the maximum length. This calls into question the viability of word-length analysis as a tool for resolving disputed documents. Word length may be a part of the analysis needed to determine unknown authors, but other variables should also be taken into consideration.

4. Other information that would be useful to identify A MOURNER would be the style of the rhetoric, vocabulary choices, use of particular phrases, and the overall flow of the writing. In other words, identifying an unknown author requires qualitative analysis in addition to quantitative analysis.

Chapter 4
Describing the Relation between Two Variables

Section 4.1

1. Univariate data measures the value of a single variable for each individual in the study. Bivariate data measures values of two variables for each individual.

3. Scatter diagram

5. −1

7. Lurking

9. Nonlinear

11. Linear; positive

13. (a) III   (b) IV   (c) II   (d) I

15. (a) Looking at the scatter diagram, the points tend to increase at a relatively consistent rate from left to right. So, there appears to be a positive, linear association between the percentage of the population with at least a bachelor's degree and median income.

    (b) The point with roughly the coordinates (55, 55000) appears to stick out. Reasons may vary. One possibility:
    A high concentration of government jobs that require a bachelor's degree and the high proportion of lobbyists and attorneys who live in Washington, D.C.

    (c) Yes, there is linear relationship because $|0.854| > 0.361$ (the critical value from Table II with $n = 30$).

17. (a)

    (b) We compute the mean and standard deviation for both variables: $\bar{x} = 5$, $s_x = 2$, $\bar{y} = 11$, and $s_y = 6$. We determine $\dfrac{x_i - \bar{x}}{s_x}$ and $\dfrac{y_i - \bar{y}}{s_y}$ in columns 3 and 4. We multiply these entries to obtain the entries in column 5.

x_i	y_i	$\dfrac{x_i - \bar{x}}{s_x}$	$\dfrac{y_i - \bar{y}}{s_y}$	$\left(\dfrac{x_i - \bar{x}}{s_x}\right)\left(\dfrac{y_i - \bar{y}}{s_y}\right)$
2	4	−1.5	−1.16667	1.75
4	8	−0.5	−0.5	0.25
6	10	0.5	−0.16667	−0.08334
6	13	0.5	0.33333	0.16667
7	20	1	1.5	1.5

We add the entries in column 5 to obtain

$$\sum \left(\frac{x_i - \bar{x}}{s_x}\right)\left(\frac{y_i - \bar{y}}{s_y}\right) = 3.58333 .$$

Finally, we use this result to compute r:

$$r = \frac{\sum \left(\dfrac{x_i - \bar{x}}{s_x}\right)\left(\dfrac{y_i - \bar{y}}{s_y}\right)}{n - 1} = \frac{3.58333}{5 - 1} \approx 0.896$$

(c) A linear relation exists between x and y.

19. (a)

    (b) We compute the mean and standard deviation for both variables: $\bar{x} = 6$, $s_x = 2.54951$, $\bar{y} = 7$, and $s_y = 1.58114$.

    We determine $\dfrac{x_i - \bar{x}}{s_x}$ and $\dfrac{y_i - \bar{y}}{s_y}$ in columns 3 and 4. We multiply these entries to obtain the entries in column 5.

x_i	y_i	$\dfrac{x_i - \bar{x}}{s_x}$	$\dfrac{y_i - \bar{y}}{s_y}$	$\left(\dfrac{x_i - \bar{x}}{s_x}\right)\left(\dfrac{y_i - \bar{y}}{s_y}\right)$
2	8	−1.56893	0.63246	−0.99228
6	7	0	0	0
6	6	0	−0.63246	0
7	9	0.39223	1.26491	0.49614
9	5	1.17670	−1.26491	−1.48842

We add the entries in column 5 to obtain

$$\sum \left(\frac{x_i - \bar{x}}{s_x}\right)\left(\frac{y_i - \bar{y}}{s_y}\right) = -1.98456 .$$

Finally, we use this result to compute r:

$$r = \frac{\sum\left(\dfrac{x_i - \overline{x}}{s_x}\right)\left(\dfrac{y_i - \overline{y}}{s_y}\right)}{n-1} = \frac{-1.98456}{5-1}$$
$$\approx -0.496$$

(c) A moderate negative relation exists between x and y.

21. (a) Positive correlation. The more infants, the more diapers will be needed.

(b) Negative correlation. The lower the interest rates, the more people can afford to buy a car.

(c) Negative correlation. More exercise is associated with lower cholesterol.

(d) Negative correlation. The higher the price of a Big Mac, the fewer Big Macs and French fries will be sold.

(e) No correlation. There is no correlation between shoe size and intelligence.

23. Yes; $0.79 > 0.361$ (critical value from Table II for $n = 30$). The correlation coefficient is high, suggesting that there is a linear relation between student task persistence and achievement score. This means that if a student has high task persistence, they will tend to have a high achievement score. Students with low task persistence will tend to have a low achievement score.

25. (a) Explanatory variable: commute time Response: well-being score

(b) The relationship between the commute time and the well-being score is illustrated below:

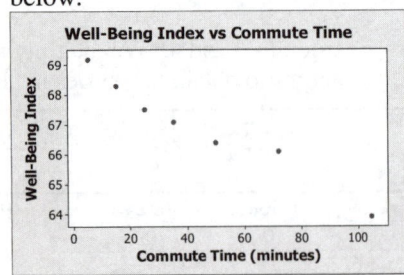

(c) We compute the mean and standard deviation for both variables:
$\overline{x} = 43.85714$, $s_x = 34.98775$,
$\overline{y} = 66.92857$, and $s_y = 1.70950$. We
determine $\dfrac{x_i - \overline{x}}{s_x}$ and $\dfrac{y_i - \overline{y}}{s_y}$ in columns
3 and 4. We multiply these entries to obtain the entries in column 5.

x_i	y_i	$\dfrac{x_i - \overline{x}}{s_x}$	$\dfrac{y_i - \overline{y}}{s_y}$	$\left(\dfrac{x_i - \overline{x}}{s_x}\right)\left(\dfrac{y_i - \overline{y}}{s_y}\right)$
5	69.2	−1.11059	1.32871	−1.47566
15	68.3	−0.82477	0.80224	−0.66167
25	67.5	−0.53896	0.33427	−0.18016
35	67.1	−0.25315	0.10028	−0.02539
50	66.4	0.17557	−0.30920	−0.05429
72	66.1	0.80436	−0.48469	−0.38986
105	63.9	1.74755	−1.77162	−3.09599

We add the entries in column 5 to obtain
$$\sum\left(\dfrac{x_i - \overline{x}}{s_x}\right)\left(\dfrac{y_i - \overline{y}}{s_y}\right) = -5.88302. \text{ Finally,}$$
we use this result to compute r:

$$r = \frac{\sum\left(\dfrac{x_i - \overline{x}}{s_x}\right)\left(\dfrac{y_i - \overline{y}}{s_y}\right)}{n-1} = \frac{-5.88302}{7-1}$$
$$\approx -0.981$$

(d) Yes, because $|-0.981| = 0.981 > 0.754$
(0.754 is the critical value from Table II), so a negative association exists between the commute time and the well-being index score.

27. (a) Explanatory variable: height; Response variable: head circumference

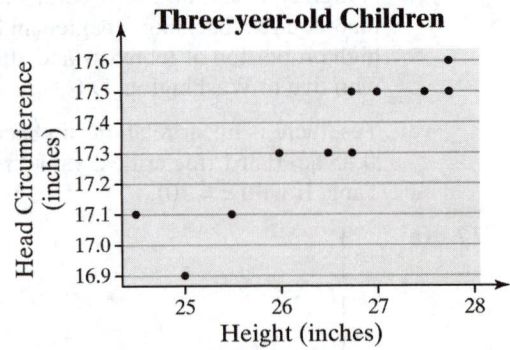

(c) To calculate the correlation coefficient, we use the computational formula:

x_i	y_i	x_i^2	y_i^2	$x_i y_i$
27.75	17.5	770.0625	306.25	485.625
24.5	17.1	600.25	292.41	418.95
25.5	17.1	650.25	292.41	436.05
26	17.3	676	299.29	449.8
25	16.9	625	285.61	422.5
27.75	17.6	770.0625	309.76	488.4
26.5	17.3	702.25	299.29	458.45
27	17.5	729	306.25	472.5
26.75	17.3	715.5625	299.29	462.775
26.75	17.5	715.5625	306.25	468.125
27.5	17.5	756.25	306.25	481.25
291	190.6	7710.25	3303.06	5044.425

From the table, we have $n = 11$, $\sum x_i = 291$, $\sum y_i = 190.6$, $\sum x_i^2 = 7710.25$, $\sum y_i^2 = 3303.06$, and $\sum x_i y_i = 5044.425$. So, the correlation coefficient is:

$$r = \frac{\sum x_i y_i - \dfrac{\sum x_i \sum y_i}{n}}{\sqrt{\left(\sum x_i^2 - \dfrac{(\sum x_i)^2}{n}\right)\left(\sum y_i^2 - \dfrac{(\sum y_i)^2}{n}\right)}}$$

$$= \frac{5044.425 - \dfrac{(291)(190.6)}{11}}{\sqrt{\left(7710.25 - \dfrac{(291)^2}{11}\right)\left(3303.06 - \dfrac{(190.6)^2}{11}\right)}}$$

$$\approx 0.911$$

(d) There is a strong positive linear association between the height and head circumference of a child.

(e) Converting the data to centimeters will have no effect on the linear correlation coefficient.

29. (a) Explanatory variable: weight
Response variable: gas mileage

(b)

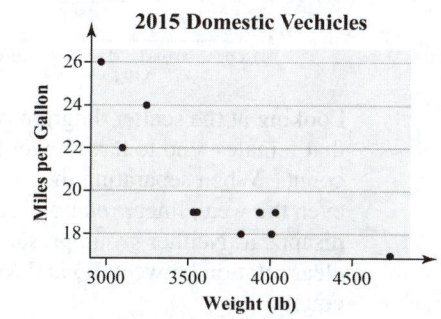

2015 Domestic Vechicles

(c) To calculate the correlation coefficient, we use the computational formula:

x_i	y_i	x_i^2	y_i^2	$x_i y_i$
4,724	17	22,316,176	289	80,308
4,006	18	16,048,036	324	72,108
3,097	22	9,591,409	484	68,134
3,555	19	12,638,025	361	67,545
4,029	19	16,232,841	361	76,551
3,934	19	15,476,356	361	74,746
3,242	24	10,510,564	576	77,808
2,960	26	8,761,600	676	76,960
3,530	19	12,460,900	361	67,070
3,823	18	14,615,329	324	68,814
36,900	201	138,651,236	4,117	730,044

From the table, we have $n = 10$, $\sum x_i = 36,900$, $\sum y_i = 201$, $\sum x_i^2 = 138,651,236$, $\sum y_i^2 = 4,117$, and

$\sum x_i y_i = 730,044$. So, the correlation coefficient is:

$$r = \frac{\sum x_i y_i - \dfrac{\sum x_i \sum y_i}{n}}{\sqrt{\left(\sum x_i^2 - \dfrac{(\sum x_i)^2}{n}\right)\left(\sum y_i^2 - \dfrac{(\sum y_i)^2}{n}\right)}}$$

$$= \frac{730,044 - \dfrac{(36,900)(201)}{10}}{\sqrt{\left(138,651,236 - \dfrac{(36,900)^2}{10}\right)\left(4,117 - \dfrac{(201)^2}{10}\right)}}$$

$$\approx -0.842$$

(d) Yes, there is a strong negative linear association between the weight of a car and its miles per gallon in the city because $|-0.842| > 0.632$ (the critical value from Table II for $n = 10$).

31. (a) The explanatory variable is stock return.

(b)

CEO Compensation and Stock Return

(c) To calculate the correlation coefficient, we use the computational formula:

x_i	y_i	x_i^2	y_i^2	$x_i y_i$
75.43	14.53	5689.685	211.1209	1095.9979
64.01	4.09	4097.280	16.7281	261.8009
142.07	7.11	20,183.885	50.5521	1010.1177
32.72	1.05	1070.598	1.1025	34.356
10.64	1.97	113.210	3.8809	20.9608
30.66	3.76	940.040	14.1376	115.2816
0.77	12.06	0.593	145.4436	9.2862
69.39	7.62	4814.972	58.0644	528.7518
58.69	8.47	3444.516	71.7409	497.1043
55.93	4.04	3128.165	16.3216	225.9572
24.28	20.87	589.518	435.5569	506.7236
32.21	6.63	1037.484	43.9569	213.5523
596.8	92.2	45109.946	1068.6064	4519.8903

From the table, we have $n = 12$, $\sum x_i = 596.8$, $\sum y_i = 92.2$, $\sum x_i^2 = 45,109.942$, $\sum y_i^2 = 1068.6064$, and $\sum x_i y_i = 4,519.8903$. So, the correlation coefficient is:

$$r = \frac{\sum x_i y_i - \frac{\sum x_i \sum y_i}{n}}{\sqrt{\left(\sum x_i^2 - \frac{(\sum x_i)^2}{n}\right)\left(\sum y_i^2 - \frac{(\sum y_i)^2}{n}\right)}}$$

$$= \frac{4,519.8903 - \frac{(596.8)(92.2)}{12}}{\sqrt{\left(45,109.942 - \frac{(596.8)^2}{12}\right)\left(1,068.6064 - \frac{(92.2)^2}{12}\right)}}$$

$$\approx -0.02779$$

(d) No, there is not a linear relation between compensation and stock return because $|{-0.02779}| < 0.576$ (critical value from Table II; Stock performance does not appear to play a role in determining the compensation of a CEO.

33. (a)

Brain Size vs. IQ

There is a moderate positive linear relation between the MRI count and the IQ.

(b) To calculate the correlation coefficient, we use the computational formula:

x_i	y_i	x_i^2	y_i^2	$x_i y_i$
816,932	133	667,377,892,624	17,689	108,651,956
951,545	137	905,437,887,025	18,769	130,361,665
991,305	138	982,685,603,025	19,044	136,800,090
833,868	132	695,335,841,424	17,424	110,070,576
856,472	140	733,544,286,784	19,600	119,906,080
852,244	132	726,319,835,536	17,424	112,496,208
790,619	135	625,078,403,161	18,225	106,733,565
866,662	130	751,103,022,244	16,900	112,666,060
857,782	133	735,789,959,524	17,689	114,085,006
948,066	133	898,829,140,356	17,689	126,092,778
949,395	140	901,350,866,025	19,600	132,915,300
1,001,121	140	1,002,243,256,641	19,600	140,156,940
1,038,437	139	1,078,351,402,969	19,321	144,342,743
965,353	133	931,906,414,609	17,689	128,391,949
955,466	133	912,915,277,156	17,689	127,076,978
1,079,549	141	1,165,426,043,401	19,881	152,216,409
924,059	135	853,885,035,481	18,225	124,747,965
955,003	139	912,030,730,009	19,321	132,745,417
935,494	141	875,149,024,036	19,881	131,904,654
949,589	144	901,719,268,921	20,736	136,740,816
18,518,961	2,728	17,256,479,190,951	372,396	2,529,103,155

From the table, we have $n = 20$, $\sum x_i = 18,518,961$, $\sum y_i = 2,728$, $\sum x_i^2 = 17,256,479,190,951$, $\sum y_i^2 = 372,396$, and $\sum x_i y_i = 2,529,103,155$. So, the correlation coefficient is:

$$r = \frac{\sum x_i y_i - \frac{\sum x_i \sum y_i}{n}}{\sqrt{\left(\sum x_i^2 - \frac{(\sum x_i)^2}{n}\right)\left(\sum y_i^2 - \frac{(\sum y_i)^2}{n}\right)}}$$

$$= \frac{2,529,103,155 - \frac{(18,518,961)(2,728)}{20}}{\sqrt{\left(17,256,479,190,951 - \frac{(18,518,961)^2}{20}\right)\left(372,396 - \frac{(2,728)^2}{20}\right)}}$$

$$\approx 0.548$$

A moderate positive linear relation exists between MRI count and IQ.

(c)

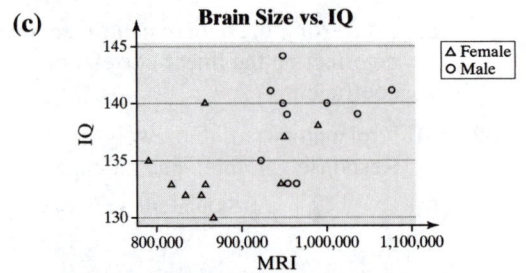

Brain Size vs. IQ

△ Female
○ Male

Looking at the scatter diagram, we can see that females tend to have lower MRI counts. When separating the two groups, even the weak linear relation seems to disappear. Neither group presents any clear relation between IQ and MRI counts.

(d) We first use the computational formula on the data from the female participants:

x_i	y_i	x_i^2	y_i^2	$x_i y_i$
816,932	133	667,377,892,624	17,689	108,651,956
951,545	137	905,437,887,025	18,769	130,361,665
991,305	138	982,685,603,025	19,044	136,800,090
833,868	132	695,335,841,424	17,424	110,070,576
856,472	140	733,544,286,784	19,600	119,906,080
852,244	132	726,319,835,536	17,424	112,496,208
790,619	135	625,078,403,161	18,225	106,733,565
866,662	130	751,103,022,244	16,900	112,666,060
857,782	133	735,789,959,524	17,689	114,085,006
948,066	133	898,829,140,356	17,689	126,092,778
8,765,495	1,343	7,721,501,871,703	180,453	1,177,863,984

From the table, we have $n = 10$,
$\sum x_i = 8,765,495$, $\sum y_i = 1,343$,
$\sum x_i^2 = 7,721,501,871,703$,
$\sum y_i^2 = 180,453$, and
$\sum x_i y_i = 1,177,863,984$.

So, the correlation coefficient for the female data is:

$$r = \frac{\sum x_i y_i - \frac{\sum x_i \sum y_i}{n}}{\sqrt{\left(\sum x_i^2 - \frac{(\sum x_i)^2}{n}\right)\left(\sum y_i^2 - \frac{(\sum y_i)^2}{n}\right)}}$$

$$= \frac{1,177,863,984 - \frac{(8,765,495)(1,343)}{10}}{\sqrt{\left(7,721,501,871,703 - \frac{(8,765,495)^2}{10}\right)\left(180,453 - \frac{(1,343)^2}{10}\right)}}$$

$$\approx 0.359$$

We next use the computational formula on the data from the male participants:

x_i	y_i	x_i^2	y_i^2	$x_i y_i$
949,395	140	901,350,866,025	19,600	132,915,300
1,001,121	140	1,002,243,256,641	19,600	140,156,940
1,038,437	139	1,078,351,402,969	19,321	144,342,743
965,353	133	931,906,414,609	17,689	128,391,949
955,466	133	912,915,277,156	17,689	127,076,978
1,079,549	141	1,165,426,043,401	19,881	152,216,409
924,059	135	853,885,035,481	18,225	124,747,965
955,003	139	912,030,730,009	19,321	132,745,417
935,494	141	875,149,024,036	19,881	131,904,654
949,589	144	901,719,268,921	20,736	136,740,816
9,753,466	1,385	9,534,977,319,248	191,943	1,351,239,171

From the table, we have $n = 10$,
$\sum x_i = 9,753,466$, $\sum y_i = 1,385$,
$\sum x_i^2 = 9,534,977,319,248$,
$\sum y_i^2 = 191,943$, and
$\sum x_i y_i = 1,351,239,171$. So, the correlation coefficient for the male data is:

$$r = \frac{\sum x_i y_i - \frac{\sum x_i \sum y_i}{n}}{\sqrt{\left(\sum x_i^2 - \frac{(\sum x_i)^2}{n}\right)\left(\sum y_i^2 - \frac{(\sum y_i)^2}{n}\right)}}$$

$$= \frac{1,351,239,171 - \frac{(9,753,466)(1,385)}{10}}{\sqrt{\left(9,534,977,319,248 - \frac{(9,753,466)^2}{10}\right)\left(191,943 - \frac{(1,385)^2}{10}\right)}}$$

$$\approx 0.236$$

There is no linear relation between MRI count and IQ. It is important to beware of lurking variables. Mixing distinct populations can produce misleading results that result in incorrect conclusions.

35. (a)

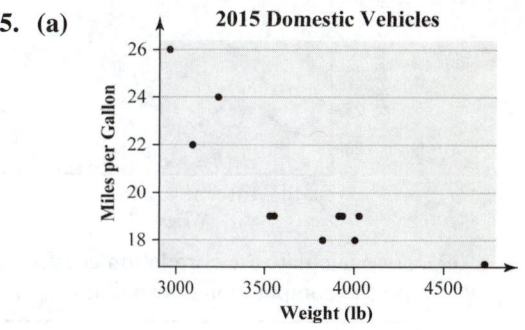

(b) To calculate the correlation coefficient, we use the computational formula:

x_i	y_i	x_i^2	y_i^2	$x_i y_i$
4,724	17	22,316,176	289	80,308
4,006	18	16,048,036	324	72,108
3,097	22	9,591,409	484	68,134
3,555	19	12,638,025	361	67,545
4,029	19	16,232,841	361	76,551
3,934	19	15,476,356	361	74,746
3,242	24	10,510,564	576	77,808
2,960	26	8,761,600	676	76,960
3,530	19	12,460,900	361	67,070
3,823	18	14,615,329	324	68,814
3,917	19	1,534,889	361	74,423
40,817	220	153,994,125	4478	804467

From the table, we have $n = 11$, $\sum x_i = 40,817$, $\sum y_i = 220$, $\sum x_i^2 = 153,994,125$, $\sum y_i^2 = 4478$, and $\sum x_i y_i = 804,467$. So, the correlation coefficient is:

$$r = \frac{\sum x_i y_i - \frac{\sum x_i \sum y_i}{n}}{\sqrt{\left(\sum x_i^2 - \frac{(\sum x_i)^2}{n}\right)\left(\sum y_i^2 - \frac{(\sum y_i)^2}{n}\right)}}$$

$$= \frac{804,467 - \frac{(40,817)(220)}{11}}{\sqrt{\left(153,994,125 - \frac{(40,817)^2}{11}\right)\left(4478 - \frac{(220)^2}{11}\right)}}$$

$$\approx -0.844$$

(c) The scatter diagram in part (a) looks very similar to the one from problem 29, and the correlation coefficient from part (b) is similar to the one computed in problem 29 (-0.842). These results are reasonable because the Taurus follows the overall pattern of the data.

(d)

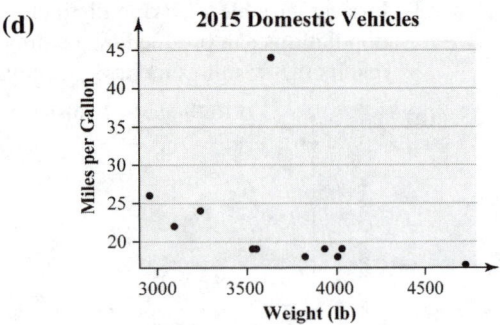

2015 Domestic Vehicles

(e) To recalculate the correlation coefficient, we use the computational formula:

x_i	y_i	x_i^2	y_i^2	$x_i y_i$
4,724	17	22,316,176	289	80,308
4,006	18	16,048,036	324	72,108
3,097	22	9,591,409	484	68,134
3,555	19	12,638,025	361	67,545
4,029	19	16,232,841	361	76,551
3,934	19	15,476,356	361	74,746
3,242	24	10,510,564	576	77,808
2,960	26	8,761,600	676	76,960
3,530	19	12,460,900	361	67,070
3,823	18	14,615,329	324	68,814
3,639	44	13,242,321	1936	160,116
40,539	245	151,893,557	6053	890,160

From the table, we have $n = 11$, $\sum x_i = 40,539$, $\sum y_i = 245$, $\sum x_i^2 = 151,893,557$, $\sum y_i^2 = 6,053$, and $\sum x_i y_i = 890,160$. So, the correlation coefficient is:

$$r = \frac{\sum x_i y_i - \dfrac{\sum x_i \sum y_i}{n}}{\sqrt{\left(\sum x_i^2 - \dfrac{(\sum x_i)^2}{n}\right)\left(\sum y_i^2 - \dfrac{(\sum y_i)^2}{n}\right)}}$$

$$= \frac{890,160 - \dfrac{(40,539)(245)}{11}}{\sqrt{\left(151,893,557 - \dfrac{(40,539)^2}{11}\right)\left(6,053 - \dfrac{(245)^2}{11}\right)}}$$

$$\approx -0.331$$

The correlation coefficient, with the Ford Fusion included. changed dramatically (from -0.842 to -0.331).

(f) The Ford Fusion is a hybrid car, so it gets much better gas mileage than the other cars which are not hybrids. As seen on the scatterplot, the Ford Fusion does not follow the pattern of the remaining cars.

37. (a) Data Set 1:

x_i	y_i	x_i^2	y_i^2	$x_i y_i$
10	8.04	100	64.6416	80.40
8	6.95	64	48.3025	55.60
13	7.58	169	57.4564	98.54
9	8.81	81	77.6161	79.29
11	8.33	121	69.3889	91.63
14	9.96	196	99.2016	139.44
6	7.24	36	52.4176	43.44
4	4.26	16	18.1476	17.04
12	10.84	144	117.5056	130.08
7	4.82	49	23.2324	33.74
5	5.68	25	32.2624	28.40
99	82.51	1001	660.1727	797.60

From the table, we have $n = 11$, $\sum x_i = 99$, $\sum y_i = 82.51$, $\sum x_i^2 = 1001$, $\sum y_i^2 = 660.1727$, and $\sum x_i y_i = 797.60$. So, the correlation coefficient is:

$$r = \frac{\sum x_i y_i - \dfrac{\sum x_i \sum y_i}{n}}{\sqrt{\left(\sum x_i^2 - \dfrac{(\sum x_i)^2}{n}\right)\left(\sum y_i^2 - \dfrac{(\sum y_i)^2}{n}\right)}}$$

$$= \frac{797.60 - \dfrac{(99)(82.51)}{11}}{\sqrt{\left(1001 - \dfrac{(99)^2}{11}\right)\left(660.1727 - \dfrac{(82.51)^2}{11}\right)}}$$

$$\approx 0.816$$

Data Set 2:

x_i	y_i	x_i^2	y_i^2	$x_i y_i$
10	9.14	100	83.5396	91.40
8	8.14	64	66.2596	65.12
13	8.74	169	76.3876	113.62
9	8.77	81	76.9129	78.93
11	9.26	121	85.7476	101.86
14	8.10	196	65.6100	113.40
6	6.13	36	37.5769	36.78
4	3.10	16	9.6100	12.40
12	9.13	144	83.3569	109.56
7	7.26	49	52.7076	50.82
5	4.47	25	19.9809	22.35
99	82.24	1001	657.6896	796.24

From the table, we have $n = 11$,

$\sum x_i = 99$, $\sum y_i = 82.24$, $\sum x_i^2 = 1001$,

$\sum y_i^2 = 657.6896$, and $\sum x_i y_i = 796.24$.

So, the correlation coefficient is:

$$r = \frac{\sum x_i y_i - \dfrac{\sum x_i \sum y_i}{n}}{\sqrt{\left(\sum x_i^2 - \dfrac{(\sum x_i)^2}{n}\right)\left(\sum y_i^2 - \dfrac{(\sum y_i)^2}{n}\right)}}$$

$$= \frac{796.24 - \dfrac{(99)(82.24)}{11}}{\sqrt{\left(1001 - \dfrac{(99)^2}{11}\right)\left(657.6896 - \dfrac{(82.24)^2}{11}\right)}}$$

≈ 0.817

Data Set 3:

x_i	y_i	x_i^2	y_i^2	$x_i y_i$
10	7.46	100	55.6516	74.60
8	6.77	64	45.8329	54.16
13	12.74	169	162.3076	165.62
9	7.11	81	50.5521	63.99
11	7.81	121	60.9961	85.91
14	8.84	196	78.1456	123.76
6	6.08	36	36.9664	36.48
4	5.39	16	29.0521	21.56
12	8.15	144	66.4225	97.80
7	6.42	49	41.2164	44.94
5	5.73	25	32.8329	28.65
99	82.50	1001	659.9762	797.47

From the table, we have $n = 11$,

$\sum x_i = 99$, $\sum y_i = 82.50$, $\sum x_i^2 = 1001$,

$\sum y_i^2 = 659.9762$, and $\sum x_i y_i = 797.47$.

So, the correlation coefficient is:

$$r = \frac{\sum x_i y_i - \dfrac{\sum x_i \sum y_i}{n}}{\sqrt{\left(\sum x_i^2 - \dfrac{(\sum x_i)^2}{n}\right)\left(\sum y_i^2 - \dfrac{(\sum y_i)^2}{n}\right)}}$$

$$= \frac{797.47 - \dfrac{(99)(82.50)}{11}}{\sqrt{\left(1001 - \dfrac{(99)^2}{11}\right)\left(659.9762 - \dfrac{(82.50)^2}{11}\right)}}$$

≈ 0.816

Data Set 4:

x_i	y_i	x_i^2	y_i^2	$x_i y_i$
8	6.58	64	43.2964	52.64
8	5.76	64	33.1776	46.08
8	7.71	64	59.4441	61.68
8	8.84	64	78.1456	70.72
8	8.47	64	71.7409	67.76
8	7.04	64	49.5616	56.32
8	5.25	64	27.5625	42.00
8	5.56	64	30.9136	44.48
8	7.91	64	62.5681	63.28
8	6.89	64	47.4721	55.12
19	12.5	361	156.2500	237.50
99	82.51	1001	660.1325	797.58

From the table, we have $n = 11$,

$\sum x_i = 99$, $\sum y_i = 82.51$, $\sum x_i^2 = 1001$,

$\sum y_i^2 = 660.1325$, and $\sum x_i y_i = 797.58$.

So, the correlation coefficient is:

$$r = \frac{\sum x_i y_i - \dfrac{\sum x_i \sum y_i}{n}}{\sqrt{\left(\sum x_i^2 - \dfrac{(\sum x_i)^2}{n}\right)\left(\sum y_i^2 - \dfrac{(\sum y_i)^2}{n}\right)}}$$

$$= \frac{797.58 - \dfrac{(99)(82.51)}{11}}{\sqrt{\left(1001 - \dfrac{(99)^2}{11}\right)\left(660.1325 - \dfrac{(82.51)^2}{11}\right)}}$$

≈ 0.817

(b)

Data Set I:

Data Set II:

Data Set III:

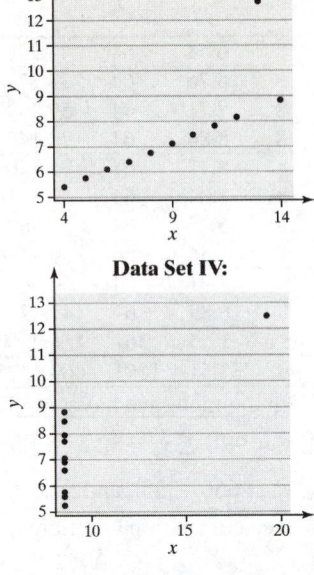

Data Set IV:

Even though the correlation coefficients for the four data sets are roughly equal, the scatter plots are clearly different. Thus, linear correlation coefficients and scatter diagrams must be used together in a statistical analysis of bivariate data to determine whether a linear relations exists.

39. If the goal is to have the lowest correlation between two stocks (i.e., a correlation close to zero), then invest in First Energy and General Electric, since their correlation coefficient is only -0.025. However, you could also invest in Walt Disney and First Energy, or Johnson & Johnson and Cisco Systems, which also have a low correlation coefficient (-0.03 and 0.058, respectively).

If the goal is to have one stock go up when the other goes down (i.e., a negative relation), then invest in First Energy and Cisco Systems since they have the strongest negative relation, -0.248.

41. $r = 0.599$ implies that a positive linear relation exists between the number of television stations and life expectancy. However, this is correlation, not causation. The more television stations a country has, the more affluent it is. The more affluent, the better the healthcare is, which in turn helps increase the life expectancy. So, wealth is a likely lurking variable.

43. No, increasing the percentage of the population that has a cell phone will not decrease the violent crime rate. A likely lurking variable is the economy. In a strong economy, crime rates tend to decrease, and consumers are better able to afford cell phones.

45. **(a)**

To calculate the correlation coefficient, we use the computational formula:

x_i	y_i	x_i^2	y_i^2	$x_i y_i$
2.2	3.9	4.84	15.21	8.58
3.7	4.0	13.69	16.00	14.80
3.9	1.4	15.21	1.96	5.46
4.1	2.8	16.81	7.84	11.48
2.6	1.5	6.76	2.25	3.90
4.1	3.3	16.81	10.89	13.53
2.9	3.6	8.41	12.96	10.44
4.7	4.9	22.09	24.01	23.03
28.2	25.4	104.62	91.12	91.22

From the table, we have $n = 8$, $\sum x_i = 28.2$, $\sum y_i = 25.4$, $\sum x_i^2 = 104.62$, $\sum y_i^2 = 91.12$, and $\sum x_i y_i = 91.22$. So, the correlation coefficient is:

$$r = \frac{\sum x_i y_i - \frac{\sum x_i \sum y_i}{n}}{\sqrt{\left(\sum x_i^2 - \frac{(\sum x_i)^2}{n}\right)\left(\sum y_i^2 - \frac{(\sum y_i)^2}{n}\right)}}$$

$$= \frac{91.22 - \frac{(28.2)(25.4)}{8}}{\sqrt{\left(104.62 - \frac{(28.2)^2}{8}\right)\left(91.12 - \frac{(25.4)^2}{8}\right)}}$$

$$\approx 0.228$$

(b)

To calculate the correlation coefficient, we use the computational formula:

x_i	y_i	x_i^2	y_i^2	$x_i y_i$
10.4	9.3	108.16	86.49	96.72
2.2	3.9	4.84	15.21	8.58
3.7	4.0	13.69	16.00	14.80
3.9	1.4	15.21	1.96	5.46
4.1	2.8	16.81	7.84	11.48
2.6	1.5	6.76	2.25	3.90
4.1	3.3	16.81	10.89	13.53
2.9	3.6	8.41	12.96	10.44
4.7	4.9	22.09	24.01	23.03
38.6	34.7	212.78	177.61	187.94

From the table, we have $n = 9$, $\sum x_i = 38.6$, $\sum y_i = 34.7$, $\sum x_i^2 = 212.78$, $\sum y_i^2 = 177.61$, and $\sum x_i y_i = 187.94$. So, the correlation coefficient is:

$$r = \frac{\sum x_i y_i - \dfrac{\sum x_i \sum y_i}{n}}{\sqrt{\left(\sum x_i^2 - \dfrac{(\sum x_i)^2}{n}\right)\left(\sum y_i^2 - \dfrac{(\sum y_i)^2}{n}\right)}}$$

$$= \frac{187.94 - \dfrac{(38.6)(34.7)}{9}}{\sqrt{\left(212.78 - \dfrac{(38.6)^2}{9}\right)\left(177.61 - \dfrac{(34.7)^2}{9}\right)}}$$

$$\approx 0.860$$

The additional data point increases r from a value that suggests no linear correlation to one that suggests a fairly strong linear correlation. However, the second scatter diagram shows that the new data point is located very far away from the rest of the data, so this new data point has a strong influence on the value of r, even though there is no apparent correlation between the variables. Therefore, correlations should always be reported along with scatter diagrams in order to check for potentially influential observations.

47.
(a)

Graduation Rate and Four-Year Cost of College

The scatterplot indicates a slight positive association between cost and graduation rate.

(b) The correlation coefficient is 0.556, which is high enough to suggest a positive association between cost and graduation rate.

(c)

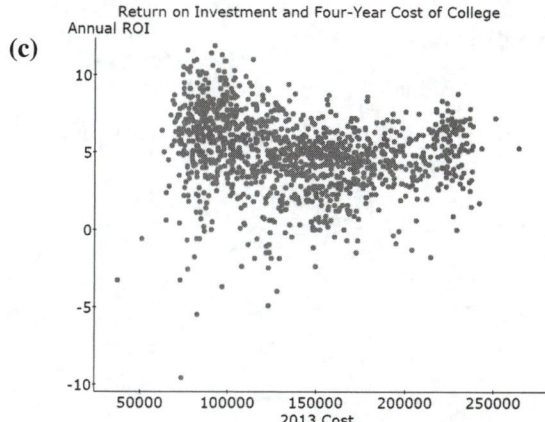

Return on Investment and Four-Year Cost of College

The correlation between cost and return on investment (ROI) is -0.201. The scatterplot shows a slight negative association, which is confirmed by the correlation coefficient. This suggests the more expensive the school, the lower the ROI tends to be, on average.

49. If the correlation coefficient equals 1, then a perfect positive linear relation exists between the variables, and the points of the scatter diagram lie exactly on a straight line with a positive slope.

51. The linear correlation coefficient can only be calculated from bivariate *quantitative* data, and the gender of a driver is a qualitative variable. A better way to write the sentence is: "Gender is associated with the rate of automobile accidents."

53. Correlation describes a relation between two variables in an observational study. Causation describes a conditional (if –then) relation in an experimental study.

55. Answers will vary. The points should all be on a line that passes through the origin and has a slope of 15 dollars per hour.

This is a deterministic relation.

Section 4.2

1. Residual

3. True; this is a property of the least-squares regression line.

5. (a)

There appears to be a negative linear relation between the x and y.

(b) $b_1 = r \cdot \dfrac{s_y}{s_x} = -0.9477\left(\dfrac{1.8239}{2.4221}\right) \approx -0.7136$

$b_0 = \overline{y} - b_1\overline{x}$

$= 3.9333 - (-0.7136)(3.6667)$

≈ 6.5496

So, the least-squares regression line is:

$\hat{y} = -0.7136x + 6.5496$

(c)

7. (a)

(b) Using the points $(3,4)$ and $(8,14)$:

$m = \dfrac{14-4}{8-3} = 2$

$y - y_1 = m(x - x_1)$

$y - 4 = 2(x - 3)$

$y - 4 = 2x - 6$

$y = 2x - 2$

(c)

(d) We compute the mean and standard deviation for each variable and the correlation coefficient to be: $\overline{x} = 5.4$, $\overline{y} = 8.6$, $s_x \approx 2.07364$, $s_y \approx 4.21901$, and $r \approx 0.99443$. Then, the slope and intercept for the least-squares regression line are:

$b_1 = r \cdot \dfrac{s_y}{s_x} = 0.99443\left(\dfrac{4.21901}{2.07364}\right)$

≈ 2.02326

$b_0 = \overline{y} - b_1\overline{x} = 8.6 - (2.02326)(5.4)$

≈ -2.3256

Rounding to four decimal places, the least-squares regression line is:

$\hat{y} = 2.0233x - 2.3256$

(e)

(f)

x	y	$\hat{y} = 2x - 2$	$y - \hat{y}$	$(y - \hat{y})^2$
3	4	4	0	0
4	6	6	0	0
5	7	8	−1	1
7	12	12	0	0
8	14	14	0	0

Total = 1

Sum of squared residuals (computed line):
1.0

(g)

x	y	$\hat{y}$	$y - \hat{y}$	$(y - \hat{y})^2$
3	4	3.7443	0.2557	0.0654
4	6	5.7676	0.2324	0.0540
5	7	7.7909	−0.7909	0.6255
7	12	11.8375	0.1625	0.0264
8	14	13.8608	0.1392	0.0194

Total = 0.7907

Sum of squared residuals (regression line):
0.7907

(h) Answers will vary. The regression line gives a smaller sum of squared residuals, so it has a better fit.

9. (a)

(b) Using the points $(-2, -4)$ and $(2, 5)$:

$$m = \frac{5 - (-4)}{2 - (-2)} = \frac{9}{4}$$

$$y - y_1 = m(x - x_1)$$

$$y - 5 = \frac{9}{4}(x - 2)$$

$$y - 5 = \frac{9}{4}x - \frac{9}{2}$$

$$y = \frac{9}{4}x + \frac{1}{2} \quad \text{or} \quad y = 2.25x + 0.5$$

(c)

(d) We compute the mean and standard deviation for each variable and the correlation coefficient to be: $\bar{x} = 0$, $\bar{y} = 1.2$, $s_x \approx 1.58114$, $s_y \approx 3.56371$, and $r \approx 0.97609$. Then, the slope and intercept for the least-squares regression line are:

$$b_1 = r \cdot \frac{s_y}{s_x} = 0.97609\left(\frac{3.56371}{1.58114}\right) \approx 2.20000$$

$$b_0 = \bar{y} - b_1\bar{x} = 1.2 + (2.20000)(0) = 1.2$$

So, the least-squares regression line is:
$\hat{y} = 2.2x + 1.2$

(e)

(f)

x	y	$y = \frac{9}{4}x + \frac{1}{2}$	$y - \hat{y}$	$(y - \hat{y})^2$
−2	−4	−4.00	0.00	0.0000
−1	0	−1.75	1.75	3.0625
0	1	0.50	0.50	0.2500
1	4	2.75	1.25	1.5625
2	5	5.00	0.00	0.0000

Total = 4.8750

Sum of squared residuals (calculated line):
4.8750

(g)

x	y	$\hat{y} = 2.2x + 1.2$	$y - \hat{y}$	$(y - \hat{y})^2$
-2	-4	-3.2	-0.8	0.64
-1	0	-1.0	1.0	1.00
0	1	1.2	-0.2	0.04
1	4	3.4	0.6	0.36
2	5	5.6	-0.6	0.36

Total = 2.40

Sum of squared residuals (regression line):
2.4

(h) Answers will vary. The regression line gives a smaller sum of squared residuals, so it has a better fit.

11. (a)

(b) Using the points $(30, 95)$ and $(60, 70)$:

$$m = \frac{70 - 95}{60 - 30} = \frac{-25}{30} = -\frac{5}{6}$$

$$y - y_1 = m(x - x_1)$$

$$y - 95 = -\frac{5}{6}(x - 30)$$

$$y - 95 = -\frac{5}{6}x + 25$$

$$y = -\frac{5}{6}x + 120$$

(c)

(d) We compute the mean and standard deviation for each variable and the correlation coefficient to be: $\overline{x} = 40$, $\overline{y} = 87.8$, $s_x \approx 15.81139$, $s_y \approx 11.73456$, and $r \approx -0.97014$. Then, the slope and intercept for the least-squares regression line are:

$$b_1 = r \cdot \frac{s_y}{s_x} = -0.97014 \left(\frac{11.73456}{15.81139} \right)$$

$$\approx -0.72000$$

$$b_0 = \overline{y} - b_1\overline{x} = 87.8 - (-0.72000)(40)$$

$$\approx 116.6$$

So, the least-squares regression line is:

$$\hat{y} = -0.72x + 116.6$$

(e)

(f)

x	y	$\hat{y} = -\dfrac{5}{6}x + 120$	$y - \hat{y}$	$(y - \hat{y})^2$
20	100	103.333	-3.333	11.1111
30	95	95.000	0.000	0.0000
40	91	86.667	4.333	18.7778
50	83	78.333	4.667	21.7778
60	70	70.000	0.000	0.0000

Total = 51.6667

Sum of squared residuals (calculated line):
51.6667.

(g)

x	y	$\hat{y}$	$y - \hat{y}$	$(y - \hat{y})^2$
20	100	102.2	-2.2	4.84
30	95	95.0	0.0	0.00
40	91	87.8	3.2	10.24
50	83	80.6	2.4	5.76
60	70	73.4	-3.4	11.56

Total = 32.40

Sum of squared residuals (regression line):
32.4

(h) Answers will vary. The regression line gives a smaller sum of squared residuals, so it has a better fit.

13. (a) Let $x = 25$ in the regression equation.

$$\hat{y} = 703.5(25) + 14.920 = \$32,507.50$$

So, we predict a median income of $32,507.50 for a state in which 25% of the adults 25 years and older have at least a bachelor's degree.

(b) Let $x = 27.1$ in the regression equation:
$$\hat{y} = 703.5(27.1) + 14.920 = \$33,984.85$$
So, we predict a median income of $33,984.85 for a state in which 27.1% of the adults 25 years and older have at least a bachelor's degree. The median income of North Dakota is $37,193, which is higher than would be expected.

(c) The slope is 703.5. The median income of a state will increase by $703.50 for each one percent increase in percentage of the adults 25 years and older that have at least a bachelor's degree, on average.

(d) It does not make sense to interpret the y-intercept in the context of this problem because there are no values near 0%, so it is outside the scope of the problem.

15. (a) The slope is -0.0527. If literacy increases by 1 percent, the age difference decreases by 0.0527 years, on average.

(b) The y-intercept is 7.1. This is outside the scope of the model. A literacy percentage of 0 is not reasonable for this model, since the model only applies to percentages between 18% and 100%.

(c) Let $x = 25$ in the regression equation:
$$\hat{y} = -0.0527(25) + 7.1 \approx 5.78$$
We predict the age difference between a husband/wife will be 5.78 years in a country where the literacy percentage is 25%.

(d) No, it would not be a good idea to use this model to predict the age difference between a husband/wife in a country where the literacy percentage is 10% because 10% is outside the scope of the model.

(e) Let $x = 99$ in the regression equation:
$$\hat{y} = -0.0527(99) + 7.1 \approx 1.88$$
So, the residual is:
$$y - \hat{y} = 2.00 - 1.88 = 0.12$$
This residual indicates that the United States has an age difference that is slightly above average.

17. (a) $\hat{y} = -0.0479x + 69.0296$

(b) Slope: For each one-minute increase in commute time, the index score decreases by 0.0479, on average; Y-intercept: The mean index score for a person with a 0-minute commute is 69.0296. A person who works from home would have a 0-minute commute so the y-intercept has a meaningful interpretation.

(c) If the commute time is 30 minutes, the predicted well-being index score would be
$$\hat{y} = -0.0479(30) + 69.0296 \approx 67.6$$

(d) The mean well-being index for a person with a 20 minute commute is:
$$\hat{y} = -0.0479(20) + 69.0296 \approx 68.1$$ So, she is less well off than would be expected.

19. (a) In Problem 27, Section 4.1, we computed the correlation coefficient. Rounded to six decimal places, it is $r = 0.911073$. We compute the mean and standard deviation for each variable to be: $\overline{x} \approx 26.454545$, $s_x \approx 1.094407$, $\overline{y} \approx 17.327273$, and $s_y \approx 0.219504$.
$$b_1 = r \cdot \frac{s_y}{s_x} = 0.911073\left(\frac{0.219504}{1.094407}\right)$$
$$\approx 0.182733$$
$$b_0 = \overline{y} - b_1\overline{x}$$
$$= 17.327273 - (0.182733)(26.454545)$$
$$\approx 12.4932$$
Rounding to four decimal places, the least-squares regression line is:
$$\hat{y} = 0.1827x + 12.4932$$

(b) If height increases by 1 inch, head circumference increases by about 0.1827 inches, on average. It is not appropriate to interpret the y-intercept since a height of 0 is outside the scope of the model. In addition, it makes no sense to consider the head circumference of a child with a height of 0 inches.

(c) Let $x = 25$ in the regression equation:
$$\hat{y} = 0.1827(25) + 12.4932 \approx 17.06$$
We predict the head circumference of a child who is 25 inches tall is 17.06 inches.

(d) $y - \hat{y} = 16.9 - 17.06 = -0.16$ inches
This indicates that the head circumference for this child is below average.

(e)

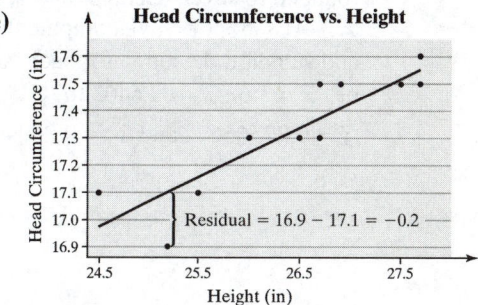

Head Circumference vs. Height

(f) The head circumferences of children who are 26.75 inches tall naturally vary.

(g) No, a height of 32 inches is well outside the scope of the model.

21. (a) In Problem 29, Section 4.1, we computed the correlation coefficient. Rounded to seven decimal places, it is $r = -0.8415760$. We compute the mean and standard deviation for each variable to be: $\bar{x} = 3690$, $s_x \approx 526.0160538$, $\bar{y} = 20.1$, and $s_y \approx 2.9230882$.

$$b_1 = r \cdot \frac{s_y}{s_x} = -0.8415760\left(\frac{2.9230882}{526.0160538}\right)$$
$$\approx -0.00467667$$
$$b_0 = \bar{y} - b_1\bar{x}$$
$$= 20.1 - (-0.00467667)(3690)$$
$$\approx 37.3569$$

Rounding to four decimal places, the least-squares regression line is
$$\hat{y} = -0.0047x + 37.3569$$

(b) For every pound added to the weight of the car, gas mileage in the city will decrease by about 0.0047 miles per gallon, on average. It is not appropriate to interpret the y-intercept because it is well beyond the scope of the model.

(c) Let $x = 3649$ in the regression equation:
$$\hat{y} = -0.0047(3649) + 37.3569 \approx 20.2066$$
mpg (or 20.2917 mpg when software is used). The mileage for the Cadillac CTS is only slightly below average for cars of this weight.

(d) No, it is not reasonable to use this least-squares regression line to predict the miles per gallon of a Toyota Prius. The data given are for domestic cars with traditional internal combustion engines. The Toyota Prius is a foreign-made hybrid car.

23. (a) We compute the correlation coefficient, rounded to seven decimal places, to be $r = -0.8056468$. We compute the mean and standard deviation for each variable to be: $\bar{x} = 3.6$, $s_x \approx 2.6403463$, $\bar{y} \approx 0.8756667$, and $s_y \approx 0.0094693$.

$$b_1 = r \cdot \frac{s_y}{s_x} = -0.8056468\left(\frac{0.0094693}{2.6403463}\right)$$
$$\approx -0.0028893$$
$$b_0 = \bar{y} - b_1\bar{x}$$
$$= 0.8756667 - (-0.0028893)(3.6)$$
$$\approx 0.8861$$

Rounding to four decimal places, the least-squares regression line is:
$$\hat{y} = -0.0029x + 0.8861$$

(b) For each additional cola consumed per week, bone mineral density will decrease by $0.0029 \text{ g}/\text{cm}^2$, on average.

(c) For a woman who does not drink cola, the mean bone mineral density will be $0.8861 \text{ g}/\text{cm}^2$.

(d) Let $x = 4$ in the regression equation:
$$\hat{y} = -0.0029(4) + 0.8861 = 0.8745$$
We predict the bone mineral density of the femoral neck of a woman who consumes four colas per week is $0.8745 \text{ g}/\text{cm}^2$.

(e) Since 0.873 is smaller than the result in part (d), this woman's bone mineral density is below average among women who consume four colas per week.

(f) No. Two cans of soda per day equates to 14 cans of soda per week, which is outside the scope of the model.

25. In Problem 31, Section 4.1, we computed the correlation coefficient: $r \approx -0.02779$. Since this value is so close to zero, we conclude that there is not a linear relation between the CEO's compensation and the stock return. For both $15 million and $25 million, $\hat{y} = \bar{y} = 49.7\%$.

27. (a) Males: In Problem 34(c), Section 4.1, we computed the correlation coefficient for males. Unrounded, it is $r \approx 0.8833643940$. We compute the mean and standard deviation for each variable for males to be: $\bar{x} = 11{,}009.555...$, $s_x \approx 7358.553596175$, $\bar{y} = 4772.111...$, and $s_y \approx 2855.260997021$.

$$b_1 = r \cdot \frac{s_y}{s_x} \approx 0.342762456$$
$$b_0 = \bar{y} - b_1\bar{x} \approx 998.4488$$

Rounding to four decimal places, the least-squares regression line for males is:
$\hat{y} = 0.3428x + 998.4488$

Females: In Problem 34(d), Section 4.1, we computed the correlation coefficient for females. Unrounded, it is
$r \approx 0.8361242533$. We compute the mean and standard deviation for each variable for males to be $\overline{x} \approx 10,988.222...$,
$s_x \approx 7240.2546532870$, $\overline{y} = 1662.222...$,
and $s_y \approx 904.7405398480$.

$b_1 = r \cdot \dfrac{s_y}{s_x} \approx 0.1044818925$

$b_0 = \overline{y} - b_1\overline{x} \approx 514.1520$

Rounding to four decimal places, the least-squares regression line is:
$\hat{y} = 0.1045x + 514.1520$

(b) Males: If the number of licensed drivers increases by 1 (thousand), then the number of fatal crashes increases by 0.3428 (thousand), on average.

Females: If the number of licensed drivers increases by 1 (thousand), then the number of fatal crashes increases by 0.1045 (thousand), on average.

Since females tend to be involved in fewer fatal crashes, an insurance company may use this information to argue for higher rates for male customers.

(c) 16 to 20 year old males: Let $x = 6424$ in the regression equation for males:
$\hat{y} = 0.3428(6424) + 998.4488 \approx 3200.6$
The actual number of fatal crashes (5180) is above this result, so it is above average.

21 to 24 year old males: Let $x = 6941$ in the regression equation for males:
$\hat{y} = 0.3428(6941) + 998.4488 \approx 3377.8$
The actual number of fatal crashes (5016) is above this result, so it is above average.

≥ 74 year old males: Let $x = 4803$ in the regression equation for males:
$\hat{y} = 0.3428(4803) + 998.4488 \approx 2644.9$
The actual number of fatal crashes (2022) is below this result, so it is below average.

An insurance company might use these results to argue for higher rates for younger drivers and lower rates for older drivers.

16 to 20 year old females: Let $x = 6139$ in the regression equation for females:
$\hat{y} = 0.1045(6139) + 514.1520 \approx 1155.7$
The actual number of fatal crashes (2113) is above this result, so it is above average.

21 to 24 year old females: Let $x = 6816$ in the regression equation for females:
$\hat{y} = 0.1045(6816) + 514.1520 \approx 1226.4$
The actual number of fatal crashes (1531) is above this result, so it is above average.

≥ 74 year old females: Let $x = 5375$ in the regression equation for females:
$\hat{y} = 0.1045(5375) + 514.1520 \approx 1075.8$
The actual number of fatal crashes (980) is below this result, so it is below average.

The same relationship holds for female drivers as for male drivers.

29. (a) Square footage is the explanatory variable.

(b)

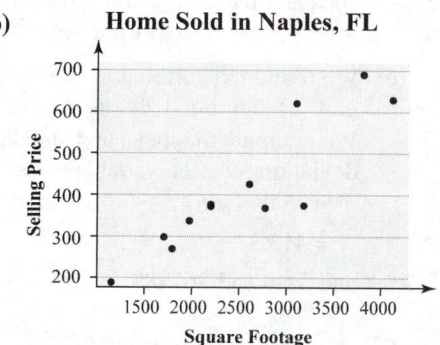

Home Sold in Naples, FL

(c) To calculate the correlation coefficient, we use the computational formula:

x_i	y_i	x_i^2	y_i^2	$x_i y_i$
2204	379.9	4,857,616	144,324.01	837,299.6
3183	375	10,131,489	140,625	1,193,625
1128	189.9	1272384	36,062.01	214,207.2
1975	338	3900625	114,244	667,550
3101	619.9	9616201	384276.01	1922309.9
2769	370	7,667,361	136,900	1,024,530
4113	627.7	16,916,769	394,007.29	2,581,730.1
2198	375	4,831,204	14,0625	824,250
2609	425	6,806,881	18,0625	1,108,825
1708	298.1	2,917,264	88,863.61	509,154.8
1786	271	3,189,796	73441	484,006
3813	690.1	14,538,969	476238.01	2,631,351.3
30,587	4959.6	86646559	2,310,230.9	13998838.9

From the table, we have $n = 12$,
$\sum x_i = 30{,}587$, $\sum y_i = 4{,}959.6$,
$\sum x_i^2 = 86{,}646{,}559$, $\sum y_i^2 = 2{,}310{,}230.94$,
and $\sum x_i y_i = 13{,}998{,}838.9$. So, the
correlation coefficient is:

$$r = \frac{\sum x_i y_i - \dfrac{\sum x_i \sum y_i}{n}}{\sqrt{\left(\sum x_i^2 - \dfrac{(\sum x_i)^2}{n}\right)\left(\sum y_i^2 - \dfrac{(\sum y_i)^2}{n}\right)}}$$

$$= \frac{13{,}998{,}838.9 - \dfrac{(30{,}587)(4{,}959.6)}{12}}{\sqrt{\left(86{,}646{,}559 - \dfrac{(30{,}587)^2}{12}\right)\left(2{,}310{,}230.94 - \dfrac{(4{,}959.6)^2}{12}\right)}}$$

$$\approx 0.9026$$

(d) Yes, a linear relation exists between square footage and asking price. The correlation coefficient $0.9026 > 0.576$, which is the critical value from Table II.

(e) We found the correlation coefficient in part (b). Unrounded, it is $r \approx 0.90256596$. We compute the mean and standard deviation for each variable to be
$\bar{x} = 2548.91667$, $s_x \approx 888.453453$,
$\bar{y} = 413.3$,
and $s_y \approx 153.867789$.

$$b_1 = r \cdot \frac{s_y}{s_x} = 0.90256596\left(\frac{153.867789}{888.453453}\right)$$
$$\approx 0.156312$$
$$b_0 = \bar{y} - b_1 \bar{x}$$
$$= 413.3 - (0.156312)(2548.91667)$$
$$\approx 14.8737$$

Rounding to four decimal places, the least-squares regression line is
$$\hat{y} = 0.1563x + 14.8737$$

(f) For each square foot added to the area, the asking price of the house will increase by $156.30 (that is 0.15630 thousand dollars), on average.

(g) It is not reasonable to interpret the y-intercept because a house with an area of 0 square feet is outside the scope of the model.

(h) Let $x = 1465$ in the regression equation:
$$\hat{y} = 0.1563(1465) + 14.8737 \approx 243.8532$$
The average price of a home with 1,465 square feet is about $243,853. Since the

actual list price of this particular home is $285,000, it is above average.

Reasons provided may vary. Some factors that could affect the price include location and updates such as thermal windows or new siding.

31. A residual is the difference between the observed value of y and the predicted value of y. If the residual is positive, the observed value is above average for the given value of the explanatory variable.

33. Answers will vary.

35. Answers will vary. Discussions should involve the scope of the model and the idea of being "outside the scope."

Section 4.3

1. Coefficient of determination

3. Residual plot

5. The variance of the residuals is not constant. The variance of the residuals increases as the value of x increases.

7. There is an outlier in the residual plot (a residual of 4).

9. **(a)** III   **(b)** II
    (c) IV   **(d)** I

11. Yes, the point is influential because the two lines have substantially different slopes and y-intercepts.

13. No, the point is not influential because there is very little change in either the slope or the y-intercept. The lines are almost identical.

15. **(a)** From the scatter diagram, a positive linear relation appears to exist between time between eruptions and length of eruption.

     (b) Yes, the residual plot does not show any obvious departures from the requirements of the model.

     (c) 83.0% of the variation in the length of eruption is explained by the least-squares regression equation.

17. (a)

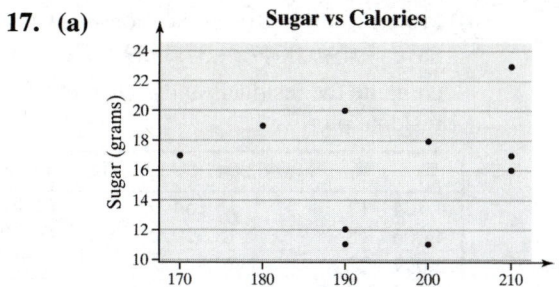

No relation appears to exist between calories and sugar content.

(b) To calculate the correlation coefficient, we use the computational formula:

x_i	y_i	x_i^2	y_i^2	$x_i y_i$
200	18	40,000	324	3600
210	23	44,100	529	4830
170	17	28,900	289	2890
190	20	36,100	400	3800
200	18	40,000	324	3600
180	19	32,400	361	3420
210	23	44,100	529	4830
210	16	44,100	256	3360
210	17	44,100	289	3570
190	12	36,100	144	2280
190	11	36,100	121	2090
200	11	40,000	121	2200
2360	205	466,000	3687	40,470

From the table, we have $n = 12$,

$\sum x_i = 2306$, $\sum y_i = 205$, $\sum x_i^2 = 466,000$,

$\sum y_i^2 = 3687$, and $\sum x_i y_i = 40,470$. So, the correlation coefficient is:

$$r = \frac{\sum x_i y_i - \dfrac{\sum x_i \sum y_i}{n}}{\sqrt{\left(\sum x_i^2 - \dfrac{(\sum x_i)^2}{n}\right)\left(\sum y_i^2 - \dfrac{(\sum y_i)^2}{n}\right)}}$$

$$= \frac{40,470 - \dfrac{(2360)(205)}{12}}{\sqrt{\left(466,000 - \dfrac{(2360)^2}{12}\right)\left(3687 - \dfrac{(205)^2}{12}\right)}}$$

$$\approx 0.261$$

No, there is no significant linear relation between calories and sugar content.

(c)

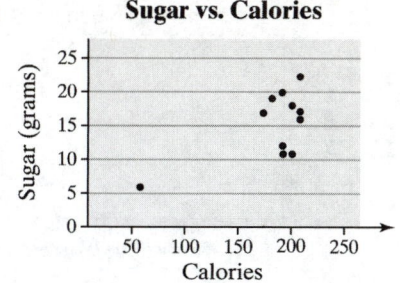

We add the following values to the table in part (a): $x_{13} = 80$, $y_{13} = 6$, $x_{13}^2 = 6400$, $y_{13}^2 = 36$, and $x_{13} y_{13} = 480$. We now have $n = 13$, $\sum x_i = 2440$, $\sum y_i = 211$, $\sum x_i^2 = 472,400$, $\sum y_i^2 = 3723$, and $\sum x_i y_i = 40,950$. So, the new correlation coefficient is:

$$r = \frac{40,950 - \dfrac{(2440)(211)}{13}}{\sqrt{\left(472,400 - \dfrac{(2440)^2}{13}\right)\left(3723 - \dfrac{(211)^2}{13}\right)}}$$

$$\approx 0.649$$

Yes, according to the new correlation coefficient, there is a linear relation between calories and sugar content. All-Bran is an influential observation. It substantially changed the correlation coefficient.

(d) It is necessary to show a scatter diagram with the correlation coefficient. Influential observations can cause the correlation coefficient to increase substantially, thereby increasing the apparent strength of the linear relation between the two variables.

19. (a) From Problem 25(c) in Section 4.1, we have $r = -0.980501$ (unrounded). So, $R^2 = (-0.980501)^2 \approx 0.961 = 96.1\%$.

(b) From Problem 17(a) in Section 4.2, we have $\hat{y} = -0.0479x + 69.0296$. We compute the residuals and construct the residual plot.

x_i	y_i	$\hat{y} = -0.0479x + 69.0296$	$y - \hat{y}$
5	69.2	68.79	0.41
15	68.3	68.31	−0.01
25	67.5	67.83	−0.33
35	67.1	67.35	−0.25
50	66.4	66.63	−0.23
72	66.1	65.58	0.52
105	63.9	64.00	−0.10

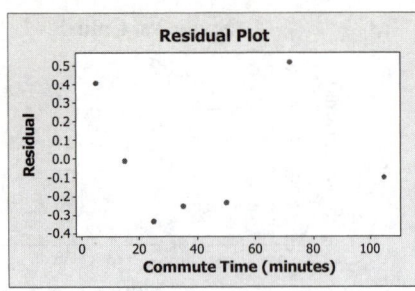

Residual Plot

(c) 96.1% of the variation in well-being score is explained by the least-squares regression line. The residual plot does not reveal any problems, so the linear model appears to be appropriate.

21. (a) From Problem 27(c) in Section 4.1, we have $r = 0.911073$ (unrounded). So,
$$R^2 = (0.911073)^2 \approx 0.830 = 83.0\%.$$

(b) From Problem 19(a) in Section 4.2, we have $\hat{y} = 0.1827x + 12.4932$. We compute the residuals and construct the residual plot.

x_i	y_i	$\hat{y} = 0.1827x + 12.4932$	$y - \hat{y}$
27.75	17.5	17.56	−0.06
24.50	17.1	16.97	0.13
25.50	17.1	17.15	−0.05
26.00	17.3	17.24	0.06
25.00	16.9	17.06	−0.16
27.75	17.6	17.56	0.04
26.50	17.3	17.33	−0.03
27.00	17.5	17.43	0.07
26.75	17.3	17.38	−0.08
26.75	17.5	17.38	0.12
27.50	17.5	17.52	−0.02

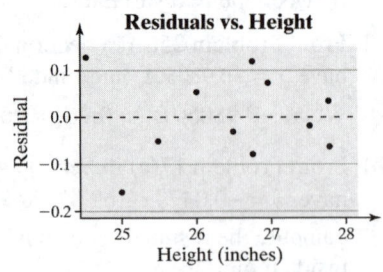

Residuals vs. Height

(c) 83.0% of the variation in head circumference is explained by the least-squares regression line. The residual plot does not reveal any problems, so the linear model appears to be appropriate.

23. (a) From Problem 29(c) in Section 4.1, we have $r \approx -0.841576$. So,
$$R^2 \approx (-0.841576)^2 \approx 0.70825 = 70.8\%.$$

(b) From Problem 21(a) in Section 4.2, we have $\hat{y} = -0.0047x + 37.3569$. We compute the residuals and construct the residual plot.

x_i	y_i	$\hat{y} = -0.0047x + 37.3569$	$y - \hat{y}$
4,724	17	15.154	1.846
4,006	18	18.529	−0.529
3,097	22	22.801	-0.801
3,555	19	20.648	-1.648
4,029	19	18.421	0.579
3,934	19	18.867	0.133
3,242	24	22.120	1.880
2,960	26	23.445	2.555
3,530	19	20.766	-1.766
3,823	18	19.389	-1.389

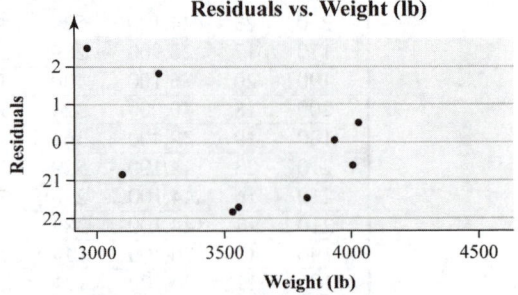

Residuals vs. Weight (lb)

(c) 70.8% of the variation in gas mileage is explained by the least-squares regression line. The residual plot does not reveal any problems, so the linear model appears to be appropriate.

25. (a)

Planetary Motion

(b) To calculate the correlation coefficient, we use the computational formula.

x_i	y_i	x_i^2	y_i^2	$x_i y_i$
36	0.24	1,296	0.0576	8.64
67	0.62	4,489	0.3844	41.54
93	1.00	8,649	1	93
142	1.88	20,164	3.5344	266.96
483	11.9	233,289	141.61	5,747.7
887	29.5	786,769	870.25	26,166.5
1,785	84	3,186,225	7,056	149,940
2,797	165	7,823,209	27,225	461,505
3,675	248	13,505,625	61,504	911,400
9,965	542.14	25,569,715	96,801.8364	1,555,169.34

From the table, we have $n = 9$,
$\sum x_i = 9,965$, $\sum y_i = 542.14$,
$\sum x_i^2 = 25,569,715$, $\sum y_i^2 = 96,801.8364$,
and $\sum x_i y_i = 1,555,169.34$. So, the
correlation coefficient is:

$$r = \frac{\sum x_i y_i - \dfrac{\sum x_i \sum y_i}{n}}{\sqrt{\left(\sum x_i^2 - \dfrac{(\sum x_i)^2}{n}\right)\left(\sum y_i^2 - \dfrac{(\sum y_i)^2}{n}\right)}}$$

$$= \frac{1,555,169.346 - \dfrac{(9965)(542.14)}{9}}{\sqrt{\left(25,569,715 - \dfrac{(9965)^2}{9}\right)\left(96,801.8364 - \dfrac{(542.14)^2}{9}\right)}}$$

$$= 0.9888989 \approx 0.989$$

Yes, there appears to be a linear relation
between distance and sidereal year.

(c) We compute the mean and standard
deviation for each variable to be:
$\bar{x} \approx 1107.2222222$, $s_x \approx 1347.972809$,
$\bar{y} \approx 60.237778$, and $s_y \approx 89.543654$.

$$b_1 = r \cdot \frac{s_y}{s_x} = 0.9888989 \left(\frac{89.543654}{1347.972809}\right)$$
$$\approx 0.06569096$$
$$b_0 = \bar{y} - b_1 \bar{x}$$
$$= 60.237778 - (0.06569095)(1107.222222)$$
$$\approx -12.4967$$

Rounding to four decimal places, the least-
squares regression line is:
$$\hat{y} = 0.0657x - 12.4967$$

(d) We compute the residuals and construct
the residual plot.

x_i	y_i	$\hat{y} = 0.0657x - 12.4967$	$y - \hat{y}$
36	0.24	−10.132	10.372
67	0.62	−8.095	8.715
93	1.00	−6.387	7.387
142	1.88	−3.167	5.047
483	11.9	19.236	−7.336
887	29.5	45.779	−16.279
1,785	84	104.778	−20.778
2,797	165	171.266	−6.266
3,675	248	228.951	19.049

Residuals vs. Distance

(e) No, the U-shaped pattern in the residual
plot indicates that a linear model is not
appropriate.

27. **(a)** From Problem 29(c) in Section 4.1, we
found $r = -0.842$. In Problem 23 of
Section 4.3, we found $R^2 = 70.8\%$.

Including the Dodge Viper data
($x_{11} = 3425$, $y_{11} = 11$), we obtain
$n = 11$, $\sum x_i = 40,325$, $\sum y_i = 212$,
$\sum x_i^2 = 150,381,861$, $\sum y_i^2 = 4,238$, and
$\sum x_i y_i = 767,719$. So, the new
correlation coefficient is:

$$r = \frac{\sum x_i y_i - \dfrac{\sum x_i \sum y_i}{n}}{\sqrt{\left(\sum x_i^2 - \dfrac{(\sum x_i)^2}{n}\right)\left(\sum y_i^2 - \dfrac{(\sum y_i)^2}{n}\right)}}$$

$$= \frac{767,719 - \dfrac{(40,325)(212)}{11}}{\sqrt{\left(150,381,861 - \dfrac{(40,325)^2}{11}\right)\left(4,238 - \dfrac{(212)^2}{11}\right)}}$$

$$\approx -0.4795$$

Therefore, the coefficient of
determination, with the Dodge Viper
included is:
$$R^2 = (-0.4795)^2 \approx 0.2299 = 22.99\%.$$
Adding the Viper reduces the amount of
variability explained by the model by
approximately 47.8%.

(b) With the Viper included, we recomputed
the means and standard deviations for
each variable to be $\bar{x} = 3665.909091$,
$s_x \approx 505.3788$, $\bar{y} \approx 19.27273$, and
$s_y \approx 3.90105$.

$b_1 = r \cdot \dfrac{s_y}{s_x} = -0.4795 \left(\dfrac{3.90105}{505.3788} \right)$

≈ -0.0037

$b_0 = \bar{y} - b_1 \bar{x}$

$= 19.27273 - (-0.0037)(3665.90901)$

≈ 32.8366

Rounding to four decimal places, the new least-squares regression line with the Viper included is $\hat{y} = -0.0037x + 32.8366$.

Since the slope and y-intercept are altered significantly from the original equation found in Problem 23(a) of Section 4.2, the Viper is an outlier and an influential point.

29. Using all 14 data points, we compute the regression line for carbon dioxide emissions, y, versus energy generated, x, to be: $\hat{y} = 0.9245x - 64.075$. The correlation coefficient is $r = 0.964$ and the coefficient of determination is $R^2 = 92.9\%$.

Excluding the Boardman Power Plant, the regression line for the remaining 13 data points becomes $\hat{y} = 0.5029x + 22.938$. The correlation coefficient is $r = 0.989$, and the coefficient of determination is $R^2 = 97.7\%$.

By removing the Boardman Power Plant, the slope and y-intercept both change substantially. Likewise, removing this plant increases the amount of variability in carbon emissions, that is explained by the model, from 92.9% to 97.7%.

The Boardman Power Plant is unusual in that it emits more carbon that expected from the overall trend. The observation is an outlier and it is influential. We illustrate this further with the following scatter plot showing the original regression line (solid) and the new regression line with Boardman Power Plant removed (dashed).

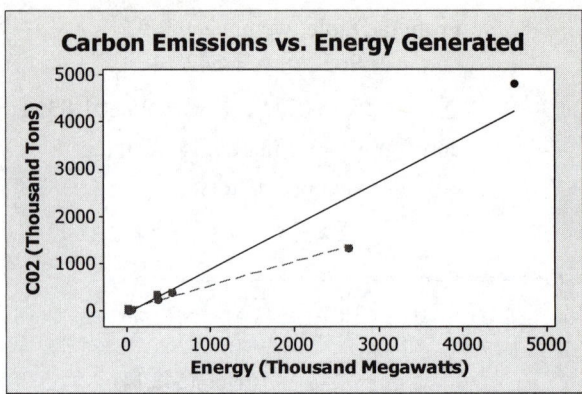

Excluding the Hermiston power plant (and putting back in the Boardman Power Plant), the regression line for the remaining 13 data points becomes $\hat{y} = 1.0455x - 39.208$. The correlation coefficient is $r = 0.999$, and the coefficient of determination is $R^2 = 99.8\%$.

By removing the Hermiston power plant, the slope and the y-intercept change slightly. Removing this plant substantially increases the amount of variability in carbon emissions that is explained by the model.

The carbon emissions for the Hermiston power plant are not unusual in the sense that they follow the overall trend of the data. Although it has a large residual when all the data are included, this effect goes away if the Boardman observation is removed from the data. The following scatter plot shows the original regression line (solid) and the new regression line with Hermiston Power Plant removed (dashed).

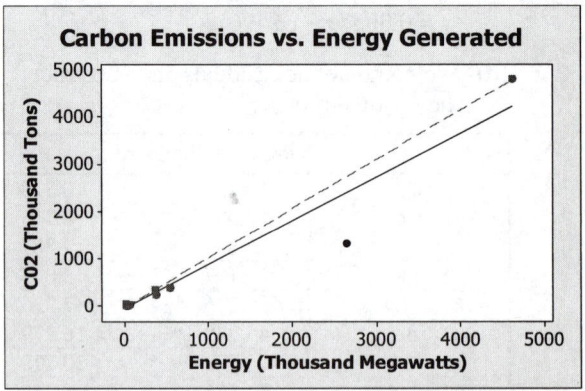

We also show a comparison of the regression line with only the Boardman plant removed (solid), and the regression line with both the Boardman plant and the Hermiston plant removed (dashed).

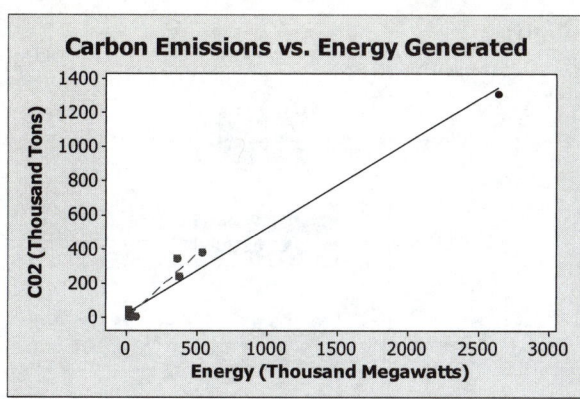

31. **(a)** The explanatory variable is the width of the tornado.

(b) Both the width and the length of the tornado are quantitative variables, so they should be analyzed as bivariate quantitative variables.

(c)

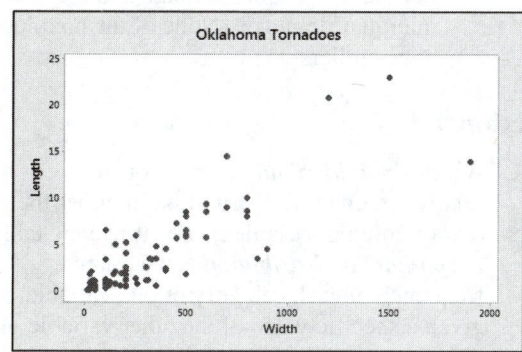

The scatterplot shows there is a positive linear relationship between width and length of tornados.

(d) The correlation coefficient is 0.851.

(e) Yes, since $0.851 > 0.361$ (critical value from Table II), a linear relation exists between the width and the length of tornados.

(f) Rounding to four decimal places, the least-squares regression line is:
$\hat{y} = 0.0111x + 0.0575$

(g) Let $x = 500$ in the regression equation:
$\hat{y} = 0.0111(500) + 0.0575 \approx 5.6075$
We predict the length of the tornado will be 5.6075 miles if the width is 500 yards.

(h) Let $x = 180$ in the regression equation:
$\hat{y} = 0.0111(180) + 0.0575 \approx 2.0555$
We predict the width of the tornado will be approximately 2.06 miles if the width

is 180 yards. Thus, a tornado of length 1.9 miles would be shorter than expected.

(i) The slope indicates that, as the width of a tornado increases by 1 yard, the length of the tornado will increase by 0.0111 miles, on average.

(j) It does not make sense to interpret the y-intercept because a tornado of width 0 yards would mean there was no tornado.

(k) $R^2 = (0.851)^2 \approx 0.7242 = 72.42\%$.
Approximately 72.42% of the variability of in tornado length can be explained by the width of the tornado.

(l)

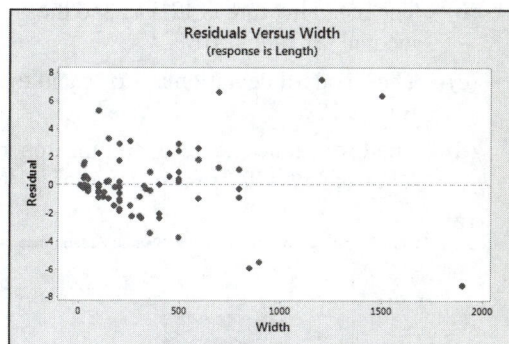

Overall, the residual plot supports the linear relation. However, there may be a slight concern of a nonconstant variance.

(m)

The boxplot shows there are a large number of outliers.

(n) Yes, the tornado was an influential point. With the additional tornado, the slope decreased from 0.0111 to 0.0058, and the y-intercept increased from 0.0575 to 1.491.

33.(a)

What Percent of Income Should Be Paid in Federal Income Tax?
Relative Frequency

The histogram is skewed to the right. The two classes with the highest relative frequency are 10 to 14.9 and 15 to 19.9.

(b) The mean tax rate is 12.3%, and the median tax rate is 10%.

(c) The standard deviation is 7.6%, and the IQR is 12%.

(d) The lower fence is -13%, and the upper fence is 35%. There is one outlier at 40%.

(e)

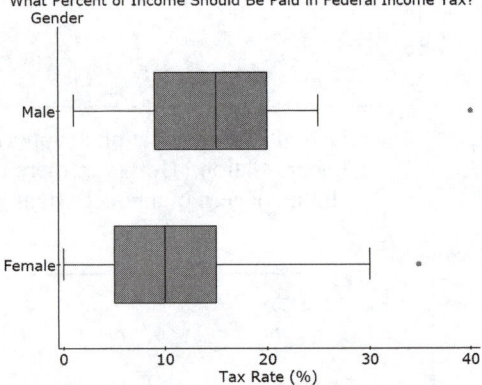

What Percent of Income Should Be Paid in Federal Income Tax?
Gender

The boxplots show that males appear to be willing to pay more in federal income tax. The median for males appears to be about 15%, while the median for females appears to be about 10%. The range of responses for females is greater than that of males, but both genders have similar interquartile ranges. Both genders have an outlier.

(f)

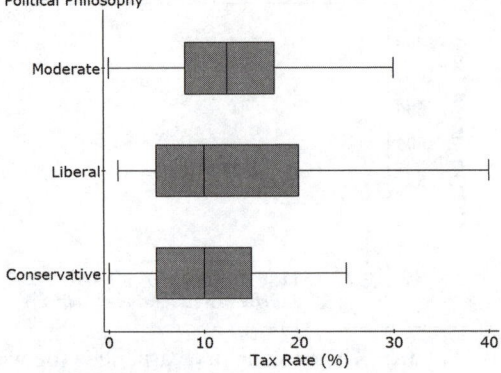

What Percent of Income Should Be Paid in Federal Income Tax?
Political Philosophy

Answers will vary. The boxplots show that moderates have the largest median at approximately 12%. Liberals have the highest tax rate at approximately 40%. Liberals have the most dispersion, and conservatives have the least dispersion as measured by both the range and the interquartile range. None of the boxplots have outliers.

Section 4.4

1. A *marginal distribution* is a frequency, or relative frequency, distribution of either the row or column variable in a contingency table. A *conditional distribution* is the relative frequency of each category of one variable, given a specific value of the other variable in a contingency table.

3. The term correlation is used with quantitative variables. In this section, we are considering the relation between qualitative variables (variables which take non-numeric values).

5. (a) We find the frequency marginal distribution for the row variable, y, by finding the total of each row. We find the frequency marginal distribution for the column variable, x, by finding the total for each column.

	x_1	x_2	x_3	Frequency Marg. Dist.
y_1	20	25	30	75
y_2	30	25	50	105
Frequency Marg. Dist.	50	50	80	180

(b) We find the relative frequency marginal distribution for the row variable, y, by dividing the row total for each y_i by the total table, 180. For example, for the relative frequency for y_1 is $\frac{75}{180} \approx 0.417$. We find the relative frequency marginal distribution for the column variable, x, by dividing the column total for each x_i by the table total. For example, for the relative frequency for x_1 is $\frac{50}{180} \approx 0.278$.

	x_1	x_2	x_3	Rel. Freq. Marg. Dist.
y_1	20	25	30	0.417
y_2	30	25	50	0.583
Rel. Freq. Marg. Dist.	0.278	0.278	0.444	1

(c) Beginning with x_1, we determine the relative frequency of each y_i given x_1 by dividing each frequency by the column total for x_1. For example, the relative frequency of y_1 given x_1 is $\frac{20}{50} = 0.4$. Next, we compute each relative frequency of each y_i given x_2. For example, the relative frequency of y_1 given x_2 is $\frac{25}{50} = 0.5$. Finally, we compute the relative frequency of each y_i given x_3.

	x_1	x_2	x_3
y_1	$\frac{20}{50} = 0.4$	$\frac{25}{50} = 0.5$	$\frac{30}{80} = 0.375$
y_2	$\frac{30}{50} = 0.6$	$\frac{25}{50} = 0.5$	$\frac{50}{80} = 0.625$
Total	1	1	1

(d) We draw two bars, side by side, for each y_i. The horizontal axis represents x_i and the vertical axis represents the relative frequency.

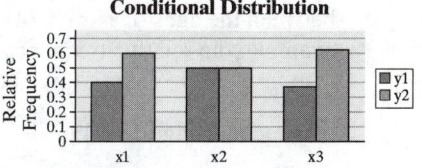

Conditional Distribution

7. (a) 2160 Americans were surveyed. 536 were 55 and older.

(b) The relative frequency marginal distribution for the row variable, response to immigration question, is found by dividing the row total for each answer choice by the table total, 2160. For example, the relative frequency for the response "More Likely" is $\frac{1329}{2160} \approx 0.615$. The relative frequency marginal distribution for the column variable, age, is found by dividing the column total for each age group by the table total. For example, the relative frequency for 18-34 year-olds is $\frac{542}{2160} \approx 0.251$.

	Age				
Likely to Buy	18-34	$35-44$	$45-54$	55+	Rel. Freq. Marg. Dist.
More Likely	238	329	360	402	0.615
Less Likely	22	6	22	16	0.031
Neither	282	201	164	118	0.354
Rel. Freq. Marg. Dist.	0.251	0.248	0.253	0.248	1

(c) The proportion of Americans who are more likely to buy a product when it says "Made in America" is 0.615.

(d) Beginning with respondents between the ages of 18 and 34, we determine the relative frequency for each response to the "Likely to Buy" question by dividing each frequency by the column total for respondents who are between the ages of 18 and 34. For example, the relative frequency for the response "More Likely" given that the respondent is age 18-34 is $\dfrac{238}{542} \approx 0.439$.

We continue with this calculation for each of the answers to the "Likely to Buy" question. Next, we compute the each relative frequency for each response given the respondent is between the ages of 35 and 44. We continue in this pattern until we have completed this for all ages.

	Age			
Likely to Buy	18-34	35−44	45−54	55+
More Likely	0.439	0.614	0.659	0.750
Less Likely	0.041	0.011	0.040	0.030
Neither	0.520	0.375	0.300	0.220
Total	1	1	1	1

(e) We draw four bars, side by side, for each response to the age question. The horizontal axis represents the likelihood to buy question, and the vertical axis represents the relative frequency.

Likelihood to Buy "Made In America"

(f) The number of people more likely to buy a product because it is made in America increases with age. On the other hand, age does not seem to be a significant factor in whether a person is less likely to buy a product because it is made in America.

9. (a) We find the frequency marginal distribution for the row variable, party affiliation, by finding the total of each row. We find the frequency marginal distribution for the column variable, gender, by finding the total for each column.

	Female	Male	Frequency Marg. Dist.
Republican	105	115	220
Democrat	150	103	253
Independent	150	179	329
Frequency Marg. Dist.	405	397	802

(b) The relative frequency marginal distribution for the row variable, party affiliation, is found by dividing the row total for each party by the table total, 802. For example, the relative frequency for Republican is $\dfrac{220}{802} \approx 0.274$. The relative frequency marginal distribution for the column variable, gender, is found by dividing the column total for each gender by the table total. For example, the relative frequency for female is $\dfrac{405}{802} \approx 0.505$.

	Female	Male	Rel. Freq. Marg. Dist.
Republican	105	115	0.274
Democrat	150	103	0.315
Independent	150	179	0.410
Rel. Freq. Marg. Dist.	0.505	0.495	1

(c) The proportion of registered voters who consider themselves Independent is 0.410.

(d) Beginning with females, we determine the relative frequency for each political party by dividing each frequency by the column total for females. For example, the relative frequency for Republicans given that the voter is female is $\dfrac{105}{405} \approx 0.259$.

We then compute each relative frequency for each political party given that the voter is male.

	Female	Male
Republican	$\dfrac{105}{405} \approx 0.259$	$\dfrac{115}{397} \approx 0.290$
Democrat	$\dfrac{150}{405} \approx 0.370$	$\dfrac{103}{397} \approx 0.259$
Independent	$\dfrac{150}{405} \approx 0.370$	$\dfrac{179}{397} \approx 0.451$
Total	1	1

(e) We draw three bars, side by side, for each political affiliation. The horizontal axis represents gender, and the vertical axis represents the relative frequency.

Party Affiliation

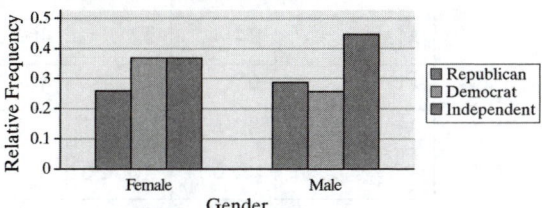

(f) Yes, gender is associated with party affiliation. Male voters are more likely to be Independents, and less likely to be Democrats.

11. To determine whether healthier people tend to also be happier, we construct a conditional distribution of people's happiness by health, and draw a bar graph of the conditional distribution.

	Poor Health	Fair Health	Good Health	Excellent Health
Not too happy	$\frac{696}{1,996} \approx 0.349$	$\frac{1,386}{6,585} \approx 0.210$	$\frac{1,629}{15,791} \approx 0.103$	$\frac{732}{11,022} \approx 0.066$
Pretty happy	$\frac{950}{1,996} \approx 0.476$	$\frac{3,817}{6,585} \approx 0.580$	$\frac{9,642}{15,791} \approx 0.611$	$\frac{5,195}{11,022} \approx 0.471$
Very happy	$\frac{350}{1,996} \approx 0.175$	$\frac{1,382}{6,585} \approx 0.210$	$\frac{4,520}{15,791} \approx 0.286$	$\frac{5,095}{11,022} \approx 0.462$
Total	1	1	1	1

We draw three bars, side by side, for each level of happiness. The horizontal axis represents level of health, and the vertical axis represents the relative frequency.

Happiness

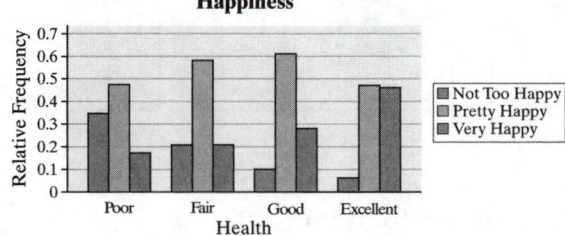

There is a relation between health and happiness. Based on the conditional distribution by health, we can see that healthier people tend to be happier. As health increases, the proportion who are very happy increases, while the proportion who are not happy decreases.

13. (a) Of the 582 smokers in the study, 139 were dead after 20 years, so the proportion of smokers who were dead after 20 years was $\frac{139}{582} \approx 0.239$. Of the 732 nonsmokers in the study, 230 were dead after 20 years, so the proportion of nonsmokers who were dead after 20 years was $\frac{230}{732} \approx 0.314$. This result implies that it is healthier to smoke.

(b) Of the 2 + 53 = 55 smokers in the 18- to 24-year-old category, 2 were dead after 20 years, so the proportion who were dead was $\frac{2}{55} \approx 0.036$. Of the 1 + 61 = 62 nonsmokers in the 18-to 24-year old category, 1 was dead after 20 years, so the proportion was $\frac{1}{62} \approx 0.016$.

(c) We repeat the procedure from part (b) for each age category, and organize the results in the table that follows:

Age	Smoker	Nonsmoker
18-24	$\frac{2}{2+53} \approx 0.036$	$\frac{1}{1+61} \approx 0.016$
25-34	$\frac{3}{3+121} \approx 0.024$	$\frac{5}{5+152} \approx 0.032$
35-44	$\frac{14}{14+95} \approx 0.128$	$\frac{7}{7+114} \approx 0.058$
45-54	$\frac{27}{27+103} \approx 0.208$	$\frac{12}{12+66} \approx 0.154$
55-64	$\frac{51}{51+64} \approx 0.443$	$\frac{40}{40+81} \approx 0.331$
65-74	$\frac{29}{29+7} \approx 0.806$	$\frac{101}{101+28} \approx 0.783$
75+	$\frac{13}{13+0} \approx 1$	$\frac{64}{64+0} \approx 1$

(d) We draw two bars, side by side, for smokers and nonsmokers. The horizontal axis represents age, and the vertical axis represents the relative frequency of deaths.

Death Rate After 20 Years

(e) Reports will vary. When taking age into account, the direction of relation changed. In almost all age groups, smokers had a higher death rate than nonsmokers. The most notable exception is for the 25 to 34 age group, the largest age group for the nonsmokers. A possible explanation could be rigorous physical activity (e.g., rock climbing) that nonsmokers are more likely to participate in than smokers.

15. (a) The relative frequency distribution is:

Political Party	Relative Frequency
Conservative	0.2761
Liberal	0.2164
Moderate	0.5075

(b)

(c)

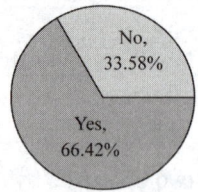

The pie chart shows that a majority of respondents (66.42%) believe there is a gender income inequality.

(d) The contingency table for gender and response including marginal distribution is:

Gender	Yes	No	Marginal Distribution
Male	24	22	46/134=0.343
Female	65	23	88/134=0.657
Marginal Distribution	89/134 =0.664	45/134 =0.336	

(e) The following conditional distribution by gender shows there is a gender gap when it comes to the belief there is gender inequality. There is a higher proportion of females who believe there is a gender inequality.

Gender	Yes	No
Male	24/46=0.522	22/46=0.478
Female	65/88=0.739	23/88=0.261

(f)

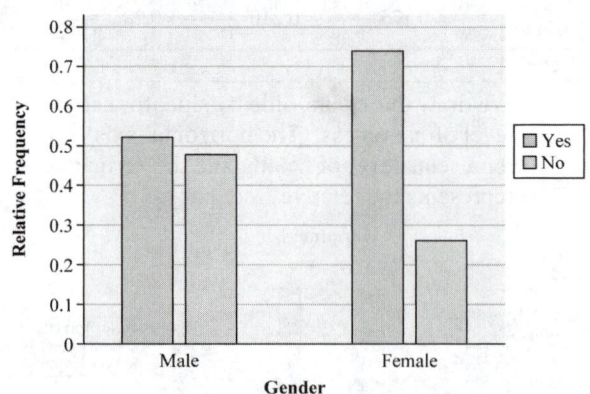

Section 4.5 (online)

1. The logarithm to the base a, where $a > 0$, $a \neq 1$, is $y = \log_a x$ if and only if $x = a^y$. The logarithm is only defined for $x > 0$.

3. $\log y = \log\left(ab^x\right)$

$\qquad = \log a + \log b^x$

$\qquad = \log a + x \log b$

The linear form of $y = ab^x$

is: $\log y = \log a + x \log b$.

If we let $Y = \log y$, $B = \log b$ and $A = \log a$,

then the linear form is $Y = A + Bx$.

5. $4^2 = 16 \quad \Leftrightarrow \quad 2 = \log_4 16$

7. $x^5 = 18 \quad \Leftrightarrow \quad 5 = \log_x 18$

9. $7^x = 15 \quad \Leftrightarrow \quad x = \log_7 15$

11. $5^3 = y \quad \Leftrightarrow \quad 3 = \log_5 y$

13. $\log_3 81 = 4 \quad \Leftrightarrow \quad 3^4 = 81$

15. $\log_a 14 = 3 \quad \Leftrightarrow \quad a^3 = 14$

17. $\log_4 c = 3 \quad \Leftrightarrow \quad 4^3 = c$

19. $\log_2 5 = k \quad \Leftrightarrow \quad 2^k = 5$

21. $\log_3(uv) = \log_3 u + \log_3 v$ (Property 1)

23. $\log_5 x^4 = 4\log_5 x$ (Property 2)

25. $\log_a\left(uv^2\right) = \log_a u + \log_a v^2$ (Property 1)

$\qquad = \log_a u + 2\log_a v$ (Property 2)

27. $\log\left(x^4 y^3\right) = \log x^4 + \log y^3$ (Property 1)

$\qquad = 4\log x + 3\log y$ (Property 2)

29. $\log 34 \approx 1.531$

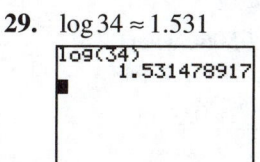

31. $\log 13.2 \approx 1.121$

```
log(13.2
      1.120573931
```

33. $10^{1.6} \approx 39.811$

```
10^1.6
      39.81071706
```

35. $10^{0.36} \approx 2.291$

37. (a)

E-coli Population vs. Time

(b)

x	y	$Y = \log y$
0	0.09	$\log 0.09 \approx -1.0458$
2.5	0.18	$\log 0.18 \approx -0.7447$
3.5	0.26	$\log 0.26 \approx -0.5850$
4.5	0.35	$\log 0.35 \approx -0.4559$
6	0.50	$\log 0.50 \approx -0.3010$

Logarithm of E-coli
Population vs. Time

(c) Using technology, we find the least-squares regression line of the transformed data to be: $\hat{Y} = 0.1266x - 1.0443$

(d) $\hat{Y} = 0.1266x - 1.0443$, where
$A = -1.0443$ and $B = 0.1266$
$a = 10^A = 10^{-1.0443} \approx 0.0903$
$b = 10^B = 10^{0.1266} \approx 1.3384$
$\hat{y} = a \cdot (b)^x$
$\hat{y} = 0.0903(1.3384)^x$

(e) $y = 0.0903(1.3384)^7 = 0.69$
The predicted population after 7 hours is 0.69.

39. (a)

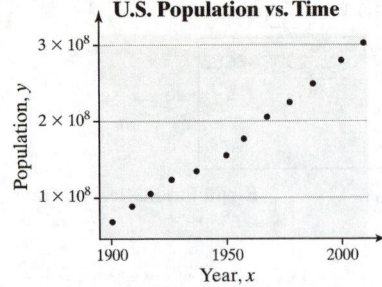

(b)

x	y	$Y = \log y$
1900	76,212,168	7.8820
1910	92,228,496	7.9649
1920	106,021,537	8.0254
1930	123,202,624	8.0906
1940	132,164,569	8.1211
1950	151,325,798	8.1799
1960	179,323,175	8.2536
1970	203,302,031	8.3081
1980	226,542,203	8.3551
1990	248,709,873	8.3957
2000	281,421,906	8.4494
2010	308,745,538	8.4896

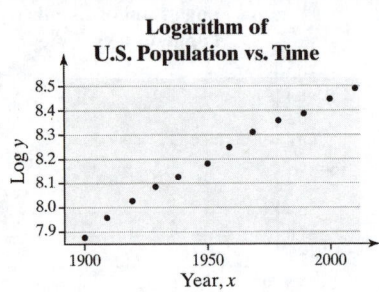

(c) Using technology, we find the least-squares regression line of the transformed data to be: $\hat{Y} = 0.0055x - 2.4495$

(d) $\hat{Y} = 0.0055x - 2.4495$, where
$A = -2.4495$ and $B = 0.0055$
$a = 10^A = 10^{-2.4955} \approx 0.0032$
$b = 10^B = 10^{0.0055} \approx 1.013$
$\hat{y} = a \cdot (b)^x$
$\hat{y} = 0.0032(1.013)^x$

(e) $\hat{y} = 0.0032(1.013)^{2020} \approx 685,850,731$.
The predicted population for the U.S. in 2020 is 685,850,731, or about 686 million.

(f) Answers may vary.

41. (a)

(b)

x	y	$X = \log x$	$Y = \log y$
36	0.24	1.5563	−0.6198
67	0.62	1.8261	−0.2076
93	1.00	1.9685	0
142	1.88	2.1523	0.2742
483	11.9	2.6839	1.0755
887	29.5	2.9479	1.4698
1785	84.0	3.2516	1.9243
2797	165.0	3.4467	2.2175
3675	248.0	3.5653	2.3945

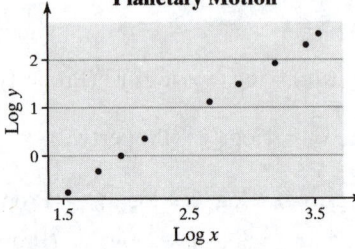

(c) Using technology, we find the least-squares regression line of the transformed data to be: $\hat{Y} = 1.4994X - 2.9506$

(d) $\hat{Y} = 1.4994X - 2.9506$, where
$A = -2.9506$ and $B = 1.4994$
$a = 10^A = 10^{-2.9506} \approx 0.0011$
$\hat{y} = a \cdot (x)^B$
$= 0.0011x^{1.4994}$

43. (a)

(b)

x	y	$X = \log x$	$Y = \log y$
1.0	0.29645	0	−0.5280
1.1	0.25215	0.0414	−0.5983
1.2	0.20547	0.0792	−0.6873
1.3	0.17462	0.1139	−0.7579
1.4	0.15342	0.1461	−0.8141
1.5	0.13521	0.1761	−0.8690
1.6	0.11450	0.2041	−0.9412
1.7	0.10243	0.2304	−0.9896
1.8	0.09231	0.2553	−1.0348
1.9	0.08321	0.2788	−1.0798
2.0	0.07342	0.3010	−1.1342

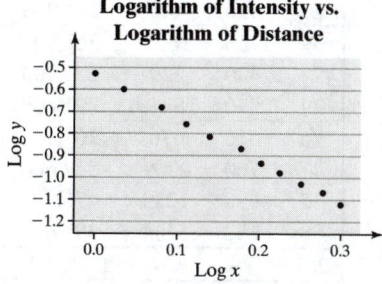

Logarithm of Intensity vs. Logarithm of Distance

(c) Using technology, we find the least-squares regression line of the transformed data to be: $\hat{Y} = -2.012X - 0.5236$

(d) $\hat{Y} = -2.012X - 0.5236$, where $A = -0.5239$ and $B = -2.012$
$a = 10^A = 10^{-0.5236} \approx 0.2995$
$\hat{y} = a \cdot (x)^B$
$\hat{y} = 0.2995x^{-2.012}$

(e) $\hat{y} = 0.2995(2.3)^{-2.012} = 0.05605$
The predicted intensity of a 100-watt light bulb that is 2.3 meters away is 0.05605..

45. (a) Using technology, we find the least-squares regression line to be:
$\hat{y} = -3711.8x + 30984.3$

We compute the residuals and construct the residual plot.

x	y	$\hat{y} = -3.711.8x + 30,984.3$	$y - \hat{y}$
1	27,417	27272.5	144.5
1	28,199	27272.5	927
2	22,995	23560.7	−565.7
2	23,195	23560.7	−365.7
3	18,499	19848.9	−1,349.9
4	16,995	16137.1	857.9
4	16,490	16137.1	352.9

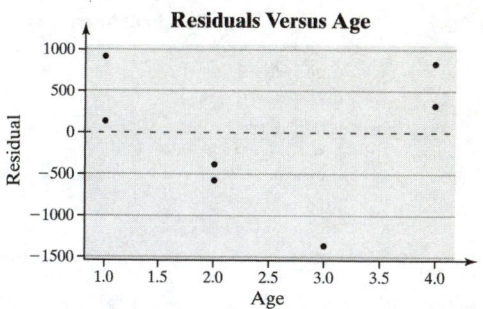

Residuals Versus Age

The residual plot shows a clear U-shaped pattern, so the relation between asking price and age is not linear.

(b) To transform the data, we determine the logarithm of the y-values.

x	y	$Y = \log y$
1	27,417	4.43802
1	28,199	4.45023
2	22,995	4.36163
2	23,195	4.36539
3	18,499	4.26715
4	16,995	4.23032
4	16,490	4.21722

Using technology, we find the least-squares regression line of the transformed data to be:

$\hat{Y} = -0.074589216x + 4.513998311$.
So, the exponential equation of best fit is:
$\hat{Y} = -0.074589216x + 4.513998311$,
where $A = 4.513998311$ and
$B = -0.074589216$
$a = 10^A = 10^{4.513998311} \approx 32658.6562$
$b = 10^B = 10^{-0.074589216} \approx 0.8422$
$\hat{y} = a \cdot (b)^x$
$y = 32658.6562(0.8422)^x$

We compute the residuals and construct the residual plot.

x	$Y = \log y$	$\hat{Y} = -0.074589x + 4.513998$	$Y - \hat{Y}$
1	4.43802	4.43941	−0.00139
1	4.45023	4.43941	0.01082
2	4.36163	4.36482	−0.00319
2	4.36539	4.36482	0.00057
3	4.26715	4.29023	−0.02308
4	4.23032	4.21564	0.01468
4	4.21722	4.21564	0.00158

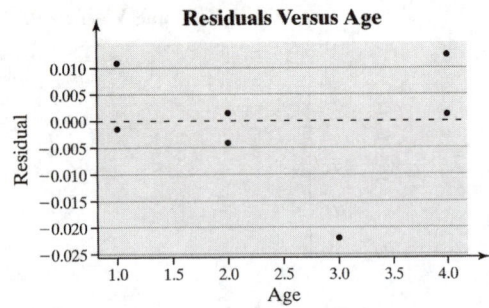

The relation between asking price and age could be exponential since the residual plot does not show a discernable pattern.

(c) The coefficient of determination for the linear model found in part (a) is $R^2 \approx 0.971$. The coefficient of determination for the exponential model found in part (b) is $R^2 \approx 0.984$. Thus, the exponential model does a better job of explaining the variation in asking price. In fact, 98.4% of the variation in asking price is explained by the age of the car in the exponential model.

Chapter 4 Review Exercises

1. (a) The predicted winning margin is $\hat{y}=1.007(3)-0.012=3.009$ points

(b) The predicted winning margin is $\hat{y}=1.007(-7)-0.012=-7.061$ points, suggesting that the visiting team is predicted to win by 7.061 points.

(c) For each 1-point increase in the spread, the winning margin increases by 1.007 points, on average.

(d) If the spread is 0, the home team is expected to lose by 0.012 points, on average.

(e) 39% of the variation in winning margins can be explained by the least-squares regression equation.

2. (a) The explanatory variable is fat content.

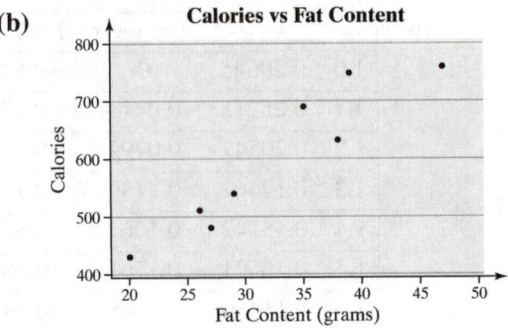

(c) To calculate the correlation coefficient, we use the computational formula:

x_i	y_i	x_i^2	y_i^2	$x_i y_i$
20	430	400	184,900	8,600
39	750	1521	562,500	29,250
27	480	729	230,400	12,960
29	540	841	291,600	15,660
26	510	676	260,100	13,260
47	760	2209	577,600	35,720
35	690	1225	476,100	24,150
38	632	1444	399,424	24,016
261	4792	9045	2,982,624	163,616

From the table, we have $n=8$, $\sum x_i = 261$, $\sum y_i = 4792$, $\sum x_i^2 = 9045$, $\sum y_i^2 = 2,982,624$, and $\sum x_i y_i = 163,616$.

So, the correlation coefficient is:

$$r = \frac{\sum x_i y_i - \dfrac{\sum x_i \sum y_i}{n}}{\sqrt{\left(\sum x_i^2 - \dfrac{(\sum x_i)^2}{n}\right)\left(\sum y_i^2 - \dfrac{(\sum y_i)^2}{n}\right)}}$$

$$= \frac{163,616 - \dfrac{(261)(4792)}{8}}{\sqrt{\left(9045 - \dfrac{(261)^2}{8}\right)\left(2,982,624 - \dfrac{(4792)^2}{8}\right)}}$$

$$\approx 0.944$$

(d) Yes, a strong positive linear relation exists between fat content and calories in fast-food restaurant sandwiches.

3. (a)

(b) To calculate each correlation coefficient, we use the computational formula:

Queens (New York City)

x_i	y_i	x_i^2	y_i^2	$x_i y_i$
500	650	250,000	422,500	325,000
588	1215	345,744	1,476,225	714,420
1000	2000	1,000,000	4,000,000	2,000,000
688	1655	473,344	2,739,025	1,138,640
825	1250	680,625	1,562,500	1,031,250
460	1805	211,600	3,258,025	830,300
1259	2700	1,585,081	7,290,000	3,399,300
650	1200	422,500	1,440,000	780,000
560	1250	313,600	1,562,500	700,000
1073	2350	1,151,329	5,522,500	2,521,550
1452	3300	2,108,304	10,890,000	4,791,600
1305	3100	1,703,025	9,610,000	4,045,500
10,360	22,475	10,245,152	49,773,275	22,277,560

From the table for the Queens (New York City) apartments, we have $n = 12$,
$\sum x_i = 10,360$, $\sum y_i = 22,475$,
$\sum x_i^2 = 10,245,152$, $\sum y_i^2 = 49,773,275$,
and $\sum x_i y_i = 22,277,560$.

So, the correlation coefficient for Queens is:

$$r = \frac{\sum x_i y_i - \frac{\sum x_i \sum y_i}{n}}{\sqrt{\left(\sum x_i^2 - \frac{(\sum x_i)^2}{n}\right)\left(\sum y_i^2 - \frac{(\sum y_i)^2}{n}\right)}}$$

$$= \frac{22,277,560 - \frac{(10,360)(22,475)}{12}}{\sqrt{\left(10,245,152 - \frac{(10,360)^2}{12}\right)\left(49,773,275 - \frac{(22,475)^2}{12}\right)}}$$

$$\approx 0.909$$

Nassau County (Long Island)

x_i	y_i	x_i^2	y_i^2	$x_i y_i$
1100	1875	1,210,000	3,515,625	2,062,500
588	1075	345,744	1,155,625	632,100
1250	1775	1,562,500	3,150,625	2,218,750
556	1050	309,136	1,102,500	583,800
825	1300	680,625	1,690,000	1,072,500
743	1475	552,049	2,175,625	1,095,925
660	1315	435,600	1,729,225	867,900
975	1400	950,625	1,960,000	1,365,000
1429	1900	2,042,041	3,610,000	2,715,100
800	1650	640,000	2,722,500	1,320,000
1906	4625	3,632,836	21,390,625	8,815,250
1077	1395	1,159,929	1,946,025	1,502,415
11,909	20,835	13,521,085	46,148,375	24,251,240

From the table for the Nassau County (Long Island) apartments, we have $n = 12$,
$\sum x_i = 11,909$, $\sum y_i = 20,835$,
$\sum x_i^2 = 13,521,085$, $\sum y_i^2 = 46,148,375$, and
$\sum x_i y_i = 24,251,240$. So, the correlation coefficient for Nassau County is:

$$r = \frac{\sum x_i y_i - \frac{\sum x_i \sum y_i}{n}}{\sqrt{\left(\sum x_i^2 - \frac{(\sum x_i)^2}{n}\right)\left(\sum y_i^2 - \frac{(\sum y_i)^2}{n}\right)}}$$

$$= \frac{24,251,240 - \frac{(11,909)(20,835)}{12}}{\sqrt{\left(13,521,085 - \frac{(11,909)^2}{12}\right)\left(46,148,375 - \frac{(20,835)^2}{12}\right)}}$$

$$\approx 0.867$$

(c) Yes, both locations have a positive linear relation between square footage and monthly rent.

(d) For small apartments (those less than 1000 square feet in area), there seems to be no difference in rent between Queens and Nassau County. In larger apartments, however, Queens seems to have higher rents than Nassau County.

4. **(a)** In Problem 2, we computed the correlation coefficient. Rounded to eight decimal places, it is $r = 0.94370937$. We compute the mean and standard deviation for each variable to be: $\bar{x} = 32.625$, $s_x \approx 8.7003695$, $\bar{y} = 599$, and $s_y \approx 126.6130212$.

$$b_1 = r \cdot \frac{s_y}{s_x} = 0.94370937 \left(\frac{126.6130212}{8.7003695} \right)$$
$$\approx 13.7334276$$
$$b_0 = \bar{y} - b_1 \bar{x}$$
$$= 599 - (13.7334276)(32.625)$$
$$\approx 150.9469$$

Rounding to four decimal places, the least-squares regression line is:
$$\hat{y} = 13.7334x + 150.9469$$

(b)

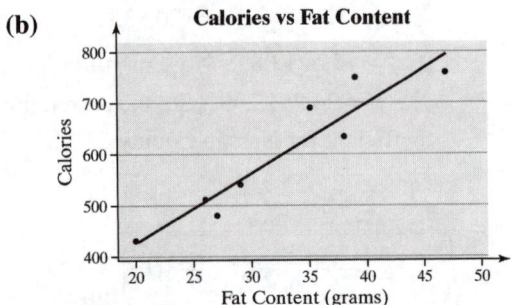

Calories vs Fat Content

(c) The slope indicates that each additional gram of fat in a sandwich adds 13.73 calories, on average. The y-intercept indicates that a sandwich with no fat will contain about 151 calories.

(d) Let $x = 30$ in the regression equation:
$$\hat{y} = 13.7334(30) + 150.9469 \approx 562.9$$
We predict that a sandwich with 30 grams of fat will have 562.9 calories.

(e) Let $x = 42$ in the regression equation:
$$\hat{y} = 13.7334(42) + 150.9469 \approx 727.7$$
Sandwiches with 42 grams of fat have an average of 727.7 calories. So, the number of calories for this cheeseburger from Sonic is below average.

5. **(a)** In Problem 3, we computed the correlation coefficient. Rounded to eight decimal places, it is $r = 0.90928694$. We compute the mean and standard deviation for each variable to be: $\bar{x} \approx 863.333333$, $s_x \approx 343.9104887$, $\bar{y} \approx 1872.916667$, and $s_y \approx 835.5440752$.

$$b_1 = r \cdot \frac{s_y}{s_x} = 0.90928694 \left(\frac{835.5440751}{343.9104887} \right)$$
$$\approx 2.20914843$$
$$b_0 = \bar{y} - b_1 \bar{x}$$
$$= 1872.916667 - (2.2091484)(863.333333)$$
$$\approx -34.3148$$

Rounding to four decimal places, the least-squares regression line is;
$$\hat{y} = 2.2091x - 34.3148.$$

(b) The slope of the least-squares regression equation indicates that, for each additional square foot of floor area, the rent increases by $2.21, on average. It is not appropriate to interpret the y-intercept, since it is not possible to have an apartment with an area of 0 square feet. The rent value of $ -34.3148 is outside the scope of the model.

(c) Let $x = 825$ in the regression equation:
$$\hat{y} = 2.2091(825) - 34.3148 \approx 1788$$
Apartments in Queens with 825 square feet have an average rent of $1788. Since the rent of the apartment in the data set with 825 square feet is $1250, it is below average.

6. **(a)**

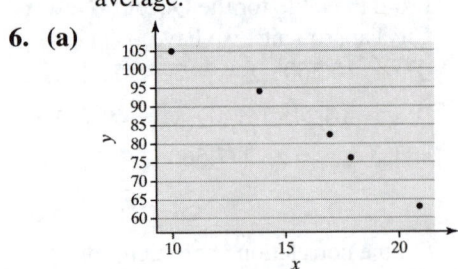

(b) Using the two points $(10, 105)$ and $(18, 76)$ gives:
$$m = \frac{76 - 105}{18 - 10} = \frac{-29}{8} = -\frac{29}{8}$$
$$y - 105 = -\frac{29}{8}(x - 10)$$
$$y - 105 = -\frac{29}{8}x + \frac{145}{4}$$
$$y = -\frac{29}{8}x + \frac{565}{4}$$

(c)

(d) We compute the mean and standard deviation for each variable and the correlation coefficient to be: $\bar{x} = 16$, $\bar{y} = 84$, $s_x \approx 4.18330$, $s_y \approx 16.20185$, and $r \approx -0.99222$. Then the slope and intercept for the least-squares regression line are:

$$b_1 = r \cdot \frac{s_y}{s_x} = -0.992221\left(\frac{16.20185}{4.18330}\right)$$
$$\approx -3.842855$$
$$b_0 = \bar{y} - b_1\bar{x} = 84 - (-3.842855)(16)$$
$$\approx 145.4857$$

So, rounding to four decimal places, the least-squares regression line is:
$$\hat{y} = -3.8429x + 145.4857 .$$

(e)

(f)

x	y	$\hat{y} = -\dfrac{29}{8}x + \dfrac{565}{4}$	$y - \hat{y}$	$(y - \hat{y})^2$
10	105	105	0	0
14	94	90.5	3.5	12.25
17	82	79.625	2.375	5.6406
18	76	76	0	0
21	63	65.125	−2.125	4.5156

Total = 22.4062

(g)

x	y	$\hat{y}$	$y - \hat{y}$	$(y - \hat{y})^2$
10	105	107.0571	−2.0571	4.2318
14	94	91.6857	2.3143	5.3559
17	82	80.1571	1.8429	3.3961
18	76	76.3143	−0.3143	0.0988
21	63	64.7857	−1.7857	3.1888

Total = 16.2714

(h) The regression line gives a smaller sum of squared residuals, so it is a better fit.

7. A linear model is appropriate. There is no pattern to the residuals and the variability stays the same.

8. A linear model is not appropriate because a discernible pattern is present in the plot.

9. A linear model is not appropriate because the amount of scatter in the residuals is clearly decreasing, indicating that the error variance is not constant.

10. (a) In Problem 1(c), we found $r \approx 0.944$.

So, $R^2 = (0.944)^2 \approx 0.891 = 89.1\%$.

This means that 89.1% of the variation in calories is explained by the least-squares regression line.

(b) From Problem 3(a), we have $\hat{y} = 13.7334x + 150.9469$. We compute the residuals and construct the residual plot.

x_i	y_i	$\hat{y} = 13.7334x + 150.9469$	$y - \hat{y}$
20	430	425.6	4.4
39	750	686.5	63.5
27	480	521.7	−41.7
29	540	549.2	−9.2
26	510	508.0	2.0
47	760	796.4	−36.4
35	690	631.6	58.4
38	632	672.8	−40.8

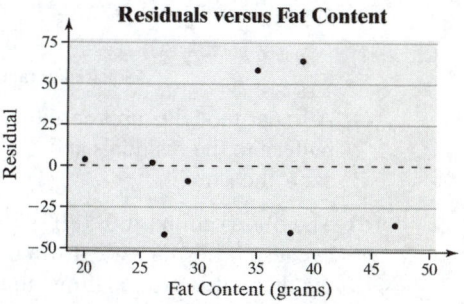

Residuals versus Fat Content

(c) Yes, based on the residual plot, a linear model is appropriate.

(d) There are no outliers or influential observations present.

(e) The observation (9, 80) is an outlier since it is far away from the rest of the data. Adding it to the data set, the regression equation becomes $\hat{y} = 17.7154x + 9.8713$.

Since there is a significant change in both the slope and y-intercept, the observation is influential.

11. (a) From Problem 2(b), the correlation coefficient for the Queens apartments is $r = 0.90928694$ (unrounded). Therefore, $R^2 = (0.90928694)^2 \approx 0.827 = 82.7\%$. This means that 82.7% of the variation in monthly rent is explained by the least-squares regression equation.

(b) From Problem 4(a), we have $\hat{y} = 2.2091x - 34.3148$. We compute the residuals and construct the residual plot.

x_i	y_i	$\hat{y} = 2.2091x - 34.3148$	$y - \hat{y}$
500	650	1070.2	−420.2
588	1215	1264.6	−49.6
1000	2000	2174.8	−174.8
688	1655	1485.5	169.5
825	1250	1788.2	−538.2
460	1805	981.9	823.1
1259	2700	2746.9	−46.9
650	1200	1401.6	−201.6
560	1250	1202.8	47.2
1073	2350	2336.0	14.0
1452	3300	3173.3	126.7
1305	3100	2848.6	251.4

Apartments

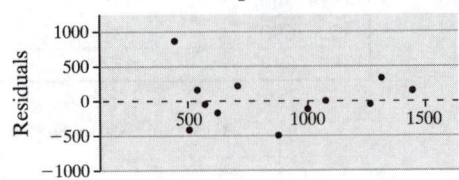

(c) A linear model is appropriate. There is no pattern to the residuals and the variability stays the same.

(d) The observation (460, 1805) is an outlier because it lies far away from the rest of the data. It is also an influential point because, if it is removed from the data set, the linear regression equation becomes $\hat{y} = 2.5315x - 399.2496$ in which the y-intercept is significantly different.

12. (a) In Problem 2(b), we computed the correlation coefficient. Rounded to eight decimal places, it is $r = 0.86741573$. We compute the mean and standard deviation for each variable to be: $\bar{x} \approx 992.416667$, $s_x \approx 393.399510$, $\bar{y} = 1736.25$, and $s_y \approx 952.203485$.

$$b_1 = r \cdot \frac{s_y}{s_x} = 0.86741573 \left(\frac{952.203485}{393.399510} \right)$$
$$\approx 2.0995356$$

$$b_0 = \bar{y} - b_1 \bar{x}$$
$$= 1736.25 - (2.0995356)(992.416667)$$
$$\approx -347.3641$$

Rounding to four decimal places, the least-squares regression line is: $\hat{y} = 2.0995x - 347.3641$.

(b) The observation (1906, 4625) is an outlier because it lies far away from the rest of the data. It is also influential point because, if it is removed from the data set, the linear regression equation becomes $\hat{y} = 0.8779x + 675.2945$ in which both the slope and y-intercept are significantly different.

13. (a)

Camaro Depreciation

(b) To calculate the correlation coefficient, we use the computational formula:

x_i	y_i	x_i^2	y_i^2	$x_i y_i$
2	15,900	4	252,810,000	31,800
5	10,988	25	120,736,144	54,940
2	16,980	4	288,320,400	33,960
5	9,995	25	99,900,025	49,975
4	11,995	16	143,880,025	47,980
5	10,995	25	120,890,025	54,975
1	20,365	1	414,733,225	20,365
2	16,463	4	271,030,369	32,926
6	10,824	36	117,158,976	64,944
1	19,995	1	399,800,025	19,995
1	18,650	1	347,822,500	18,650
4	10,488	16	109,998,144	41,952
38	173,638	158	2,687,079,858	472,462

From the table, we have $n = 12$, $\sum x_i = 38$, $\sum y_i = 173,638$, $\sum x_i^2 = 158$, $\sum y_i^2 = 2,687,079,858$, and $\sum x_i y_i = 472,462$. So, the correlation coefficient is:

$$r = \frac{472,462 - \frac{(38)(173,638)}{12}}{\sqrt{\left(158 - \frac{(38)^2}{12}\right)\left(2,687,079,858 - \frac{(173,638)^2}{12}\right)}}$$
$$\approx -0.95440918 \approx -0.954$$

Yes, there is strong negative linear relation between age and price.

(c) We compute the mean and standard deviation for each variable: $\bar{x} \approx 3.1666667$, $\bar{y} \approx 14,469.8333333$, $s_x \approx 1.85047087$, and $s_y \approx 3983.68191577$. Then the slope and intercept for the least-squares regression line are:

$$b_1 = r \cdot \frac{s_y}{s_x}$$

$$= -0.95440918\left(\frac{3983.68191577}{1.85047087}\right)$$

$$\approx -2054.646011$$

$$b_0 = \bar{y} - b_1\bar{x}$$

$$= 14,469.8333333 - (-2054.646011)(3.1666667)$$

$$\approx 20,976.2124$$

So, rounding to four decimal places, the least-squares regression line is:
$\hat{y} = -2054.6460x + 20,976.2124$.

(d) We compute the residuals and construct the residual plot.

x_i	y_i	$\hat{y} = -2054.6460x + 20,976.2124$	$y - \hat{y}$
2	15,900	16,866.9	−966.9
5	10,988	10,703.0	285.0
2	16,980	16,866.9	113.1
5	9,995	10,703.0	−708.0
4	11,995	12,757.6	−762.6
5	10,995	10,703.0	292.0
1	20,365	18,921.6	1,443.4
2	16,463	16,866.9	−403.9
6	10,824	8,648.3	2,175.7
1	19,995	18,921.6	1,073.4
1	18,650	18,921.6	−271.6
4	10,488	12,757.6	−2,269.6

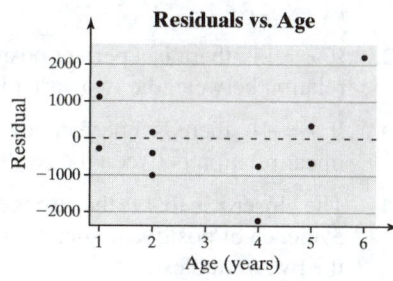

Residuals vs. Age

(e) No, the residuals follow a U-shaped pattern indicating that there is a nonlinear relation.

14. No. Correlation does not imply causation. Florida has a large number of tourists in warmer months, times when more people will be in the water to cool off. The larger number of people in the water splashing around lends to a larger number of shark attacks.

15. (a) 203 buyers were extremely satisfied with their automobile purchase.

(b) The relative frequency marginal distribution for the row variable, level of satisfaction, is found by dividing the row total for each level of satisfaction by the table total, 396. For example, the relative frequency for "Not too satisfied" is $\frac{36}{396} \approx 0.091$. The relative frequency marginal distribution for the column variable, purchase type (new versus used), is found by dividing the column total for each purchase type by the table total. For example, the relative frequency for "New" is $\frac{207}{396} \approx 0.523$.

	New	Used	Rel. Freq. Marg. Dist.
Not Too Satisfied	11	25	$\frac{36}{396} \approx 0.091$
Pretty Satisfied	78	79	$\frac{157}{396} \approx 0.396$
Extremely Satisfied	118	85	$\frac{203}{396} \approx 0.513$
Rel. Freq. Marg. Dist.	$\frac{207}{396} \approx 0.523$	$\frac{189}{396} \approx 0.477$	1

(c) The proportion of consumers who were extremely satisfied with their automobile purchase was 0.513.

(d) Beginning with new automobiles, we determine the relative frequency for level of satisfaction by dividing each frequency by the column total for new automobiles. For example, the relative frequency for "Not Too Satisfied" given that the consumer purchased a new automobile is $\frac{11}{207} \approx 0.053$. We compute each relative frequency for the used automobiles.

	New	Used
Not Too Satisfied	$\frac{11}{207} \approx 0.053$	$\frac{25}{189} \approx 0.132$
Pretty Satisfied	$\frac{78}{207} \approx 0.377$	$\frac{79}{189} \approx 0.418$
Extremely Satisfied	$\frac{118}{207} \approx 0.570$	$\frac{85}{189} \approx 0.450$
Total	1	1

(e) We draw three bars, side by side, for each level of satisfaction. The horizontal axis represents purchase type (new versus used), and the vertical axis represents the relative frequency.

Customer Satisfaction

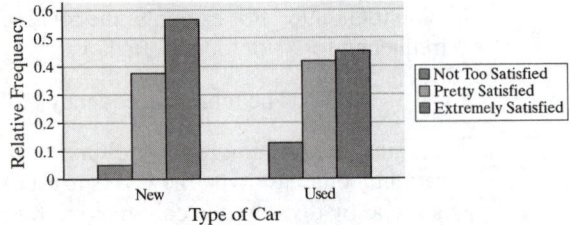

(f) Yes, there appears to be some association between purchase type (new versus used) and level of satisfaction. Buyers of used cars are more likely to be dissatisfied and are less likely to be extremely satisfied than buyers of new cars.

16. (a) In 1982, the unemployment rate can be found by dividing the number of people who are unemployed by the total number of people who are either employed or unemployed

$$\frac{11.3 \text{ thousand}}{(99.1+11.3) \text{ thousand}} \approx 0.102 = 10.2\%$$

In 2009, the unemployment rate was:

$$\frac{14.5 \text{ thousand}}{(130.1+14.5) \text{ thousand}} \approx 0.100 = 10.0\%$$

(b) To find the unemployment rate for the recession of 1982 for those who do not have a high school diploma, for example, we divide the number in the group who are unemployed by the total number of people in the group.

$$\frac{3.9 \text{ thousand}}{(20.3+3.9) \text{ thousand}} \approx 0.161 = 16.1\%$$

	High School or Less	High School	Bachelor's or Higher
Recession of 1982	$\frac{3.9}{24.2} \approx 16.1\%$	$\frac{6.6}{64.8} \approx 10.2\%$	$\frac{0.8}{21.2} \approx 3.8\%$
Recession of 2009	$\frac{2.0}{12.0} \approx 16.7\%$	$\frac{10.3}{87.0} \approx 11.8\%$	$\frac{2.2}{45.6} \approx 4.8\%$

(c) We draw two bars, side by side, for each of the two years. The horizontal axis represents the level of education, and the vertical axis represents the relative frequency.

Unemployment by Level of Education

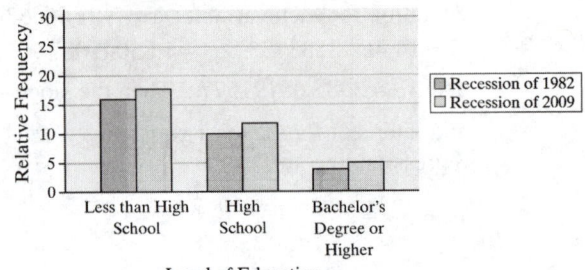

(d) Answers will vary. The discussion should include the observation that, although the overall unemployment rate was higher in 1982, the unemployment rate within each level of education was higher in 2009.

17. (a) A positive linear relation appears to exist between number of marriages and number unemployed.

(b) Population is highly correlated with both the number of marriages and the number unemployed. The size of the population affects both variables.

(c) It appears that no relation exists between the two variables.

(d) Answers may vary. A strong correlation between two variables may be due to a third variable that is highly correlated with the two original variables.

18. The eight properties of a linear correlation coefficient are:

1. The linear correlation coefficient is always between -1 and 1, inclusive. That is, $-1 \le r \le 1$.

2. If $r = +1$, there is a perfect positive linear relation between the two variables.

3. If $r = -1$, there is a perfect negative linear relation between the two variables.

4. The closer r is to $+1$, the stronger is the evidence of positive association between the two variables.

5. The closer r is to -1, the stronger is the evidence of negative association between the two variables.

6. If r is close to 0, there is no evidence of a *linear* relation between the two variables. Because the linear correlation coefficient is a measure of the strength of the linear relation, an r close to 0 does not imply no relation, just no linear relation.

7. The linear correlation coefficient is a unitless measure of association. So, the unit of measure for x and y plays no role in the interpretation of r.

8. The correlation coefficient is not resistant.

19. (a) Answers will vary.

(b) The slope can be interpreted as "the school day decreases by about 0.01 hours for every 1 percent increase in the percentage of the district with low income", on average. The y-intercept can be interpreted as the length of the school day for a district with no low income families.

(c) $\hat{y} = -0.0102(20) + 7.11 \approx 6.91$

The school day is predicted to be about 6.91 hours long if 20% of the district is low income.

(d) Answers will vary. Because so many of the data values are clumped together, it is difficult to see any patterns. As shown, it does not appear that any of the model requirements are badly violated.

(e) Answers will vary. The likely candidates will be the three points with the highest percentage of low income population.

(f) Answers will vary.

(g) Answers will vary. There is some indication that there is a positive association between length of school day and PSAE score.

(h) Answers will vary. The scatter diagram indicates that there is some negative association between PSAE score and percentage of population as low income. However, it is likely not as strong as the correlation coefficient indicates. A few influential observations could be inflating the correlation coefficient.

(i) Answers will vary.

Chapter 4 Test

1. (a) The likely explanatory variable is temperature because crickets would likely chirp more frequently in warmer temperatures and less frequently in colder temperatures, so the temperature could be seen as explaining the number of chirps.

(b)

Cricket Chirp Rates

(c) To calculate the correlation coefficient, we use the computational formula:

x_i	y_i	x_i^2	y_i^2	$x_i y_i$
88.6	20.0	7849.96	400.00	1772.00
93.3	19.8	8704.89	392.04	1847.34
80.6	17.1	6496.36	292.41	1378.26
69.7	14.7	4858.09	216.09	1024.59
69.4	15.4	4816.36	237.16	1068.76
79.6	15.0	6336.16	225.00	1194.00
80.6	16.0	6496.36	256.00	1289.60
76.3	14.4	5821.69	207.36	1098.72
71.6	16.0	5126.56	256.00	1145.60
84.3	18.4	7106.49	338.56	1551.12
75.2	15.5	5655.04	240.25	1165.60
82.0	17.1	6724.00	292.41	1402.20
83.3	16.2	6938.89	262.44	1349.46
82.6	17.2	6822.76	295.84	1420.72
83.5	17.0	6972.25	289.00	1419.50
1200.6	249.8	96,725.86	4200.56	20,127.47

From the table, we have $n = 15$, $\sum x_i = 1200.6$, $\sum y_i = 249.8$, $\sum x_i^2 = 96{,}725.86$, $\sum y_i^2 = 4200.56$, and $\sum x_i y_i = 20{,}127.47$. So, the correlation coefficient is:

$$r = \frac{\sum x_i y_i - \frac{\sum x_i \sum y_i}{n}}{\sqrt{\left(\sum x_i^2 - \frac{(\sum x_i)^2}{n}\right)\left(\sum y_i^2 - \frac{(\sum y_i)^2}{n}\right)}}$$

$$= \frac{20{,}127.47 - \frac{(1200.6)(249.8)}{15}}{\sqrt{\left(96{,}725.86 - \frac{(1200.6)^2}{15}\right)\left(4200.56 - \frac{(249.8)^2}{15}\right)}}$$

≈ 0.8351437868

≈ 0.835

(d) Based on the scatter diagram and the linear correlation coefficient, a positive linear relation between temperature and chirps per second.

2. (a) We compute the mean and standard deviation for each variable: $\bar{x} = 80.04$, $\bar{y} \approx 16.6533333$, $s_x \approx 6.70733074$, and $s_y \approx 1.70204359$. The slope and intercept for the least-squares regression line are:

$$b_1 = r \cdot \frac{s_y}{s_x} = 0.8351437868 \left(\frac{1.70204359}{6.70733074} \right)$$
$$\approx 0.21192501$$
$$b_0 = \bar{y} - b_1 \bar{x}$$
$$= 16.6533333 - (0.21192501)(80.04)$$
$$\approx -0.3091$$

So, rounding to four decimal places, the least-squares regression line is:
$$\hat{y} = 0.2119x - 0.3091.$$

(b) If the temperature increases $1°F$, the number of chirps per second increases by 0.2119, on average. Since there are no observations near $0°F$, it is outside the scope of the model. So, it does not make sense to interpret the y-intercept.

(c) Let $x = 83.3$ in the regression equation:
$$\hat{y} = 0.2119(83.3) - 0.3091 \approx 17.3$$
We predict that, if the temperature is $83.3°F$, then there will be 17.3 chirps per second.

(d) Let $x = 82$ in the regression equation:
$$\hat{y} = 0.2119(82) - 0.3091 \approx 17.1$$
There will be an average of 17.1 chirps per second when the temperature is $82°F$. Therefore, 15 chirps per second is below average.

(e) No, we should not use this model to predict the number of chirps when the temperature is $55°F$ because $55°F$ is outside the scope of the model.

3. (a) $R^2 = (0.8351437868)^2 \approx 0.697 = 69.7\%$
69.7% of the variation in number of chirps per second is explained by the least-squares regression line.

(b) Yes, a linear model is appropriate. There is no pattern to the residuals and the variability stays the same.

4. If we add the point (92, 14.5) to the data set, then the linear regression equation becomes $\hat{y} = 0.1431x + 4.9565$. Since both the slope and y-intercept change significantly, the observation is influential.

5. We compute the correlation coefficient to be $r \approx 0.992$, so from this it appears that there is a strong positive linear relation between speed and braking distance. However, we must analyze this further by computing the least-squares regression equation and construct a residual plot. The least-squares regression line is $\hat{y} = 5.3943x - 131.3524$. We compute the residuals and construct the residual plot.

x_i	y_i	$\hat{y} = 5.3943x - 131.3524$	$y - \hat{y}$
30	48	30.5	17.5
40	79	84.4	−5.4
50	123	138.4	−15.4
60	184	192.3	−8.3
70	243	246.2	−3.2
80	315	300.2	14.8

The residual plot has a clear U-shaped pattern, so a linear model is not appropriate despite the correlation coefficient.

6. There is a problem with the politician's reasoning. Correlation does not imply causation. It is possible that a lurking variable, such as income level or educational level, is affecting both the explanatory and response variables.

7. (a) The relative frequency marginal distribution for the row variable, level of education, is found by dividing the row total for each level of satisfaction by the table total, 2375. For example, the relative frequency for "Less than high school" is $\frac{412}{2375} \approx 0.173$. The relative frequency marginal distribution for the column variable, response concerning belief in Heaven, is found by dividing the column total for each response by the table total. For example, the relative frequency for "Yes, Definitely" is $\dfrac{1539}{2375} = 0.648$.

	Yes, Definitely	Yes, Probably	No, Probably Not	No, Definitely Not	Rel. Freq. Marg. Dist.
Less than High School	316	66	21	9	0.173
High School	956	296	122	65	0.606
Bachelor's	267	131	62	64	0.221
Rel. Freq. Marg. Dist.	0.648	0.208	0.086	0.058	1

(b) The proportion of adult Americans in the survey who definitely believe in Heaven is 0.648.

(c) Beginning with "Less than high school", we determine the relative frequency for responses concerning belief in Heaven by dividing each frequency by the row total for "Less than high school." For example, the relative frequency for "Yes, Definitely" given that the respondent has less than a high school education is $\dfrac{316}{412} \approx 0.767$. We then compute each relative frequency for the other levels of education.

	Yes, Definitely	Yes, Probably	No, Probably Not	No, Definitely Not	Total
< HS	$\dfrac{316}{412} \approx 0.767$	$\dfrac{66}{412} \approx 0.160$	$\dfrac{21}{412} \approx 0.051$	$\dfrac{9}{412} \approx 0.022$	1
HS	$\dfrac{956}{1439} \approx 0.664$	$\dfrac{296}{1439} \approx 0.206$	$\dfrac{122}{1439} \approx 0.085$	$\dfrac{65}{1439} \approx 0.045$	1
BS	$\dfrac{267}{524} \approx 0.510$	$\dfrac{131}{524} \approx 0.250$	$\dfrac{62}{524} \approx 0.118$	$\dfrac{64}{524} \approx 0.122$	1

(d) We draw four bars, side by side, for each response concerning belief in Heaven. The horizontal axis represents level of education, and the vertical axis represents the relative frequency.0

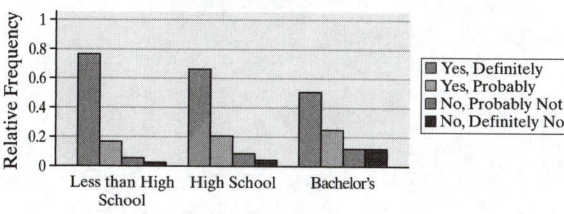

Belief in Heaven

(e) Yes, as education level increases, the proportion who definitely believe in

Heaven decreases (i.e., doubt in Heaven increases).

8. (a) For each gender, we determine the proportions for acceptance status by dividing the number of applicants in each acceptance status by the total number of applicants of that gender:

	Accepted	Denied
Male	$\dfrac{98}{98+522} \approx 0.158$	$\dfrac{522}{98+522} \approx 0.842$
Female	$\dfrac{90}{90+200} \approx 0.310$	$\dfrac{200}{90+200} \approx 0.690$

(b) The proportion of males that was accepted is 0.158. The proportion of females that was accepted is 0.310.

(c) The college accepted a higher proportion of females than males.

(d) The proportion of male applicants accepted into the business school was:
$$\frac{90}{90+510} = 0.15$$
The proportion of female applicants accepted into the business school was:
$$\frac{10}{10+60} \approx 0.143$$

(e) The proportion of male applicants accepted into the social work school was:
$$\frac{8}{8+12} = 0.4$$
The proportion of male applicants accepted into the social work school was:
$$\frac{80}{80+140} \approx 0.364$$

(f) Answers will vary. A larger number of males applied to the business school, which has an overall lower acceptance rate than the social work school, so more male applicants were declined.

9. A set of quantitative bivariate data whose linear coefficient is -1 would have a perfect negative linear relation. A scatter diagram would show all the observations being collinear (falling on the same line) with a negative slope.

10. If the slope of the least-squares regression line is negative, then the correlation between the explanatory and response variables is also negative.

11. If a linear correlation is close to zero, then this means that there is no linear relation between the explanatory and response variables. This does not necessarily mean that there is no relation at all; however, just no *linear* relation.

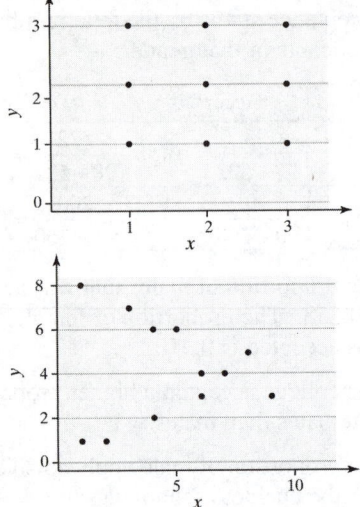

Case Study: Thomas Malthus, Population, and Subsistence

(b)

$$\log P = 0.00877t + 6.78385$$

$$a = 10^{6.78385} \approx 6,079,249.95$$

$$\log b = 0.00877$$
$$b = 10^{0.00877} \approx 1.0204$$

The annual growth rate is
$$b - 1 = 1.0204 - 1 = 0.0204 = 2.04\%.$$

Thus, $P = 6,079,249.95(1.0204)^t$

(c) $A = 1.2P$
$$= 1.2\left(6,079,249.95(1.0204)^t\right)$$
$$= 7,295,099.94(1.0204)^t$$

(d) Answers will vary. The following is for the years 1995–2003.

	t (years after 1790)	Farm Acres (millions)
1995	205	963
1996	206	959
1997	207	956
1998	208	952
1999	209	948
2000	210	945
2001	211	942
2002	212	940
2003	213	939

$$\hat{y} = -3.1333t + 1604.2$$

(e) Since the number of acres is in millions, we need to divide our model for A by one million to keep the same units.

$$\hat{y}_1 = \frac{A}{1,000,000} = 7.295099(1.0204)^t$$
$$\hat{y}_2 = -3.1333t + 1604.2$$

Based on this information, the United States will no longer be able to provide a diverse diet for its entire population $t = 237$ years after 1790. That is, in the year 2027.

(f) Answers will vary. As technology improves, particularly in regards to agriculture, it becomes possible to generate a higher yield from a smaller piece of land. Therefore, food production may actually increase even if the number of acres of farmland is decreasing.

Chapter 5
Probability

Section 5.1

1. The probability of an impossible event is zero. No. An event with probability approximately zero is a very unlikely event, but it is not necessarily an impossible event.

3. True

5. experiment

7. Rule 1 is satisfied since all of the probabilities in the model are greater than or equal to zero and less than or equal to one. Rule 2 is satisfied since the sum of the probabilities in the model is one: $0.3 + 0.15 + 0 + 0.15 + 0.2 + 0.2 = 1$. In this model, the outcome "blue" is an impossible event.

9. This cannot be a probability model because $P(\text{green}) < 0$.

11. Probabilities must be between 0 and 1, inclusive, so the only values which could be probabilities are: 0, 0.01, 0.35, and 1.

13. The probability of 0.89 means that in 100 elections, where a senate candidate is winning his/her election with a 5% lead in the average of the polls with a week until the election, we would expect the leading candidate to win approximately 89 of the elections. Probability deals with long-term behavior so while we would expect to see about 89 such wins in every 100, on average, this does not mean we will always get exactly 89 wins out of every 100 elections.

15. The empirical probability that the next flip would result in a head is $\dfrac{95}{100} \approx 0.95$.

17. $P(2) \neq \dfrac{1}{11}$ because the 11 possible outcomes are not equally likely.

19. The sample space is $S = \{$1H, 2H, 3H, 4H, 5H, 6H, 1T, 2T, 3T, 4T, 5T, 6T$\}$

21. The probability that a randomly selected three-year-old is enrolled in daycare is $P = 0.428$.

23. Event E contains 3 of the 10 equally likely outcomes, so $P(E) = \dfrac{3}{10} = 0.3$.

25. There are 10 equally likely outcomes and 4 are even numbers less than 9, so
$$P(E) = \frac{4}{10} = \frac{2}{5} = 0.4 .$$

27. (a) $P(\text{plays organized sports}) = \dfrac{288}{500} = 0.576$

    (b) If we sampled 1000 high school students, we would expect that about 576 of the students play organized sports.

29. (a) Since 85 of the 1000 homeruns were caught by the fans,
$$P(\text{caught}) = \frac{85}{1000} = 0.085$$

    (b) Since 296 of the 1000 homeruns were dropped by the fans,
$$P(\text{dropped}) = \frac{296}{1000} = 0.296$$

    (c) Since 2 of the 85 homeruns that were caught by the fans were caught in hat,
$$P(\text{hat}) = \frac{2}{85} = 0.024 \text{ If 85 caught}$$
homeruns were selected, we would expect about 2 to have been caught in a hat.

    (d) Since 8 of the 296 dropped homeruns were a failed hat attempt,
$$P(\text{failed hat}) = \frac{8}{296} = 0.027 . \text{ If 296}$$
dropped homeruns were selected, we would expect about 8 to have been a failed hat attempt to catch the ball.

31. (a) The sample space is $S = \{0, 00, 1, 2, 3, 4,\ldots, 36\}$.

(b) Since the slot marked 8 is one of the 38 equally likely outcomes,

$P(8) = \dfrac{1}{38} \approx 0.0263$. This means that, in many spins of such a roulette wheel, the long-run relative frequency of the ball landing on "8" will be close to

$\dfrac{1}{38} \approx 0.0263 = 2.63\%$. That is, if we spun the wheel 100 times, we would expect about 3 of those times to result in the ball landing in slot 8.

(c) Since there are 18 odd slots (1, 3, 5, 7, 9, 11, 13, 15, 17, 19, 21, 23, 25, 27, 29, 31, 33, 35) in the 38 equally likely outcomes,

$P(\text{odd}) = \dfrac{18}{38} = \dfrac{9}{19} \approx 0.4737$. This means that, in many spins of such a roulette wheel, the long-run relative frequency of the ball landing in an odd slot will be

close to $\dfrac{9}{19} \approx 0.4737 = 47.37\%$. That is,

if we spun the wheel 100 times, we would expect about 47 of those times to result in an odd number.

33. (a) The sample space of possible genotypes is $\{SS, Ss, sS, ss\}$.

(b) Only one of the four equally likely genotypes gives rise to sickle-cell anemia, namely ss. Thus, the probability

is $P(ss) = \dfrac{1}{4} = 0.25$. This means that of

the many children who are offspring of two Ss parents, approximately 25% will have sickle-cell anemia.

(c) Two of the four equally likely genotypes result in a carrier, namely Ss and sS. Thus, the probability of this is

$P(Ss \text{ or } sS) = \dfrac{2}{4} = \dfrac{1}{2} = 0.5$. This means

that of the many children who are offspring of two Ss parents, approximately 50% will not themselves have sickle-cell anemia, but will be carriers of sickle-cell anemia.

35. (a) There are
$125 + 324 + 552 + 1257 + 2518 = 4776$
college students in the survey. The individuals can be thought of as the trials of the probability experiment. The relative frequency of "Never" is

$\dfrac{125}{4776} \approx 0.026$. We compute the relative

frequencies of the other outcomes similarly and obtain the probability model below.

Response	Probability
Never	0.026
Rarely	0.068
Sometimes	0.116
Most of the time	0.263
Always	0.527

(b) Yes, it is unusual to find a college student who never wears a seatbelt when riding in a car driven by someone else. The approximate probability of this is only 0.026, which is less than 0.05.

37. (a) There are
$4 + 6 + 133 + 219 + 90 + 42 + 143 + 5 = 642$
police records included in the survey. The individuals can be thought of as the trials of the probability experiment. The relative frequency of "Pocket picking" is

$\dfrac{4}{642} \approx 0.006$. We compute the relative

frequencies of the other outcomes similarly and obtain the probability model below.

Type of Larceny Theft	Probability
Pocket picking	0.006
Purse snatching	0.009
Shoplifting	0.207
From motor vehicles	0.341
Motor vehicle accessories	0.140
Bicycles	0.065
From buildings	0.223
From coin-operated machines	0.008

(b) Yes, purse-snatching larcenies are unusual since the probability is only $0.009 < 0.05$.

(c) No, bicycle larcenies are not unusual since the probability is $0.065 > 0.05$.

39. Assignments A, B, C, and F are consistent with the definition of a probability model. Assignment D cannot be a probability model because it contains a negative probability, and Assignment E cannot be a probability model because is does not add up to 1.

41. Assignment B should be used if the coin is known to always come up tails.

43. **(a)** The sample space is $S = \{$John-Roberto; John-Clarice; John-Dominique; John-Marco; Roberto-Clarice; Roberto-Dominique; Roberto-Marco; Clarice-Dominique; Clarice-Marco; Dominique-Marco$\}$.

(b) Clarice-Dominique is one of the ten possible samples from part (a). Thus,

$$P(\text{Clarice and Dominique}) = \frac{1}{10} = 0.1.$$

(c) Four of the ten samples from part (a) include Clarice. Thus,

$$P(\text{Clarice attends}) = \frac{4}{10} = \frac{2}{5} = 0.4.$$

(d) Six of the ten samples from part (a) do not include John. Thus,

$$P(\text{John stays home}) = \frac{6}{10} = \frac{3}{5} = 0.6.$$

45. **(a)** Since 24 of the 73 homeruns went to right field, $P(\text{right field}) = \frac{24}{73} \approx 0.329$.

(b) Since 2 of the 73 homeruns went to left field, $P(\text{left field}) = \frac{2}{73} \approx 0.027$.

(c) Yes, it was unusual for Barry Bonds to hit a homerun to left field. The probability is below 0.05.

47. **(a)–(d)** Answers will vary depending on the results from the simulation.

49. If the dice were fair, then each outcome should occur approximately $\frac{400}{6} \approx 67$ times.

Since 1 and 6 occurred with considerably higher frequency, the dice appear to be loaded.

51. Half of all families are above the median and half are below, so

$$P(\text{Income greater than \$58,500}) = \frac{1}{2} = 0.5.$$

53. The relative frequency distribution for the position of 2015 NFL players is:

Position	Relative Frequency
C	0.015
CB	0.120
DE	0.081
DT	0.089
FB	0.015
ILB	0.046
LS	0.004
OG	0.062
OLB	0.081
OT	0.081
P	0.004
QB	0.046
RB	0.100
S	0.062
TE	0.042
WR	0.154

If a player is randomly selected from the 2015 NFL combine, a Wide Receiver (WR) has the highest probability of being selected because that position has the highest relative frequency.

It would be surprising if a center was randomly selected since this should occur with probability 0.015 meaning only about 15 out of 1000 selections would be a center.

55. The Law of Large Numbers states that as the number of repetitions of a probability experiment increases (in the long term), the proportion with which a certain outcome is observed (i.e. the relative frequency) gets closer to the probability of the outcome. The games at a gambling casino are designed to benefit the casino in the long run; the risk to the casino is minimal because of the large number of gamblers.

57. An event is unusual if it has a low probability of occurring. The same "cut-off" should not always be used to identify unusual events. Selecting a "cut-off" is subjective and should take into account the consequences of incorrectly identifying an event as unusual.

59. Empirical probability is based on the outcomes of a probability experiment and is the relative frequency of the event. Classical probability is based on counting techniques and is equal to the ratio of the number of ways an event can occur to the number of possible outcomes of the experiment.

61. It is impossible to be absolutely certain, but due to the law of large numbers it is most likely the smaller hospital. The larger hospital likely has more births so it less likely that the births would deviate from the expected proportion of girls.

Section 5.2

1. Two events are disjoint (mutually exclusive) if they have no outcomes in common.

3. $P(E) + P(F) - P(E \text{ and } F)$

5. E and $F = \{5, 6, 7\}$. No, E and F are not mutually exclusive because they have simple events in common.

7. F or $G = \{5, 6, 7, 8, 9, 10, 11, 12\}$.
$P(F \text{ or } G) = P(F) + P(G) - P(F \text{ and } G)$
$= \dfrac{5}{12} + \dfrac{4}{12} - \dfrac{1}{12} = \dfrac{8}{12} = \dfrac{2}{3}$.

9. E and $G = \{\ \}$. Yes, E and G are mutually exclusive because they have no simple events in common.

11. $E^c = \{1, 8, 9, 10, 11, 12\}$.
$P(E^c) = 1 - P(E) = 1 - \dfrac{6}{12} = \dfrac{1}{2}$

13. $P(E \text{ or } F) = P(E) + P(F) - P(E \text{ and } F)$
$= 0.25 + 0.45 - 0.15 = 0.55$

15. $P(E \text{ or } F) = P(E) + P(F) = 0.25 + 0.45 = 0.7$

17. $P(E^c) = 1 - P(E) = 1 - 0.25 = 0.75$

19. $P(E \text{ or } F) = P(E) + P(F) - P(E \text{ and } F)$
$0.85 = 0.60 + P(F) - 0.05$
$P(F) = 0.85 - 0.60 + 0.05 = 0.30$

21. $P(\text{Titleist or Maxfli}) = \dfrac{9+8}{20} = \dfrac{17}{20} = 0.85$

23. $P(\text{not Titleist}) = 1 - P(\text{Titleist})$
$= 1 - \dfrac{9}{20} = \dfrac{11}{20} = 0.55$

25. (a) Rule 1 is satisfied since all of the probabilities in the model are between 0 and 1.
Rule 2 is satisfied since the sum of the probabilities in the model is one: $0.472 + 0.023 + 0.025 + 0.170 + 0.122 + 0.056 + 0.132 = 1$.

(b) $P(\text{rifle or shotgun}) = 0.023 + 0.025$
$= 0.048$.
If 1000 murders in 2009 were randomly selected, we would expect 48 of them to be committed with a rifle or a shotgun.

(c) $P(\text{handgun, rifle, or shotgun})$
$= 0.472 + 0.023 + 0.025$
$= 0.520$.
If 100 murders in 2009 were randomly selected, we would expect 52 of them to be committed with a handgun, a rifle, or a shotgun.

(d) To find the probability that a randomly selected murder was committed with a weapon other than a gun, we subtract the probability that a gun was used from 1.
$P(\text{not a gun}) = 1 - P(\text{a gun was used})$
$= 1 - (0.472 + 0.023 + 0.025 + 0.170)$
$= 1 - 0.690 = 0.310$
This means that there is a 31.0% probability of randomly selecting a murder that was not committed with a gun.

(e) Yes, murders with a shotgun are unusual since the probability is $0.025 < 0.05$.

27. No; for example, on one draw of a card from a standard deck, let $E = \text{diamond}$, $F = \text{club}$, and $G = \text{red card}$. Here, E and F are disjoint, as are F and G. However, E and G are *not* disjoint since diamond cards are red.

29. (a) Since there are 23 scores out of 125 that are between 5 and 5.9 then
$P(5 - 5.9) = \dfrac{23}{125} \approx 0.184$

(b) Use the complementation rule:
$P(\text{not } 5 - 5.9) = 1 - P(5 - 5.9)$
$\approx 1 - 0.184 = 0.816$

(c) Use the complementation rule:
$P(\text{less than } 9) = 1 - P(9 - 10)$
$= 1 - \dfrac{5}{125}$
$= 0.96$

(d) In order for a hospital to receive reduced Medicare payments their score must be at least 8. To have a score of at least 8 means a score between 8 and 8.9 or between 9 and 10.

$P(\text{reduced}) = P(\text{at least } 8)$

$= P(8 - 8.9) + P(9 - 10)$

$= \dfrac{5}{125} + \dfrac{5}{125} = 0.08$

This means that if 100 hospitals in Illinois are randomly selected, we would expect 8 to receive reduced Medicare payments. Since this is expected to happen at 8% of the hospitals, it is not a very unusual result.

31. (a) $P(\text{Heart or Club}) = P(\text{Heart}) + P(\text{Club})$

$= \dfrac{13}{52} + \dfrac{13}{52}$

$= \dfrac{1}{2} = 0.5$

(b) $P(\text{Heart or Club or Diamond})$

$= P(\text{Heart}) + P(\text{Club}) + P(\text{Diamond})$

$= \dfrac{13}{52} + \dfrac{13}{52} + \dfrac{13}{52}$

$= \dfrac{3}{4} = 0.75$

(c) $P(\text{Ace or Heart})$

$= P(\text{Ace}) + P(\text{Heart}) - P(\text{Ace of Hearts})$

$= \dfrac{4}{52} + \dfrac{13}{52} - \dfrac{1}{52}$

$= \dfrac{4}{13} \approx 0.308$

33. (a) $P(\text{not on Nov. } 8) = 1 - P(\text{on Nov. } 8)$

$= 1 - \dfrac{1}{365}$

$= \dfrac{364}{365} \approx 0.997$

(b) $P(\text{not on the } 1^{\text{st}}) = 1 - P(\text{on the } 1^{\text{st}})$

$= 1 - \dfrac{12}{365}$

$= \dfrac{353}{365} \approx 0.967$

(c) $P(\text{not on the } 31^{\text{st}}) = 1 - (\text{on the } 31^{\text{st}})$

$= 1 - \dfrac{7}{365}$

$= \dfrac{358}{365} \approx 0.981$

(d) $P(\text{not in Dec.}) = 1 - P(\text{in Dec.})$

$= 1 - \dfrac{31}{365}$

$= \dfrac{334}{365} \approx 0.915$

35. No, we cannot compute the probability of randomly selecting a citizen of the U.S. who has hearing problems or vision problems by adding the given probabilities because the events "hearing problems" and "vision problems" are not disjoint. That is, some people have both vision and hearing problems, but we do not know the proportion.

37. (a) $P(\text{only English or only Spanish})$
$= P(\text{only English}) + P(\text{only Spanish})$
$= 0.81 + 0.12 = 0.93$

(b) $P(\text{neither only English nor only Spanish})$
$= 1 - P(\text{only English or only Spanish})$
$= 1 - 0.93 = 0.07$

(c) $P(\text{not only English}) = 1 - P(\text{only English})$
$= 1 - 0.81 = 0.19$

(d) No, the probability that a randomly selected household speaks Polish at home cannot equal 0.08 because the sum of the probabilities would be more than 1 and there would be no probability model.

39. (a) Of the 137,243 men included in the study, $782 + 91 + 141 = 1014$ died from cancer. Thus,

$P(\text{died from cancer}) = \dfrac{1014}{137,243} \approx 0.007$.

(b) Of the 137,243 men included in the study, $141 + 7725 = 7866$ were current cigar smokers. Thus,

$P(\text{current cigar smoker}) = \dfrac{7866}{137,243}$
≈ 0.057.

(c) Of the 137,243 men included in the study, 141 were current cigar smokers who died from cancer. Thus,

$P(\text{died from cancer and current smoker})$

$= \dfrac{141}{137,243} \approx 0.001$.

(d) Of the 137,243 men included in the study, 1014 died from cancer, 7866 were current cigar smokers, and 141 were current cigar smokers who died from cancer. Thus,

$P(\text{died from cancer or current smoker})$

$= P(\text{died from cancer}) + P(\text{current smoker})$

$\quad - P(\text{died from cancer and current smoker})$

$= \dfrac{1014}{137,243} + \dfrac{7866}{137,243} - \dfrac{141}{137,243}$

$= \dfrac{8739}{137,243} \approx 0.064.$

41. (a) Of the 250 study participants, 100 were given the placebo. Thus,

$$P(\text{placebo}) = \frac{100}{250} = \frac{2}{5} = 0.4 \ .$$

(b) Of the 250 study participants, 188 reported that the headache went away within 45 minutes. Thus,

$$P(\text{headache went away}) = \frac{188}{250} = \frac{94}{125}$$
$$= 0.752$$

(c) Of the 250 study participants, 56 were given the placebo and reported that the headache went away. Thus,

$$P\left(\begin{array}{c}\text{placebo and headache}\\ \text{went away}\end{array}\right) = \frac{56}{250} = \frac{28}{125}$$
$$= 0.224$$

(d) Of the 250 study participants, 100 were given the placebo, 188 reported that the headache went away, and 56 were both given the placebo and reported that the headache went away. Thus,

$P(\text{placebo or headache went away})$

$= P(\text{placebo}) + P(\text{headache went away})$

$\quad - P(\text{placebo and headache went away})$

$= \dfrac{100}{250} + \dfrac{188}{250} - \dfrac{56}{250}$

$= \dfrac{232}{250} = \dfrac{116}{125}$

$= 0.928$

43. (a) Of the 39,900 drivers involved in fatal crashes, 25,911 were male. Thus,

$$P(\text{Male}) = \frac{25,911}{39,900} \approx 0.649 \ .$$

(b) Of the 39,900 fatal crashes, 6430 occurred on Sunday. Thus,

$$P(\text{Sunday}) = \frac{6,430}{39,900} \approx 0.161 \ .$$

(c) Of the 39,900 fatal crashes, 4,143 occurred on Sunday and involved a male. Thus,

$$P(\text{Sunday and male}) = \frac{4,143}{39,900} \approx 0.104 \ .$$

(d) Of the 39,900 fatal crashes, 25,911 involved a male, 6,430 occurred on Sunday, and 4,143 occurred on Sunday and involved a male. Thus,

$P(\text{Sunday or male})$

$= P(\text{Sunday}) + P(\text{male})$

$\quad - P(\text{Sunday and male})$

$= \dfrac{6,430}{39,900} + \dfrac{25,911}{39,900} - \dfrac{4,143}{39,900}$

$= \dfrac{28,198}{39,900} \approx 0.707$

(e) Of the 39,900 fatal crashes, 1,729 occurred on a Wednesday and involved a female.

$P(\text{Wednesday and female})$

$= \dfrac{1,729}{39,900} \approx 0.043$

Yes, a fatality on Wednesday involving a female driver is unusual, $0.043 < 0.05$.

45. (a) The variables presented in the table are crash type, system, and number of crashes.

(b) Crash type is qualitative because it describes an attribute or characteristic.

System is qualitative because it describes an attribute or characteristic.

Number of crashes is quantitative because it is a numerical measure. It is discrete because the numerical measurement is the result of a count.

(c) Of the 1,121 projected crashes under the current system, 289 were projected to have reported injuries. Of the 922 projected crashes under the new system, 221 were projected to have reported injuries. The relative frequency for reported injury crashes would be

$\dfrac{289}{1,121} \approx 0.26$ under the current system

and $\dfrac{221}{922} \approx 0.24$ under the new system.

Similar computations can be made for the remaining crash types for both systems.

Crash Type	Current System	w/Red-Light Cameras
Reported Injury	0.26	0.24
Reported Property Damage Only	0.35	0.36
Unreported Injury	0.07	0.07
Unreported Property Damage Only	0.32	0.33
Total	1	1

(d)

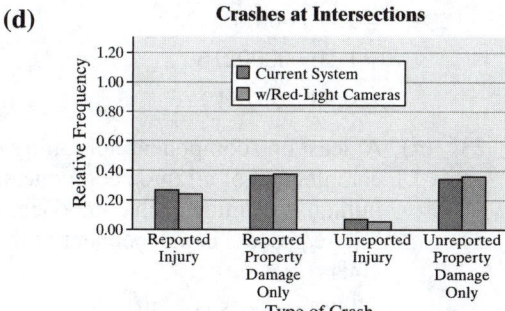

Crashes at Intersections

(e) Mean for Current System:

$\dfrac{1121}{13} \approx 86.2$ crashes per intersection

Mean for Red-light Camera System:

$\dfrac{922}{13} \approx 70.9$ crashes per intersection

(f) It is not possible to compute the standard deviation because we do not know the number of crashes at each intersection.

(g) Since the mean number of crashes is lower with the cameras, it appears that the program will be beneficial.

(h) There are 1,121 crashes under the current system and 289 with reported injuries. Thus,

$$P(\text{reported injuries}) = \frac{289}{1121} \approx 0.258.$$

(i) There are 922 crashes under the camera system. Of these, 333 had reported property damage only and 308 had unreported property damage only. Thus,

$$P\left(\begin{array}{c}\text{property damage}\\ \text{only}\end{array}\right) = \frac{333 + 308}{922}$$
$$= \frac{641}{922} \approx 0.695$$

(j) Simpson's Paradox refers to situations when conclusions reverse or change direction when accounting for a specific variable. When accounting for the cause of the crash (rear-end vs. red-light running), the camera system does not reduce all types of accidents. Under the camera system, red-light running crashes decreased, but rear-end crashes increased.

(k) Recommendations may vary. The benefits of the decrease in red-light running crashes must be weighed against the negative of increased rear-end crashes. Seriousness of injuries and amount of property damage may need to be considered.

Section 5.3

1. independent

3. Addition

5. $P(E) \cdot P(F)$

7. (a) Dependent. Speeding on the interstate increases the probability of being pulled over by a police officer.

(b) Dependent: Eating fast food affects the probability of gaining weight.

(c) Independent: Your score on a statistics exam does not affect the probability that the Boston Red Sox win a baseball game.

9. Since E and F are independent,
$$P(E \text{ and } F) = P(E) \cdot P(F)$$
$$= (0.3)(0.6)$$
$$= 0.18$$

11. $P(5 \text{ heads in a row}) = \left(\dfrac{1}{2}\right)\left(\dfrac{1}{2}\right)\left(\dfrac{1}{2}\right)\left(\dfrac{1}{2}\right)\left(\dfrac{1}{2}\right)$

$$= \left(\dfrac{1}{2}\right)^5 = \dfrac{1}{32} = 0.03125$$

If we flipped a coin five times, 100 different times, we would expect to observe 5 heads in a row about 3 times.

13. $P(2 \text{ left-handed people}) = (0.13)(0.13)$

$$= 0.0169$$

$$P\left(\begin{matrix}\text{At least 1 is}\\ \text{right-handed}\end{matrix}\right) = 1 - P\left(\begin{matrix}2 \text{ left-handed}\\ \text{people}\end{matrix}\right)$$

$$= 1 - 0.0169 = 0.9831$$

15. **(a)** $P(\text{all 5 negative})$

$$= (0.995)(0.995)(0.995)(0.995)(0.995)$$

$$= (0.995)^5 \approx 0.9752$$

(b) $P(\text{at least one positive})$

$$= 1 - P(\text{all 5 negative})$$

$$= 1 - 0.9752$$

$$= 0.0248$$

17. **(a)** $P(\text{Both live to be 41}) = (0.99757)(0.99757)$

$$\approx 0.99515$$

(b) $P(\text{All 5 live to be 41}) = (0.99757)^5$

$$\approx 0.98791$$

(c) This is the complement of the event in (b), so the probability is
$1 - 0.98791 = 0.01209$ which is unusual since $0.01209 < 0.05$.

19. **(a)** Using the complementation rule,
$$P(\text{not default}) = 1 - P(\text{default})$$
$$= 1 - 0.01 = 0.99$$

(b) Assuming that the likelihood of default is independent,
$$P(\text{All 5 will not default}) = (0.99)^5$$
$$= 0.951$$

(c) Probability the derivative is worthless is the probability that at least one of the mortgages defaults,
$$P(\text{At least 1 defaults})$$
$$= 1 - P(\text{None default})$$
$$= 1 - P(\text{All 5 will not default})$$
$$= 1 - (0.99)^5 = 1 - 0.951$$
$$= 0.049$$

(d) The assumption that the likelihood of default is independent is probably not reasonable. Economic conditions (such as recessions) will impact all mortgages. Thus, if one mortgage defaults, the likelihood of a second mortgage defaulting may be higher.

21. **(a)** Assuming each component's failure/success is independent of the others,
$$P(\text{all three fail}) = (0.006)^3$$
$$\approx 0.000000216$$

(b) At least one component not failing is the complement of all three components failing, so
$$P(\text{at least one does not fail})$$
$$= 1 - P(\text{all 3 fail})$$
$$= 1 - (0.006)^3$$
$$= 1 - 0.00000216$$
$$= 0.999999784$$

23. **(a)** At least one component not failing is the complement of all three components failing. Assuming each component's failure/success is independent of the others,
$$P(\text{system does not fail})$$
$$P(\text{at least one component works})$$
$$= 1 - P(\text{no components work})$$
$$= 1 - (0.003)^3$$
$$= 0.9999973$$

(b) From part (a) we know that three components are not enough, so we increase the number.
4 components:
$$P(\text{system succeeds}) = 1 - (0.03)^4$$
$$\approx 0.99999919$$

5 components:

$$P(\text{system succeeds}) = 1 - (0.03)^5$$
$$= 0.9999999757$$

6 components:

$$P(\text{system succeeds}) = 1 - (0.03)^6$$
$$= 0.9999999993$$

Therefore, 6 components would be needed to make the probability of the system succeeding greater than 0.99999999.

25. (a) $P\left(\begin{array}{c}\text{two strikes}\\\text{in a row}\end{array}\right) = (0.3)(0.3) = 0.09$

(b) $P(\text{turkey}) = (0.3)^3 = 0.027$

(c) $P\left(\begin{array}{c}\text{gets a turkey}\\\text{but fails to get}\\\text{a clover}\end{array}\right) = P\left(\begin{array}{c}\text{three strikes}\\\text{followed by}\\\text{a non strike}\end{array}\right)$

$$= P\left(\begin{array}{c}\text{three strikes}\\\text{in a row}\end{array}\right) \cdot P(\text{non-strike})$$

$$= (0.3)^3 (0.7) = 0.0189$$

27. The probability that an individual satellite will not detect a missile is $1 - 0.9 = 0.1$, so

$$P\left(\begin{array}{c}\text{none of the 4}\\\text{will detect the missile}\end{array}\right) = (0.1)^4 = 0.0001.$$

Thus,

$$P\left(\begin{array}{c}\text{at least one of}\\\text{the 4 satellites}\\\text{will detect the}\\\text{missile}\end{array}\right) = 1 - 0.0001 = 0.9999.$$

Answer will vary. Generally, one would probably feel safe since only 1 in 10,000 missiles should go undetected.

29. Since, the events are independent, the probability that a randomly selected pregnancy will result in a girl and weight gain over 40 pounds is:

$$P(\text{girl and weight gain over 40 pounds})$$
$$= P(\text{girl}) \cdot P(\text{weight gain of 40 pounds})$$
$$= (0.495)(0.201) \approx 0.099$$

31. (a) $P\left(\begin{array}{c}\text{male and bets on}\\\text{professional sports}\end{array}\right)$

$$= P(\text{male}) \cdot P\left(\begin{array}{c}\text{bets on}\\\text{professional}\\\text{sports}\end{array}\right)$$

$$= (0.484)(0.170)$$
$$\approx 0.0823$$

(b) $P(\text{male or bets})$

$$= P(\text{male}) + P(\text{bets}) - P(\text{male and bets})$$
$$= 0.17 + 0.484 - 0.0823$$
$$= 0.5717$$

(c) Since $P\left(\begin{array}{c}\text{male and bets on}\\\text{professional sports}\end{array}\right) = 0.106,$

but we computed it as 0.0823 assuming independence, it appears that the independence assumption is not correct.

(d) $P(\text{male or bets})$

$$= P(\text{male}) + P(\text{bets}) - P(\text{male and bets})$$
$$= 0.17 + 0.484 - 0.106$$
$$= 0.548$$

The actual probability is lower than we computed assuming independence.

33. Assuming that gender of children for different births are independent then the fact that the mother already has three girls does not affect the likelihood of having a fourth girl.

Section 5.4

1. $F; E$

3. $P(F \mid E) = \dfrac{P(E \text{ and } F)}{P(E)} = \dfrac{0.6}{0.8} = 0.75$

5. $P(F \mid E) = \dfrac{N(E \text{ and } F)}{N(E)} = \dfrac{420}{740} = 0.568$

7. $P(E \text{ and } F) = P(E) \cdot P(F \mid E)$
$$= (0.8)(0.4)$$
$$= 0.32$$

9. No, the events "earn more than \$100,000 per year" and "earned a bachelor's degree" are not independent because $P(\text{"earn more than \$100,000 per year"} \mid \text{"earned a bachelor's degree"}) \neq P(\text{"earn more than \$100,000 per year"})$.

11. $P(\text{club}) = \dfrac{13}{52} = \dfrac{1}{4};$

$$P(\text{club} \mid \text{black card}) = \dfrac{13}{26} = \dfrac{1}{2}$$

13. $P(\text{rainy} \mid \text{cloudy}) = \dfrac{P(\text{rainy and cloudy})}{P(\text{cloudy})}$

$$= \dfrac{0.21}{0.37} \approx 0.568$$

15. $P(\text{unemployed} \mid \text{high school dropout})$

$$= \frac{P(\text{umployed and high school dropout})}{P(\text{high school dropout})}$$

$$= \frac{0.021}{0.080} \approx 0.263$$

17. (a) $P(\text{age } 35\text{–}44 \mid \text{more likely})$

$$= \frac{N(\text{age } 35\text{–}44 \text{ and more likely})}{N(\text{more likely})}$$

$$= \frac{329}{1329} \approx 0.248$$

(b) $P(\text{more likely} \mid \text{age } 35\text{–}44)$

$$= \frac{N(\text{more likely and age } 35\text{–}44)}{N(\text{age } 35\text{–}44)}$$

$$= \frac{329}{536} \approx 0.614$$

(c) For 18–34 year olds, the probability that they are more likely to buy a product that is 'Made in America' is:

$$P\left(\text{More likely} \mid 18\text{–}34 \text{ years old}\right)$$

$$= \frac{238}{542} \approx 0.439$$

For individuals in general, the probability is:

$$P\left(\text{More likely}\right) = \frac{1329}{2160} \approx 0.615$$

18–34 year olds are less likely to buy a product that is labeled 'Made in America' than individuals in general.

19. (a) $P\left(\text{female} \mid \text{Sunday}\right)$

$$= \frac{N\left(\text{female and Sunday}\right)}{N\left(\text{Sunday}\right)}$$

$$= \frac{2,287}{6,430} \approx 0.356$$

(b) $P\left(\text{Sunday} \mid \text{female}\right)$

$$= \frac{N\left(\text{Sunday and female}\right)}{N\left(\text{female}\right)}$$

$$= \frac{2,287}{13,989} \approx 0.163$$

(c) $P\left(\text{male} \mid \text{Sunday}\right)$

$$= \frac{N\left(\text{male and Sunday}\right)}{N\left(\text{Sunday}\right)}$$

$$= \frac{4,143}{6,430} \approx 0.644$$

$P\left(\text{male} \mid \text{Monday}\right)$

$$= \frac{N\left(\text{male and Monday}\right)}{N\left(\text{Monday}\right)}$$

$$= \frac{3,178}{4,883} \approx 0.651$$

$P\left(\text{male} \mid \text{Tuesday}\right)$

$$= \frac{N\left(\text{male and Tuesday}\right)}{N\left(\text{Tuesday}\right)}$$

$$= \frac{3,280}{5,109} \approx 0.654$$

$P\left(\text{male} \mid \text{Wednesday}\right)$

$$= \frac{N\left(\text{male and Wednesday}\right)}{N\left(\text{Wednesday}\right)}$$

$$= \frac{3,197}{4,926} \approx 0.649$$

$P\left(\text{male} \mid \text{Thursday}\right)$

$$= \frac{N\left(\text{male and Thursday}\right)}{N\left(\text{Thursday}\right)}$$

$$= \frac{3,389}{5,228} \approx 0.648$$

$P\left(\text{male} \mid \text{Friday}\right)$

$$= \frac{N\left(\text{male and Friday}\right)}{N\left(\text{Friday}\right)}$$

$$= \frac{3,975}{6,154} \approx 0.646$$

$P\left(\text{male} \mid \text{Saturday}\right)$

$$= \frac{N\left(\text{male and Saturday}\right)}{N\left(\text{Saturday}\right)}$$

$$= \frac{4,749}{7,260} \approx 0.654$$

Each of the probabilities is about the same, so there does not appear to be any day where male fatalities are more likely than other days.

21. P(both televisions work)

$\quad = P(\text{1st works}) \cdot P(\text{2nd works} \mid \text{1st works})$

$\quad = \dfrac{4}{6} \cdot \dfrac{3}{5} = 0.4$

P(at least one television does not work)

$\quad = 1 - P(\text{both televisions work})$

$\quad = 1 - 0.4 = 0.6$

23. (a) $P\big(\text{both kings}\big)$

$\quad = P\big(\text{first king}\big) \cdot P\big(\text{2nd king} \mid \text{first king}\big)$

$\quad = \dfrac{4}{52} \cdot \dfrac{3}{51} = \dfrac{1}{221} \approx 0.005$

(b) $P\big(\text{both kings}\big)$

$\quad = P\big(\text{first king}\big) \cdot P\big(\text{2nd king} \mid \text{first king}\big)$

$\quad = \dfrac{4}{52} \cdot \dfrac{4}{52} = \dfrac{1}{169} \approx 0.006$

25. P(Dave 1st and Neta 2nd)

$\quad = P(\text{Dave 1st}) \cdot P(\text{Neta 2nd} \mid \text{Dave 1st})$

$\quad = \dfrac{1}{5} \cdot \dfrac{1}{4} = \dfrac{1}{20} = 0.05$

27. (a) P(like both songs)

$\quad = P(\text{like 1st}) \cdot P(\text{like 2nd} \mid \text{like 1st})$

$\quad = \dfrac{5}{13} \cdot \dfrac{4}{12} = \dfrac{5}{39} \approx 0.128$

The probability is greater than 0.05. This is not a small enough probability to be considered unusual.

(b) P(dislike both songs)

$\quad = P(\text{dislike 1st}) \cdot P(\text{dislike 2nd} \mid \text{dislike 1st})$

$\quad = \dfrac{8}{13} \cdot \dfrac{7}{12} = \dfrac{14}{39} \approx 0.359$

(c) Since you either like both or neither or exactly one (and these are disjoint) then the probability that you like exactly one is given by

$\quad P\big(\text{like exactly one song}\big)$

$\quad = 1 - \big(P\big(\text{like both}\big) + P\big(\text{dislike both}\big)\big)$

$\quad = 1 - \left(\dfrac{5}{39} + \dfrac{14}{39}\right) = \dfrac{20}{39} \approx 0.513$

(d) P(like both songs)

$\quad = P(\text{like 1st}) \cdot P(\text{like 2nd} \mid \text{like 1st})$

$\quad = \dfrac{5}{13} \cdot \dfrac{5}{13} = \dfrac{25}{169} \approx 0.148$

The probability is greater than 0.05. This is not a small enough probability to be considered unusual.

$\quad P$(dislike both songs)

$\quad = P(\text{dislike 1st}) \cdot P(\text{dislike 2nd} \mid \text{dislike 1st})$

$\quad = \dfrac{8}{13} \cdot \dfrac{8}{13} = \dfrac{64}{169} \approx 0.379$

$\quad P\big(\text{like exactly one song}\big)$

$\quad = 1 - \big(P\big(\text{like both}\big) + P\big(\text{dislike both}\big)\big)$

$\quad = 1 - \left(\dfrac{25}{169} + \dfrac{64}{169}\right) = \dfrac{80}{169} \approx 0.473$

29.

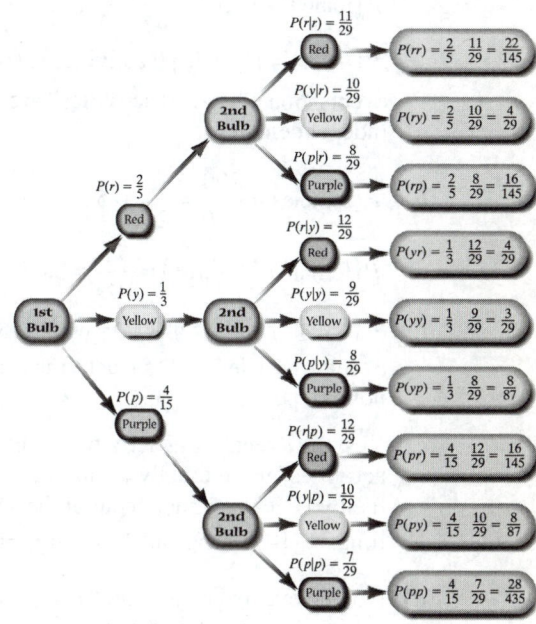

(a) $P\big(\text{both red}\big) = \dfrac{22}{145} \approx 0.152$

(b) $P\big(\text{1st red and 2nd yellow}\big) = \dfrac{4}{29} \approx 0.138$

(c) $P\big(\text{1st yellow and 2nd red}\big) = \dfrac{4}{29} \approx 0.138$

(d) Since one each of red and yellow must be either 1st red, 2nd yellow or vice versa, by the addition rule this probability is

$\quad P\big(\text{one red and one yellow}\big) = \dfrac{4}{29} + \dfrac{4}{29}$

$\quad\quad\quad\quad\quad\quad\quad\quad\quad\quad = \dfrac{8}{29} \approx 0.276.$

31. $P(\text{female and smoker})$

$= P(\text{female} \mid \text{smoker}) \cdot P(\text{smoker})$

$= (0.445)(0.203) \approx 0.090$

The probability is greater than 0.05. It is not unusual to select a female who smokes.

33. (a) $P(10 \text{ different birthdays})$

$= \dfrac{365}{365} \cdot \dfrac{364}{365} \cdot \dfrac{363}{365} \cdots \dfrac{358}{365} \cdot \dfrac{357}{365} \cdot \dfrac{356}{365}$

≈ 0.883

(b) $P(\text{at least } 2)$

$= 1 - P(\text{none})$

$\approx 1 - 0.883$

$= 0.117$

35. (a) $P(\text{male}) = \dfrac{200}{400} = \dfrac{1}{2}$

$P(\text{male} \mid 0 \text{ activ.}) = \dfrac{21}{42} = \dfrac{1}{2}$

$P(\text{male}) = P(\text{male} \mid 0 \text{ activities})$, so the events "male" and "0 activities" are independent.

(b) $P(\text{female}) = \dfrac{200}{400} = \dfrac{1}{2} = 0.5$

$P(\text{female} \mid 5+ \text{ activ.}) = \dfrac{71}{109} \approx 0.651$

$P(\text{female}) \neq P(\text{female} \mid 5+ \text{ activ.})$, so the events "female" and "5+ activities" are not independent.

(c) Yes, the events "1–2 activities" and "3–4 activities" are mutually exclusive because the two events cannot occur at the same time. $P(1\text{–}2 \text{ activ. and } 3\text{–}4 \text{ activ.}) = 0$

(d) No, the events "male" and "1–2 activities" are not mutually exclusive because the two events can happen at the same time.

$P(\text{male and } 1\text{–}2 \text{ activ.}) = \dfrac{81}{400}$

$= 0.2025 \neq 0$

37. (a) $P(\text{being dealt 5 clubs})$

$= \dfrac{13}{52} \cdot \dfrac{12}{51} \cdot \dfrac{11}{50} \cdot \dfrac{10}{49} \cdot \dfrac{9}{48} = \dfrac{33}{66,640}$

≈ 0.000495

(b) $P(\text{bing dealt a flush})$

$= 4\left(\dfrac{33}{66,640}\right) = \dfrac{33}{16,660} \approx 0.002$

39. (a) $P(\text{selecting two defective chips})$

$= \dfrac{50}{10,000} \cdot \dfrac{49}{9,999} \approx 0.0000245$

(b) Assuming independence,

$P(\text{selecting two defective chips})$

$\approx (0.005)^2$

$= 0.000025$

The difference in the results of parts (a) and (b) is only 0.0000005, so the assumption of independence did not significantly affect the probability.

41. $P(45\text{–}54 \text{ yrs old}) = \dfrac{546}{2,160} \approx 0.253$

$P(45\text{–}54 \text{ years old} \mid \text{more likely}) = \dfrac{360}{1,329}$

≈ 0.271

No, the events "45–54 years old" and "more likely" are not independent since the preceding probabilities are not equal.

43. (a)-(d) Answers will vary.

For part (d), of the 18 boxes in the outermost ring, 12 indicate you win if you switch while 6 indicate you lose if you switch. Assuming random selection so each box is equally likely, we get $P(\text{win if switch}) = \dfrac{12}{18} = \dfrac{2}{3} \approx 0.667$.

Section 5.5

1. permutation

3. True

5. $5! = 5 \cdot 4 \cdot 3 \cdot 2 \cdot 1 = 120$

7. $10! = 10 \cdot 9 \cdot 8 \cdot 7 \cdot 6 \cdot 5 \cdot 4 \cdot 3 \cdot 2 \cdot 1 = 3,628,800$

9. $0! = 1$

11. $_6P_2 = \dfrac{6!}{(6-2)!} = \dfrac{6!}{4!} = 6 \cdot 5 = 30$

13. $_4P_4 = \dfrac{4!}{(4-4)!} = \dfrac{4!}{0!} = \dfrac{24}{1} = 24$

15. $_5P_0 = \dfrac{5!}{(5-0)!} = \dfrac{5!}{5!} = 1$

17. $_8P_3 = \dfrac{8!}{(8-3)!} = \dfrac{8!}{5!} = 8 \cdot 7 \cdot 6 = 336$

19. $_8C_3 = \dfrac{8!}{3!(8-3)!} = \dfrac{8!}{3!5!} = \dfrac{8 \cdot 7 \cdot 6}{3 \cdot 2 \cdot 1} = 56$

21. $_{10}C_2 = \dfrac{10!}{2!(10-2)!} = \dfrac{10!}{2!8!} = \dfrac{10 \cdot 9}{2 \cdot 1} = 45$

23. $_{52}C_1 = \dfrac{52!}{1!(52-1)!} = \dfrac{52!}{1!51!} = \dfrac{52}{1} = 52$

25. $_{48}C_3 = \dfrac{48!}{3!(48-3)!} = \dfrac{48!}{3!45!} = \dfrac{48 \cdot 47 \cdot 46}{3 \cdot 2 \cdot 1}$
$= 17,296$

27. *ab*, *ac*, *ad*, *ae*, *ba*, *bc*, *bd*, *be*, *ca*, *cb*, *cd*, *ce*, *da*, *db*, *dc*, *de*, *ea*, *eb*, *ec*, *ed*
Since there are 20 permutations, $_5P_2 = 20$.

29. *ab*, *ac*, *ad*, *ae*, *bc*, *bd*, *be*, *cd*, *ce*, *de*
Since there are 10 combinations, $_5C_2 = 10$.

31. Here we use the Multiplication Rule of Counting. There are six shirts and four ties, so there are $6 \cdot 4 = 24$ different shirt-and-tie combinations the man can wear.

33. There are 12 ways Dan can select the first song, 11 ways to select the second song, etc. From the Multiplication Rule of Counting, there are $12 \cdot 11 \cdot \ldots \cdot 2 \cdot 1 = 12! = 479,001,600$ ways that Dan can arrange the 12 songs.

35. There are 8 ways to pick the first city, 7 ways to pick the second, etc. From the Multiplication Rule of Counting, there are $8 \cdot 7 \cdot \ldots \cdot 2 \cdot 1 = 8! = 40,320$ different routes possible for the salesperson.

37. Since the company name can be represented by 1, 2, or 3 letters, we find the total number of 1 letter abbreviations, 2 letter abbreviations, and 3 letter abbreviations, then we sum the results to obtain the total number of abbreviations possible. There are 26 letters that can be used for abbreviations, so there are 26 one-letter abbreviations. Since repetitions are allowed, there are $26 \cdot 26 = 26^2$ different two-letter abbreviations, and $26 \cdot 26 \cdot 26 = 26^3$ different three-letter abbreviations. Therefore, the maximum number of companies that can be listed on the New York Stock Exchange is 26 (one letter) $+ 26^2$ (two letters)
$+ 26^3$ (three letters) $= 18,278$ companies.

39. (a) $10 \cdot 10 \cdot 10 \cdot 10 = 10^4 = 10,000$ different codes are possible.

(b) $P(\text{guessing the correct code}) = \dfrac{1}{10,000}$
$= 0.0001$

41. Since lower and uppercase letters are considered the same, each of the 8 letters to be selected has 26 possibilities. Therefore, there are $26^8 = 208,827,064,576$ different usernames possible for the local area network.

43. (a) $50 \cdot 50 \cdot 50 = 50^3 = 125,000$ different combinations are possible.

(b) $P(\text{guessing combination}) = \dfrac{1}{50^3} = \dfrac{1}{125,000}$
$= 0.000008$

45. Since order matters, we use the permutation formula $_nP_r$.
$_{40}P_3 = \dfrac{40!}{(40-3)!} = \dfrac{40!}{37!} = 40 \cdot 39 \cdot 38 = 59,280$
There are 59,280 ways in which the top three cars can result.

47. Since the order of selection determines the office of the member, order matters. Therefore, we use the permutation formula $_nP_r$.
$_{20}P_4 = \dfrac{20!}{(20-4)!} = \dfrac{20!}{16!} = 20 \cdot 19 \cdot 18 \cdot 17$
$= 116,280$
There are 116,280 different leadership structures possible.

49. Since the problem states the numbers must be matched in order, this is a permutation problem.
$_{25}P_4 = \dfrac{25!}{(25-4)!} = \dfrac{25!}{21!} = 25 \cdot 24 \cdot 23 \cdot 22$
$= 303,600$
There are 303,600 different outcomes possible for this game.

51. Since order of selection does not matter, this is a combination problem.
$_{50}C_5 = \dfrac{50 \cdot 49 \cdot 48 \cdot 47 \cdot 46}{5 \cdot 4 \cdot 3 \cdot 2 \cdot 1} = 2,118,760$
There are 2,118,760 different simple random samples of size 5 possible.

53. There are 6 children and we need to determine the number of ways two can be boys. The order of the two boys does not matter, so this is a combination problem. There are

$$_6C_2 = \frac{6!}{2!(6-2)!} = \frac{6!}{2!\,4!} = \frac{6\cdot5}{2\cdot1} = 15 \text{ ways to}$$

select 2 of the 6 children to be boys (the rest are girls), so there are 15 different birth and gender orders possible.

55. Since there are three A's, two C's, two G's, and three T's from which to form a DNA sequence, we can make $\dfrac{10!}{3!2!2!3!} = 25{,}200$ distinguishable DNA sequences.

57. Arranging the trees involves permutations with repetitions. Using Formula (3), we find that there are $\dfrac{11!}{4!5!2!} = 6930$ different ways the landscaper can arrange the trees.

59. Since the order of the balls does not matter, this is a combination problem. Using Formula (2), we find there are $_{39}C_5 = 575{,}757$ possible choices (without regard to order), so

$$P(\text{winning}) = \frac{1}{575{,}757} \approx 0.00000174 \,.$$

61. **(a)** $P(\text{all students}) = \dfrac{_8C_5}{_{18}C_5}$

$$= \frac{8\cdot7\cdot6}{3\cdot2\cdot1} \cdot \frac{5\cdot4\cdot3\cdot2\cdot1}{18\cdot17\cdot16\cdot15\cdot14}$$

$$= \frac{1}{153} \approx 0.0065$$

(b) $P(\text{all faculty}) = \dfrac{_{10}C_5}{_{18}C_5}$

$$= \frac{10\cdot9\cdot8\cdot7\cdot6}{5\cdot4\cdot3\cdot2\cdot1} \cdot \frac{5\cdot4\cdot3\cdot2\cdot1}{18\cdot17\cdot16\cdot15\cdot14}$$

$$= \frac{1}{34} \approx 0.0294$$

(c) $P(\text{2 students and 3 faculty}) = \dfrac{_8C_2 \cdot {}_{10}C_3}{_{18}C_5}$

$$= \frac{8\cdot7}{2\cdot1} \cdot \frac{10\cdot9\cdot8}{3\cdot2\cdot1} \cdot \frac{5\cdot4\cdot3\cdot2\cdot1}{18\cdot17\cdot16\cdot15\cdot14}$$

$$= \frac{20}{51} \approx 0.3922$$

63. $P(\text{one or more defective})$

$= 1 - P(\text{none defective})$

$$= 1 - \frac{_{116}C_4}{_{120}C_4}$$

$$= 1 - \frac{116\cdot115\cdot114\cdot113}{4\cdot3\cdot2\cdot1} \cdot \frac{4\cdot3\cdot2\cdot1}{120\cdot119\cdot118\cdot117}$$

≈ 0.1283

There is a probability of 0.1283 that the shipment is rejected.

65. **(a)** $P(\text{you like 2 of the 4 songs}) = \dfrac{_5C_2 \cdot {}_8C_2}{_{13}C_4}$

$$\approx 0.3916$$

There is a probability of 0.3916 that you will like 2 of the 4 songs played.

(b) $P(\text{you like 3 of the 4 songs}) = \dfrac{_5C_3 \cdot {}_8C_1}{_{13}C_4}$

$$\approx 0.1119$$

There is a probability of 0.1119 that you will like 3 of the 4 songs played.

(c) $P(\text{you like all 4 songs}) = \dfrac{_5C_4 \cdot {}_8C_0}{_{13}C_4}$

$$\approx 0.0070$$

There is a probability of 0.007 that you will like all 4 songs played.

67. **(a)** Five cards can be selected from a deck in $_{52}C_5 = 2{,}598{,}960$ ways.

(b) There are $_4C_3 = 4$ ways of choosing 3 two's, and so on for each denomination. Hence, there are $13\cdot4 = 52$ ways of choosing three of a kind.

(c) There are $_{12}C_2 = 66$ choices of two additional denominations (different from that of the three of a kind) and 4 choices of suit for the first remaining card and then, for each choice of suit for the first remaining card, there are 4 choices of suit for the last card. This gives a total of $66\cdot4\cdot4 = 1056$ ways of choosing the last two cards.

(d) $P(\text{three of a kind}) = \dfrac{52\cdot1056}{2{,}598{,}960} \approx 0.0211$

69. $P(\text{all 4 modems work}) = \dfrac{17}{20}\cdot\dfrac{16}{19}\cdot\dfrac{15}{18}\cdot\dfrac{14}{17}$

$$\approx 0.4912$$

There is a probability of 0.4912 that the shipment will be accepted.

71. (a) Using the Multiplication Rule of Counting and the given password format, there are
$21 \cdot 5 \cdot 21 \cdot 21 \cdot 5 \cdot 21 \cdot 10 \cdot 10 = 486,202,500$
different passwords that are possible.

(b) If letters are case sensitive, there are
$42 \cdot 10 \cdot 42 \cdot 42 \cdot 10 \cdot 42 \cdot 10 \cdot 10 = 420^4 =$
$31,116,960,000$ different passwords that are possible.

Section 5.6

1. In a permutation, the order in which the objects are chosen matters; in a combination, the order in which the objects are chosen is unimportant.

3. 'AND' is generally associated with multiplication, while 'OR' is generally associated with addition.

5. $P(E) = \dfrac{4}{10} = \dfrac{2}{5} = 0.4$

7. *abc, acb, abd, adb, abe, aeb, acd, adc, ace, aec, ade, aed, bac, bca, bad, bda, bae, bea, bcd, bdc, bce, bec, bde, bed, cab, cba, cad, cda, cae, cea, cbd, cdb, cbe, ceb, cde, ced, dab, dba, dac, dca, dae, dea, dbc, dcb, dbe, deb, dce, dec, eab, eba, eac, eca, ead, eda, ebc, ecb, ebd, edb, ecd, edc*

9. $P(E \text{ or } F) = P(E) + P(F) - P(E \text{ and } F)$
$= 0.7 + 0.2 - 0.15$
$= 0.75$

11. ${}_7P_3 = \dfrac{7!}{(7-3)!} = \dfrac{7!}{4!} = 7 \cdot 6 \cdot 5 = 210$

13. $P(E \text{ and } F) = P(E) \cdot P(F)$
$= (0.8)(0.5)$
$= 0.4$

15. $P(E \text{ and } F) = P(E) \cdot P(F \mid E)$
$= (0.9)(0.3)$
$= 0.27$

17. $P(\text{soccer}) \approx \dfrac{22}{500} = \dfrac{11}{250} = 0.044$

19. Since the actual order of the men in their three seats matters, there are ${}_3P_3 = 6$ ways to arrange the men. Similarly, there are ${}_3P_3 = 6$ ways to arrange the women among their seats.

Therefore, there are $6 \cdot 6 = 36$ ways to arrange the men and women.

Alternatively, using the Multiplication Rule of Counting, and starting with a female, we get:
$\underset{F}{\dfrac{3}{}} \cdot \underset{M}{\dfrac{3}{}} \cdot \underset{F}{\dfrac{2}{}} \cdot \underset{M}{\dfrac{2}{}} \cdot \underset{F}{\dfrac{1}{}} \cdot \underset{M}{\dfrac{1}{}} = 3 \cdot 3 \cdot 2 \cdot 2 \cdot 1 \cdot 1 = 36$
Again, there are 36 ways to arrange the men and women.

21. (a) $P(\text{survived}) = \dfrac{711}{2224} \approx 0.320$

(b) $P(\text{female}) = \dfrac{425}{2224} \approx 0.191$

(c) $P(\text{female or child}) = \dfrac{425 + 109}{2224} = \dfrac{534}{2224}$
$= \dfrac{267}{1112} \approx 0.240$

(d) $P(\text{female and survived}) = \dfrac{316}{2224}$
$= \dfrac{79}{556} \approx 0.142$

(e) $P(\text{female or survived})$
$= P(\text{female}) + P(\text{survived})$
$\quad - P(\text{female and survived})$
$= \dfrac{425}{2224} + \dfrac{711}{2224} - \dfrac{316}{2224}$
$= \dfrac{820}{2224} = \dfrac{205}{556} \approx 0.369$

(f) $P(\text{survived} \mid \text{female}) = \dfrac{316}{425} \approx 0.744$

(g) $P(\text{survived} \mid \text{child}) = \dfrac{57}{109} \approx 0.523$

(h) $P(\text{survived} \mid \text{male}) = \dfrac{338}{1690} = \dfrac{1}{5} = 0.2$

(i) Yes; the survival rate was much higher for women and children than for men.

(j) $P(\text{both females survived})$
$= P(\text{1st surv.}) \cdot P(\text{2nd surv.} \mid \text{1st surv.})$
$= \dfrac{316}{425} \cdot \dfrac{315}{424} \approx 0.552$

23. Since the order in which the colleges are selected does not matter, there are ${}_{12}C_3 = 220$ different ways to select 3 colleges from the 12 he is interested in.

25. **(a)** Since the games are independent of each other, we get

$$P(\text{win both}) = \frac{1}{5,200,000} \cdot \frac{1}{705,600}$$
$$\approx 0.00000000000027$$

(b) Since the games are independent of each other, we get

$$P\left(\begin{array}{c}\text{win Jubilee}\\\text{twice}\end{array}\right) = \left(\frac{1}{705,600}\right)^2$$
$$\approx 0.000000000002.$$

27. **(a)** The events "bachelor's degree" and "never married" are not independent because

$$P(\text{Bachelor's} \mid \text{never married}) = 0.228$$

and P(bachelor's) = 0.202 are not equal.

(b) Using the multiplication Rule,
P(Bachelor's and never married)

$$= P(\text{bachelor's}) \cdot P(\text{never married}|\text{bachelor's})$$

$$= 0.202(0.186)$$

$$\approx 0.038$$

Approximately 3.8% of American women aged 25 years or older have a bachelor's degree and have never been married.

29. The order in which the questions are answered does not matter so there are $_{12}C_8 = 495$ different sets of questions that could be answered.

31. Using the Multiplication Rule of Counting, there are $2 \cdot 2 \cdot 3 \cdot 8 \cdot 2 = 192$ different Hyundai Genesis cars that are possible.

Section 5.7 (online)

1.

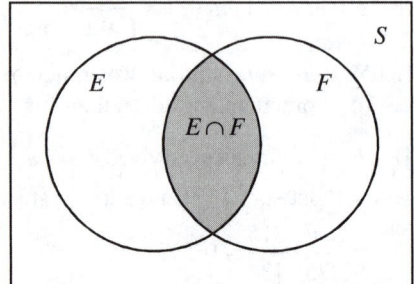

3. To partition a sample space S into three subsets A_1, A_2, and A_3 means that S is separated into three nonempty subsets where the subsets are all pairwise-disjoint. That is, $A_1 \cup A_2 \cup A_3 = S$.

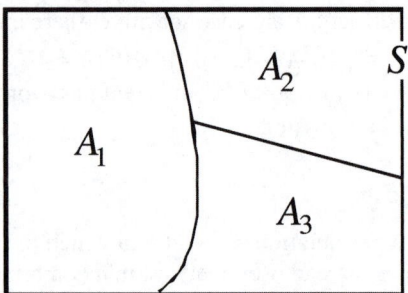

5. Answers will vary. The Rule of Total Probability expresses a probability as a weighted average of conditional probabilities.

7. From the figure, $P(E \mid A) = 0.6$.

9. From the figure, $P(E^c \mid A) = 0.4$.

11. From the figure, $P(E \mid C) = 0.4$.

13. $P(E) = P(A) \cdot P(E \mid A)$
$$+ P(B) \cdot P(E \mid B)$$
$$+ P(B) \cdot P(E \mid B)$$
$$= 0.2(0.6) + 0.55(0.7) + 0.25(0.4)$$
$$= 0.12 + 0.385 + 0.1$$
$$= 0.605$$

15. $P(A \mid E) = \dfrac{P(A) \cdot P(E \mid A)}{P(E)}$

$$= \frac{0.2(0.6)}{0.2(0.6) + 0.55(0.7) + 0.25(0.4)}$$

$$= \frac{0.12}{0.605} \approx 0.198$$

17. $P(C \mid E) = \dfrac{P(C) \cdot P(E \mid C)}{P(E)}$

$$= \frac{0.25(0.4)}{0.2(0.6) + 0.55(0.7) + 0.25(0.4)}$$

$$= \frac{0.1}{0.605} \approx 0.165$$

19.
$$P(B\,|\,E) = \frac{P(B) \cdot P(E\,|\,B)}{P(E)}$$

$$= \frac{0.55(0.7)}{0.2(0.6) + 0.55(0.7) + 0.25(0.4)}$$

$$= \frac{0.385}{0.605} \approx 0.636$$

21.
$$P(E) = P(A_1) \cdot P(E\,|\,A_1) + P(A_2) \cdot P(E\,|\,A_2)$$

$$= 0.55(0.06) + 0.45(0.08)$$

$$= 0.069$$

23.
$$P(E) = P(A_1) \cdot P(E\,|\,A_1) + P(A_2) \cdot P(E\,|\,A_2)$$
$$+ P(A_3) \cdot P(E\,|\,A_3)$$

$$= 0.35(0.25) + 0.45(0.18) + 0.2(0.14)$$

$$= 0.1965$$

25. **(a)**
$$P(A_1\,|\,E) = \frac{P(A_1) \cdot P(E\,|\,A_1)}{P(E)}$$

$$= \frac{0.55(0.06)}{0.069} \approx 0.4783$$

(b)
$$P(A_2\,|\,E) = \frac{P(A_2) \cdot P(E\,|\,A_2)}{P(E)}$$

$$= \frac{0.45(0.08)}{0.069} \approx 0.5217$$

27. **(a)**
$$P(A_1\,|\,E) = \frac{P(A_1) \cdot P(E\,|\,A_1)}{P(E)}$$

$$= \frac{0.35(0.25)}{0.1965} \approx 0.4453$$

(b)
$$P(A_2\,|\,E) = \frac{P(A_2) \cdot P(E\,|\,A_2)}{P(E)}$$

$$= \frac{0.45(0.18)}{0.1965} \approx 0.4122$$

(c)
$$P(A_3\,|\,E) = \frac{P(A_3) \cdot P(E\,|\,A_3)}{P(E)}$$

$$= \frac{0.2(0.14)}{0.1965} \approx 0.1425$$

29. **(a)**

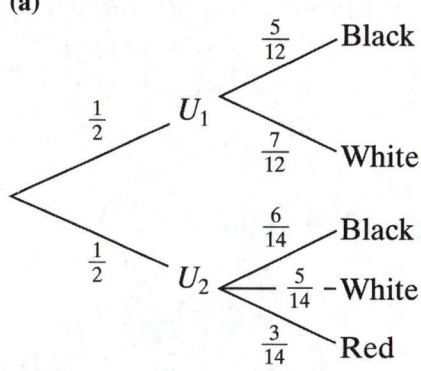

$$P(B) = B \cap U_1 + B \cap U_2$$

$$= \frac{1}{2}\left(\frac{5}{12}\right) + \frac{1}{2}\left(\frac{6}{14}\right) = \frac{71}{168} \approx 0.4226$$

(b)
$$P(B) = P(U_1) \cdot P(B\,|\,U_1)$$
$$+ P(U_2) \cdot P(B\,|\,U_2)$$

$$= \frac{1}{2}\left(\frac{5}{12}\right) + \frac{1}{2}\left(\frac{6}{14}\right)$$

$$= \frac{71}{168} \approx 0.4226$$

(c)
$$P(U_1\,|\,B) = \frac{P(U_1) \cdot P(B\,|\,U_1)}{P(B)}$$

$$= \frac{(1/2)(5/12)}{71/168} = \frac{35}{71} \approx 0.4930$$

(d) $P(U_1\,|\,R) = 0$ since there are no red balls in urn I.

$P(U_2\,|\,R) = 1$ since the only red balls are in urn II.

31. **(a)**

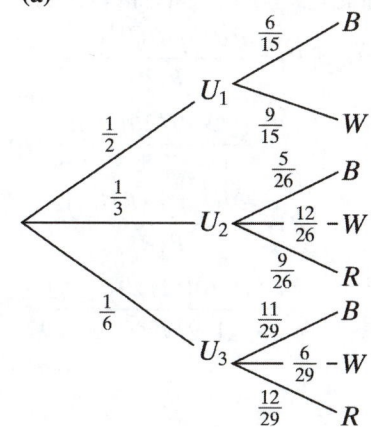

$$P(B) = B \cap U_1 + B \cap U_2 + B \cap U_3$$
$$= \frac{3}{6}\left(\frac{6}{15}\right) + \frac{2}{6}\left(\frac{5}{26}\right) + \frac{1}{6}\left(\frac{11}{29}\right)$$
$$= \frac{617}{1885} \approx 0.3273$$

(b) $P(B) = P(U_1) \cdot P(B \mid U_1)$
$$+ P(U_2) \cdot P(B \mid U_2)$$
$$+ P(U_3) \cdot P(B \mid U_3)$$
$$= \frac{3}{6}\left(\frac{6}{15}\right) + \frac{2}{6}\left(\frac{5}{26}\right) + \frac{1}{6}\left(\frac{11}{29}\right)$$
$$= \frac{617}{1885} \approx 0.3273$$

(c) $P(U_1 \mid B) = \dfrac{P(U_1) \cdot P(B \mid U_1)}{P(B)}$
$$= \frac{(1/2)(6/15)}{617/1885} = \frac{377}{617} \approx 0.6110$$

(d) $P(R) = P(U_1) \cdot P(R \mid U_1)$
$$+ P(U_2) \cdot P(R \mid U_2)$$
$$+ P(U_3) \cdot P(R \mid U_3)$$
$$= \frac{3}{6}(0) + \frac{2}{6}\left(\frac{9}{26}\right) + \frac{1}{6}\left(\frac{12}{29}\right)$$
$$= \frac{139}{754} \approx 0.1844;$$
$$P(U_1 \mid R) = \frac{P(U_1) \cdot P(R \mid U_1)}{P(R)}$$
$$= \frac{(1/2)(0)}{139/754} = 0$$
(note that no red balls are in urn I);
$$P(U_2 \mid R) = \frac{P(U_2) \cdot P(R \mid U_2)}{P(R)}$$
$$= \frac{(1/3)(9/26)}{139/754} = \frac{87}{139} \approx 0.6259$$
$$P(U_3 \mid R) = \frac{P(U_3) \cdot P(R \mid U_3)}{P(R)}$$
$$= \frac{(1/6)(12/29)}{139/754} = \frac{52}{139} \approx 0.3741$$

33. (a) $P(\text{colorblind})$
$$= P(\text{male}) \cdot P(\text{colorblind} \mid \text{male})$$
$$+ P(\text{female}) \cdot P(\text{colorblind} \mid \text{female})$$
$$= 0.491(0.08) + 0.509(0.0064)$$
$$= 0.0425376 \approx 0.0425$$

(b) $P(\text{female} \mid \text{colorblind})$
$$= \frac{P(\text{female}) \cdot P(\text{colorblind} \mid \text{female})}{P(\text{colorblind})}$$
$$= \frac{0.509(0.0064)}{0.0425376} \approx 0.0766$$

35. (a) $P(M) = P(A_1) \cdot P(M \mid A_1)$
$$+ P(A_2) \cdot P(M \mid A_2)$$
$$+ P(A_3) \cdot P(M \mid A_3)$$
$$+ P(A_4) \cdot P(M \mid A_4)$$
$$+ P(A_5) \cdot P(M \mid A_5)$$
$$+ P(A_6) \cdot P(M \mid A_6)$$
$$= 0.118(0.501) + 0.298(0.486)$$
$$+ 0.168(0.474) + 0.099(0.431)$$
$$+ 0.201(0.478) + 0.116(0.496)$$
$$= 0.479861 \approx 0.4799$$

(b) $P(A_6 \mid M) = \dfrac{P(A_6) \cdot P(M \mid A_6)}{P(M)}$
$$= \frac{0.116(0.496)}{0.479861} \approx 0.1199$$

(c) $P(A_1 \mid M) = \dfrac{P(A_1) \cdot P(M \mid A_1)}{P(M)}$
$$= \frac{0.118(0.501)}{0.479861} \approx 0.1232$$

37. (a)

$$P(D) = P(A_1) \cdot P(D \mid A_1)$$
$$+ P(A_2) \cdot P(D \mid A_2)$$
$$+ P(A_3) \cdot P(D \mid A_3)$$
$$= 0.062(0.64) + 0.572(0.51) + 0.366(0.47)$$
$$= 0.50342 \approx 0.503$$

(b) $P(A_3 \mid D) = \dfrac{P(A_3) \cdot P(D \mid A_3)}{P(D)}$

$= \dfrac{0.366(0.47)}{0.50342} \approx 0.342$

(c) $P(A_1 \mid D) = \dfrac{P(A_1) \cdot P(D \mid A_1)}{P(D)}$

$= \dfrac{0.062(0.64)}{0.50342} \approx 0.079$

39. $P(\text{found } G) = P(G) \cdot P(\text{found } G \mid G)$

$\qquad\qquad\quad + P(I) \cdot P(\text{found } G \mid I)$

$\qquad = 0.95(0.90) + 0.05(0.01)$

$\qquad = 0.8555$

$P(I \mid \text{found } G) = \dfrac{P(I) \cdot P(\text{found } G \mid I)}{P(\text{found } G)}$

$\qquad\qquad = \dfrac{0.05(0.01)}{0.8555} \approx 0.00058$

Chapter 5 Review Exercises

1. **(a)** Probabilities must be between 0 and 1, inclusive, so the possible probabilities are: 0, 0.75, 0.41.

(b) Probabilities must be between 0 and 1, inclusive, so the possible probabilities are: $\dfrac{2}{5}, \dfrac{1}{3}, \dfrac{6}{7}$.

2. Event E contains 1 of the 5 equally likely outcomes, so $P(E) = \dfrac{1}{5} = 0.2$.

3. Event F contains 2 of the 5 equally likely outcomes, so $P(F) = \dfrac{2}{5} = 0.4$.

4. Event E contains 3 of the 5 equally likely outcomes, so $P(E) = \dfrac{3}{5} = 0.6$.

5. Since $P(E) = \dfrac{1}{5} = 0.2$, we have

$P(E^c) = 1 - \dfrac{1}{5} = \dfrac{4}{5} = 0.8$.

6. $P(E \text{ or } F) = P(E) + P(F) - P(E \text{ and } F)$

$\qquad = 0.76 + 0.45 - 0.32 = 0.89$

7. Since events E and F are mutually exclusive,

$P(E \text{ or } F) = P(E) + P(F)$

$\qquad = 0.36 + 0.12 = 0.48$.

8. Since events E and F are independent,

$P(E \text{ and } F) = P(E) \cdot P(F) = 0.45 \cdot 0.2 = 0.09$.

9. No, events E and F are not independent because

$P(E) \cdot P(F) = 0.8 \cdot 0.5 = 0.40 \neq P(E \text{ and } F)$.

10. $P(E \text{ and } F) = P(E) \cdot P(F \mid E)$

$\qquad = 0.59 \cdot 0.45 = 0.2655$

11. $P(E \mid F) = \dfrac{P(E \text{ and } F)}{P(F)} = \dfrac{0.35}{0.7} = 0.5$

12. **(a)** $7! = 7 \cdot 6 \cdot 5 \cdot 4 \cdot 3 \cdot 2 \cdot 1 = 5040$

(b) $0! = 1$

(c) $_9C_4 = \dfrac{9!}{4!(9-4)!} = \dfrac{9!}{4!5!} = \dfrac{9 \cdot 8 \cdot 7 \cdot 6}{4 \cdot 3 \cdot 2 \cdot 1} = 126$

(d) $_{10}C_3 = \dfrac{10!}{3!(10-3)!} = \dfrac{10!}{3!7!} = \dfrac{10 \cdot 9 \cdot 8}{3 \cdot 2 \cdot 1} = 120$.

(e) $_9P_2 = \dfrac{9!}{(9-2)!} = \dfrac{9!}{7!} = 9 \cdot 8 = 72$

(f) $_{12}P_4 = \dfrac{12!}{(12-4)!} = \dfrac{12!}{8!}$

$\qquad = 12 \cdot 11 \cdot 10 \cdot 9 = 11,880$.

13. **(a)** $P(\text{green}) = \dfrac{2}{38} = \dfrac{1}{19} \approx 0.0526$. If the wheel is spun 100 times, we would expect about 5 spins to end with the ball in a green slot.

(b) $P(\text{green or red}) = \dfrac{2+18}{38} = \dfrac{20}{38}$

$\qquad = \dfrac{10}{19} \approx 0.5263$

If the wheel is spun 100 times, we would expect about 53 spins to end with the ball in a green or a red slot.

(c) $P(00 \text{ or red}) = \dfrac{1+18}{38} = \dfrac{19}{38} = \dfrac{1}{2} = 0.5$

If the wheel is spun 100 times, we would expect about 50 spins to end with the ball in either a 00 or a red slot.

(d) Since 31 is an odd number and the odd slots are colored red,
$P(31 \text{ and black}) = 0$. This is called an impossible event.

14. (a) Of the 32,719 accidents, 10,076 were alcohol related, so
$$P(\text{alcohol related}) = \frac{10,076}{32,719} \approx 0.308 .$$

(b) Of the 32,719 accidents, 32,719 – 10,076 = 22,643 were not alcohol related, so
$$P(\text{not alcohol related}) = \frac{22,643}{32,719} \approx 0.692$$

(c) $P(\text{both alcohol related}) = \frac{10,076}{32,719} \cdot \frac{10,076}{32,719}$
$$\approx 0.095$$

(d) $P(\text{neither was alcohol related})$
$$= \frac{22,643}{32,719} \cdot \frac{22,643}{32,719} \approx 0.479$$

(e) $P(\text{at least one of the two was alcohol related})$
$$= 1 - P(\text{neither was alcohol related})$$
$$= 1 - \frac{22,643}{32,719} \cdot \frac{22,643}{32,719} \approx 0.521$$

15. (a) There are 126 + 262 + 263 + 388 = 1039 people who were surveyed. The individuals can be thought of as the trials of the probability experiment. The relative frequency of "Age 18–79" is
$$\frac{126}{1,039} \approx 0.121 .$$
We compute the relative frequencies of the other outcomes similarly and obtain the probability model:

Age	Probability
18–79	0.121
80–89	0.252
90–99	0.253
100+	0.373

(b) No, since the probability that an individual will want to live between 18 and 79 years is 0.121 (which is greater than 0.05), this is not unusual.

16. (a) Of the 3,948,828 births included in the table, 225,044 are postterm. Thus,
$$P(\text{postterm}) = \frac{225,044}{3,948,828} \approx 0.057.$$

(b) Of the 3,948,828 births included in the table, 2,596,802 weighed 3000 to 3999 grams. Thus, $P(\text{3000 to 3999 grams})$
$$= \frac{2,596,802}{3,948,828} \approx 0.658 .$$

(c) Of the 3,948,828 births included in the table, 160,354 both weighed 3000 to 3999 grams and were postterm. Thus, $P(\text{3000 to 3999 grams and postterm})$
$$= \frac{160,354}{3,948,828} \approx 0.041 .$$

(d) $P(\text{3000 to 3999 grams or postterm})$
$= P(\text{3000 to 3999 grams}) + P(\text{postterm})$
$- P(\text{3000 to 3999 grams and postterm})$
$$= \frac{2,596,802}{3,948,828} + \frac{225,044}{3,948,828} - \frac{160,354}{3,948,828}$$
$$= \frac{2,661,492}{3,948,828} \approx 0.674$$

(e) Of the 3,948,828 births included in the table, 26 both weighed less than 1000 grams and were postterm. Thus, $P(\text{less than 1000 grams and postterm})$
$$= \frac{26}{3,948,828} \approx 0.000007 .$$
This event is highly unlikely, but not impossible.

(f) $P(\text{3000 to 3999 grams} \mid \text{postterm})$
$$= \frac{N(\text{3000 to 3999 grams and postterm})}{N(\text{postterm})}$$
$$= \frac{160,354}{225,044} \approx 0.713$$

(g) No, the events "postterm baby" and "weighs 3000 to 3999 grams" are not independent since
$P(\text{3000 to 3999 grams}) \cdot P(\text{postterm})$
$\approx 0.658(0.057) = 0.038$
$\neq P(\text{3000 to 3999 grams and postterm})$
$\approx 0.041 .$

17. **(a)** $P(\text{trust}) = 0.18$

(b) $P(\text{not trust}) = 1 - 0.18 = 0.82$

(c) $P(\text{all 3 trust}) = (0.18)^3 \approx 0.006$. This is surprising, since the probability is less than 0.05.

(d) $P(\text{at least one of the three does not trust})$
$= 1 - P(\text{all 3 trust})$
$= 1 - 0.006 = 0.994$

(e) $P(\text{none of the 5 trust}) = (0.82)^5 \approx 0.371$. It is not surprising.

(f) $P(\text{at least one of the five trust})$
$= 1 - P(\text{none of 5 trust})$
$= 1 - 0.371 = 0.629$

18. $P(\text{matching the winning PICK 3 numbers})$
$= \dfrac{1}{10} \cdot \dfrac{1}{10} \cdot \dfrac{1}{10} = \dfrac{1}{1000} = 0.001$

19. $P(\text{matching the winning PICK 4 numbers})$
$= \dfrac{1}{10} \cdot \dfrac{1}{10} \cdot \dfrac{1}{10} \cdot \dfrac{1}{10} = \dfrac{1}{10,000} = 0.0001$

20. $P(\text{three aces}) = \dfrac{4}{52} \cdot \dfrac{3}{51} \cdot \dfrac{2}{50} \approx 0.00018$

21. $26 \cdot 26 \cdot 10^4 = 6,760,000$ license plates are possible.

22. $_{10}P_4 = 10 \cdot 9 \cdot 8 \cdot 7 = 5040$ seating arrangements are possible.

23. $\dfrac{10!}{4!3!2!} = 12,600$ different vertical arrangements of flags are possible.

24. $_{55}C_8 = 1,217,566,350$ simple random samples are possible.

25. $P\left(\begin{array}{c}\text{winning Arizona's} \\ \text{Pick 5}\end{array}\right) = \dfrac{1}{_{35}C_5}$
$= \dfrac{1}{324,632} \approx 0.000003$

26. **(a)** $P(\text{all three are Merlot}) = \dfrac{5}{12} \cdot \dfrac{4}{11} \cdot \dfrac{3}{10}$
$= \dfrac{1}{22} \approx 0.0455$

(b) $P\left(\begin{array}{c}\text{exactly two} \\ \text{are Merlot}\end{array}\right) = \dfrac{_5C_2 \cdot {_7C_1}}{_{12}C_3}$
$= \dfrac{5 \cdot 4}{2 \cdot 1} \cdot \dfrac{7}{1} \cdot \dfrac{3 \cdot 2 \cdot 1}{12 \cdot 11 \cdot 10}$
$= \dfrac{7}{22} \approx 0.3182$

(c) $P(\text{none are Merlot}) = \dfrac{7}{12} \cdot \dfrac{6}{11} \cdot \dfrac{5}{10}$
$= \dfrac{7}{44} \approx 0.1591$

27. **(a), (b)** Answers will vary depending on the results of the simulation. Results should be reasonably close to $\frac{1}{38}$ for part (a) and $\frac{1}{19}$ for part (b).

28. Subjective probabilities are probabilities based on personal experience or intuition. Examples will vary. Some examples are the likelihood of life on other planets or the chance that the Packers will make it to the NFL playoffs next season.

29. **(a)** There are 13 clubs in the deck.

(b) There are 37 cards remaining in the deck. There are also 4 cards unknown by you. So, there are $37 + 4 = 41$ cards not known to you. Of the unknown cards, $13 - (3 + 2) = 8$ are clubs.

(c) $P(\text{next card dealt is a club}) = \dfrac{8}{41}$
≈ 0.1951

(d) $P(\text{two clubs in a row}) = \dfrac{8}{41} \cdot \dfrac{7}{40} \approx 0.0341$

(e) Answers will vary. One possibility follows: No, you should not stay in the game because the probability of completing the flush is low (0.0341 is less than 0.05).

30. **(a)** Since 20 of the 70 home runs went to left field, $P(\text{left field}) = \dfrac{34}{70} = \dfrac{17}{35} \approx 0.486$. Mark McGwire hit 48.6% of his home runs to left field that year.

(b) Since none of the 70 home runs went to right field, $P(\text{right field}) = \dfrac{0}{70} = 0$.

(c) No. While Mark McGwire did not hit any home runs to right field in 1998, this does not imply that it is impossible for him to hit a right-field home run. He just never did it.

31. Someone winning a lottery twice is not that unlikely considering millions of people play lotteries who have already won a lottery (sometimes more than once) each week, and many lotteries have multiple drawings each week.

32. (a) $P(\text{Bryce}) = \dfrac{119}{1009} \approx 0.118$

This is not unusual since the probability is greater than 0.05.

(b) $P(\text{Gourmet}) = \dfrac{264}{1009} \approx 0.262$

(c) $P(\text{Mallory} \mid \text{Single Cup}) = \dfrac{75}{625} = 0.12$

(d) $P(\text{Bryce} \mid \text{Gourmet}) = \dfrac{9}{264} = \dfrac{3}{88} \approx 0.034$

This is unusual since the probability is less than 0.05.

(e) While it is not unusual for Bryce to sell a case, it is unlikely that he will sell a Gourmet case.

(f) $P(\text{Mallory}) = \dfrac{186}{1009} \approx 0.184$

$P(\text{Mallory} \mid \text{Filters}) = \dfrac{40}{120} = \dfrac{1}{3} \approx 0.333$

No, the events 'Mallory' and 'Filters' are not independent since $P(\text{Mallory}) \neq P(\text{Mallory} \mid \text{Filters})$.

(g) No, the events 'Paige' and 'Gourmet' are not mutually exclusive because the two events can happen at the same time.

$P(\text{Paige and Gourmet}) = \dfrac{42}{1009} \approx 0.042 \neq 0$

Chapter 5 Test

1. Probabilities must be between 0 and 1, inclusive, so the possible probabilities are: 0.23, 0, and $\frac{3}{4}$.

2. Event E contains 1 of the 5 equally likely outcomes, so $P(E) = \dfrac{1}{5} = 0.2$.

3. Event F contains 2 of the 5 equally likely outcomes, so $P(F) = \dfrac{2}{5} = 0.4$.

4. Since $P(E) = \dfrac{1}{5} = 0.2$, we have

$P(E^c) = 1 - \dfrac{1}{5} = \dfrac{4}{5} = 0.8$.

5. (a) Since events E and F are mutually exclusive, $P(E \text{ or } F) = P(E) + P(F)$
$= 0.37 + 0.22$
$= 0.59.$

(b) Since events E and F are independent, $P(E \text{ and } F) = P(E) \cdot P(F)$
$= (0.37)(0.22)$
$= 0.0814.$

6. (a) $P(E \text{ and } F) = P(E) \cdot P(F \mid E)$
$= 0.15 \cdot 0.70$
$= 0.105$

(b) $P(E \text{ or } F) = P(E) + P(F) - P(E \text{ and } F)$
$= 0.15 + 0.45 - 0.105$
$= 0.495$

(c) $P(E \mid F) = \dfrac{P(E \text{ and } F)}{P(F)} = \dfrac{0.105}{0.45}$
≈ 0.233

(d) $P(E \mid F) \approx 0.233$ and $P(E) = 0.15$. Since $P(E \mid F) \neq F(E)$, the events E and F are not independent.

7. (a) $8! = 8 \cdot 7 \cdot 6 \cdot 5 \cdot 4 \cdot 3 \cdot 2 \cdot 1 = 40{,}320$

(b) ${}_{12}C_6 = \dfrac{12!}{6!(12-6)!} = \dfrac{12!}{6!6!}$
$= \dfrac{12 \cdot 11 \cdot 10 \cdot 9 \cdot 8 \cdot 7}{6 \cdot 5 \cdot 4 \cdot 3 \cdot 2 \cdot 1}$
$= 924$

(c) ${}_{14}P_8 = \dfrac{14!}{(14-8)!} = \dfrac{14!}{6!}$
$= 14 \cdot 13 \cdot 12 \cdot 11 \cdot 10 \cdot 9 \cdot 8 \cdot 7$
$= 121{,}080{,}960$

8. (a) A "7" can occur in six ways: (1, 6), (2, 5), (3, 4), (4,3), (5, 2), or (6, 1). An "11" can occur two ways: (6, 5) or (5, 6). Thus, 8 of the 36 possible outcomes of the first roll results in a win, so

$$P(\text{wins on first roll}) = \frac{8}{36} = \frac{2}{9} \approx 0.2222.$$

If the dice are thrown 100 times, we would expect that the player will win on the first roll about 22 times.

(b) A "2" can occur in only one way: (1, 1). A "3" can occur in two ways: (1, 2) or (2, 1). A "12" can occur in only one way: (6, 6). Thus, 4 of the 36 possible outcomes of the first roll results in a loss, so

$$P(\text{loses on first roll}) = \frac{4}{36} = \frac{1}{9} \approx 0.1111.$$

If the dice are thrown 100 times, we would expect that the player will lose on the first roll about 11 times.

9. (a) $P(\text{healthy}) = 0.26$ In a random selection of 100 adult Americans, we expect 26 to believe his/her diet is healthy

(b) Using the complementation rule,
$$P(\text{not healthy}) = 1 - P(\text{healthy})$$
$$= 1 - 0.26$$
$$= 0.74$$

10. (a) Rule 1 is satisfied since all of the probabilities in the model are between 0 and 1.
Rule 2 is satisfied since the sum of the probabilities in the model is one:
$0.25 + 0.19 + 0.13 + 0.11 + 0.09 + 0.23 = 1.$

(b) $P\left(\begin{array}{l}\text{PB Patties/Tagalongs or}\\\text{PB Sandwich/Do-si-dos}\end{array}\right) = 0.13 + 0.11$
$$= 0.24$$

(c) $P\left(\begin{array}{l}\text{Thin Mints, Samoas,}\\\text{Shortbread}\end{array}\right) = 0.25 + 0.19 + 0.09$
$$= 0.53$$

(d) $P(\text{not Thin Mints}) = 1 - 0.25 = 0.75$

11. (a) Of the 297 people surveyed, 155 thought the ideal number of children is 2. Thus,
$$P(\text{ideal is 2}) = \frac{155}{297} \approx 0.522.$$

(b) Of the 297 people surveyed, 87 were females and thought the ideal number of

children is 2. Thus,
$$P(\text{female and ideal is 2}) = \frac{87}{297} \approx 0.293.$$

(c) Of the 297 people surveyed, 155 thought the ideal number of children is 2; 188 were females, and 87 were females and thought the ideal number of children is 2. Thus,
$P(\text{female or ideal is 2})$
$= P(\text{female}) + P(\text{ideal is 2})$
$\quad - P(\text{female and ideal is 2})$
$= \frac{188}{297} + \frac{155}{297} - \frac{87}{297} = \frac{256}{297} \approx 0.862$

(d) Of the 188 females, 87 thought the ideal number of children is 2. Thus,
$$P(\text{ideal is 2 | female}) = \frac{87}{188} \approx 0.463.$$

(e) Of the 36 people who said the ideal number of children was 4, 8 were males. Thus,
$$P(\text{male | ideal is 4}) = \frac{8}{36} \approx 0.222.$$

12. (a) Assuming the outcomes are independent, we get $P(\text{win two games in a row})$
$$= (0.58)^2 \approx 0.336.$$

(b) Assuming the outcomes are independent, we get $P(\text{win seven games in a row})$
$$= (0.58)^7 \approx 0.022.$$

(c) $P(\text{lose at least one of next seven games})$
$P(\text{lose at least one of next seven games})$
$= 1 - P(\text{win seven games in a row})$
$= 1 - (0.58)^7 \approx 1 - 0.022 = 0.978$

13. $P(\text{accept}) = P(\text{1st works}) \cdot P(\text{2nd works | 1st works})$
$$= \frac{9}{10} \cdot \frac{8}{9} = 0.8$$

14. $6! = 720$ different arrangements are possible for the letters in the 'word' LINCEY.

15. $_{29}C_5 = 118,755$ different subcommittees are possible.

16. $P(\text{winning Pennsylvania's Cash 5})$
$$= \frac{1}{_{43}C_5} = \frac{1}{962,598} \approx 0.00000104$$

17. There are 26 ways to select the first character, and 36 ways to select each of the final 7 characters. Therefore, using the Multiplication Rule of Counting, there are $26 \cdot 36^7 = 2.04 \times 10^{12}$ different passwords that are possible for the local area network.

18. Subjective probability assignment was used. It is not possible to repeat probability experiments to estimate the probability of life on Mars.

19. Since there are two A's, four C's, four G's, and five T's from which to form a DNA sequence, there are $\dfrac{15!}{2!4!4!5!} = 9,459,450$ distinguishable DNA sequences possible using the 15 letters.

20. There are 40 digits, of which 9 are either a 0 or a 1. Therefore,
$$P(\text{guess correctly}) \approx \frac{9}{40} = 0.225 .$$

Case Study: The Case of the Body in the Bag

1. Of the 7285 offenders included in the "Victim-Offender Relationship by Age" table, 6327 are known to be at least 18. Thus,
$$P(\text{offender is at least } 18) = \frac{6327}{7285} \approx 0.8685 .$$

2. Of the 7285 offenders included in the "Victim-Offender Relationship by Race" table, 3318 are known to be white. Thus,
$$P(\text{offender is white}) = \frac{3318}{7285} \approx 0.4555 .$$

3. Of the 7285 offenders included in the "Sex of Victim by Sex of Offender" table, 6495 are known to be male. Thus,
$$P(\text{offender is male}) = \frac{6495}{7285} \approx 0.8916 .$$

4. Of the 14,990 victims included in the "Murder Victims by Race and Sex" table, 1883 are known to be white females. Thus,
$$P(\text{victim is white and female}) = \frac{1883}{14,990}$$
$$\approx 0.1256.$$

5. $P(\text{victim is white or female})$
$= P(\text{victim is white}) + P(\text{victim is female})$
$\quad - P(\text{victim is white and female})$
$$= \frac{6956}{14490} + \frac{3156}{14490} - \frac{1883}{14490} == \frac{8229}{14490} \approx 0.5490$$

6. $P\left(\begin{array}{l}\text{victim and} \\ \text{offender are} \\ \text{from same} \\ \text{age category}\end{array}\right)$
$= P\left(\begin{array}{l}\text{both are less} \\ \text{than 18}\end{array}\right) + P\left(\begin{array}{l}\text{both are at} \\ \text{least 18}\end{array}\right) + P\left(\begin{array}{l}\text{both are} \\ \text{unknown}\end{array}\right)$
$$= \frac{124}{7285} + \frac{5628}{7285} + \frac{21}{7285} = \frac{5773}{7285} \approx 0.7925$$

7. $P\left(\begin{array}{l}\text{victim and offender} \\ \text{are from different} \\ \text{age categories}\end{array}\right) = 1 - P\left(\begin{array}{l}\text{victim and} \\ \text{offender are} \\ \text{from same} \\ \text{age categories}\end{array}\right)$
$$\approx 1 - 0.7925 = 0.2075$$

8. $P\left(\begin{array}{l}\text{victim and} \\ \text{offender are} \\ \text{from same} \\ \text{race categories}\end{array}\right)$
$= P(\text{both are white}) + P(\text{both are black})$
$\quad + P(\text{both are other}) + P(\text{both are unknown})$
$$= \frac{3026}{7285} + \frac{3034}{7285} + \frac{97}{7285} + \frac{26}{7285}$$
$$= \frac{6183}{7285} \approx 0.8487$$

9. $P\left(\begin{array}{l}\text{victim and offender} \\ \text{are from different} \\ \text{race categories}\end{array}\right)$
$= 1 - P\left(\begin{array}{l}\text{victim and offender} \\ \text{are from same race} \\ \text{categories}\end{array}\right)$
$$\approx 1 - 0.8487 = 0.1513$$

10. $P\left(\begin{array}{l}\text{victim and} \\ \text{offender are} \\ \text{of the same} \\ \text{sex categories}\end{array}\right)$
$= P(\text{both are male}) + P(\text{both are female})$
$\quad + P(\text{both are unknown})$
$$= \frac{4710}{7285} + \frac{150}{7285} + \frac{26}{7285} = \frac{4886}{7285} \approx 0.6707$$

11. $$P\begin{pmatrix}\text{victim and}\\\text{offender are}\\\text{from different}\\\text{sex categories}\end{pmatrix}=1-P\begin{pmatrix}\text{victim and}\\\text{offender are}\\\text{from same}\\\text{sex categories}\end{pmatrix}$$
$$\approx 1-0.6707=0.3293$$

12. Answers will vary. One possibility is that the victim is more likely to be at least 18, white, and male. The reason is that the probabilities of each outcome is higher than the other outcomes for each category:

$$P(\text{victim is at least 18})=\frac{6398}{7285}\approx 0.8782\ ;$$

$$P(\text{victim is white})=\frac{3709}{7285}\approx 0.5091\ ;$$

$$P(\text{victim is male})=\frac{5289}{7285}\approx 0.7260\ .$$

13. Answers will vary. One possibility is that the offender is more likely to be at least 18, black, and male. The reason is that the probabilities of each outcome is higher than the other outcomes for given category:

$$P(\text{offender is at least 18})=\frac{6327}{7285}\approx 0.8685\ ;$$

$$P(\text{offender is black})=\frac{3662}{7285}\approx 0.5027\ ;$$

$$P(\text{offender is male})=\frac{6495}{7285}\approx 0.8916\ .$$

14. We have no information to relate age of the offender to the race or sex of the victim. The best we can determine is:

$$P(\text{offender is at least 18}\mid\text{victim is at least 18})$$
$$=\frac{5628}{290+5628+480}=\frac{5628}{6398}\approx 0.8796$$

15. We have no information to relate race of the offender to the age of the victim. The best we can determine is:

$$P(\text{offender is white}\mid\text{victim is white})$$
$$=\frac{3026}{3026+573+53+57}=\frac{3026}{3709}\approx 0.8159$$

$$P(\text{offender is white}\mid\text{victim is female})$$
$$=\frac{1126}{1126+700+59+24}=\frac{1126}{1909}\approx 0.5898$$

Note: We cannot determine the probability that the offender is white given that the victim is a white female because necessary information is not provided.

16. We have no information to relate sex of the offender to the age of the victim. The best we can determine is:

$$P(\text{offender is male}\mid\text{victim is white})$$
$$=\frac{3322}{3322+330+57}=\frac{3322}{3709}\approx 0.8957$$

$$P(\text{offender is male}\mid\text{victim is female})$$
$$=\frac{1735}{1735+150+24}=\frac{1735}{1909}\approx 0.9089$$

Note: We cannot determine the probability that the offender is male given that the victim is a white female because necessary information is not provided.

17. Since the victim's age category is "at least 18," we find:

$$P(\text{offender is at least 18}\mid\text{victim is at least 18})$$
$$=\frac{2528}{290+5628+480}=\frac{2528}{6398}\approx 0.8796$$

18. Since the victim's age category is "at least 18," we find:

$$P(\text{offender is not "at least 18"}\mid\text{victim is at least 18})$$
$$\approx 1-0.8796=0.1204$$

19. Since the victim is white, we find:

$$P(\text{offender is white}\mid\text{victim is white})$$
$$=\frac{3026}{3026+573+53+57}=\frac{3026}{3709}\approx 0.8159$$

20. Since the victim is white, we find:

$$P(\text{offender is not white}\mid\text{victim is white})$$
$$\approx 1-0.8159=0.1841$$

21. Since the victim is female, we find:

$$P(\text{offender is female}\mid\text{victim is female})$$
$$=\frac{150}{1735+150+24}=\frac{150}{1909}\approx 0.0786$$

22. Since the victim is female, we find:

$$P(\text{offender is not female}\mid\text{victim is female})$$
$$\approx 1-0.0786=0.9214$$

23. Answers will vary. One possibility is that, since we know the victim is a 30-year-old, white female, the offender is more likely to be at least 18, white, and male. The reason is that, given the restrictions for the victim, the probabilities of each outcome is higher than the other outcomes for given category:

$P($offender is at least 18 | victim is at least 18$)$

$= 0.8796$

$P($offender is white | victim is female$)$

$= 0.5898$

$P($offender is male | victim is female$)$

$= 0.9089$

$P($offender is male | victim is white$)$

$= 0.8957.$

Note: we do not have sufficient information to determine the probability that the offender is a white male who is at least 18 given that the victim is a 30-year-old white female.

24. Answers will vary. In the provided example, the profile of the offender changed from a black male who was at least 18 to a white male who was at least 18 once the age, race, and sex of the victim was known.

Final profiles will vary. One possibility follows: We expect that the offender in this case is a white male who is at least 18 years old. Some key suspects would be any men who have a relationship with the victim, such as husband or boyfriend. We would also want to look at other males who were related to or acquainted with the victim.

Chapter 6
Discrete Probability Distributions

Section 6.1

1. A random variable is a numerical measure of the outcome of a probability experiment, so its value is determined by chance.

3. For a discrete probability distribution, each probability must be between 0 and 1 (inclusive) and the sum of the probabilities must equal one.

5. (a) The number of light bulbs that burn out, X, is a discrete random variable because the value of the random variable results from counting.
   Possible values: $x = 0, 1, 2, 3, ..., 20$

   (b) The time it takes to fly from New York City to Los Angeles is a continuous random variable because time is measured. If we let the random variable T represent the time it takes to fly from New York City to Los Angeles, the possible values for T are all positive real numbers; that is $t > 0$.

   (c) The number of hits to a web site in a day is a discrete random variable because the value of the random variable results from counting. If we let the random variable X represent the number of hits, then the possible values of X are $x = 0, 1, 2, ...$

   (d) The amount of snow in Toronto during the winter is a continuous random variable because the amount of snow is measured. If we let the random variable S represent the amount of snow in Toronto during the winter, the possible values for S are all nonnegative real numbers; that is $s \geq 0$.

7. (a) The amount of rain in Seattle during April is a continuous random variable because the amount of rain is measured. If we let the random variable R represent the amount of rain, the possible values for R are all nonnegative real numbers; that is, $r \geq 0$.

   (b) The number of fish caught during a fishing tournament is a discrete random variable because the value of the random variable results from counting. If we let the random variable X represent the number of fish caught, the possible values of X are $x = 0, 1, 2, ...$

   (c) The number of customers arriving at a bank between noon and 1:00 P.M. is a discrete random variable because the value of the random variable results from counting. If we let the random variable X represent the number of customers arriving at the bank between noon and 1:00 P.M., the possible values of X are $x = 0, 1, 2, ...$

   (d) The time required to download a file from the Internet is a continuous random variable because time is measured. If we let the random variable T represent the time required to download a file, the possible values of T are all positive real numbers; that is, $t > 0$.

9. Yes, because $\sum P(x) = 1$ and $0 \leq P(x) \leq 1$ for all x.

11. No, because $P(50) < 0$.

13. No, because $\sum P(x) = 0.95 \neq 1$.

15. We need the sum of all the probabilities to equal 1. For the given probabilities, we have $0.4 + 0.1 + 0.2 = 0.7$. For the sum of the probabilities to equal 1, the missing probability must be $1 - 0.7 = 0.3$. That is, $P(4) = 0.3$.

17. **(a)** This is a discrete probability distribution because all the probabilities are between 0 and 1 (inclusive) and the sum of the probabilities is 1.

(b)

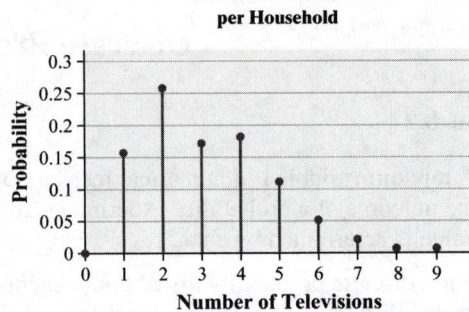

Number of Televisions per Household

(c) $\mu_X = \sum\left[x \cdot P(x)\right] = 0(0) + 1(0.161) + \ldots + 9(0.010) = 3.210 \approx 3.2$

If we surveyed many households, we would expect the mean number of televisions per household to be about 3.2.

(d) $\sigma_X{}^2 = \sum\left[(x - \mu_X)^2 \cdot P(x)\right]$

$= (0 - 3.210)^2 (0) + (1 - 3.210)^2 (0.161) + \ldots + (9 - 3.210)^2 (0.010)$

≈ 3.0159

$\sigma_X = \sqrt{\sigma_X{}^2} = \sqrt{3.0159} \approx 1.7366$ or about 1.7 televisions per household.

(e) $P(3) = 0.176$

(f) $P(3 \text{ or } 4) = P(3) + P(4) = 0.176 + 0.186 = 0.362$

(g) $P(0) = 0$; This is not an impossible event, but it is very unlikely.

19. **(a)** This is a discrete probability distribution because all the probabilities are between 0 and 1 (inclusive) and the sum of the probabilities is 1.

(b)

Ichiro's Hit Parade

(c) $\mu_X = \sum\left[x \cdot P(x)\right] = 0(0.1677) + 1(0.3354) + \ldots + 5(0.0248) = 1.6273 \approx 1.6$

Over many games, Ichiro is expected to average about 1.6 hits per game.

(d) $\sigma_X{}^2 = \sum\left[(x - \mu_X)^2 \cdot P(x)\right]$

$= (0 - 1.6273)^2 (0.1677) + (1 - 1.6273)^2 (0.3354) + \ldots + (5 - 1.6273)^2 (0.0248)$

≈ 1.389

$\sigma_X = \sqrt{\sigma^2{}_X} \approx \sqrt{1.389} \approx 1.179$ or about 1.2 hits

(e) $P(2) = 0.2857$

(f) $P(X > 1) = 1 - P(X \leq 1) = 1 - P(0 \text{ or } 1) = 1 - (0.1677 + 0.3354) = 1 - 0.5031 = 0.4969$

21. (a) Total number of World Series
$= 18 + 18 + 20 + 35 = 91.$

x (games played)	$P(x)$
4	$\dfrac{18}{91} \approx 0.1978$
5	$\dfrac{18}{91} \approx 0.1978$
6	$\dfrac{20}{91} \approx 0.2198$
7	$\dfrac{35}{91} \approx 0.3846$

(b)

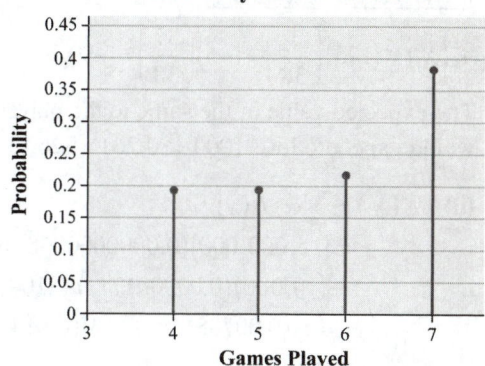

Games Played in World Series

(c) $\mu_X = \sum [x \cdot P(x)]$
$\approx 4 \cdot (0.1978) + 5 \cdot (0.1978) + 6 \cdot (0.2198) + 7 \cdot (0.3846) \approx 5.79$ or about 5.8 games

The World Series, if played many times, would be expected to last about 5.8 games on average.

(d) $\sigma_X^2 = \sum (x - \mu_x)^2 \cdot P(x)$
$\approx (4 - 5.7912)^2 \cdot 0.1978 + (5 - 5.7912)^2 \cdot 0.1978 + (6 - 5.7912)^2 \cdot 0.2198 + (7 - 5.7912)^2 \cdot 0.3846$
≈ 1.3300
$\sigma_X = \sqrt{\sigma_X^2} \approx \sqrt{1.3300} \approx 1.153$ or about 1.2 games

23. (a) $P(4) = 0.096$

(b) $P(4 \text{ or } 5) = P(4) + P(5) = 0.096 + 0.047 = 0.143$

(c) $P(X \geq 6) = P(6) + P(7) + P(8) + P(9) + p(10)$
$= 0.054 + 0.018 + 0.020 + 0.016 + 0.011 = 0.119$

(d) $E(X) = \mu_X = \sum x \cdot P(x) = 1(0.352) + 2(0.262) + 3(0.124) + ... + 10(0.011) = 2.731$ or about 2.7

We would expect the mother to have had 2.7 live births, on average.

25. $E(X) = \sum x \cdot P(x)$
$= (200)(0.999544) + (200 - 250,000)(0.000456) = \$86.00$

If the company sells many of these policies to 20-year old females, then they will make an average of \$86.00 per policy.

27. (a) $E(X) = \sum x \cdot P(x)$
$= (0)(0.0982) + (30)(0.0483) + (20)(0.389275) + (-20)(0.464225) = -0.05$

If you play black jack a large number of times with a \$20 bet, then they will lose an average of \$0.05 or 5¢ per game.

(b) If a player was dealt 40 hands then the player can expect to lose \$0.05 on each of the 40 hands for a total loss of (0.05)(40) = \$2 per hour. Thus, after three hours the total loss would be (3)(\$2)=\$6.

29. Let X = player winnings for \$5 bet on a single number.

Winnings, x (\$)	175	–5
Probability	$\dfrac{1}{38}$	$\dfrac{37}{38}$

$$E(X) = (175)\left(\frac{1}{38}\right) + (-5)\left(\frac{37}{38}\right) = -\$0.26$$

The expected value of the game to the player is a loss of \$0.26. If you played the game 1000 times, you would expect to lose $1000 \cdot (\$0.26) = \$260$.

31. (a) $E(X) = \sum x \cdot P(x)$
$$= (15,000,000)(0.00000000684) + (200,000)(0.00000028)$$
$$+ (10,000)(0.000001711) + (100)(0.000153996) + (7)(0.004778961)$$
$$+ (4)(0.007881463) + (3)(0.01450116) + (0)(0.9726824222)$$
$$\approx 0.30$$

After many \$1 plays, you would expect to win an average of \$0.30 per play. That is, you would lose an average of \$0.70 per \$1 play for a net profit of $-\$0.70$.

(Note: the given probabilities reflect changes made in April 2005 to create larger jackpots that are built up more quickly. It is interesting to note that prior to the change, the expected cash prize was still \$0.30)

(b) We need to find the break-even point. That is, the point where we expect to win the same amount that we pay to play. Let x be the grand prize. Set the expected value equation equal to 1 (the cost for one play) and then solve for x.

$$E(X) = \sum x \cdot P(x)$$
$$1 = (x)(0.00000000684) + (200,000)(0.00000028)$$
$$+ (10,000)(0.000001711) + (100)(0.000153996) + (7)(0.004778961)$$
$$+ (4)(0.007881463) + (3)(0.01450116) + (0)(0.9726824222)$$
$$1 = 0.196991659 + 0.00000000684x$$
$$0.803008341 = 0.00000000684x$$
$$117,398,880.3 = x$$
$$118,000,000 \approx x$$

The grand prize should be at least \$118,000,000 for you to expect a profit after many \$1 plays.

(Note: prior to the changes mentioned in part (a), the grand prize only needed to be about \$100 million to expect a profit after many \$1 plays)

(c) No, the size of the grand prize does not affect your chance of winning. Your chance of winning the grand prize is determined by the number of balls that are drawn and the number of balls from which they are drawn. However, the size of the grand prize will impact your expected winnings. A larger grand prize will increase your expected winnings.

33. Answers will vary. The simulations illustrate the Law of Large Numbers.

35. (a) The mean is the sum of the values divided by the total number of observations.

The mean is: $\mu_X = \frac{529}{160} \approx 3.3063$ or about 3.3 credit cards

(b) The standard deviation is computed by subtracting the mean from each value, squaring the result, and summing. Then, to get the population variance, we divide the sum by the number of observations. The square root of the variance is the standard deviation.

$$\sigma_X^2 = \frac{\sum (x - \mu_x)^2}{N} = \frac{(3 - 3.3063)^2 + (2 - 3.3063)^2 + \ldots + (5 - 3.3063)^2}{160} \approx 5.3585$$

$$\sigma_X = \sqrt{\sigma_X^2} \approx \sqrt{5.3585} \approx 2.3148 \text{ or about 2.3}$$

(c)

x (Cards)	P(x)	x (Cards)	P(x)	x (Cards)	P(x)
1	$\frac{23}{160} = 0.14375$	5	$\frac{13}{160} = 0.08125$	9	$\frac{2}{160} = 0.0125$
2	$\frac{44}{160} = 0.275$	6	$\frac{7}{160} = 0.04375$	10	$\frac{2}{160} = 0.0125$
3	$\frac{43}{160} = 0.26875$	7	$\frac{4}{160} = 0.025$	20	$\frac{1}{160} = 0.00625$
4	$\frac{18}{160} = 0.1125$	8	$\frac{3}{160} = 0.01875$		

(d)

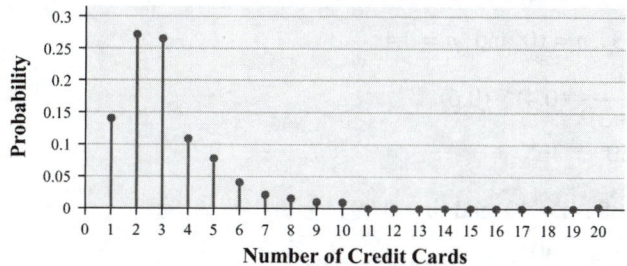

Credit Cards

(e) $\mu_X = \sum \left[x \cdot P(x) \right]$

$\approx 1 \cdot (0.14375) + 2 \cdot (0.275) + 3 \cdot (0.26875) + \cdots + 10 \cdot (0.0125) + 20 \cdot (0.00625)$

≈ 3.3063 or about 3.3 credit cards

$\sigma_X^2 = \sum (x - \mu_x)^2 \cdot P(x)$

$= (1 - 3.3063)^2 \cdot 0.14375 + (2 - 3.3063)^2 \cdot 0.275 + \ldots + (20 - 3.3063)^2 \cdot 0.00625$

≈ 5.234961

$\sigma_X = \sqrt{\sigma_X^2} \approx \sqrt{5.234961} \approx 2.3076$ or about 2.3 credit cards

(f) $\mu - 2\sigma = 3.3 - 2 \cdot 2.3 = -1.3$

$\mu + 2\sigma = 3.3 + 2 \cdot 2.3 = 7.9$

So, the probability of selecting an individual whose number of credit cards is more than two standard deviations from the mean is the same as picking someone who has at least 8 credit cards.

$P(X \geq 8) = P(X = 8) + P(X = 9) + P(X = 10 + P(X = 20)$

$= 0.01875 + 0.0125 + 0.0125 + 0.00625$

$= 0.05$

This result is right on the boundary between being unusual and not unusual. We can say that this result is a little unusual.

(g) $P(\text{Choose 2 with exactly 2 cards}) = \left(\frac{44}{160} \right) \left(\frac{43}{159} \right) = 0.0744$

Section 6.2

1. trial

3. True

5. np

7. This is not a binomial experiment because there are more than two possible values for the random variable 'age.'

9. This is a binomial experiment. There is a fixed number of trials $(n = 100$ where each trial corresponds to administering the drug to one of the 100 individuals), the trials are independent (due to random selection), there are two outcomes (favorable or unfavorable response), and the probability of a favorable response is assumed to be constant for each individual.

11. This is not a binomial experiment because the trials (cards) are not independent and the probability of getting an ace changes for each trial (card). Because the cards are not replaced, the probability of getting an ace on the second card depends on what was drawn first.

13. This is not a binomial experiment because the number of trials is not fixed.

15. This is a binomial experiment. There is a fixed number of trials ($n = 100$ where each trial corresponds to selecting one of the 100 parents), the trials are independent (due to random selection), there are two outcomes (spanked or never spanked), and there is a fixed probability (for a given population) that a parent has ever spanked their child.

---

17. Using $P(x) = {}_nC_x \, p^x (1-p)^{n-x}$ with $x = 3$, $n = 10$ and $p = 0.4$:

$$P(3) = {}_{10}C_3 \cdot (0.4)^3 \cdot (1-0.4)^{10-3} = \frac{10!}{3!(10-3)!} \cdot (0.4)^3 \cdot (0.6)^7$$

$$= 120 \cdot (0.064) \cdot (0.0279936) \approx 0.2150$$

19. Using $P(x) = {}_nC_x \, p^x (1-p)^{n-x}$ with $x = 38$, $n = 40$ and $p = 0.99$:

$$P(38) = {}_{40}C_{38} \cdot (0.99)^{38} \cdot (1-0.99)^{40-38} = \frac{40!}{38!(40-38)!} \cdot (0.99)^{38} \cdot (0.01)^2$$

$$= 780 \cdot (0.6825\ldots) \cdot (0.0001) \approx 0.0532$$

21. Using $P(x) = {}_nC_x \, p^x (1-p)^{n-x}$ with $x = 3$, $n = 8$ and $p = 0.35$:

$$P(3) = {}_8C_3 \cdot (0.35)^3 \cdot (1-0.35)^{8-3} = \frac{8!}{3!(8-3)!} \cdot (0.35)^3 \cdot (0.65)^5$$

$$= 56 \cdot (0.42875)(0.116029\ldots) \approx 0.2786$$

23. Using $n = 9$ and $p = 0.2$:

$$P(X \le 3) = P(0) + P(1) + P(2) + P(3)$$
$$= {}_9C_0 \cdot (0.2)^0 (0.8)^9 + {}_9C_1 \cdot (0.2)^1 (0.8)^8 + {}_9C_2 \cdot (0.2)^2 (0.8)^7 + {}_9C_3 \cdot (0.2)^3 (0.8)^6$$
$$\approx 0.134218 + 0.301990 + 0.301990 + 0.176161 \approx 0.9144$$

25. Using $n = 7$ and $p = 0.5$:

$$P(X > 3) = P(X \ge 4)$$
$$= P(4) + P(5) + P(6) + P(7)$$
$$= {}_7C_4 \cdot (0.5)^4 (0.5)^3 + {}_7C_5 \cdot (0.5)^5 (0.5)^2 + {}_7C_6 \cdot (0.5)^6 (0.5)^1 + {}_7C_7 \cdot (0.5)^7 (0.5)^0$$
$$= 0.2734375 + 0.1640625 + 0.0546875 + 0.0078125 = 0.5$$

27. Using $n = 12$ and $p = 0.35$:

$$P(X \le 4) = P(0) + P(1) + P(2) + P(3) + P(4)$$
$$= {}_{12}C_0 \cdot (0.35)^0 (0.65)^{12} + {}_{12}C_1 \cdot (0.35)^1 (0.65)^{11} + {}_{12}C_2 \cdot (0.35)^2 (0.65)^{10}$$
$$+ {}_{12}C_3 \cdot (0.35)^3 (0.65)^9 + {}_{12}C_4 \cdot (0.35)^4 (0.65)^8$$
$$\approx 0.005688 + 0.036753 + 0.108846 + 0.195365 + 0.236692 \approx 0.5833$$

29. (a)

Distribution

x	$P(x)$
0	0.1176
1	0.3025
2	0.3241
3	0.1852
4	0.0595
5	0.0102
6	0.0007

	$x \cdot P(x)$	$(x - \mu_x)^2 \cdot P(x)$
	0.0000	0.3812
	0.3025	0.1936
	0.6483	0.0130
	0.5557	0.2667
	0.2381	0.2881
	0.0510	0.1045
	0.0044	0.0129
Σ	1.8000	1.2600

(b) $\mu_X = 1.8$ (from first column in table above and to the right)

$\sigma_X = \sqrt{\sigma_X^2} = \sqrt{1.26} \approx 1.1$ (from second column in table above and to the right)

(c) $\mu_X = n \cdot p = 6 \cdot (0.3) = 1.8$ and $\sigma_X = \sqrt{n \cdot p \cdot (1-p)} = \sqrt{(6) \cdot (0.3) \cdot (0.7)} \approx 1.1$

(d)

The distribution is skewed right.

31. (a)

Distribution

x	$P(x)$
0	0.0000
1	0.0001
2	0.0012
3	0.0087
4	0.0389
5	0.1168
6	0.2336
7	0.3003
8	0.2253
9	0.0751

	$x \cdot P(x)$	$(x - \mu_x)^2 \cdot P(x)$
	0.0000	0.0002
	0.0001	0.0034
	0.0025	0.0279
	0.0260	0.1217
	0.1557	0.2944
	0.5840	0.3577
	1.4016	0.1314
	2.1024	0.0188
	1.8020	0.3520
	0.6758	0.3801
Σ	6.7500	1.6876

(b) $\mu_X = 6.75$ (from first column in table above and to the right)

$\sigma_X = \sqrt{\sigma_X^2} = \sqrt{1.6876} \approx 1.3$ (from second column in table above and to the right)

(c) $\mu_X = n \cdot p = 9 \cdot (0.75) = 6.75$ and $\sigma_X = \sqrt{n \cdot p \cdot (1-p)} = \sqrt{(9) \cdot (0.75) \cdot (0.25)} \approx 1.3$

(d)

The distribution is skewed left.

33. (a)

Distribution

x	$P(x)$
0	0.0010
1	0.0098
2	0.0439
3	0.1172
4	0.2051
5	0.2461
6	0.2051
7	0.1172
8	0.0439
9	0.0098
10	0.0010

	$x \cdot P(x)$	$(x - \mu_x)^2 \cdot P(x)$
	0.0000	0.0250
	0.0098	0.1568
	0.0878	0.3952
	0.3516	0.4690
	0.8204	0.2053
	1.2305	0.0000
	1.2306	0.2049
	0.8204	0.4686
	0.3512	0.3950
	0.0882	0.1568
	0.0100	0.0250
Total	5.0005	2.5016

(b) $\mu_X = 5.0$ (from first column in table above and to the right)

$\sigma_X = \sqrt{\sigma_X^2} = \sqrt{2.5016} \approx 1.6$ (from second column in table above and to the right)

(c) $\mu_X = n \cdot p = 10 \cdot (0.5) = 5$ and $\sigma_X = \sqrt{n \cdot p \cdot (1-p)} = \sqrt{10 \cdot (0.5) \cdot (0.5)} \approx 1.6$

(d)

The distribution is symmetric.

35. (a) This is a binomial experiment because it satisfies each of the four requirements:
   1) There are a fixed number of trials ($n = 15$).
   2) The trials are all independent (randomly selected).
   3) For each trial, there are only two possible outcomes ('on time' and 'not on time').
   4) The probability of "success" (i.e. on time) is the same for all trials ($p = 0.80$).

   (b) We have $n = 15$, $p = 0.80$, and $x = 10$. In the binomial table, we go to the section for $n = 15$ and the column that contains $p = 0.80$. Within the $n = 15$ section, we look for the row $x = 10$.

   $P(10) = 0.1032$

   There is a probability of 0.1032 that in a random sample of 15 such flights, exactly 10 will be on time.

   (c) $P(X < 10) = P(X \le 9)$, so using the cumulative binomial table, we go to the section for $n = 15$ and the column that contains $p = 0.8$. Within the $n = 15$ section, we look for the row $x = 9$. We find

   $P(X \le 9) = 0.0611$. In a random sample of 15 such flights, there is a 0.0611 probability that less than 10 flights will be on time.

   (d) Here we wish to find $P(X \ge 10)$. Using the complement rule we can write:

   $P(X \ge 10) = 1 - P(X < 10) = 1 - P(X \le 9) = 1 - 0.0611 = 0.9389$ In a random sample of 15 such flights, there is a 0.9389 probability that at least 10 flights will be on time.

   (e) Using the binomial probability table we get:

   $P(8 \le X \le 10) = P(8) + P(9) + P(10) = 0.0138 + 0.0430 + 0.1032 = 0.1600$

   Using the cumulative binomial probability table we get:

   $P(8 \le X \le 10) = P(X \le 10) - P(X \le 7) = 0.1642 - 0.0042 = 0.1600$

   In a random sample of 15 such flights, there is a 0.1600 probability that between 8 and 10 flights, inclusive, will be on time.

37. (a) This is a binomial experiment because it satisfies each of the four requirements:
   1) There are a fixed number of trials ($n = 20$).
   2) The trials are all independent (randomly selected).
   3) For each trial, there are only two possible outcomes ('uses foot to flush public toilet' and 'does not use foot').
   4) The probability of "success" (i.e. uses foot to flush public toilet) is the same for all trials ($p = 0.64$).

   (b) We have $n = 20$, $p = 0.64$, and $x = 12$.

   $$P(12) = {}_{20}C_{12} \cdot (0.64)^{12} \cdot (1 - 0.64)^{20-12} = \frac{20!}{12!(20-12)!} \cdot (0.64)^{12} \cdot (0.36)^8$$

   $= 125970 \cdot (0.64)^{12} \cdot (0.36)^8 = 0.1678$

   There is a 0.1678 probability that in a random sample of 20 adult Americans who complete the survey, exactly 12 flush a public toilet with their foot.

   (c) Using the cumulative binomial probability table, we get: $P(X \ge 16) = 1 - P(X \le 15) = 0.1011$.

   In a random sample of 20 adult Americans, there is a 0.1011 probability that 16 or more use their foot to flush a public toilet.

   (d) $P(9 \le X \le 11) = P(9) + P(10) + P(11)$

   $= {}_{20}C_9 \cdot (0.64)^9 \cdot (1 - 0.64)^{20-9} + {}_{20}C_{10} \cdot (0.64)^{10} \cdot (1 - 0.64)^{20-10} + {}_{20}C_{11} \cdot (0.64)^{11} \cdot (1 - 0.64)^{20-11}$

   $\approx 0.039825 + 0.077880 + 0.125866 \approx 0.2436$

   In a random sample of 20 adult Americans, there is a 0.2436 probability that between 9 and 11, inclusive, use their foot to flush a public toilet.

(e) $P(X > 17) = P(18) + P(19) + P(20)$

$$= {}_{20}C_{18} \cdot (0.64)^{18} \cdot (1-0.64)^{20-18} + {}_{20}C_{19} \cdot (0.64)^{19} \cdot (1-0.64)^{20-19} + {}_{20}C_{20} \cdot (0.64)^{20} \cdot (1-0.64)^{20-20}$$

$$\approx 0.007991 + 0.001495 + 0.000133 \approx 0.0096$$

In a random sample of 20 adult Americans, there is a 0.0096 probability that more than 17 use their foot to flush a public toilet. These would be very unusual results.

39. (a) Calculating this probability by hand, we get:

$$P(X = 4) = {}_{10}C_4 \cdot (0.267)^4 \cdot (1-0.733)^{10-4} = \frac{10!}{4!(10-4)!} \cdot (0.267)^4 \cdot (0.733)^6$$

$$= 210 \cdot (0.267)^4 \cdot (0.733)^6 = 0.1655$$

In a random sample of 10 people, there is a probability of 0.1655 that exactly 4 will cover their mouth when sneezing.

(b) $P(X < 3) = P(0) + P(1) + P(2)$

$$= {}_{10}C_0 \cdot (0.267)^{10} \cdot (0.733)^0 + {}_{10}C_1 \cdot (0.267)^9 \cdot (0.733)^1 + {}_{10}C_2 \cdot (0.267)^8 \cdot (0.733)^2$$

$$= 1 \cdot (0.267)^{10} \cdot (0.733)^0 + 10 \cdot (0.267)^9 \cdot (0.733)^1 + 45 \cdot (0.267)^8 \cdot (0.733)^2$$

$$= 0.0448 + 0.1631 + 0.2673$$

$$= 0.4752$$

We expect that, in a random sample of 10 adult Americans, there is a probability of 0.4752 that less than 3 will cover their mouth when sneezing.

(c) If fewer than half covered their mouth, then at least half did not cover their mouth. In other words, 5 or more out of the 10 people did not cover their mouth.

$$P(X \geq 5) = P(5) + P(6) + P(7) + P(8) + P(9) + P(10)$$

$$= {}_{10}C_5 \cdot (0.267)^5 \cdot (0.733)^5 + {}_{10}C_6 \cdot (0.267)^6 \cdot (0.733)^4 + {}_{10}C_7 \cdot (0.267)^7 \cdot (0.733)^3$$

$$+ {}_{10}C_8 \cdot (0.267)^8 \cdot (0.733)^2 + {}_{10}C_9 \cdot (0.267)^9 \cdot (0.733)^1 + {}_{10}C_{10} \cdot (0.267)^{10} \cdot (0.733)^0$$

$$= 252 \cdot (0.267)^5 \cdot (0.733)^5 + 210 \cdot (0.267)^6 \cdot (0.733)^4 + 120 \cdot (0.267)^7 \cdot (0.733)^3$$

$$+ 45 \cdot (0.267)^8 \cdot (0.733)^2 + 10 \cdot (0.267)^9 \cdot (0.733)^1 + 1 \cdot (0.267)^{10} \cdot (0.733)^0$$

$$= 0.0724 + 0.0220 + 0.0046 + 0.0006 + 0.0001 + 0.0000$$

$$= 0.0996 \text{ or } 0.0997, \text{ depending upon rounding}$$

In a random sample of 10 people, there is a probability of 0.0997 that fewer than half will cover their mouth when sneezing. This is not unusual.

41. (a) The proportion of the jury that is Hispanic is: $\frac{2}{12} = 0.1667$. So, about 16.67% of the jury is Hispanic.

(b) We can compute this by hand or use the binomial probability table. By hand, we have:

$$P(X \leq 2) = P(X = 0) + P(X = 1) + P(X = 2)$$

$$= {}_{12}C_0 \cdot (0.45)^0 \cdot (1-0.45)^{12-0} + {}_{12}C_1 \cdot (0.45)^1 \cdot (1-0.45)^{12-1} + {}_{12}C_2 \cdot (0.45)^2 \cdot (1-0.45)^{12-2}$$

$$= \frac{12!}{0!(12-0)!} \cdot (0.45)^0 \cdot (0.55)^{12} + \frac{12!}{1!(12-1)!} \cdot (0.45)^1 \cdot (0.55)^{11} + \frac{12!}{2!(12-2)!} \cdot (0.45)^2 \cdot (0.55)^{10}$$

$$= 1 \cdot (0.45)^0 \cdot (0.55)^{12} + 12 \cdot (0.45)^1 \cdot (0.55)^{11} + 66 \cdot (0.45)^2 \cdot (0.55)^{10}$$

$$= 0.0008 + 0.0075 + 0.0339$$

$$= 0.0421 \text{ or } 0.0422, \text{ depending on rounding}$$

In a random sample of 12 jurors, there is a probability of 0.421 that 2 or fewer would be Hispanic.

(c) The probability in part (b) is less than 0.05, so this is an unusual event. I would argue that Hispanics are underrepresented on the jury and that the jury selection process was not fair.

43. (a) We have $n = 100$ and $p = 0.80$.

$\mu_X = n \cdot p = 100(0.80) = 80$ flights; $\sigma_X = \sqrt{np(1-p)} = \sqrt{100(.80)(1-.80)} = \sqrt{16} = 4$ flights

(b) We expect that, in a random sample of 100 flights from Orlando to Los Angeles, 80 will be on time.

(c) Since $np(1-p) = 16 \geq 10$, the distribution is approximately bell shaped (approximately normal) and we can apply the Empirical Rule.

$\mu_X - 2\sigma_X = 80 - 2(4) = 72$

Since 75 is less than 2 standard deviations below the mean, we would conclude that it would not be unusual to observe 75 on-time flights in a sample of 100 flights.

45. (a) We have $n = 500$ and $p = 0.64$.

$\mu_X = n \cdot p = 500(0.64) = 320$ adults who use their foot to flush public toilets;

$\sigma_X = \sqrt{np(1-p)} = \sqrt{(500) \cdot (0.64) \cdot (0.36)} = \sqrt{115.2} \approx 10.733$

(b) Since $np(1-p) = 500(0.64)(0.36) = 115.2 \geq 10$, we can use the Empirical Rule to check for unusual observations.

280 is below the mean, and we have $\mu_X + 2\sigma_X = 320 - 2(10.733) = 298.534$.

This indicates that 280 is more than two standard deviations below the mean. Therefore, it would be considered unusual to find 280 adult Americans who flush the toilet with their foot, from a sample of 500 adult Americans.

47. We have $n = 1030$ and $p = 0.80$.

$E(X) = \mu_X = n \cdot p = 1030(0.80) = 824$. If attitudes have not changed, we expect 824 of the parents surveyed to spank their children.

Since $np(1-p) = 164.8 \geq 10$, we can use the Empirical rule to check if 781 would be an unusual observation.

$\sigma_X = \sqrt{np(1-p)} = \sqrt{1030(0.80)(1-0.80)} = \sqrt{164.8} \approx 12.8$

781 is below the mean and we have $\mu_X - 2\sigma_X = 824 - 2(12.8) = 798.4$. Since 781 is more than two standard deviations away from the mean, it would be considered unusual if 781 parents from a sample of 1030 said they spank their children. This suggests that parents' attitudes have changed since 1995.

49. We would expect $500,000(0.56) = 280,000$ of the stops to be pedestrians who are nonwhite. Because

$np(1-p) = 500,000(0.56)(1-0.56) = 123,200 \geq 10$, we can use the Empirical Rule to identify cutoff points

for unusual results. The standard deviation number of stops is $\sigma_x = \sqrt{500,000(0.56)(1-0.56)} \approx 351.0$. If the

number of stops of nonwhites exceeds $\mu_X + 2\sigma_X = 280,000 + 2(351) = 280,702$, we would say the result is

unusual. The actual number of stops is $500,000(0.89) = 445,000$, which is definitely unusual. A potential

criticism of this analysis is the use of 0.44 as the proportion of whites, since the actual proportion of whites may be different due to individuals commuting back and forth to the city. Additionally, there could be confounding due to the part of the city where stops are made. If the distribution of nonwhites is not uniform across the city, then location should be taken into account. See Simpson's Paradox in chapter 4.

51. (a) Using n =56 and p =1- 0.0995=0.9005 which is the probability that a passenger does not miss his/her flight.

$P(55 \text{ or } 56) = P(55) + P(56)$

$= {}_{56}C_{55} \cdot (0.9005)^{55}(0.0995)^1 + {}_{56}C_{56} \cdot (0.9005)^{56}(0.0995)^0$

$\approx 0.01748 + 0.00283$

≈ 0.0203

(b) The probability of being bumped would be the probability that more than 54 passengers out of the 60 passengers showed up for the flight. $P(X > 54) = P(X \ge 55) = 0.4423$. If 60 tickets are sold there is a probability of 0.4423 that a passenger will be bumped.

(c) We would like the probability, $P(X \ge 251)$, to be 0.01 or less, with $p = 0.9005$ and n to be determined.

$P(X \ge 251) = 1 - P(X \le 250) = 1 - (P(0) + \ldots + P(250))$ and using technology we find that:

when $n = 267$, $P(X \ge 251) = 1 - P(X \le 250) = 1 - 0.98491 = 0.01509$

when $n = 266$, $P(X \ge 251) = 1 - P(X \le 250) = 1 - 0.99149 = 0.0085$

Thus the maximum number of seats that should be booked is 266 if the probability of being bumped is less than 0.01.

53. (a) Using $P(x) = p \cdot (1-p)^{x-1}$ with $p = 0.524$ gives:

$P(3) = (0.524) \cdot (1 - 0.524)^{3-1} = 0.1187$

(b)

x	$P(x) = p \cdot (1-p)^{x-1}$
1	0.5240
2	0.2494
3	0.1187
4	0.0565
5	0.0269
6	0.0128
7	0.0061
8	0.0029
9	0.0014
10	0.0007

(c)

$x \cdot P(x)$
0.5240
0.4988
0.3562
0.2261
0.1345
0.0768
0.0427
0.0232
0.0124
0.0066
$\mu_X = \sum xP(x) \approx 1.9$

(d) $\dfrac{1}{p} = \dfrac{1}{0.524} \approx 1.9$ free throws

We would expect Shaq to take about 1.9 free throws before making his first one. This result is the same as that obtained in part (c).

55. (a) We expect np=5000(0.5) = 2500 investment advisors across the country to beat the market in any given year.

(b) To beat market for two years means beating the market the first year and the second year. Thus,

$P(\text{beat two years}) = P(\text{beat first year})P(\text{beat second year}) = (0.50)(0.50) = 0.25$

We expect np=5000(0.25) = 1250 investment advisors across the country to beat the market for two years.

(c) To beat market for five years means beating the market each of the five years. Thus,

$P(\text{beat five years}) = (0.50)^5 = 0.03125$

We expect np = 5000(0.03125) = 156.25 investment advisors across the country to beat the market for five years.

(d) To beat market for ten years means beating the market each of the ten years. Thus,

$P(\text{beat ten years}) = (0.50)^{10} = 0.00098$

We expect np=5000(0.00098) =4.9 investment advisors across the country to beat the market for ten years.

(e) This is a binomial experiment if all of the following conditions are satisfied:

(1) the experiment consists of a fixed number of trials, where $n = 5000$;
(2) the trials are independent because the sample size is small relative to the size of the population (assuming there are tens of thousands of investment advisors in the population);
(3) each trial has two possible, mutually exclusive, outcomes: either the investment advisor beats the market or not;
(4) the probability of beating the market, $p=0.50$, remains constant.

(f) Using technology, the probability of 6 investment advisors beating the market for ten years is:

$P(\text{at least 6 beat market for ten years}) = 0.3635.$ This means if we randomly select 5000 investment advisors 100 different times and the probability any individual advisor beats the market in a single year is 0.5 we would expect approximately 36 of the time for at least 6 advisors to beat the market for 10 years in a row. This is not an unusual result and suggests that those advisors who beat the market are not necessarily better than those who do not. That is we expect these types of results simply by randomness.

57. The term "success" indicates the outcome that you are observing.

59. When n is small, the shape of the binomial distribution is heavily affected by p. If p is close to zero, the distribution is skewed right; if p is close to 0.5, the distribution is approximately symmetric; if p is close to 1, the distribution is skewed left.

Section 6.3

1. A random variable X, measuring the number of successes in a fixed interval, follows a Poisson process provided that the following conditions are met:
   (1) the probability of two or more successes in any sufficiently small subinterval is 0;
   (2) the probability of success is the same for any two intervals of equal length;
   (3) the number of successes in any interval is independent of the number of successes in any other non-overlapping (disjoint) interval.

3. $\lambda = 10$ hits per minute, $t = 5$ minutes (from 7:30 P.M. to 7:35 P.M.).

5. $\lambda = 0.07$ flaws per linear foot, $t = 20$ linear feet of timber.

7. Use $P(x) = \dfrac{\mu^x}{x!} \cdot e^{-\mu}$ to find the following, assuming $\mu = 5$:

(a) $P(6) = \dfrac{5^6}{6!} \cdot e^{-5} = 0.1462$

(b) $P(X < 6) = P(X \le 5) = P(0) + P(1) + P(2) + P(3) + P(4) + P(5)$

$= \dfrac{5^0}{0!} \cdot e^{-5} + \dfrac{5^1}{1!} \cdot e^{-5} + \dfrac{5^2}{2!} \cdot e^{-5} + \dfrac{5^3}{3!} \cdot e^{-5} + \dfrac{5^4}{4!} \cdot e^{-5} + \dfrac{5^5}{5!} \cdot e^{-5} = 0.6160$

(c) $P(X \ge 6) = 1 - P(X < 6) = 1 - 0.6160 = 0.3840$

(d) $P(2 \le X \le 4) = P(2) + P(3) + P(4) = \dfrac{5^2}{2!} \cdot e^{-5} + \dfrac{5^3}{3!} \cdot e^{-5} + \dfrac{5^4}{4!} \cdot e^{-5} = 0.4001$

9. Use $P(x) = \dfrac{(\lambda t)^x}{x!} \cdot e^{-\lambda t}$ to find the following, given $\lambda = 0.07$, $t = 10$, and $\lambda t = 0.7$:

(a) $P(4) = \dfrac{0.7^4}{4!} \cdot e^{-0.7} = 0.0050$

(b) $P(X < 4) = P(X \le 3) = P(X = 0) + P(1) + P(2) + P(3)$

$$= \frac{0.7^0}{0!} \cdot e^{-0.7} + \frac{0.7^1}{1!} \cdot e^{-0.7} + \frac{0.7^2}{2!} \cdot e^{-0.7} + \frac{0.7^3}{3!} \cdot e^{-0.7} = 0.9942$$

(c) $P(X \ge 4) = 1 - P(X < 4) = 1 - 0.9942 = 0.0058$

(d) $P(4 \le X \le 6) = P(4) + P(5) + P(6) = \frac{0.7^4}{4!} \cdot e^{-0.7} + \frac{0.7^5}{5!} \cdot e^{-0.7} + \frac{0.7^6}{6!} \cdot e^{-0.7} = 0.0057$

(e) $\mu_X = \lambda t = (0.07)(10) = 0.7$ and $\sigma_X = \sqrt{\lambda t} = \sqrt{0.7} \approx 0.8$

11. We are given that the hits occur at a rate of 1.4 per minute, so $\lambda = 1.4$; and the interval of time we are interested in is 5 minutes (from 7:30 P.M. to 7:35 P.M.), so $t = 5$. Therefore, $\lambda t = (1.4)(5) = 7$.

(a) $P(7) = \frac{7^7}{7!} e^{-7} \approx 0.1490$

On about 15 out of every 100 days, exactly 7 hits will occur between 7:30 P.M. and 7:35 P.M.

(b) $P(X < 7) = P(X \le 6) = P(0) + \ldots + P(6)$

$$= \frac{7^0}{0!} e^{-7} + \frac{7^1}{1!} e^{-7} + \frac{7^2}{2!} e^{-7} + \frac{7^3}{3!} e^{-7} + \frac{7^4}{4!} e^{-7} + \frac{7^5}{5!} e^{-7} + \frac{7^6}{6!} e^{-7}$$

$\approx 0.00091 + 0.00638 + 0.02234 + 0.05213 + 0.09123 + 0.12772 + 0.14900$

≈ 0.44971

On about 45 out of every 100 days, fewer than 7 hits will occur between 7:30 P.M. and 7:35 P.M.

(c) Use the Complement Rule: $P(X \ge 7) = 1 - P(X < 7) = 1 - 0.4497 = 0.5503$.

On about 55 out of every 100 days, at least 7 hits will occur between 7:30 P.M. and 7:35 P.M.

13. We are given that FDAL level for insect filth in peanut butter is 0.3 insect fragments per gram, so $\lambda = 0.3$; and the interval we are interested in is 5 grams (the 5-gram sample), so $t = 5$. Therefore, $\lambda t = (0.3)(5) = 1.5$.

(a) $P(2) = \frac{(1.5)^2}{2!} e^{-1.5} \approx 0.2510$

In about 25 out of every hundred 5-gram samples of this peanut butter, exactly 2 insect fragments will be found.

(b) $P(X < 2) = P(X \le 1) = P(0) + P(1) \approx 0.22313 + 0.33470 \approx 0.55783$

In about 56 out of every hundred 5-gram samples of this peanut butter, fewer than 2 insect fragments will be found.

(c) Use the Complement Rule: $P(X \ge 2) = 1 - P(X < 2) = 1 - 0.5578 = 0.4422$.

In about 44 out of every hundred 5-gram samples of this peanut butter, at least 2 insect fragments will be found.

(d) Use the Complement Rule: $P(X \ge 1) = 1 - P(X < 1) = 1 - P(0) = 1 - 0.2231 = 0.7769$

In about 78 out of every hundred 5-gram samples of this peanut butter, at least 1 insect fragment will be found.

(e) $P(X \ge 4) = 1 - P(X < 4) = 1 - P(X \le 3)$

$$= 1 - \left[P(0) + P(1) + P(2) + P(3) \right]$$

$$\approx 1 - \left[0.22313 + 0.33470 + 0.25102 + 0.12551 \right]$$

$$= 1 - 0.93436 = 0.06564$$

Since $P(X \ge 4) > 0.05$, it would not be unusual for a 5-gram sample of peanut butter to contain four or more insect fragments. About 7 out of every hundred 5-gram samples of this peanut butter will contain at least 4 fragments.

15. (a) We are given that fatal accidents occur at the rate of 0.05 per 100 million miles. If we think in terms of 100 million miles, we have $\lambda = 0.05$. The interval we are interested in is 100 million miles, so $t = 1$ (that is, 1 one hundred million mile unit). Therefore, $\lambda t = (0.05)(1) = 0.05$.

$$P(0) = \frac{(0.05)^0}{0!} e^{-0.05} = e^{-0.05} \approx 0.9512$$

In about 95 out of every 100 one hundred million-mile samples, no fatal accidents will occur.

(b) $P(X \geq 1) = 1 - P(X < 1) = 1 - P(0) = 1 - 0.9512 = 0.0488$

In about 5 out of every 100 one hundred million-mile samples, at least one fatal accident will occur.

(c) $P(X > 1) = 1 - P(X \leq 1) = 1 - \left[P(0) + P(1) \right]$

$$= 1 - \left[\frac{(0.05)^0}{0!} e^{-0.01} + \frac{(0.05)^1}{1!} e^{-0.01} \right] \approx 1 - 0.9988 = 0.0012$$

In about 1 out of every 1,000 one hundred million-mile samples, more than one fatal accident will occur.

17. We are given that major hurricanes strike Florida at a rate of 31 per 115 year interval. That is,

$\lambda = \dfrac{31}{115} \approx 0.2696$ major hurricanes per year. The time interval we are interested in is one year, so $t = 1$.

(a) $P(X = 0) = \dfrac{\left(\frac{31}{115}\right)^0}{0!} e^{-\frac{31}{115}} \approx 0.7637$

(b) $P(X \geq 1) = 1 - P(X = 0) = 1 - 0.7637 = 0.2363$

Since this probability is greater than 0.05 it would not be unusual for Florida to be hit by at least one major hurricane in a single year.

(c) $P(X \geq 3) = 1 - P(X < 3) = 1 - P(X \leq 2)$

$\qquad = 1 - \left[P(0) + P(1) + P(2) \right]$

$\qquad = 1 - \left[\dfrac{\left(\frac{31}{115}\right)^0}{0!} e^{-\frac{31}{115}} + \dfrac{\left(\frac{31}{115}\right)^1}{1!} e^{-\frac{31}{115}} + \dfrac{\left(\frac{31}{115}\right)^2}{2!} e^{-\frac{31}{115}} \right]$

$\qquad \approx 1 - 0.9973$

$\qquad = 0.0027$

Since this probability is less than 0.05, it would be unusual for Florida to be hit by at least 3 major hurricanes in a single year. Thus, the 2004 hurricane season in Florida was extremely unusual.

(d) $P(X \geq 2) = 1 - P(X < 2) = 1 - P(X \leq 1)$

$\qquad = 1 - \left[P(0) + P(1) \right]$

$\qquad = 1 - \left[\dfrac{\left(\frac{31}{115}\right)^0}{0!} e^{-\frac{31}{115}} + \dfrac{\left(\frac{31}{115}\right)^1}{1!} e^{-\frac{31}{115}} \right]$

$\qquad \approx 1 - 0.9696$

$\qquad = 0.0304$

Since this probability is less than 0.05, it would be unusual for Florida to be hit by at least 2 major hurricanes in a single year. Thus, the 2005 hurricane season in Florida was unusual.

19. (a) and (b)

Use technology and the given parameters $\lambda = 0.2$ and $t = 30$ to construct a table:

x	$P(x) = \dfrac{(\lambda t)^x}{x!} \cdot e^{-\lambda t}$	Expected # of restaurants having x arrivals $= 200 \cdot P(x)$	Observed number of restaurants having x arrivals
0	0.0025	0.50	0
1	0.0149	2.97	4
2	0.0446	8.92	5
3	0.0892	17.85	13
4	0.1339	26.77	23
5	0.1606	32.12	25
6	0.1606	32.12	28
7	0.1377	27.54	25
8	0.1033	20.65	27
9	0.0688	13.77	21
10	0.0413	8.26	15
11	0.0225	4.51	5
12	0.0113	2.25	3
13	0.0052	1.04	2
14	0.0022	0.45	2
15	0.0009	0.18	0
16	0.0003	0.07	2

(c) The observed frequencies are lower for fewer cars (x small) and higher for more cars (x large), so the advertising campaign appears to be effective.

21. (a) and (c)

Use technology to construct a table:

x	Observed Frequency	Proportion of years $=$ Frequency $\div 200$	$x \cdot$ proportion of years	$P(x) = \dfrac{(0.61)^x}{x!} \cdot e^{-0.61}$
0	109	0.545	0.000	0.5434
1	65	0.325	0.325	0.3314
2	22	0.110	0.220	0.1011
3	3	0.015	0.045	0.0206
4	1	0.005	0.020	0.0031
		Total $=$	0.610	

(b) From the data in the table, the mean number of deaths per year is $\mu_X = 0.610$.

(c) Using $\mu = 0.61$ we calculate the Poisson probabilities as shown in the last column.

(d) Answers will vary. These probabilities are very close to the relative frequencies, so the random variable X appears to follow a Poisson distribution.

23. (a) $E(X) = \lambda t = \dfrac{23.6}{100} \cdot (500) = 118$ colds.

(b) – (g) Answers will vary.

Section 6.4 (online)

1. In both cases, the random variable X represents the number of successes in n trials of an experiment, where each trial results in one of two mutually exclusive (disjoint) outcomes classified as "success" or "failure". The binomial probability distribution requires that each trial be independent of the others and that the probability of success remain the same for each trial. This is not the case for the hypergeometric probability distribution. For the hypergeometric probability distribution, the population size is finite and there are a fixed number of successes in the population. The sample is obtained without replacement, so the trials are not independent and the probability of success does not remain constant. If the population size is infinite, there is no difference between the hypergeometric distribution and the binomial distribution.

3. n is the sample size and there cannot be more successes than trials. Therefore, we must have $x \le n$. In turn, the population only contains k successes. Therefore, we must also have $x \le k$. If the sample size n is larger than the number of successes in the population (that is, $n > k$), we are still restricted by the fact that there are only k successes in the population. Therefore, the possible values of the random variable must be less than or equal to the smaller of the sample size (n) and the total number of successes in the population (k).

5. This is a hypergeometric probability experiment because:
 1) The population is finite ($N = 49$ numbers).
 2) There are only two possible outcomes (number matches or does not match).
 3) The sample size is $n = 6$ and is drawn without replacement.
 There are only 6 matching numbers so $k = 6$. The possible values for X are $x = 0, 1, 2, ..., 6$.

7. This is a hypergeometric probability experiment because:
 1) The population is finite ($N = 250$ numbers).
 2) There are only two possible outcomes (chip is defective or chip is not defective).
 3) The sample size is $n = 20$ and is drawn without replacement.
 There are only 12 chips that are defective so $k = 12$. The possible values for X are $x = 0, 1, 2, ..., 12$.

9. Given $N = 150$, $n = 20$, $k = 30$, and $x = 5$:

$$P(5) = \frac{\left({}_{30}C_5\right)\left({}_{150-30}C_{20-5}\right)}{{}_{150}C_{20}} = \frac{\left({}_{30}C_5\right)\left({}_{120}C_{15}\right)}{{}_{150}C_{20}} = 0.1856$$

11. Given $N = 230$, $n = 15$, $k = 200$, and $x = 12$:

$$P(12) = \frac{\left({}_{200}C_{12}\right)\left({}_{230-200}C_{15-12}\right)}{{}_{230}C_{15}} = \frac{\left({}_{200}C_{12}\right)\left({}_{30}C_3\right)}{{}_{230}C_{15}} = 0.1939$$

13. Given $N = 150$, $n = 20$, and $k = 30$:

$$\mu_X = n \cdot \frac{k}{N} = 20 \cdot \frac{30}{150} = 4\ ;$$

$$\sigma_X = \sqrt{\left(\frac{N-n}{N-1}\right) \cdot n \cdot \frac{k}{N} \cdot \frac{N-k}{N}} = \sqrt{\left(\frac{150-20}{150-1}\right) \cdot 20 \cdot \frac{30}{150} \cdot \frac{150-30}{150}} \approx 1.67$$

15. Given $N = 230$, $n = 15$, and $k = 200$:

$$\mu_X = n \cdot \frac{k}{N} = 15 \cdot \frac{200}{230} \approx 13.0\ ;$$

$$\sigma_X = \sqrt{\left(\frac{N-n}{N-1}\right) \cdot n \cdot \frac{k}{N} \cdot \frac{N-k}{N}} = \sqrt{\left(\frac{230-15}{230-1}\right) \cdot 15 \cdot \frac{200}{230} \cdot \frac{230-200}{230}} \approx 1.26$$

17. We are given $N = 47$, $n = 6$, and $k = 6$:

(a) $P(3) = \dfrac{\left(_6C_3\right)\left(_{47-6}C_{6-3}\right)}{_{47}C_6} = \dfrac{\left(_6C_3\right)\left(_{41}C_3\right)}{_{47}C_6} = 0.01986$

There is a 0.01986 probability of matching 3 numbers.

(b) $P(4) = \dfrac{\left(_6C_4\right)\left(_{47-6}C_{6-4}\right)}{_{47}C_6} = \dfrac{\left(_6C_4\right)\left(_{41}C_2\right)}{_{47}C_6} = 0.001146$

There is a 0.001146 probability of matching 4 numbers.

(c) $P(5) = \dfrac{\left(_6C_5\right)\left(_{47-6}C_{6-5}\right)}{_{47}C_6} = \dfrac{\left(_6C_5\right)\left(_{41}C_1\right)}{_{47}C_6} = 0.0000229$

There is a 0.0000229 probability of matching 5 numbers.

(d) $P(6) = \dfrac{\left(_6C_6\right)\left(_{47-6}C_{6-6}\right)}{_{47}C_6} = \dfrac{\left(_6C_6\right)\left(_{41}C_0\right)}{_{47}C_6} = 0.000000093$

There is a 0.000000093 probability of matching 6 numbers.

(e) $P(\text{win}) = P(3) + P(4) + P(5) + P(6)$

$= 0.01986 + 0.001146 + 0.0000229 + 0.000000093$

$= 0.02103$

There is a 0.02103 probability of purchasing a winning ticket. This probability is less than 0.05, so it would be considered unusual to purchase a winning ticket.

(f) $\mu_X = n \cdot \dfrac{k}{N} = 6 \cdot \dfrac{6}{47} \approx 0.766$;

$\sigma_X = \sqrt{\left(\dfrac{N-n}{N-1}\right) \cdot n \cdot \dfrac{k}{N} \cdot \dfrac{N-k}{N}} = \sqrt{\left(\dfrac{47-6}{47-1}\right) \cdot 6 \cdot \dfrac{6}{47} \cdot \dfrac{47-6}{47}} \approx 0.772$

$E(X) = \mu_x \approx 0.766$. For a randomly selected ticket, we expect to match 0.766 numbers.

19. We are given $N = 250$, $n = 20$, and $k = 12$:

(a) $P(4) = \dfrac{\left(_{12}C_4\right)\left(_{250-12}C_{20-4}\right)}{_{250}C_{20}} = \dfrac{\left(_{12}C_4\right)\left(_{238}C_{16}\right)}{_{250}C_{20}} = 0.0087$

There is a 0.0087 probability of obtaining 4 defective chips.

(b) $P(3) = \dfrac{\left(_{12}C_3\right)\left(_{250-12}C_{20-3}\right)}{_{250}C_{20}} = \dfrac{\left(_{12}C_3\right)\left(_{238}C_{17}\right)}{_{250}C_{20}} = 0.0507$

There is a 0.0507 probability of obtaining 3 defective chips.

(c) $P(0) = \dfrac{\left(_{12}C_0\right)\left(_{250-12}C_{20-0}\right)}{_{250}C_{20}} = \dfrac{\left(_{12}C_0\right)\left(_{238}C_{20}\right)}{_{250}C_{20}} = 0.3590$

There is a 0.3590 probability of obtaining no defective chips.

(d) $P(14) = 0$ because the random variable cannot take on the value 14. The minimum of n and k is 12. Therefore, the possible values for the random variable are $x = 0, 1, 2, \ldots, 12$.

(e) $E(X) = \mu_X = n \cdot \dfrac{k}{N} = 20 \cdot \dfrac{12}{250} = 0.96$

In a random sample of 20 chips, we expect to obtain 0.96 defective chips on average.

21. Let the random variable X represent the number of jurors on the jury who will never convict. From the problem statement, we get $N = 30$, $n = 12$, and $k = 2$.

Here it will take less work if we find the probability of avoiding a hung jury and subtracting from 1. To avoid a hung jury, we need $X = 0$.

$$P(X = 0) = \frac{\left(_2C_0\right)\left(_{30-2}C_{12-0}\right)}{_{30}C_{12}} = \frac{\left(_2C_0\right)\left(_{28}C_{12}\right)}{_{30}C_{12}} = 0.3517$$

Since, $P(X \geq 1) = 1 - P(X = 0) = 1 - 0.3517 = 0.6483$, there is a 0.6483 probability of having a hung jury.(Note: Here we are assuming that there will never be a hung jury unless one of the two potential jurors who will never convict is on the jury.)

23. Let the random variable X represent the number of cylinders that pass inspection. From the problem statement we have $N = 200$ and $n = 4$. To accept the shipment, we must obtain 0 defective cylinders. That is, we want $X = 0$.

(a) If 10% are defective, we have $k = (0.1)(200) = 20$.

$$P(0) = \frac{\left(_{20}C_0\right)\left(_{200-20}C_{4-0}\right)}{_{200}C_4} = \frac{\left(_{20}C_0\right)\left(_{180}C_4\right)}{_{200}C_4} = 0.6539$$

If 10% are defective, there is a 0.6539 probability that the shipment is accepted.

(b) If 20% are defective, we have $k = (0.2)(200) = 40$.

$$P(0) = \frac{\left(_{40}C_0\right)\left(_{200-40}C_{4-0}\right)}{_{200}C_4} = \frac{\left(_{40}C_0\right)\left(_{160}C_4\right)}{_{200}C_4} = 0.4065$$

If 20% are defective, there is a 0.4065 probability that the shipment is accepted.

(c) If 40% are defective, we have $k = (0.4)(200) = 80$.

$$P(0) = \frac{\left(_{80}C_0\right)\left(_{200-80}C_{4-0}\right)}{_{200}C_4} = \frac{\left(_{80}C_0\right)\left(_{120}C_4\right)}{_{200}C_4} = 0.1270$$

If 40% are defective, there is a 0.1270 probability that the shipment is accepted.

(d) If 60% are defective, we have $k = (0.6)(200) = 120$.

$$P(0) = \frac{\left(_{120}C_0\right)\left(_{200-120}C_{4-0}\right)}{_{200}C_4} = \frac{\left(_{120}C_0\right)\left(_{80}C_4\right)}{_{200}C_4} = 0.0245$$

If 60% are defective, there is a 0.0245 probability that the shipment is accepted.

(e) If 80% are defective, we have $k = (0.8)(200) = 160$.

$$P(0) = \frac{\left(_{160}C_0\right)\left(_{200-160}C_{4-0}\right)}{_{200}C_4} = \frac{\left(_{160}C_0\right)\left(_{40}C_4\right)}{_{200}C_4} = 0.0014$$

If 80% are defective, there is a 0.0014 probability that the shipment is accepted.

(f)

Chapter 6 Review Exercises

1. **(a)** The number of students in a randomly selected classroom is a discrete random variable because its value results from counting. If we let the random variable S represent the number of students, the possible values for S are all nonnegative integers (up to the capacity of the room). That is, $s = 0, 1, 2, ...$.

   (b) The number of inches of snow that falls in Minneapolis during the winter season is a continuous random variable because its value results from a measurement. If we let the random variable S represent the number of inches of snow, the possible values for S are all nonnegative real numbers. That is, $s \geq 0$.

   (c) The amount of flight time accumulated is a continuous random variable because its value results from a measurement. If we let the random variable H represent the accumulated flight time, the possible values for H are all nonnegative real numbers. That is, $h \geq 0$.

   (d) The number of points scored by the Miami Heat in a game is a discrete random variable because its value results from a count. If we let the random variable X represent the number of points scored by the Miami Heat in a game, the possible values for X are nonnegative integers. That is, $x = 0, 1, 2, ...$.

2. **(a)** This is not a valid probability distribution because the sum of the probabilities does not equal 1.
   ($\sum P(x) = 0.73$)

   (b) This is a valid probability distribution because $\sum P(x) = 1$ and $0 \leq P(x) \leq 1$ for all x.

3. **(a)** Total number of Stanley Cups
   $= 20 + 18 + 22 + 16 = 76$.

x (games played)	$P(x)$
4	$\frac{20}{76} \approx 0.2632$
5	$\frac{18}{76} \approx 0.2368$
6	$\frac{22}{76} \approx 0.2895$
7	$\frac{16}{76} \approx 0.2105$

   (b)
   Stanley Cup Finals

   (c) $\mu_X = \sum [x \cdot P(x)] \approx 4 \cdot (0.2632) + 5 \cdot (0.2368) + 6 \cdot (0.2895) + 7 \cdot (0.2105)$
   $= 5.4473$ or about 5.4 games
   We expect the Stanley Cup to last, on average, about 5.4 games.

   (d) $\sigma_X^2 = \sum (x - \mu_x)^2 \cdot P(x)$
   $\approx (4 - 5.4473)^2 \cdot 0.2632 + (5 - 5.4473)^2 \cdot 0.2368 + (6 - 5.4473)^2 \cdot 0.2895 + (7 - 5.4473)^2 \cdot 0.2105$
   ≈ 1.19452
   $\sigma_X = \sqrt{\sigma_X^2} \approx \sqrt{1.19462} \approx 1.093$ or about 1.1 games

4. **(a)** Let X represent the profit.
   $\mu_X = E(X) = \sum [x \cdot P(x)]$
   $\approx (200)\left(\frac{12}{5525}\right) + (150)\left(\frac{1}{425}\right) + (30)\left(\frac{36}{1105}\right) + (20)\left(\frac{274}{5525}\right) + (5)\left(\frac{72}{425}\right) + (-5)\left(\frac{822}{1105}\right)$
   $\approx -0.1158 \approx -\$0.12$
   If this game is played many times, the players can expect to lose about $0.12 per game, in the long run.

   (b) In four hours, at a rate of 35 hands per hour, a total of 140 hands will be played. So, the player can expect to *lose* $(140)(\$0.1158) \approx \$16.21$.

5. (a) This is a binomial experiment. There are a fixed number of trials ($n = 10$) where each trial corresponds to a randomly chosen freshman, the trials are independent, there are only two possible outcomes (graduated or did not graduate within six years), and the probability of a graduation within six years is fixed for all trials ($p = 0.54$).

(b) This is not a binomial experiment because the number of trials is not fixed.

6. (a) We have $n = 10$, $p = 0.05$, and $x = 1$. $P(1) = {}_{10}C_1 \cdot (0.05)^1 (0.95)^9 \approx 0.3151$

In about 31.5% of random samples of 10 visitors to the ER, exactly one will die within a year.

(b) $P(X < 2) = P(X = 0) + P(X = 1) = {}_{25}C_0 \cdot (0.05)^0 (0.95)^{25} + {}_{25}C_1 \cdot (0.05)^1 (0.95)^{24}$
$= 0.2774 + 0.3650 = 0.6424$

In about 64% of random samples of 25 ER patients, fewer than 2 of the twenty-five will die within one year.

(c) $P(X \geq 2) = 1 - P(X < 2) = 1 - 0.6424 = 0.3576$

In about 36% of random samples of 25 ER patients, at least 2 will die within one year.

(d) The probability that at least 8 of 10 will not die is the same as the probability that 2 or fewer out of 10 will die.

$P(X \leq 2) = P(X = 0) + P(X = 1) + P(X = 2)$
$= {}_{10}C_0 \cdot (0.05)^0 (0.95)^{10} + {}_{10}C_1 \cdot (0.05)^1 (0.95)^9 + {}_{10}C_2 \cdot (0.05)^2 (0.95)^8$
$= 0.5987 + 0.3151 + 0.0746 = 0.9885$ or 0.9884, depending on rounding

In about 99% of random samples of 10 ER patients, at least 8 will not die.

(e) $P(X > 3) = 1 - P(X \leq 3)$
$= 1 - \left[P(X = 0) + P(X = 1) + P(X = 2) + P(X = 3) \right]$
$= 1 - \left[{}_{30}C_0 \cdot (0.05)^0 (0.95)^{30} + {}_{30}C_1 \cdot (0.05)^1 (0.95)^{29} \right.$
$\left. + {}_{30}C_2 \cdot (0.05)^2 (0.95)^{28} + {}_{30}C_3 \cdot (0.05)^2 (0.95)^{27} \right]$
$= 1 - \left[0.2146 + 0.3389 + 0.2586 + 0.1270 \right] = 0.0608$ or 0.0609, depending on rounding

Since 0.0608 is greater than 0.05, it would not be unusual for more than 3 of thirty ER patients to die within one year of their visit.

(f) $E(X) = \mu_X = n \cdot p = 1000(0.05) = 50$; $\sigma_X = \sqrt{np(1-p)} = \sqrt{1000(0.05)(0.95)} \approx 6.9$

(g) No. Since $np \cdot (1-p) = 800 \cdot (0.05) \cdot (0.95) = 38 \geq 10$, we can use the Empirical Rule to check for unusual observations. $E(X) = \mu_X = n \cdot p = 800(0.05) = 40$ and $\sigma_X = \sqrt{np(1-p)} = \sqrt{800(0.05)(0.95)} \approx 6.2$.

We have that 51 is above the mean and $\mu + 2\sigma = 40 + 2 \cdot 6.2 = 52.4$. This indicates that 51 is within two standard deviations of the mean, so observing 51 deaths after one year in an ER with 800 visitors would not be unusual.

7. (a) We have $n = 15$ and $p = 0.6$.

Using the binomial probability table, we get: $P(10) = 0.1859$

(b) Using the cumulative binomial probability table, we get: $P(X < 5) = P(X \leq 4) = 0.0093$

(c) Using the complement rule, we get: $P(X \geq 5) = 1 - P(X < 5) = 1 - 0.0093 = 0.9907$

(d) Using the binomial probability table, we get:
$P(7 \leq X \leq 12) = P(7) + P(8) + P(9) + P(10) + P(11) + P(12)$
$= 0.1181 + 0.1771 + 0.2066 + 0.1859 + 0.1268 + 0.0634 = 0.8779$

Using the cumulative binomial probability table, we get:

$$P(7 \le X \le 12) = P(X \le 12) - P(X \le 6) = 0.9729 - 0.0950 = 0.8779$$

There is a 0.8779 [Tech: 0.8778] probability that in a random sample of 15 U.S. women 18 years old or older, between 7 and 12, inclusive, would feel that the minimum driving age should be 18.

(e) $E(X) = \mu_X = np = 200(0.6) = 120$ women;

$$\sigma_X = \sqrt{np(1-p)} = \sqrt{200 \cdot (0.6) \cdot (0.4)} = \sqrt{48} \approx 6.928 \text{ or about 6.9 women}$$

(f) No; since $np(1-p) = 48 \ge 10$, we can use the Empirical Rule to check for unusual observations.

$\mu - 2\sigma \approx 120 - 13.8 = 106.2$ and $\mu + 2\sigma \approx 120 + 13.8 = 133.8$

Since 110 is within this range of values, it would not be considered an unusual observation.

8. (a)

Distribution

x	$P(x)$
0	0.00002
1	0.00037
2	0.00385
3	0.02307
4	0.08652
5	0.20764
6	0.31146
7	0.26697
8	0.10011

$x \cdot P(x)$	$(x - \mu_x)^2 \cdot P(x)$
0.0000	0.0005
0.0004	0.0092
0.0077	0.0615
0.0692	0.2076
0.3461	0.3461
1.0382	0.2076
1.8688	0.0000
1.8688	0.2670
0.8009	0.4005
Σ 6.0000	1.5000

(b) Using the formulas from Section 6.1:

$\mu_X = 6$ (from first column in table above and to the right)

$\sigma_X = \sqrt{\sigma_X^2} = \sqrt{1.5} \approx 1.2$ (from second column in table above and to the right)

Using the formulas from Section 6.2:

$\mu_X = np = 8(0.75) = 6$ and $\sigma_X = \sqrt{np \cdot (1-p)} = \sqrt{8(0.75)(0.25)} \approx 1.2$.

(c)

The distribution is skewed left.

9. If $np \cdot (1-p) \ge 10$, then the Empirical Rule can be used to check for unusual observations.

10. In sampling from large populations without replacement, the trials may be assumed to be independent provided that the sample size is small in relation to the size of the population. As a general rule of thumb, this condition is satisfied if the sample size is less than 5% of the population size.

11. A random variable X, the number of successes in a fixed interval, follows a Poisson process provided that following conditions are met:
    1) the probability of two or more successes in any sufficiently small subinterval is 0;
    2) the probability of success is the same for any two intervals of equal length;
    3) the number of successes in any interval is independent of the number of successes in any other non-overlapping interval.

12. Use $P(x) = \dfrac{(\lambda t)^x}{x!} \cdot e^{-\lambda t}$ with $\lambda = 0.05$ and $t = 30$ so that $\lambda t = 1.5$:

    (a) $P(2) = \dfrac{1.5^2}{2!} \cdot e^{-1.5} = 0.2510$

    (b) $P(X < 2) = P(0) + P(1) = 0.22313 + 0.33470 \approx 0.55783$

    (c) $P(X \geq 2) = 1 - P(X < 2) = 1 - 0.5578 = 0.4422$

    (d) $P(1 \leq X \leq 3) = P(1) + P(2) + P(3) = 0.33470 + 0.25102 + 0.12551 = 0.71123$

    (e) $\mu_X = \lambda t = (0.05) \cdot 30 = 1.5$; $\sigma_X = \sqrt{\lambda t} = \sqrt{1.5} \approx 1.2$

13. We need to find $P(X \geq 2)$ where X is the number of defects in a roll. We have $\mu = \lambda \cdot t = \dfrac{1}{500} \cdot (100) = 0.2$

    $$P(X \geq 2) = 1 - P(X \leq 1) = 1 - \left[\frac{0.2^0}{0!} \cdot e^{-0.2} + \frac{0.2^1}{1!} \cdot e^{-0.2} \right] = 1 - 0.9825 = 0.0175$$

    About 1.75% of the rolls are rejected.

14. We have that maintenance calls occur at the rate of two per month, so $\lambda = 2$. The time interval of interest is one month, so $t = 1$. Therefore, $\lambda t = (2)(1) = 2$.

    $$P(X > 4) = 1 - P(X \leq 4)$$
    $$= 1 - \left[\frac{2^0}{0!} \cdot e^{-2} + \frac{2^1}{1!} \cdot e^{-2} + \frac{2^2}{2!} \cdot e^{-2} + \frac{2^3}{3!} \cdot e^{-2} + \frac{2^4}{4!} \cdot e^{-2} \right]$$
    $$= 1 - 0.9473 = 0.0527$$

    Since $P(X > 4) > 0.05$, it would not be unusual for the copy machine to require more than 4 maintenance calls in a given month.

15. We have $n = 40$, $p = 0.17$, $x = 12$, and find $\mu_X = n \cdot p = 40(0.17) = 6.8$. Since 12 is above the mean, we compute $P(X \geq 12)$. Using technology we find $P(X \geq 12) \approx 0.0301 < 0.05$, so the result of the survey is unusual. This suggests that emotional abuse may be a factor that increases the likelihood of self-injurious behavior.

Chapter 6 Test

1. **(a)** The number of days of measurable rainfall in Honolulu during a year is a discrete random variable because its value results from counting. If we let the random variable R represent the number of days for which there is measurable rainfall, the possible values for R are integers between 0 and 365, inclusive. That is, $r = 0, 1, 2, \ldots 365$.

    (b) The miles per gallon for a Toyota Prius is a continuous random variable because its value results from a measurement. If we let the random variable M represent the miles per gallon, the possible values for M are all nonnegative real numbers. That is, $m \geq 0$.
    (Note: we need to include the possibility that $m = 0$ since the engine could be idling, in which case it would be getting 0 miles to the gallon.)

    (c) The number of golf balls hit into the ocean on the famous 18th hole at Pebble Beach is a discrete random variable because its value results from a count. If we let the random variable X represent the number of golf balls hit into the ocean, the possible values for X are nonnegative integers. That is, $x = 0, 1, 2, \ldots$

    (d) The weight of a randomly selected robin egg is a continuous random variable because its value results from a measurement. If we let the random variable W represent the weight of a robin's egg, the possible values for W are all nonnegative real numbers. That is, $w > 0$.

2. **(a)** This is a valid probability distribution because $\sum P(x) = 1$ and $0 \leq P(x) \leq 1$ for all x.

    (b) This is not a valid probability distribution because $P(4) < 0$.

3. **(a)**   Total number of Wimbledon Men's Single Finals = 20 + 14 + 14 = 48.

x (sets played)	$P(x)$
3	$\dfrac{20}{48} \approx 0.4167$
4	$\dfrac{14}{48} \approx 0.2917$
5	$\dfrac{14}{48} \approx 0.2917$

(b)

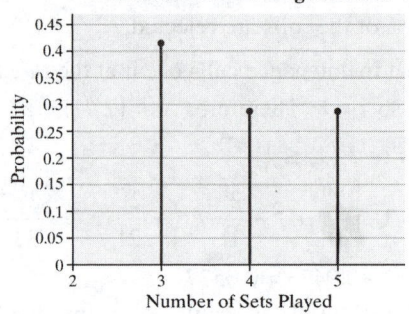

Wimbledon Men's Singles Finals

(c) $\mu_X = \sum [x \cdot P(x)] = 3 \cdot (0.4167) + 4 \cdot (0.2917) + 5 \cdot (0.2917)$
≈ 3.875 or about 3.9 sets
The Wimbledon men's single finals, if played many times, would be expected to last about 3.9 sets.

(d) $\sigma_X^2 = \sum (x - \mu_x)^2 \cdot P(x)$
$\approx (3 - 3.875)^2 \cdot 0.4167 + (4 - 3.875)^2 \cdot 0.2917 + (5 - 3.875)^2 \cdot 0.2917$
≈ 0.6928
$\sigma_X = \sqrt{\sigma_X^2} \approx \sqrt{0.6928} \approx 0.832$ or about 0.8 sets

4. $E(X) = \sum x \cdot P(x) = \$200 \cdot (0.998725) + (-\$99,800) \cdot (0.001275) = \$72.50$.

    If the company issues many \$100,000 1-year policies to 35-year-old males, then they will make an average profit of \$72.50 per male.

5. A probability experiment is said to be a binomial probability experiment provided:
   a) The experiment consists of a fixed number, n, of trials.
   b) The trials are all independent.
   c) For each trial there are two mutually exclusive (disjoint) outcomes, success or failure.
   d) The probability of success, p, is the same for each trial.

6. (a) This is not a binomial experiment because the number of trials is not fixed and there are not two mutually exclusive outcomes.

(b) This is a binomial experiment. There are a fixed number of trials ($n = 25$) where each trial corresponding to a randomly chosen property crime, the trials are independent, there are only two possible outcomes (cleared or not cleared), and the probability that the property crime is cleared is fixed for all trials ($p = 0.16$).

7. (a) We have $n = 20$ and $p = 0.80$. We can use the binomial probability table, or compute this by hand.

$$P(15) = {}_{20}C_{15} \cdot (0.8)^{15} (0.2)^5 \approx 0.1746$$

(b) $P(X \geq 19) = P(X = 19) + P(X = 20) = {}_{20}C_{19} \cdot (0.8)^{19} (0.2)^1 + {}_{20}C_1 \cdot (0.8)^{20} (0.2)^0$
$$= 0.0576 + 0.0115 = 0.0692$$

(c) Using the complement rule, we get:
$$P(X < 19) = 1 - P(X \geq 19) = 1 - 0.0692 = 0.9308$$

(d) Using the binomial probability table, we get:
$$P(15 \leq X \leq 17) = P(15) + P(16) + P(17) = 0.1746 + 0.2182 + 0.2054 = 0.5981$$

Using the cumulative binomial probability table, we get:
$$P(15 \leq X \leq 17) = P(X \leq 17) - P(X \leq 14) = 0.7939 - 0.1958 = 0.5981$$

There is a probability of 0.5981 that in a random sample of 20 couples, between 15 and 17 inclusive, would hide purchases from their mates.

8. (a) We have $n = 1200$ and $p = 0.5$. $\mu_X = np = 1200 \cdot (0.5) = 600$. We expect that 600 adult Americans would state that their taxes are too high.

(b) Since $np(1 - p) == 1200(0.5)(0.5) = 300 \geq 10$, we can use the Empirical Rule to check for unusual observations.

(c) $\mu_X = np = 1200 \cdot (0.5) = 600$ adult Americans
$$\sigma_X = \sqrt{np \cdot (1 - p)} = \sqrt{1200(0.5)(0.5)} \approx 17.3$$
$\mu - 2\sigma = 565.4$ and $\mu + 2\sigma = 634.6$. Since 640 is not between these values, a result of 640 would be unusual. This result contradicts the belief that adult Americans are equally split in their belief that the amount of tax they pay is too high.

9. (a)

Distribution		$x \cdot P(x)$	$(x - \mu_x)^2 \cdot P(x)$
x	$P(x)$	0	0.3277
0	0.3277	0.4096	0.0000
1	0.4096	0.4096	0.2048
2	0.2048	0.1536	0.2048
3	0.0512	0.0256	0.0576
4	0.0064	0.0016	0.0051
5	0.0003	Σ 1.0000	0.8000

(b) Using the formulas from Section 6.1:

$\mu_X = 1$ (from first column in table above and to the right)

$\sigma_X = \sqrt{\sigma_X^2} = \sqrt{0.8} \approx 0.9$ (from second column in table above and to the right)

Using the formulas from Section 6.2:

$\mu_X = np = 5(0.2) = 1$ and $\sigma_X = \sqrt{np \cdot (1-p)} = \sqrt{5(0.2)(0.8)} \approx 0.9$.

(c)

The distribution is skewed right.

10. Use $P(x) = \dfrac{(\lambda t)^x}{x!} \cdot e^{-\lambda t}$ with $\lambda = 0.05$ and $t = 8$ so that $\lambda t = 0.4$:

(a) $P(3) = \dfrac{0.4^3}{3!} \cdot e^{-0.4} = 0.0072$

(b) $P(X < 3) = P(0) + P(1) + P(2) = 0.67032 + 0.26813 + 0.05363 \approx 0.99208$

(c) $P(X \geq 3) = 1 - P(X < 3) = 1 - 0.99208 = 0.00792$

(d) $P(3 \leq X \leq 5) = P(3) + P(4) + P(5) = 0.00715 + 0.00072 + 0.00006 \approx 0.00793$

(e) $\mu_X = \lambda t = (0.05) \cdot 8 = 0.4$; $\sigma_X = \sqrt{\lambda t} = \sqrt{0.4} \approx 0.6$

11. We are given that cars arrive at the rate of 0.41 cars per minute, so $\lambda = 0.41$. The time interval of interest is 10 minutes (4:00 P.M. and 4:10 P.M.), so $t = 10$. Therefore, $\lambda t = 0.41(10) = 4.1$.

(a) $P(4) = \dfrac{4.1^4}{4!} \cdot e^{-4.1} \approx 0.1951$

On about 20 of every 100 randomly chosen days, exactly 4 cars will arrive at the bank's drive-through window between 4:00 P.M. and 4:10 P.M.

(b) $P(X < 4) = P(X \leq 3) = P(0) + P(1) + P(2) + P(3)$

$$= \frac{4.1^0}{0!} \cdot e^{-4.1} + \frac{4.1^1}{1!} \cdot e^{-4.1} + \frac{4.1^2}{2!} \cdot e^{-4.1} + \frac{4.1^3}{3!} \cdot e^{-4.1}$$

$$\approx 0.01657 + 0.06795 + 0.13929 + 0.19037$$

$$\approx 0.41418$$

On about 41 of every 100 randomly chosen days, fewer than 4 cars will arrive at the bank's drive-through window between 4:00 P.M. and 4:10 P.M.

(c) Using the complement rule, we get: $P(X \geq 4) = 1 - P(X < 4) = 1 - 0.4142 = 0.5858$

On about 59 of every 100 randomly chosen days, at least 4 cars will arrive at the bank's drive-through window between 4:00 P.M. and 4:10 P.M.

(d) $\mu_X = \lambda t = (0.41) \cdot 10 = 4.1$ cars

(e) $\sigma_X = \sqrt{\lambda t} = \sqrt{4.1} \approx 2.0$ cars

Case Study: The Voyage of the St. Andrew

1. <u>Distribution of the number of families per parish:</u>

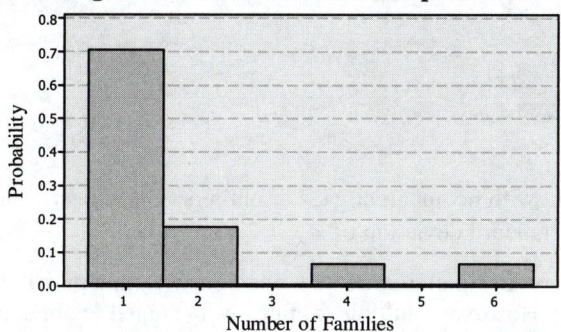

The distribution of the number of families per parish is skewed right with two gaps. The heaviest concentration occurs for 1 family per parish. The expected number of families per parish is $\mu_X \approx 1.65$ with a standard deviation of $\sigma_X \approx 1.33$.

<u>Distribution of the number of freights purchased:</u>

The distribution of the number of freights purchased is skewed right with two gaps. The most frequent purchase was 2 freights. The expected number of freights purchased is $\mu_X \approx 2.54$ with a standard deviation of $\sigma_X \approx 1.00$.

<u>Distribution of the number of family members on board:</u>

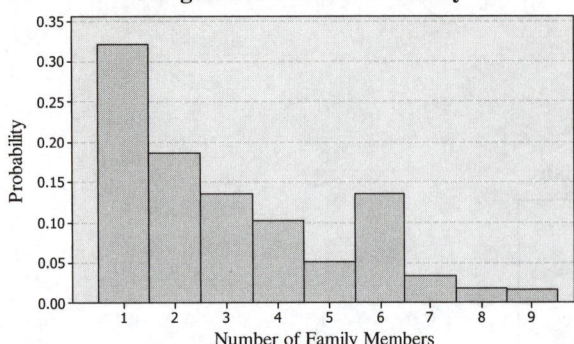

Histogram of Number in Family

The distribution of family members is skewed right with no apparent gaps or outliers. The expected number of family members on board is $\mu_X \approx 3.10$ with a standard deviation of $\sigma_X \approx 2.13$.

2. It appears that the *Neuländers* were mildly successful in signing more than one family from a parish. In the vast majority of cases, only one family was signed. However, multiple families were signed in about 30% of the cases. Since villages were likely spread out in the German countryside, it is not likely that many families knew each other prior to undertaking the voyage. When multiple families were signed, it is very likely that families from the same parish knew each other.

3. Using 2.54 as the mean number of freights, the estimated cost per family would be:

   $(2.54)(7.5) = 19.05$ or about 19 pistoles

   $(2.54)(2000) = \$5080$ or about \$5100

4. Answers will vary. It would not be appropriate to estimate the average cost of the voyage from the mean family size because the age of the family members is a major factor. The cost for a family of 6 adults would be significantly more than the cost for a family of 6 with 2 adults and 4 children.

5. Based on the distribution of freights from the *St. Andrew*, we will assume the probability is $p = 0.05$ that a family would purchase more than 4 freights. Using this value, $n = 6$ (families), and assuming each family purchases freights independently of the others, the random variable X is a binomial random variable and represents the number of families purchasing more than 4 freights.

   $$P(X \geq 5) = P(5) + P(6)$$
   $$= {}_6C_5 \cdot (0.05)^5 (0.95)^1 + {}_6C_6 \cdot (0.05)^6 (0.95)^0 \approx 0.0000018$$

   This probability is very small. Therefore, it seems very unlikely that these families came from a population similar to that of the Germans on board the *St. Andrew*.

6. Answers will vary. Topics for discussion could include issues such as independence or whether or not families were disjoint units. For example, did some families send their children with a wealthier family friend to give their children a chance at a better life?

Chapter 7
The Normal Probability Distribution

Section 7.1

1. probability density function

3. True

5. $\mu - \sigma$; $\mu + \sigma$

7. No, the graph cannot represent a normal density function because it is not symmetric.

9. No, the graph cannot represent a normal density function because it crosses below the horizontal axis. That is, it is not always greater than 0.

11. Yes, the graph can represent a normal density function.

13. (a) The figure presents the graph of the density function with the area we wish to find shaded.

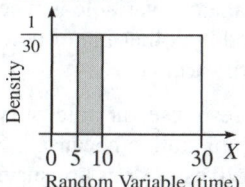
Random Variable (time)

The width of the rectangle is $10 - 5 = 5$ and the height is $\frac{1}{30}$. Thus, the area between 5 and 10 is $5\left(\frac{1}{30}\right) = \frac{1}{6}$. The probability that the friend is between 5 and 10 minutes late is $\frac{1}{6}$.

(b) 40% of 30 minutes is $0.40 \cdot 30 = 12$ minutes. So, there is a 40% probability your friend will arrive within the next 12 minutes.

15. The figure presents the graph of the density function with the area we wish to find shaded.

Random Variable (time)

The width of the rectangle is $30 - 20 = 10$ and the height is $\frac{1}{30}$. Thus, the area between 20 and 30 is $10\left(\frac{1}{30}\right) = \frac{1}{3}$. The probability that the friend is at least 20 minutes late is $\frac{1}{3}$.

17. (a)

(b) $P(0 \leq X \leq 0.2) = 1(0.2 - 0) = 0.2$

(c) $P(0.25 \leq X \leq 0.6) = 1(0.6 - 0.25) = 0.35$

(d) $P(X \geq 0.95) = 1(1 - 0.95) = 0.05$

(e) Answers will vary.

19. The histogram is symmetrical and bell-shaped, so a normal distribution can be used as a model for the variable.

21. The histogram is skewed to the right, so normal distribution cannot be used as a model for the variable.

23. Graph A matches $\mu = 10$ and $\sigma = 2$, and graph B matches $\mu = 10$ and $\sigma = 3$. We can tell because a higher standard deviation makes the graph lower and more spread out.

25. The center is at 2, so $\mu = 2$. The distance to the inflection points is ± 3, so $\sigma = 3$.

27. The center is at 100, so $\mu = 100$. The distance to the inflection points is ± 15, so $\sigma = 15$.

29.

31. (a)

(b)

(c) Interpretation 1: 15.87% of the cell phone plans in the United States are less than $44.00 per month.

Interpretation 2: The probability is 0.1587 that a randomly selected cell phone plan in the United States is less than $44.00 per month.

33. (a)

(b)

(c) Interpretation 1: 2.28% of all full-term babies have a birth weight of at least 4410 grams.

Interpretation 2: The probability is 0.0228 that the birth weight of a randomly chosen full-term baby is at least 4410 grams.

35. (a) Interpretation 1: The proportion of human pregnancies that last more than 280 days is 19.08%.

Interpretation 2: The probability is 0.1908 that a randomly selected human pregnancy lasts more than 280 days.

(b) Interpretation 1: The proportion of human pregnancies that last between 230 and 260 days is 34.16%.

Interpretation 2: The probability is 0.3416 that a randomly selected human pregnancy lasts between 230 and 260 days.

37. (a)

(b) The normal density function appears to describe the distance Michael hits a pitching wedge fairly accurately. Looking at the graph, the normal curve is a fairly good approximation to the histogram.

39. (a) This is an observational study in which the data are collected over time.

(b) The explanatory variable is whether the patient had hypothermia or not. This is a qualitative variable.

(c) The first response variable, survival status, is qualitative because the patient either survives or does not survive.

The second response variable, length of stay in the ICU, is quantitative because it counts the number of days.

The third response variable, time spent on a ventilator, is quantitative because it counts time spent in hours.

(d) Time on the ventilator is a statistic because it is calculated from a sample of male patients who had an out-of-hospital cardiac arrest, instead of the entire population of those patients.

(e) The population is male patients who have an out-of-hospital cardiac arrest.

(f) $P(\text{survive, hypothermia}) = \dfrac{37}{52} \approx 0.712$

$P(\text{survive, no hypothermia}) = \dfrac{43}{74} \approx 0.581$

Section 7.2

1. standard normal distribution

3. 0.3085

5. The standard normal table (Table V) gives the area to the left of the z-score. Thus, we look up each z-score and read the corresponding area. We can also use technology. The areas are:

(a) The area to the left of $z = -2.45$ is 0.0071.

(b) The area to the left of $z = -0.43$ is 0.3336.

(c) The area to the left of $z = 1.35$ is 0.9115.

(d) The area to the left of $z = 3.49$ is 0.9998.

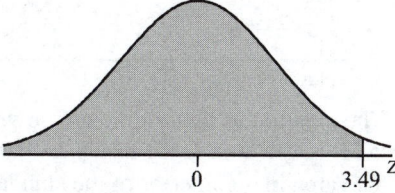

7. The standard normal table (Table V) gives the area to the left of the z-score. Thus, we look up each z-score and read the corresponding area from the table. The area to the right is one minus the area to the left. We can also use technology to find the area. The areas are:

(a) The area to the right of $z = -3.01$ is $1 - 0.0013 = 0.9987$.

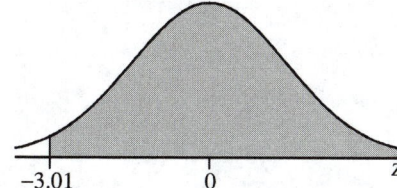

(b) The area to the right of $z = -1.59$ is $1 - 0.0559 = 0.9441$.

(c) The area to the right of $z = 1.78$ is $1 - 0.9625 = 0.0375$.

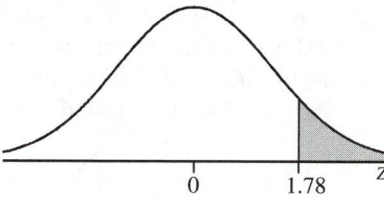

(d) The area to the right of $z = 3.11$ is $1 - 0.9991 = 0.0009$.

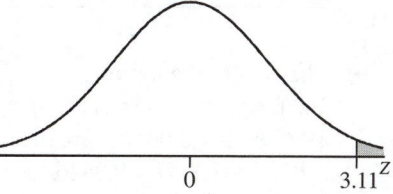

9. To find the area between two z-scores using the standard normal table (Table V), we look up the area to the left of each z-score and then we find the difference between these two. We can also use technology to find the areas:

(a) The area to the left of $z = -2.04$ is 0.0207, and the area to the left of $z = 2.04$ is 0.9793. So, the area between is $0.9793 - 0.0207 = 0.9586$.

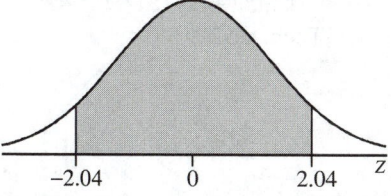

(b) The area to the left of $z = -0.55$ is 0.2912, and the area to the left of $z = 0$ is 0.5. So, the area between is $0.5 - 0.2912 = 0.2088$.

(c) The area to the left of $z = -1.04$ is 0.1492, and the area to the left of $z = 2.76$ is 0.9971. So, the area between is $0.9971 - 0.1492 = 0.8479$

11. (a) The area to the left of $z = -2$ is 0.0228, and the area to the right of $z = 2$ is $1 - 0.9772 = 0.0228$. So, the total area is $0.0228 + 0.0228 = 0.0456$. [Tech: 0.0455]

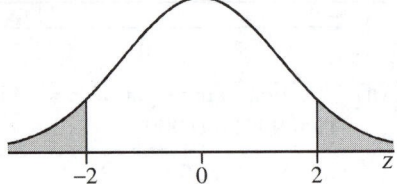

(b) The area to the left of $z = -1.56$ is 0.0594, and the area to the right of $z = 2.56$ is $1 - 0.9948 = 0.0052$. So, the total area is $0.0594 + 0.0052 = 0.0646$.

(c) The area to the left of $z = -0.24$ is 0.4052, and the area to the right of $z = 1.20$ is $1 - 0.8849 = 0.1151$. So, the total area is $0.4052 + 0.1151 = 0.5203$.
[Tech: 0.5202]

13. The area in the interior of the standard normal table (Table V) that is closest to 0.1000 is 0.1003, corresponding to $z = -1.28$.

15. The area to the left of the unknown z-score is $1 - 0.25 = 0.75$. The area in the interior of the standard normal table (Table V) that is closest to 0.7500 is 0.7486, corresponding to $z = 0.67$.

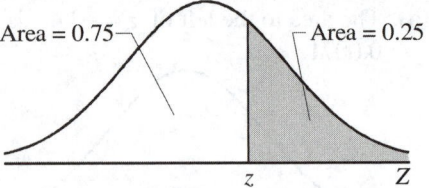

17. The z-scores for the middle 99% are the z-scores for the top and bottom 0.5%. The area to the left of z_1 is 0.005, and the area to the left of z_2 is 0.995.
The areas in the interior of the standard normal table (Table V) that are closest to 0.0050 are 0.0049 and 0.0051. So, we use the average of their corresponding z-scores: -2.58 and -2.57, respectively.
This gives $z_1 = -2.575$. The areas in the interior of the standard normal table (Table V) that are closest to 0.9950 are 0.9949 and 0.9951. So, we use the average of their corresponding z-scores: 2.57 and 2.58, respectively. This gives $z_2 = 2.575$.

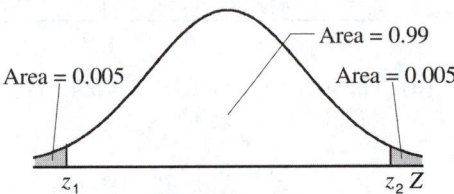

19. The area to the right of the unknown z-score is 0.01, so the area to the left is $1 - 0.01 = 0.99$. The area in the interior of the standard normal table (Table V) that is closest to 0.9900 is 0.9901, so $z_{0.01} = 2.33$.

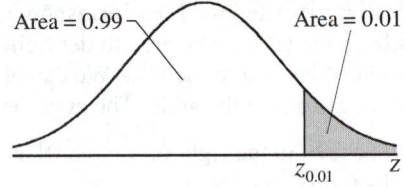

21. The area to the right of the unknown z-score is 0.025, so the area to the left is $1 - 0.025 = 0.975$. The area in the interior of the standard normal table (Table V) that is closest to 0.9750 is 0.9750, so $z_{0.025} = 1.96$.

23. $z = \dfrac{x - \mu}{\sigma} = \dfrac{35 - 50}{7} \approx -2.14$

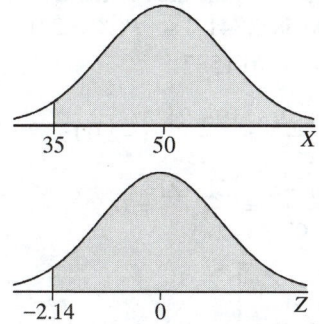

From Table V, the area to the left is 0.0162, so $P(X > 35) = 1 - 0.0162 = 0.9838$. [Tech: 0.9839]

25. $z = \dfrac{x - \mu}{\sigma} = \dfrac{45 - 50}{7} \approx -0.71$

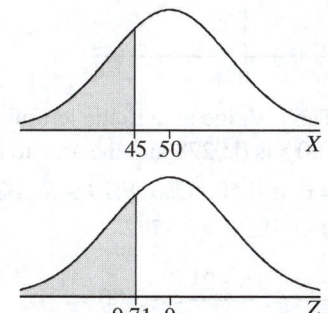

From Table V, the area to the left is 0.2389, so $P(X \le 45) = 0.2389$. [Tech: 0.2375]

27. $z_1 = \dfrac{x_1 - \mu}{\sigma} = \dfrac{40 - 50}{7} \approx -1.43$;

$z_2 = \dfrac{x_2 - \mu}{\sigma} = \dfrac{65 - 50}{7} \approx 2.14$

From Table V, the area to the left of $z_1 = -1.43$ is 0.0764 and the area to the left of $z_2 = 2.14$ is 0.9838, so $P(40 < X < 65) = 0.9838 - 0.0764 = 0.9074$. [Tech: 0.9074]

29. $z_1 = \dfrac{x_1 - \mu}{\sigma} = \dfrac{55 - 50}{7} \approx 0.71$;

$z_2 = \dfrac{x_2 - \mu}{\sigma} = \dfrac{70 - 50}{7} \approx 2.86$

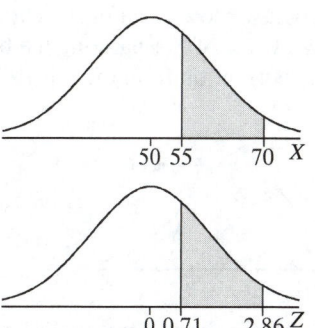

From Table V, the area to the left of $z_1 = 0.71$ is 0.7611 and the area to the left of $z_2 = 2.86$ is 0.9979, so $P(55 \le X \le 70) = 0.9979 - 0.7611 = 0.2368$. [Tech: 0.2354]

31. $z_1 = \dfrac{x_1 - \mu}{\sigma} = \dfrac{38 - 50}{7} \approx -1.71$;

$z_2 = \dfrac{x_2 - \mu}{\sigma} = \dfrac{55 - 50}{7} \approx 0.71$

From Table V, the area to the left of $z_1 = -1.71$ is 0.0436 and the area to the left of $z_2 = 0.71$ is 0.7611, so $P(38 < X \le 55) = 0.7611 - 0.0436 = 0.7175$. [Tech: 0.7192]

33. The figure below shows the normal curve with the unknown value of X separating the bottom 9% of the distribution from the top 91% of the distribution.

From Table V, the area closest to 0.09 is 0.0901. The corresponding z-score is -1.34. So, the 9th percentile for X is $x = \mu + z\sigma =$ $50 + (-1.34)(7) = 40.62$. [Tech: 40.61]

35. The figure below shows the normal curve with the unknown value of X separating the bottom 81% of the distribution from the top 19% of the distribution.

From Table V, the area closest to 0.81 is 0.8106. The corresponding z-score is 0.88. So, the 81st percentile for X is $x = \mu + z\sigma =$ $50 + 0.88(7) = 56.16$. [Tech: 56.15]

37. (a)

(b) $z = \dfrac{x - \mu}{\sigma} = \dfrac{20 - 21}{1} = -1.00$

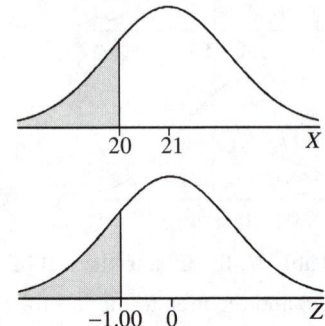

From Table V, the area to the left of $z = -1.00$ is 0.1587, so $P(X < 20) = 0.1587$.

(c) $z = \dfrac{x - \mu}{\sigma} = \dfrac{22 - 21}{1} = 1.00$

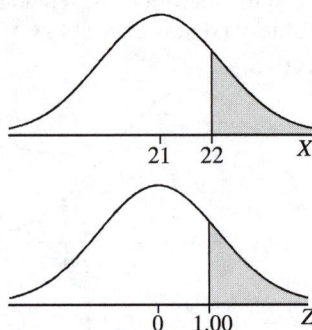

From Table V, the area to the left of $z = 1.00$ is 0.8413, so $P(X > 22) = 1 - 0.8413 = 0.1587$.

(d) $z_1 = \dfrac{x_1 - \mu}{\sigma} = \dfrac{19 - 21}{1} = -2.00$;

$z_2 = \dfrac{x_2 - \mu}{\sigma} = \dfrac{21 - 21}{1} = 0$

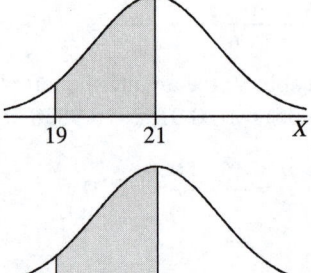

From Table V, the area to the left of $z_1 = -2.00$ is 0.0228 and the area to the left of $z_2 = 0$ is 0.5000, so $P(19 \le X \le 21) = 0.5000 - 0.0228 = 0.4772$.

(e) $z = \dfrac{x - \mu}{\sigma} = \dfrac{18 - 21}{1} = -3.00$

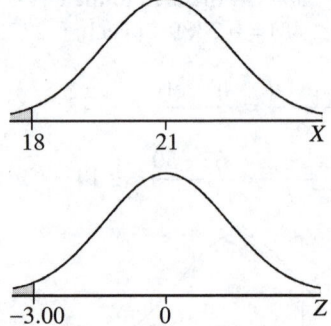

From Table V, the area to the left of $z = -3.00$ is 0.0013, so $P(X < 18) = 0.0013$.

Yes, it would be unusual for an egg to hatch in less than 18 days. Only about 1 egg in 1000 hatches in less than 18 days.

39. (a) $z_1 = \dfrac{x_1 - \mu}{\sigma} = \dfrac{1000 - 1262}{118} \approx -2.22$;

$z_2 = \dfrac{x_2 - \mu}{\sigma} = \dfrac{1400 - 1262}{118} \approx 1.17$

From Table V, the area to the left of $z_1 = -2.22$ is 0.0132 and the area to the left of $z_2 = 1.17$ is 0.8790, so
$P(1000 \le X \le 1400) = 0.8790 - 0.0132 = 0.8658$. [Tech: 0.8657]

(b) $z = \dfrac{x - \mu}{\sigma} = \dfrac{1000 - 1262}{118} \approx -2.22$

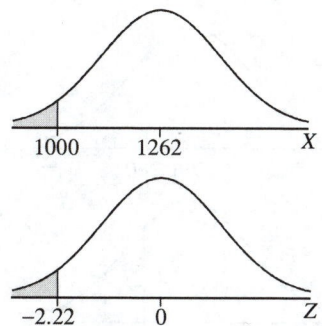

From Table V, the area to the left of $z = -2.22$ is 0.0132, so $P(X < 1000) = 0.0132$.

(c) $z = \dfrac{x - \mu}{\sigma} = \dfrac{1200 - 1262}{118} \approx -0.53$

From Table V, the area to the left of $Z = -0.53$ is 0.2981, so $P(X > 1200) = 1 - 0.2981 = 0.7019$. [Tech: 0.7004] So, the proportion of 18-ounce bags of Chips Ahoy! cookies that contains more than 1200 chocolate chips is 0.7019, or 70.19%.

(d) $z = \dfrac{x - \mu}{\sigma} = \dfrac{1125 - 1262}{118} \approx -1.16$.

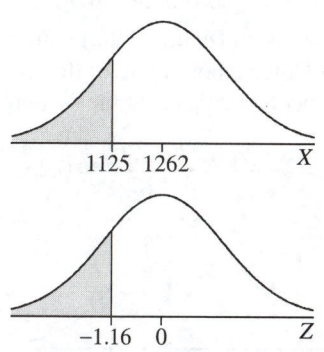

From Table V, the area to the left of $z = -1.16$ is 0.1230, so $P(X < 1125) = 0.1230$. [Tech: 0.1228] So, the proportion of 18-ounce bags of Chips Ahoy! cookies that contains fewer than 1125 chocolate chips is 0.1230, or 12.30%.

(e) $z = \dfrac{x - \mu}{\sigma} = \dfrac{1475 - 1262}{118} \approx 1.81$

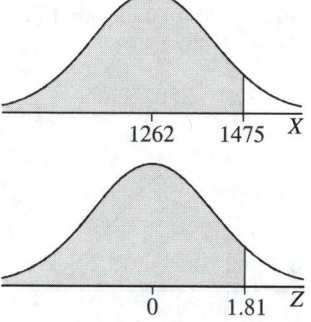

From Table V, the area to the left of $z = 1.81$ is 0.9649, so $P(X < 1475) = 0.9649$. [Tech: 0.9645] An 18-ounce bag of Chips Ahoy! Cookies that contains 1475 chocolate chips is at the 96th percentile.

(f) $z = \dfrac{x - \mu}{\sigma} = \dfrac{1050 - 1262}{118} \approx -1.80$.

From Table V, the area to the left of $z = -1.80$ is 0.0359, so $P(X < 1050) = 0.0359$. [Tech: 0.0362] An 18-ounce bag of Chips Ahoy! cookies that contains 1050 chocolate chips is at the 4th percentile.

41. (a) $z = \dfrac{x - \mu}{\sigma} = \dfrac{270 - 266}{16} = 0.25$

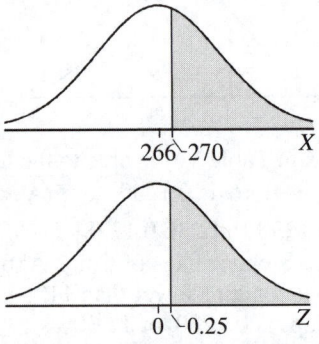

From Table V, the area to the left of $z = 0.25$ is 0.5987, so $P(X > 270) = 1 - 0.5987 = 0.4013$. So, the proportion of human pregnancies that last more than 270 days is 40.13%.

(b) $z = \dfrac{x - \mu}{\sigma} = \dfrac{250 - 266}{16} = -1$

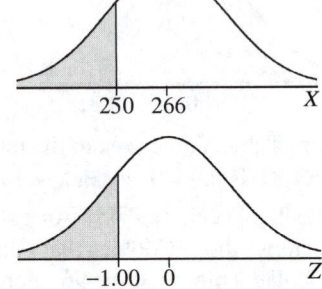

From Table V, the area to the left of $z = -1.00$ is 0.1587, so $P(X < 250) = 0.1587$. So, the proportion of human pregnancies that last less than 250 days is 0.1587, or 15.87%.

(c) $z_1 = \dfrac{x_1 - \mu}{\sigma} = \dfrac{240 - 266}{16} \approx -1.63$;

$z_2 = \dfrac{x_2 - \mu}{\sigma} = \dfrac{280 - 266}{16} \approx 0.88$

From Table V, the area to the left of $z_1 = -1.63$ is 0.0516 and the area to the left of $z_2 = 0.88$ is 0.8106, so $P(240 \le X \le 280) = 0.8106 - 0.0516 = 0.7590$. [Tech: 0.7571] So, the proportion of human pregnancies that last between 240 and 280 days is 0.7590, or 75.90%.

(d) $z = \dfrac{x - \mu}{\sigma} = \dfrac{280 - 266}{16} \approx 0.88$

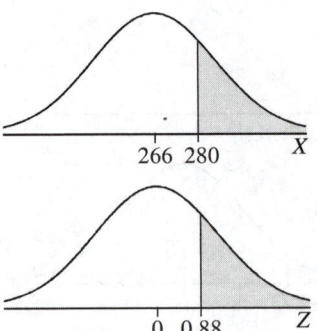

From Table V, the area to the left of $z = 0.88$ is 0.8106, so $P(X > 280) = 1 - 0.8106 = 0.1894$. [Tech: 0.1908]

(e) $z = \dfrac{x - \mu}{\sigma} = \dfrac{245 - 266}{16} \approx -1.31$

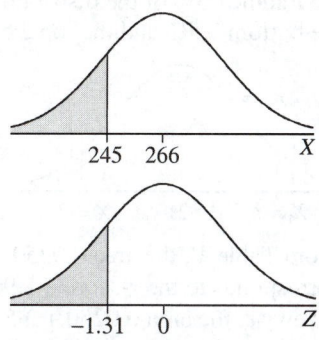

From Table V, the area to the left of $z = -1.31$ is 0.0951, so $P(X \le 245) = 0.0951$. [Tech: 0.0947]

(f) $z = \dfrac{x - \mu}{\sigma} = \dfrac{224 - 266}{16} \approx -2.63$

From Table V, the area to the left of $z = -2.63$ is 0.0043, so $P(X < 224) = 0.0043$.

Yes, "very preterm" babies are unusual. Only about 4 births in 1000 are "very preterm."

43. (a) $z = \dfrac{x - \mu}{\sigma} = \dfrac{24.9 - 25}{0.07} \approx -1.43$

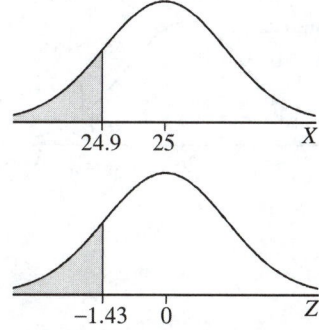

From Table V, the area to the left of $z = -1.43$ is 0.0764, so $P(X < 24.9) =$

0.0764. [Tech: 0.0766] So, the proportion of rods that has a length less than 24.9 cm is 0.0764, or 7.64%.

(b) $z_1 = \dfrac{x_1 - \mu}{\sigma} = \dfrac{24.85 - 25}{0.07} \approx -2.14$;

$z_2 = \dfrac{x_2 - \mu}{\sigma} = \dfrac{25.15 - 25}{0.07} \approx 2.14$

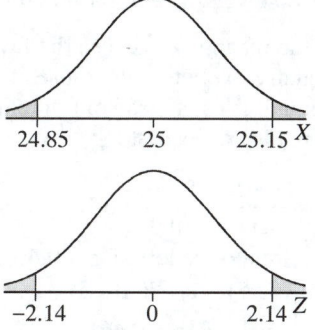

From Table V, the area to the left of $z_1 = -2.14$ is 0.0162, so $P(X < 24.85) = 0.0162$. The area to the left of $z_2 = 2.14$ is 0.9838, so $P(X > 25.15) = 1 - 0.9838 = 0.0162$. So, $P(X < 24.85 \text{ or } X > 25.15) = 2(0.0162) = 0.0324$. [Tech: 0.0321] The proportion of rods that will be discarded is 0.0324, or 3.24%.

(c) The manager should expect to discard $5000(0.0324) = 162$ of the 5000 steel rods.

(d) $z_1 = \dfrac{x_1 - \mu}{\sigma} = \dfrac{24.9 - 25}{0.07} \approx -1.43$;

$z_2 = \dfrac{x_2 - \mu}{\sigma} = \dfrac{25.1 - 25}{0.07} \approx 1.43$

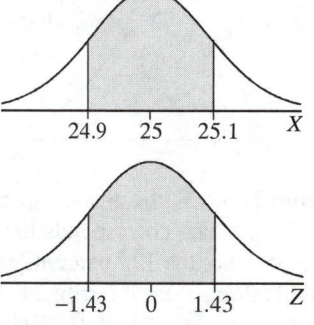

From Table V, the area to the left of $z_1 = -1.43$ is 0.0764 and the area to the left of $z_2 = 1.43$ is 0.9236, so $P(24.9 \le X \le 25.1) = 0.9236 - 0.0764 = 0.8472$. [Tech: 0.8469] So, 0.8472, or

84.72%, of the rods manufactured will be between 24.9 and 25.1 cm. Let n represent the number of rods that must be manufactured. Then, $0.8472n = 10,000$,

so $n = \dfrac{10,000}{0.8472} \approx 11,803.59$. Increase this to the next whole number: 11,804. To meet the order, the manager should manufacture 11,804 rods. [Tech: 11,808 rods]

45. (a) If the mean is zero, then the favored team is equally likely to win or lose relative to the spread. Yes, a mean of 0 implies that the spreads are accurate.

(b) $z = \dfrac{x - \mu}{\sigma} = \dfrac{5 - 0}{10.9} \approx 0.46$. From Table V, the area to the left of $z = 0.46$ is 0.6772, so

$P(X \geq 5) = 1 - P(X < 5)$

$= 1 - P(z < 0.46)$

$= 1 - 0.6772 = 0.3228$.

[Tech: 0.3232] The probability that the favored team wins by 5 or more points relative to the spread is 0.3228.

(c) $z = \dfrac{x - \mu}{\sigma} = \dfrac{-2 - 0}{10.9} \approx -0.18$. From Table V, the area to the left of $z = -0.18$ is 0.4286, so

$P(X \leq -2) = P(z \leq -0.18) = 0.4286$.

[Tech: 0.4272] The probability that the favored team loses by 2 or more points relative to the spread is 0.4286.

47. (a) The figure below shows the normal curve with the unknown value of X separating the bottom 17% of the distribution from the top 83%.

From Table V, the area closest to 0.17 is 0.1711, which corresponds to the z-score -0.95. So, the 17th percentile for incubation times of fertilized chicken eggs is $x = \mu + z\sigma = 21 + (-0.95)(1) \approx 20$ days.

(b) The figure below shows the normal curve with the unknown values of X separating the middle 95% of the distribution from the bottom 2.5% and the top 2.5%.

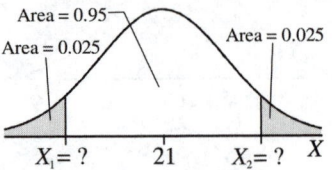

From Table V, the area 0.0250 corresponds to the z-score -1.96. Likewise, the area $0.0250 + .95 = 0.975$ corresponds to the z-score 1.96. Now, $x_1 = \mu + z_1\sigma = 21 + (-1.96)(1) \approx 19$ and $x_2 = \mu + z_2\sigma = 21 + 1.96(1) \approx 23$. Thus, the incubation times that make up the middle 95% of fertilized chicken eggs is between 19 and 23 days.

49. (a) The figure below shows the normal curve with the unknown value of X separating the bottom 30% of the distribution from the top 70%.

From Table V, the area closest to 0.30 is 0.3015, which corresponds to the z-score -0.52. So, the 30th percentile for the number of chocolate chips in an 18-ounce bag of Chips Ahoy! cookies is $x = \mu + z\sigma$

$= 1262 + (-0.52)(118) \approx 1201$ chocolate chips. [Tech: 1200 chocolate chips]

(b) The figure below shows the normal curve with the unknown values of X separating the middle 99% of the distribution from the bottom 0.5% and the top 0.5%.

From Table V, the areas 0.0049 and 0.0051 are equally close to 0.005. We average the corresponding z-scores -2.58 and -2.57 to obtain $z_1 = -2.575$. Likewise, the area $0.005 + 0.99 = 0.995$ is equally close to 0.9949 and 0.9951.
We average the corresponding z-scores 2.57 and 2.58 to obtain $z_2 = 2.575$. Now,

$x_1 = \mu + z_1\sigma = 1262 + (-2.575)(118) \approx 958$

and

$x_2 = \mu + z_2\sigma = 1262 + 2.575(118) \approx 1566$.

The number of chocolate chips that make up the middle 99% of 18-ounce bags of Chips Ahoy! cookies is 958 to 1566 chips.

51. (a) $z = \dfrac{x - \mu}{\sigma} = \dfrac{20 - 17}{2.5} = 1.2$

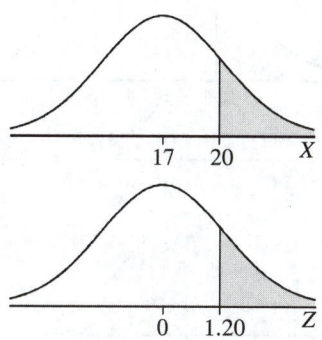

From Table V, the area to the left of $z = 1.20$ is 0.8849, so $P(X > 20) = 1 - 0.8849 = 0.1151$. So, about 11.51% of Speedy Lube's customers receive the service for half price.

(b) The figure below shows the normal curve with the unknown value of X separating the top 3% of the distribution from the bottom 97%.

From Table V, the area closest to 0.97 is 0.9699, which corresponds to the z-score 1.88. So, $x = \mu + z\sigma = 17 + 1.88(2.5) \approx 22$.

In order to discount only about 3% of its customers, Speedy Lube should make the guaranteed time limit 22 minutes.

53. The area under a normal curve can be interpreted as a probability, a proportion, or a percentile.

55. Reporting the probability as > 0.9999 accurately describes the event as highly likely, but not certain. Reporting the probability as 1.0000 might be incorrectly interpreted to mean that the event is certain.

Section 7.3

1. normal probability plot

3. (a)

(b) The correlation between the observed points and the expected z-scores is 0.991.

(c) From table VI the critical value is 0.906. Since 0.991>0.906 there is evidence that the sample data come from a population that is normally distributed.

5. (a)

(b) The correlation between the observed points and the expected z-scores is 0.873.

(c) From table VI the critical value is 0.898. Since 0.873<0.898 there is evidence that the sample data do not come from a population that is normally distributed.

7. The normal probability plot is approximately linear, so the sample data could come from a normally distributed population.

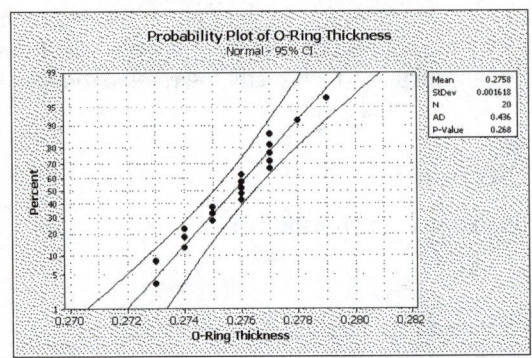

9. The normal probability plot is not approximately linear (points lie outside the provided bounds), so the sample data do not come from a normally distributed population.

11. (a)

The normal probability plot is roughly linear, and all the data lie within the provided bounds, so the sample data could come from a population that is normally distributed.

(b) $\sum x = 48,895$; $\sum x^2 = 62,635,301$; $n = 40$

$$\overline{x} = \frac{\sum x}{n} = \frac{49,895}{40} \approx 1247.4 \text{ chips};$$

$$s = \sqrt{\frac{\sum x^2 - \frac{(\sum x)^2}{n}}{n-1}}$$

$$= \sqrt{\frac{62,635,301 - \frac{(49,895)^2}{40}}{40-1}}$$

$$\approx 101.0 \text{ chips}$$

(c) $\mu - \sigma \approx \overline{x} - s = 1247.4 - 101.0 = 1146.4$;
$\mu + \sigma \approx \overline{x} + s = 1247.4 + 101.0 = 1348.4$

(d) $z = \dfrac{x - \mu}{\sigma} = \dfrac{1000 - 1247.4}{101.0} \approx -2.45$

From Table V, the area to the left of $z = -2.45$ is 0.0071, so $P(X \ge 1000) = 1 - 0.0071 = 0.9929$. [Tech: 0.9928]

(e) $z_1 = \dfrac{x_1 - \mu}{\sigma} = \dfrac{1200 - 1247.4}{101.0} \approx -0.47$;

$z_2 = \dfrac{x_2 - \mu}{\sigma} = \dfrac{1400 - 1247.4}{101.0} \approx 1.51$

From Table V, the area to the left of $z_1 = -0.47$ is 0.3192 and the area to the left of $z_2 = 1.51$ is 0.9345. So,

$P(1200 \le X \le 1400) = 0.9345 - 0.3192 = 0.6153$. [Tech: 0.6152] The proportion of 18-ounce bags of Chips Ahoy! that contains between 1200 and 1400 chips is 0.6153, or 61.53%.

13. (a)

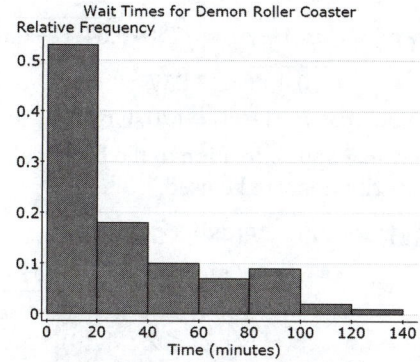

Wait Times for Demon Roller Coaster

(b) Distribution is skewed right.

(c)

Correlation = 0.922

The plotted points are not linear, so the data are not from a population that is normally distributed.

Section 7.4

1. $np(1-p) \ge 10$; np; $\sqrt{np(1-p)}$

3. $P(X \le 4.5)$; we use the continuity correction and find the area less than or equal to 4.5.

5. Approximate $P(X \ge 40)$ by computing the area under the normal curve to the right of $x = 39.5$.

7. Approximate $P(X = 8)$ by computing the area under the normal curve between $x = 7.5$ and $x = 8.5$.

9. Approximate $P(18 \le X \le 24)$ by computing the area under the normal curve between $x = 17.5$ and $x = 24.5$.

11. Approximate $P(X > 20) = P(X \ge 21)$ by computing the area under the normal curve to the right of $x = 20.5$.

13. Approximate $P(X > 500) = P(X \ge 501)$ by computing the area under the normal curve to the right of $x = 500.5$.

15. Using $P(x) = {}_nC_x p^x (1-p)^{n-x}$, with the parameters $n = 60$ and $p = 0.4$, we get

$P(20) = {}_{60}C_{20}(0.4)^{20}(0.6)^{40} \approx 0.0616$. Now $np(1-p) = 60 \cdot 0.4 \cdot (1-0.4) = 14.4 \ge 10$, so the normal approximation can be used, with

$\mu_X = np = 60(0.4) = 24$ and $\sigma_X = \sqrt{np(1-p)}$ $= \sqrt{14.4} \approx 3.795$. With continuity correction we calculate:

$P(20) \approx P(19.5 < X < 20.5)$

$= P\left(\dfrac{19.5 - 24}{\sqrt{14.4}} < Z < \dfrac{20.5 - 24}{\sqrt{14.4}}\right)$

$= P(-1.19 < Z < -0.92)$

$= 0.1788 - 0.1170$

$= 0.0618$ [Tech: 0.0603]

17. Using $P(x) = {}_nC_x p^x (1-p)^{n-x}$, with the parameters $n = 40$ and $p = 0.25$, we get

$P(30) = {}_{40}C_{30}(0.25)^{30}(0.75)^{70} \approx 4.1 \times 10^{-11}$. Now $np(1-p) = 40 \cdot 0.25 \cdot (1-0.25) = 7.5$, which is below 10, so the normal approximation cannot be used.

19. Using $P(x) = {}_nC_x p^x (1-p)^{n-x}$, with the parameters $n = 75$ and $p = 0.75$, we get

$P(60) = {}_{75}C_{60}(0.75)^{60}(0.25)^{15} \approx 0.0677$. Now $np(1-p) = 75 \cdot 0.75 \cdot (1-0.75) = 14.0625 \ge 10$, so the normal approximation can be used, with $\mu_X = 75(0.75) = 56.25$ and $\sigma_X = \sqrt{np(1-p)} =$

$\sqrt{14.0625} = 3.75$. With continuity correction we calculate:

$P(60) \approx P(59.5 < X < 60.5)$

$$= P\left(\frac{59.5 - 56.25}{3.75} < Z < \frac{60.5 - 56.25}{3.75}\right)$$

$$= P(0.87 < Z < 1.13)$$

$$= 0.8708 - 0.8078$$

$$= 0.0630 \quad [\text{Tech: } 0.0645]$$

56.25 59.5 60.5 X

21. From the parameters $n = 150$ and $p = 0.9$, we get $\mu_X = np = 150 \cdot 0.9 = 135$ and $\sigma_X = \sqrt{np(1-p)} = \sqrt{150 \cdot 0.9 \cdot (1-0.9)} = \sqrt{13.5} \approx 3.674$. Note that $np(1-p) = 13.5 \geq 10$, so the normal approximation to the binomial distribution can be used.

(a) $P(130) \approx P(129.5 < X < 130.5)$

$$= P\left(\frac{129.5 - 135}{\sqrt{13.5}} < Z < \frac{130.5 - 135}{\sqrt{13.5}}\right)$$

$$= P(-1.50 < Z < -1.22)$$

$$= 0.1112 - 0.0668$$

$$= 0.0444 \quad [\text{Tech: } 0.0431]$$

(b) $P(X \geq 130) \approx P(X \geq 129.5)$

$$= P\left(Z \geq \frac{129.5 - 135}{\sqrt{13.5}}\right)$$

$$= P(Z \geq -1.50)$$

$$= 1 - 0.0668$$

$$= 0.9332 \quad [\text{Tech: } 0.9328]$$

(c) $P(X < 125) = P(X \leq 124)$

$$\approx P(X \leq 124.5)$$

$$= P\left(Z \leq \frac{124.5 - 135}{\sqrt{13.5}}\right)$$

$$= P(Z \leq -2.86)$$

$$= 0.0021$$

(d) $P(125 \leq X \leq 135)$

$$\approx P(124.5 \leq X \leq 135.5)$$

$$= P\left(\frac{124.5 - 135}{\sqrt{13.5}} < Z < \frac{135.5 - 135}{\sqrt{13.5}}\right)$$

$$= P(-2.86 < Z < 0.14)$$

$$= 0.5557 - 0.0021$$

$$= 0.5536 \quad [\text{Tech: } 0.5520]$$

23. From the parameters $n = 740$ and $p = 0.64$ we get $\mu_X = np = 740 \cdot 0.64 = 473.6$ and

$$\sigma_x = \sqrt{np(1-p)} = \sqrt{740 \cdot 0.64 \cdot (1-0.64)}$$

$$= \sqrt{170.496} \approx 13.057$$

Note that $np(1-p) = 170.496 \geq 10$, so the normal approximation to the binomial distribution can be used.

(a) $P(490) \approx P(489.5 \leq X \leq 490.5)$

$$= P\left(\frac{489.5 - 473.6}{13.057} \leq Z \leq \frac{490.5 - 473.5}{13.057}\right)$$

$$= P(1.22 \leq Z \leq 1.29)$$

$$= 0.9015 - 0.8888$$

$$= 0.0127 \quad [\text{Tech: } 0.0139]$$

(b) $P(X \leq 490) \approx P(X \leq 489.5)$

$$= P\left(Z \leq \frac{489.5 - 473.6}{13.057}\right)$$

$$= P(Z \leq 1.22)$$

$$= 0.9015 \quad [\text{Tech: } 0.9022]$$

(c) $P(X \geq 503) = P(X \geq 502.5)$

$$= P\left(Z \geq \frac{502.5 - 473.6}{13.057}\right)$$

$$= P(Z \geq 2.21)$$

$$= 1 - 0.9864$$

$$= 0.0136 \quad [\text{Tech: } 0.0134]$$

These results would suggest that the proportion of adult women 18 to 24 years of age who flush public toilets with their foot is greater than 0.64.

25. From the parameters $n = 200$ and $p = 0.55$, we get $\mu_X = np = 200 \cdot 0.55 = 110$ and $\sigma_X = \sqrt{np(1-p)} = \sqrt{200 \cdot 0.55 \cdot (1-0.55)} = \sqrt{49.5} \approx 7.036$. Note that $np(1-p) = 49.5 \geq 10$, so the normal approximation to the binomial distribution can be used.

(a) $P(X \geq 130) \approx P(X \geq 129.5)$

$$= P\left(Z \geq \frac{129.5 - 110}{\sqrt{49.5}} \right)$$

$$= P(Z \geq 2.77)$$

$$= 1 - 0.9972$$

$$= 0.0028$$

(b) Yes, the result from part (a) contradicts the results of the *Current Population Survey* because the result from part (a) is unusual. Fewer than 3 samples in 1000 will result in 130 or more male students living at home if the true percentage is 55%.

27. From the parameters $n = 150$ and $p = 0.37$, we get $\mu_X = np = 150 \cdot 0.37 = 55.5$ and

$$\sigma_X =$$

$$\sqrt{np(1-p)} = \sqrt{150 \cdot 0.37 \cdot (1 - 0.37)} = \sqrt{34.965}$$

≈ 5.913. Note that
$np(1-p) = 34.965 \geq 10$, so the normal approximation to the binomial distribution can be used.

(a) $P(X \geq 75) \approx P(X \geq 74.5)$

$$= P\left(Z \geq \frac{74.5 - 55.5}{\sqrt{34.965}} \right)$$

$$= P(Z \geq 3.21)$$

$$= 1 - 0.9993$$

$$= 0.0007$$

(b) Yes, the result from part (a) contradicts the results of the *Current Population Survey* because the result from part (a) is unusual. Fewer than 1 sample in 1000 will result in 75 or more respondents preferring a male if the true percentage who prefer a male is 37%.

Chapter 7 Review Exercises

1. (a) μ is the center (and peak) of the normal distribution, so $\mu = 60$.

(b) σ is the distance from the center to the points of inflection, so $\sigma = 70 - 60 = 10$.

(c) Interpretation 1: The proportion of values of the random variable to the right of $x = 75$ is 6.68%.
Interpretation 2: The probability that a randomly selected value is greater than $x = 75$ is 0.0668.

(d) Interpretation 1: The proportion of values of the random variable between $x = 50$ and $x = 75$ is 77.45%.
Interpretation 2: The probability that a randomly selected value is between $x = 50$ and $x = 75$ is 0.7745.

2.

Using the standard normal table, the area to the left of $z = -1.04$ is 0.1492.

3.

Using the standard normal table, the area between $z = -0.34$ and $z = 1.03$ is $0.8485 - 0.3669 = 0.4816$.

4. If the area to the right of z is 0.483 then the area to the left z is $1 - 0.483 = 0.5170$.

The closest area to this in the interior of the normal tables is 0.5160, corresponding to $z = 0.04$.

5. The z-scores for the middle 92% are the z-scores for the top and bottom 4%. The area to the left of z_1 is 0.04, and the area to the left of z_2 is 0.96. The area in the interior of the standard normal table (Table V) that is closest to 0.0400 is 0.0401, corresponding to $z_1 = -1.75$. The area in the interior of the standard normal table (Table V) that is closest to 0.9600 is 0.9599, corresponding to $z_2 = 1.75$.

6. The area to the right of the unknown z-score is 0.20, so the area to the left is $1 - 0.20 = 0.80$. The area in the interior of the standard normal table (Table V) that is closest to 0.8000 is 0.7995, corresponding to $z = 0.84$, so $z_{0.20} = 0.84$.

7.

$$z = \frac{55 - 50}{6} \approx 0.83$$

From Table V, the area to the left of $z = 0.83$ is 0.7967, so $P(X > 55) = 1 - 0.7967 = 0.2033$. [Tech: 0.2023]

8.

$$z = \frac{23 - 30}{5} = -1.40$$

From Table V, the area to the left of $z = -1.40$ is 0.0808, so $P(X \le 23) = 0.0808$.

9.

$$z_1 = \frac{65 - 70}{10} = -0.50 \; ; \; z_2 = \frac{85 - 70}{10} = 1.50$$

From Table V, the area to the left of $z = -0.50$ is 0.3085 and the area to the left of $z = 1.50$ is 0.9332, so $P(65 < X < 85) = 0.9332 - 0.3085 = 0.6247$.

10. (a)

$$z = \frac{75,000 - 70,000}{4,400} \approx 1.14$$

From Table V, the area to the left of $z = 1.14$ is 0.8729, so $P(X \ge 75,000) = 1 - 0.8729 = 0.1271$ [Tech: 0.1279]

So, the proportion of tires that will last at least 75,000 miles is 0.1271, or 12.71%.

(b)

$$z = \frac{60,000 - 70,000}{4,400} \approx -2.27$$

From Table V, the area to the left of $z = -2.27$ is 0.0116, so $P(X \le 60,000) = 0.0116$ [Tech: 0.0115] So, the proportion of tires that will last 60,000 miles or less is 0.0116, or 1.16%.

(c)

$$z_1 = \frac{65,000 - 70,000}{4,400} \approx -1.14 \; ;$$

$$z_2 = \frac{80,000 - 70,000}{4,400} \approx 2.27$$

From Table V, the area to the left of $z_1 = -1.14$ is 0.1271 and the area to the left of $z_2 = 2.27$ is 0.9884, so $P(65,000 \le X \le 80,000) = 0.9884 - 0.1271 = 0.8613$ [Tech: 0.8606]

(d) The figure below shows the normal curve with the unknown value of X separating the bottom 2% of the distribution from the top 98% of the distribution.

From Table V, 0.0202 is the area closest to 0.02. The corresponding z-score is −2.05. So, $x = \mu + z\sigma = 70,000 + (-2.05)(4,400)$ $= 60,980$. [Tech: 60,964] In order to warrant only 2% of its tires, Dunlop should advertise its warranty mileage as 60,980 miles.

11. (a)

$$z = \frac{125 - 100}{15} \approx 1.67$$

From Table V, the area to the left of $z = 1.67$ is 0.9525, so $P(X > 125) =$ $1 - 0.9525 = 0.0475$ [Tech: 0.0478].

(b)

$$z = \frac{90 - 100}{15} \approx -0.67$$

From Table V, the area to the left of $z = -0.67$ is 0.2514, so $P(X < 90) =$ 0.2514. [Tech: 0.2525]

(c)

$$z_1 = \frac{110 - 100}{15} \approx 0.67 \; ; \; z_2 = \frac{140 - 100}{15} \approx 2.67$$

From Table V, the area to the left of $z_1 = 0.67$ is 0.7486 and the area to the left of $z_2 = 2.67$ is 0.9962, so $P(110 < X < 140) = 0.9962 - 0.7486$ $= 0.2476$ [Tech: 0.2487] So, the proportion of test takers who score between 110 and 140 is 0.2476, or 24.76%.

(d)

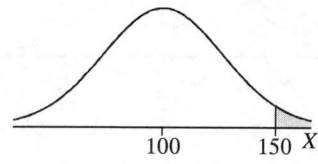

$$z = \frac{150 - 100}{15} \approx 3.33$$

From Table V, the area to the left of $z = 3.33$ is 0.9996, so $P(X > 150) =$ $= 1 - 0.9996 = 0.0004$.

(e) The figure below shows the normal curve with the unknown value of X separating the bottom 98% of the distribution from the top 2% of the distribution.

From Table V, 0.9798 is the area closest to 0.98. The corresponding z-score is 2.05. So, $x = \mu + z\sigma = 100 + 2.05(15) \approx 131$. A score of 131 places a child in the 98th percentile.

(f) The figure below shows the normal curve with the unknown values of X separating the middle 95% of the distribution from the bottom 2.5% and top 2.5% of the distribution.

From Table V, the area 0.0250 corresponds to $z_1 = -1.96$ and the area 0.9750 corresponds to $z_2 = 1.96$. So, $x_1 = \mu + z_1\sigma$ $= 100 + (-1.96)(15) \approx 71$ and $x_2 = \mu + z_2\sigma$ $= 100 + 1.96(15) \approx 129$. Thus, children of normal intelligence scores are between 71 and 129 on the Wechsler Scale.

12. (a)

$$z = \frac{5.25 - 5.11}{0.062} \approx 2.26$$

From Table V, the area to the left of $z = 2.26$ is 0.9881, so $P(X > 5.25) =$ $= 1 - 0.9881 = 0.0119$. [Tech: 0.0120] So, the proportion of baseballs produced by this factory that are too heavy for use by major league baseball is 0.0119, or 1.19%.

(b)

$z = \dfrac{5 - 5.11}{0.062} \approx -1.77$ From Table V, the area to the left of $z = -1.77$ is 0.0384, so $P(X < 5) = 0.0384$. [Tech: 0.0380] So, the proportion of baseballs produced by this factory that are too light for use by major league baseball is 0.0384, or 3.84%.

(c)

From parts (a) and (b), $z_1 \approx -1.77$ and $z_2 \approx 2.26$. The area to the left of $z_1 = -1.77$ is 0.0384 and the area to the left of $z_2 = 2.26$ is 0.9881, so $P(5 \leq X \leq 5.25) = 0.9881 - 0.0384 =$ 0.9497 [Tech: 0.9500]. So, the proportion of baseballs produced by this factory that can be used by major league baseball is 0.9497, or 94.97%.

(d) From part (c), we know that 94.97% of the baseballs can be used by major league baseball. Let n represent the number of baseballs that must be produced. Then, $0.9497n = 8,000$, so

$n = \dfrac{8,000}{0.9497} \approx 8,423.71$. Increase this to the next whole number: 8,424. To meet the order, the factory should produce 8,424 baseballs. [Tech: 8421 baseballs]

13. (a) Since $np(1 - p) = 250(0.46)(1 - 0.46) =$ $62.1 \geq 10$, the normal distribution can be used to approximate the binomial distribution. The parameters are $\mu_X = np$ $= 250(0.46) = 115$ and $\sigma_X = \sqrt{np(1 - p)}$ $= \sqrt{250(0.46)(1 - 0.46)} = \sqrt{62.1} \approx 7.880$

(b) For the normal approximation, we make corrections for continuity 124.5 and 125.5.

$P(125) \approx P(124.5 < X < 125.5)$

$= P\left(\dfrac{124.5 - 115}{\sqrt{62.1}} < Z < \dfrac{125.5 - 115}{\sqrt{62.1}}\right)$

$= P(1.21 < Z \leq 1.33)$

$= 0.9082 - 0.8869$

$= 0.0213$ [Tech: 0.0226]

Interpretation: Approximately 2 of every 100 random samples of 250 adult Americans will result in exactly 125 who state that they have read at least 6 books within the past year.

(c) $P(X < 120) = P(X \leq 119)$

For the normal approximation, we make a correction for continuity to 119.5.

$P(X < 120) = P(X \leq 119)$

$\approx P(X < 119.5)$

$= P\left(Z < \dfrac{119.5 - 115}{\sqrt{62.1}}\right)$

$= P(Z < 0.57)$

$= 0.7157$ [Tech: 0.7160]

Interpretation: Approximately 72 of every 100 random samples of 250 adult Americans will result in fewer than 120 who state that they have read at least 6 books within the past year.

(d) For the normal approximation, we make a correction for continuity to 139.5.

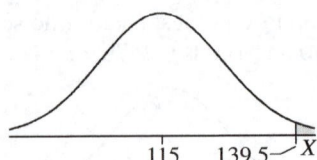

$$P(X \geq 140) \approx P(X > 139.5)$$

$$= P\left(Z > \frac{139.5 - 115}{\sqrt{62.1}} \right)$$

$$\approx P(Z > 3.11)$$

$$= 1 - 0.9991 = 0.0009$$

Interpretation: Approximately 1 of every 1000 random samples of 250 adult Americans will result in 140 or more who state that they have read at least 6 books within the past year.

(e) For the normal approximation, we make corrections for continuity 99.5 and 120.5.

$$P(100 \leq X \leq 120)$$

$$\approx P(99.5 < X < 120.5)$$

$$= P\left(\frac{99.5 - 115}{\sqrt{62.1}} < Z < \frac{120.5 - 115}{\sqrt{62.1}} \right)$$

$$\approx P(-1.97 < Z < 0.70)$$

$$= 0.7580 - 0.0244$$

$$= 0.7336 \quad [\text{Tech: } 0.7328]$$

Interpretation: Approximately 73 of every 100 random samples of 250 adult Americans will result in between 100 and 120, inclusive, who state that they have read at least 6 books within the past year.

14. (a)

(b) The correlation between the observed values and the expected z-scores is 0.986.

(c) Because 0.986> 0.898 (Critical value from Table VI), there is evidence the sample could come from a population that is normally distributed. Also, the normal probability plot is roughly linear, so the sample data could come from a normally distributed population.

15. The plotted points do not appear linear and do not lie within the provided bounds, so the sample data are not from a normally distributed population. The correlate between the age of the car and the expected z-scores is 0.914. The critical value from Table VI is 0.951. 0.914<0.951 suggests the sample data do not come from a normally distributed population.

16. The plotted points are not linear and do not lie within the provided bounds, so the data are not from a population that is normally distributed. The correlation is 0.873 which is less than 0.959 (the critical value from Table VI for n=25).

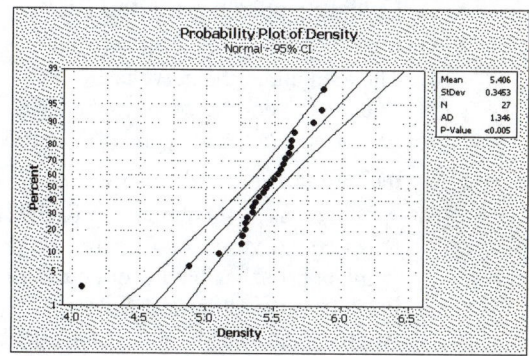

17. From the parameters $n = 250$ and $p = 0.20$, we get $\mu_X = np = 250 \cdot 0.20 = 50$ and $\sigma_X = \sqrt{np(1 - p)} = \sqrt{250(0.20)(1 - 0.20)} = \sqrt{40} \approx 6.325$. Note that $np(1 - p) = 40 \geq 10$, so the normal approximation to the binomial distribution can be used.

(a) For the normal approximation, we make corrections for continuity to 30.5.

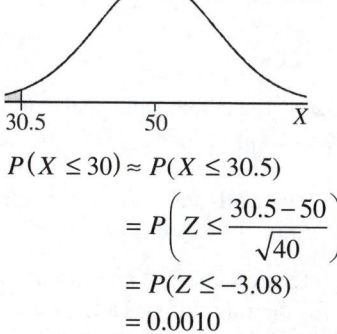

$$P(X \leq 30) \approx P(X \leq 30.5)$$

$$= P\left(Z \leq \frac{30.5 - 50}{\sqrt{40}} \right)$$

$$= P(Z \leq -3.08)$$

$$= 0.0010$$

(b) Yes, the result from part (a) contradicts the *USA Today* "Snapshot" because the result from part (a) is unusual. About 1 sample in 1000 will result in 30 or fewer who do their most creative thinking while driving, if the true percentage is 20%.

18. (a)

Density vs. Random Variable, 0.05 from 0 to 20 on X.

(b) $P(0 \le X \le 5) = 0.05 \cdot (5 - 0) = 0.25$

(c) $P(10 \le X \le 18) = 0.05 \cdot (18 - 10) = 0.4$

19. The standard normal curve has the following properties:
   (1) It is symmetric about its mean $\mu = 0$ and has standard deviation $\sigma = 1$.
   (2) Its highest point occurs at $\mu = 0$.
   (3) It has inflection points at -1 and 1.
   (4) The area under the curve is 1.
   (5) The area under the curve to the right of $\mu = 0$ equals the area under the curve to the left of $\mu = 0$, which equals 0.5.
   (6) As Z increases, the graph approaches but never equals zero. As Z decreases, the graph approaches but never equals zero.
   (7) It satisfies the Empirical Rule: Approximately 68% of the area under the standard normal curve is between -1 and 1. Approximately 95% of the area under the standard normal curve is between -2 and 2. Approximately 99.7% of the area under the standard normal curve is between -3 and 3.

20. The graph plots actual observations against expected z-scores, assuming that the data are normally distributed. If the plot is not linear, then we have evidence that that data are not from a normal distribution.

Chapter 7 Test

1. (a) μ is the center (and peak) of the normal distribution, so $\mu = 7$.

   (b) σ is the distance from the center to the points of inflection, so $\sigma = 9 - 7 = 2$.

   (c) Interpretation 1: The proportion of values for the random variable to the left of $x = 10$ is 93.32%.
   Interpretation 2: The probability that a randomly selected value is less than $x = 10$ is 0.9332.

   (d) Interpretation 1: The proportion of values for the random variable between $x = 5$ and $x = 8$ is 53.28%.

Interpretation 2: The probability that a randomly selected value is between $x = 5$ and $x = 8$ is 0.5328.

2.

Normal curve with shaded area to the right of $z = 2.04$.

Using Table V, the area to the left of $z = 2.04$ is 0.9793, so the area to the right of $z = 2.04$ is $1 - 0.9793 = 0.0207$.

3. The z-scores for the middle 88% are the z-scores for the top and bottom 6%.

Normal curve with Area = 0.88 in the middle, Area = 0.06 on each tail; $Z_1 = ?$ and $Z_2 = ?$ marked.

The area to the left of z_1 is 0.06, and the area to the left of z_2 is 0.94. From the interior of Table V, the areas 0.0606 and 0.0594 are equally close to 0.0600. These areas correspond to $z = -1.55$ and $z = -1.56$, respectively. Splitting the difference we obtain $z_1 = -1.555$. From the interior of Table V, the areas 0.9394 and 0.9406 are equally close to 0.9400. These correspond to $z = 1.55$ and $z = 1.56$, respectively. Splitting the difference, we obtain $z_2 = 1.555$.

4. The area to the right of the unknown z-score is 0.04, so the area to the left is $1 - 0.04 = 0.96$.

Normal curve with Area = 0.96 to left, Area = 0.04 shaded to right; $Z = ?$ marked.

The area in the interior of the Table V that is closest to 0.9600 is 0.9599, corresponding to $z = 1.75$, so $z_{0.04} = 1.75$.

5. (a)

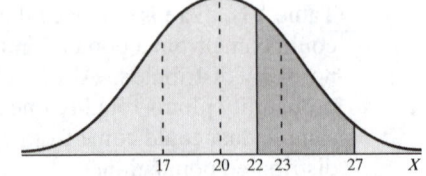

Normal curve over X with marks at 17, 20, 22, 23, 27.

(b) $z_1 = \dfrac{22-20}{3} \approx 0.67$; $z_2 = \dfrac{27-20}{3} \approx 2.33$

From Table V, the area to the left of $z_1 = 0.67$ is 0.7486 and the area to the left of $z_2 = 2.33$ is 0.9901, so $P(22 \le X \le 27)$ $= 0.9901 - 0.7486 = 0.2415$. [Tech: 0.2427]

6. (a)

$z = \dfrac{6-7}{0.8} = -1.25$

From Table V, the area to the left of $z = -1.25$ is 0.1956, so $P(X \ge 6) =$ $1 - 0.1056 = 0.8944$. So, the proportion of the time that a fully charged iPhone will last at least 6 hours is 0.8944, or 89.44%.

(b)

$z = \dfrac{5-7}{0.8} = -2.5$

From Table V, the area to the left of $z = -2.50$ is 0.0062, so $P(X < 5) =$ 0.0062 . That is, the probability that a fully charged iPhone will last less than 5 hours is 0.0062. This is an unusual result.

(c) The figure below shows the normal curve with the unknown value of X separating the top 5% of the distribution from the bottom 95% of the distribution.

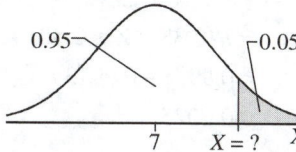

From the interior of Table V, the areas 0.9495 and 0.9505 are equally close to 0.9500. These areas correspond to $z = 1.64$ and $z = 1.65$, respectively. Splitting the difference, we obtain the z-score 1.645. So,

$x = \mu + z\sigma = 7 + 1.645(0.8) \approx 8.3$. The cutoff for the top 5% of all talk times is 8.3 hours.

(d)

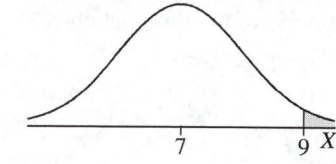

$z = \dfrac{9-7}{0.8} = 2.5$

From Table V, the area to the left of $z = 2.50$ is 0.9938, so $P(X > 9) =$ $1 - 0.9938 = 0.0062$. So, yes, it would be unusual for the iPhone to last more than 9 hours. Only about 6 out of every 1000 full charges will result in the iPhone lasting more than 9 hours.

7. (a)

$z = \dfrac{100-92.5}{13.7} \approx 0.55$

From Table V, the area to the left of $z = 0.55$ is 0.7088, so $P(X < 6) = 0.7088$. [Tech: 0.7080] So, the proportion of 20- to 29-year-old males whose waist circumferences is less than 100 cm is 0.7088, or 70.88%.

(b)

$z_1 = \dfrac{80-92.5}{13.7} = -0.91$;

$z_2 = \dfrac{100-92.5}{13.7} \approx 0.55$

From Table V, the area to the left of $z_1 = -0.91$ is 0.1814 and the area to the left of $z_2 = 0.55$ is 0.7088, so $P(80 \le X \le 100) = 0.7088 - 0.1814 =$ 0.5274 . [Tech: 0.5272] That is, the probability that a randomly selected 20- to 29-year-old male has a waist circumference between 80 and 100 cm is 0.5274.

(c) The figure that follows shows the normal curve with the unknown values of X separating the middle 90% of the distribution from the bottom 5% and the top 5% of the distribution.

From the interior of Table V, the areas 0.0495 and 0.0505 are equally close to 0.0500. These areas correspond to $z = -1.65$ and $z = -1.64$, respectively. Splitting the difference we obtain $z_1 = -1.645$. So, $x_1 = \mu + z_1\sigma =$ $= 92.5 + (-1.645)(13.7) \approx 70$. From the interior of Table V, the areas 0.9495 and 0.9505 are equally close to 0.9500. These areas correspond to $z = 1.64$ and $z = 1.65$, respectively. Splitting the difference we obtain the $z_2 = 1.645$. So, $x_2 = \mu + z_2\sigma =$ $= 92.5 + 1.645(13.7) \approx 115$. Therefore, waist circumferences between 70 and 115 cm make up the middle 90% of all waist circumferences.

(d) The figure below shows the normal curve with the unknown value of X separating the bottom 10% of the distribution from the top 90% of the distribution.

From the interior of Table V, the area closest to 0.10 is 0.1003. The corresponding z-score is -1.28. So, $x = \mu + z\sigma = 92.5 + (-1.28)(13.7) \approx 75$. Thus, a waist circumference of 75 cm is at the 10th percentile.

8. If the top 6% get an A, then 94% (or 0.9400) of the students will have grades lower than an A. Similarly, 80% (or 0.80) will have grades lower than a B. 20% (or 0.20) will have grades lower than a C. Finally, 6% (or 0.06) will have a grade lower than a D, that is, an F. To find the cutoff score corresponding to each of these values, we find the z-score for each of

the cutoffs, then use the equation $X = \mu + z \cdot \sigma$ to find the point value for each cutoff.

For the cutoff for the low A's: $z_{0.94} = 1.55$, and $X = 64 + 1.55(8) = 76.4$, so everyone who scores at least 76.4 points will get an A.
For the cutoff for the low B's: $z_{0.80} = 0.84$, and $X = 64 + 0.84(8) \approx 70.7$, so everyone who scores at least 70.7 points and less than 76.4 will get a B.
For the cutoff for the low C's: $z_{0.20} = -0.84$, and $X = 64 + (-0.84)(8) \approx 57.3$, so everyone who scores at least 57.3 and less than 70.7 points will get a C.
For the cutoff for the low D's: $z_{0.06} = -1.55$, and $X = 64 + (-1.55)(8) = 51.6$, so everyone who scores at least 51.6 and less than 57.3 points will get a D.
Students who score less than 51.6 points will get an F.

9. $np(1-p) = 500(0.16)(1-0.16) = 67.2 \geq 10$, so the normal distribution can be used to approximate the binomial distribution. The parameters are $\mu_X = np = 500(0.16) = 80$ and $\sigma_X = \sqrt{np(1-p)} = \sqrt{250(0.16)(1-0.16)} = \sqrt{67.2} \approx 8.198$.

(a) For the normal approximation of $P(100)$, we make corrections for continuity 99.5 and 100.5.

$P(100) \approx P(99.5 < X < 100.5)$
$= P\left(\dfrac{99.5 - 80}{\sqrt{67.2}} < Z < \dfrac{100.5 - 80}{\sqrt{67.2}} \right)$
$= P(2.38 < Z \leq 2.50)$
$= 0.9938 - 0.9913$
$= 0.0025$

(b) $P(X < 60) = P(X \leq 59)$
For the normal approximation, we make a correction for continuity to 59.5.

(c)

$$P(X < 60) = P(X \le 59)$$
$$\approx P(X < 59.5)$$
$$= P\left(Z < \frac{59.5 - 80}{\sqrt{67.2}}\right)$$
$$= P(Z < -2.50)$$
$$= 0.0062$$

10. The correlation is 0.974. From table VI the critical value is 0.939. Since 0.974>0.939 there is evidence that the sample data come from a population that is normally distributed. Also, the plotted points appear linear and lie within the provided bounds, so the sample data do likely come from a normally distributed population.

11. **(a)**

(b) $P(20 \le X \le 30) = \dfrac{1}{40}(30 - 20) = 0.25$

(c) $P(X < 15) = P(10 \le X \le 15)$
$$= \frac{1}{40}(15 - 10)$$
$$= 0.125$$

Case Study: A Tale of Blood Chemistry

1. $\mu \approx \text{midpoint} = \dfrac{35 + 150}{2} = 92.5 \text{ mg/dL}$

$\sigma \approx \dfrac{\text{range}}{4} = \dfrac{150 - 35}{4} = 28.75 \text{ mg/dL}$

These results agree with the entries in the table.

2. White blood cell count:
$$P(X \le 5.3) = P\left(Z \le \frac{5.3 - 7.25}{1.625}\right)$$
$$= P(Z \le -1.2)$$
$$= 0.1151$$

Red blood cell count:
$$P(X \le 4.62) = P\left(Z \le \frac{4.62 - 4.85}{0.375}\right)$$
$$\approx P(Z \le -0.61)$$
$$= 0.2709 \quad \text{[Tech: 0.2698]}$$

Hemoglobin:
$$P(X \le 14.6) = P\left(Z \le \frac{14.6 - 14.75}{1.125}\right)$$
$$\approx P(Z \le -0.13)$$
$$= 0.4483 \quad \text{[Tech: 0.4470]}$$

Hematocrit:
$$P(X \le 41.7) = P\left(Z \le \frac{41.7 - 43.0}{3.5}\right)$$
$$\approx P(Z \le -0.37)$$
$$= 0.3557 \quad \text{[Tech: 0.3552]}$$

Glucose, serum:
$$P(X \ge 95.0) = P\left(Z \ge \frac{95.0 - 87.0}{11.0}\right)$$
$$\approx P(Z \ge 0.73)$$
$$= 1 - 0.7673$$
$$= 0.2327 \quad \text{[Tech: 0.2335]}$$

Creatine, serum:
$$P(X \le 0.8) = P\left(Z \le \frac{0.8 - 1.00}{0.25}\right)$$
$$= P(Z \le -0.80)$$
$$= 0.2119$$

Sodium, serum:
$$P(X \ge 143.0) = P\left(Z \ge \frac{143.0 - 141.5}{3.25}\right)$$
$$\approx P(Z \ge 0.46)$$
$$= 1 - 0.6772$$
$$= 0.3228 \quad \text{[Tech: 0.3222]}$$

Potassium, serum:
$$P(X \ge 5.1) = P\left(Z \ge \frac{5.1 - 4.5}{0.5}\right)$$
$$= P(Z \ge 1.20)$$
$$= 1 - 0.8849$$
$$= 0.1151$$

Chloride, serum:
$$P(X \le 100.0) = P\left(Z \le \frac{100.0 - 102.5}{3.25}\right)$$
$$\approx P(Z \le -0.77)$$
$$= 0.2206 \quad \text{[Tech: 0.2209]}$$

Carbon dioxide, total:

$$P(X \leq 25.0) = P\left(Z \leq \frac{25.0 - 26.0}{3.0}\right)$$
$$\approx P(Z \leq -0.33)$$
$$= 0.3707 \quad \text{[Tech: 0.3694]}$$

Calcium, serum:

$$P(X \geq 10.1) = P\left(Z \geq \frac{10.1 - 9.55}{0.525}\right)$$
$$\approx P(Z \geq 1.05)$$
$$= 1 - 0.8531$$
$$= 0.1469 \quad \text{[Tech: 0.1474]}$$

Total cholesterol:

$$P(X \geq 253.0) = P\left(Z \geq \frac{253.0 - 149.5}{24.75}\right)$$
$$\approx P(Z \geq 4.18)$$
$$\approx 1 - 0.9998$$
$$= 0.0002 \quad \text{[Tech: <0.0001]}$$

Triglycerides:

$$P(X \geq 150.0) = P\left(Z \geq \frac{150.0 - 99.5}{49.75}\right)$$
$$\approx P(Z \geq 1.02)$$
$$\approx 1 - 0.8461$$
$$\approx 0.1539 \quad \text{[Tech: 0.1550]}$$

HDL cholesterol:

$$P(X \leq 42.0) = P\left(Z \leq \frac{42.0 - 92.5}{28.75}\right)$$
$$\approx P(Z \leq -1.76)$$
$$= 0.0392 \quad \text{[Tech: 0.0395]}$$

LDL cholesterol:

$$P(X \geq 181.0) = P\left(Z \geq \frac{181.0 - 64.5}{32.25}\right)$$
$$\approx P(Z \geq 3.61)$$
$$\approx 1 - 0.9998$$
$$= 0.0002 \quad \text{[Tech: 0.0002]}$$

LDL/HDL ratio:

$$P(X \geq 4.3) = P\left(Z \geq \frac{4.3 - 1.8}{0.72}\right)$$
$$\approx P(Z \geq 3.47)$$
$$= 1 - 0.9997$$
$$= 0.0003$$

TSH, high sensitivity, serum:

$$P(X \geq 3.15) = P\left(Z \geq \frac{3.15 - 2.925}{1.2875}\right)$$
$$\approx P(Z \geq 0.17)$$
$$= 1 - 0.5675$$
$$= 0.4325 \quad \text{[Tech: 0.4306]}$$

Using Abby's criterion, she should be concerned about her total cholesterol, her LDL cholesterol, and her LDL/HDL ratio.

Chapter 8
Sampling Distributions

Section 8.1

1. sampling distribution

3. standard error; mean

5. False; if the sample is drawn from a population that is not normal, then the distribution of the sample mean will not necessarily be normally distributed.

7. The sampling distribution is normal. The mean of the sampling distribution of $\bar{x}$ is $\mu_{\bar{x}} = \mu = 30$ and the standard deviation is given by $\sigma_{\bar{x}} = \dfrac{\sigma}{\sqrt{n}} = \dfrac{8}{\sqrt{10}}$.

9. $\mu_{\bar{x}} = \mu = 80$; $\sigma_{\bar{x}} = \dfrac{\sigma}{\sqrt{n}} = \dfrac{14}{\sqrt{49}} = \dfrac{14}{7} = 2$

11. $\mu_{\bar{x}} = \mu = 52$; $\sigma_{\bar{x}} = \dfrac{\sigma}{\sqrt{n}} = \dfrac{10}{\sqrt{21}} \approx 2.182$

13. (a) The sampling distribution is symmetric about 500, so the mean is $\mu_{\bar{x}} = 500$.

    (b) The inflection points are at 480 and 520, so
    $$\sigma_{\bar{x}} = 520 - 500 = 20 \ \text{(or}$$
    $$500 - 480 = 20) \,.$$

    (c) Since $n = 16 \le 30$, the population must be normal so that the sampling distribution of $\bar{x}$ is normal.

    (d) $\sigma_{\bar{x}} = \dfrac{\sigma}{\sqrt{n}}$
    $$20 = \dfrac{\sigma}{\sqrt{16}}$$
    $$\sigma = 20\sqrt{16} = 20(4) = 80$$
    The standard deviation of the population from which the sample is drawn is 80.

15. (a) Since $\mu = 80$ and $\sigma = 14$, the mean and standard deviation of the sampling distribution of $\bar{x}$ are given by:
    $$\mu_{\bar{x}} = \mu = 80; \ \sigma_{\bar{x}} = \dfrac{\sigma}{\sqrt{n}} = \dfrac{14}{\sqrt{49}} = \dfrac{14}{7} = 2 \,.$$
    We are not told that the population is normally distributed, but we do have a large sample size ($n = 49 \ge 30$). Therefore, we can use the Central Limit

Theorem to say that the sampling distribution of $\bar{x}$ is approximately normal.

(b) $P(\bar{x} > 83) = P\left(Z > \dfrac{83 - 80}{2}\right)$
$$= P(Z > 1.50)$$
$$= 1 - P(Z \le 1.50)$$
$$= 1 - 0.9332 = 0.0668$$
If we take 100 simple random samples of size $n = 49$ from a population with $\mu = 80$ and $\sigma = 14$, then about 7 of the samples will result in a mean that is greater than 83.

(c) $P(\bar{x} \le 75.8) = P\left(Z \le \dfrac{75.8 - 80}{2}\right)$
$$= P(Z \le -2.10) = 0.0179$$
If we take 100 simple random samples of size $n = 49$ from a population with $\mu = 80$ and $\sigma = 14$, then about 2 of the samples will result in a mean that is less than or equal to 75.8.

(d)
$$P(78.3 < \bar{x} < 85.1)$$
$$= P\left(\dfrac{78.3 - 80}{2} < Z < \dfrac{85.1 - 80}{2}\right)$$
$$= P(-0.85 < Z < 2.55)$$
$$= 0.9946 - 0.1977 = 0.7969 \ \text{[Tech: 0.7970]}$$
If we take 100 simple random samples of size $n = 49$ from a population with $\mu = 80$ and $\sigma = 14$, then about 78 of the samples will result in a mean that is between 78.3 and 85.1.

17. (a) The population must be normally distributed. If this is the case, then the sampling distribution of $\bar{x}$ is exactly normal. The mean and standard deviation of the sampling distribution are $\mu_{\bar{x}} = \mu = 64$ and $\sigma_{\bar{x}} = \dfrac{\sigma}{\sqrt{n}} = \dfrac{17}{\sqrt{12}} \approx 4.907$.

(b) $P(\bar{x} < 67.3) = P\left(Z < \dfrac{67.3 - 64}{17 / \sqrt{12}}\right)$

$= P(Z < 0.67)$

$= 0.7486 \quad [\text{Tech: } 0.7493]$

If we take 100 simple random samples of size $n = 12$ from a population that is normally distributed with $\mu = 64$ and $\sigma = 17$, then about 75 of the samples will result in a mean that is less than 67.3.

(c) $P(\bar{x} \geq 65.2) = P\left(Z \geq \dfrac{65.2 - 64}{17 / \sqrt{12}}\right)$

$= P(Z \geq 0.24)$

$= 1 - P(Z < 0.24)$

$= 1 - 0.5948$

$= 0.4052 \quad [\text{Tech: } 0.4034]$

If we take 100 simple random samples of size $n = 12$ from a population that is normally distributed with $\mu = 64$ and $\sigma = 17$ then about 41 of the samples will result in a mean that is greater than or equal to 65.2.

19. (a) $P(X < 260) = P\left(Z < \dfrac{260 - 266}{16}\right)$

$= P(Z < -0.38)$

$= 0.3520 \quad [\text{Tech: } 0.3538]$

If we select a simple random sample of $n = 100$ human pregnancies, then about 35 of the pregnancies would last less than 260 days.

(b) Since the length of human pregnancies is normally distributed, the sampling distribution of $\bar{x}$ is normal with

$\mu_{\bar{x}} = 266$ and $\sigma_{\bar{x}} = \dfrac{16}{\sqrt{20}} \approx 3.578$.

(c) $P(\bar{x} \leq 260) = P\left(Z \leq \dfrac{260 - 266}{16 / \sqrt{20}}\right)$

$= P(Z \leq -1.68)$

$= 0.0465 \quad [\text{Tech: } 0.0468]$

If we take 100 simple random samples of size $n = 20$ human pregnancies, then about 5 of the samples will result in a mean gestation period of 260 days or less.

(d) $\mu_{\bar{x}} = \mu = 266$; $\sigma_{\bar{x}} = \dfrac{\sigma}{\sqrt{n}} = \dfrac{16}{\sqrt{50}}$

$P(\bar{x} \leq 260) = P\left(Z \leq \dfrac{260 - 266}{16 / \sqrt{50}}\right)$

$= P(Z \leq -2.65)$

$= 0.0040$

If we take 1000 simple random samples of size $n = 50$ human pregnancies, then about 4 of the samples will result in a mean gestation period of 260 days or less.

(e) Answers will vary. Part (d) indicates that this result would be an unusual observation. Therefore, we would conclude that the sample likely came from a population whose mean gestation period is less than 266 days.

(f) $\mu_{\bar{x}} = \mu = 266$; $\sigma_{\bar{x}} = \dfrac{\sigma}{\sqrt{n}} = \dfrac{16}{\sqrt{15}}$

$P(256 \leq \bar{x} \leq 276)$

$= P\left(\dfrac{256 - 266}{16 / \sqrt{15}} \leq Z \leq \dfrac{276 - 266}{16 / \sqrt{15}}\right)$

$= P(-2.42 \leq Z \leq 2.42)$

$= 0.9922 - 0.0078$

$= 0.9844 \quad [\text{Tech: } 0.9845]$

If we take 100 simple random samples of size $n = 15$ human pregnancies, then about 98 of the samples will result in a mean gestation period between 256 and 276 days, inclusive.

21. (a) $P(X > 95) = P\left(Z > \dfrac{95 - 90}{10}\right)$

$= P(Z > 0.5)$

$= 1 - P(Z \leq 0.5)$

$= 1 - 0.6915$

$= 0.3085$

If we select a simple random sample of $n = 100$ second grade students, then about 31 of the students would read more than 95 words per minute.

(b) $\mu_{\bar{x}} = \mu = 90$; $\sigma_{\bar{x}} = \dfrac{\sigma}{\sqrt{n}} = \dfrac{10}{\sqrt{12}}$

$$P(\bar{x} > 95) = P\left(Z > \dfrac{95 - 90}{10/\sqrt{12}}\right)$$

$$= P(Z > 1.73)$$

$$= 1 - P(Z \leq 1.73)$$

$$= 1 - 0.9582$$

$$= 0.0418 \quad \text{[Tech: 0.0416]}$$

If we take 100 simple random samples of size $n = 12$ second grade students, then about 4 of the samples will result in a mean reading rate that is more than 95 words per minute.

(c) $\mu_{\bar{x}} = \mu = 90$; $\sigma_{\bar{x}} = \dfrac{\sigma}{\sqrt{n}} = \dfrac{10}{\sqrt{24}}$

$$P(\bar{x} > 95) = P\left(Z > \dfrac{95 - 90}{10/\sqrt{24}}\right)$$

$$= P(Z > 2.45)$$

$$= 1 - P(Z \leq 2.45)$$

$$= 1 - 0.9929$$

$$= 0.0071 \quad \text{[Tech: 0.0072]}$$

If we take 1000 simple random samples of size $n = 24$ second grade students, then about 7 of the samples will result in a mean reading rate that is more than 95 words per minute.

(d) Increasing the sample size decreases the probability that $\bar{x} > 95$. This happens because $\sigma_{\bar{x}}$ decreases as n increases.

(e) No, this result would not be unusual because, if $\mu_{\bar{x}} = \mu = 90$ and

$\sigma_{\bar{x}} = \dfrac{\sigma}{\sqrt{n}} = \dfrac{10}{\sqrt{20}}$, then

$$P(\bar{x} > 92.8) = P\left(Z > \dfrac{92.8 - 90}{10/\sqrt{20}}\right)$$

$$= P(Z > 1.25)$$

$$= 1 - P(Z \leq 1.25)$$

$$= 1 - 0.8944$$

$$= 0.1056 \quad \text{[Tech: 0.1052]}$$

If we take 100 simple random samples of size $n = 20$ second grade students, then about 11 of the samples will result in a mean reading rate that is above 92.8 words per minute.
This result does not qualify as unusual. This

means that the new reading program is not abundantly more effective than the old program.

(f) We need to find a number of words per minute c such that $P(\bar{x} > c) = 0.05$. So, we have

$$P(\bar{x} > c) = P\left(Z > \dfrac{c - 90}{10/\sqrt{20}}\right) = 0.05.$$

The area to the right of z is 0.05 if $z = 1.645$.

So, $\dfrac{c - 90}{10/\sqrt{20}} = 1.645$. Solving for c, we get

$c = 93.7$ [tech 93.7] words per minute.

23. (a) $P(X > 0) = P\left(Z > \dfrac{0 - 0.007233}{0.04135}\right)$

$$= P(Z > -0.17)$$

$$= 1 - P(Z \leq -0.17)$$

$$= 1 - 0.4325$$

$$= 0.5675 \quad \text{[Tech: 0.5694]}$$

If we select a simple random sample of $n = 100$ months, then about 57 of the months would have positive rates of return.

(b) $\mu_{\bar{x}} = \mu = 0.007233$;

$$\sigma_{\bar{x}} = \dfrac{\sigma}{\sqrt{n}} = \dfrac{0.04135}{\sqrt{12}}$$

$$P(\bar{x} > 0) = P\left(Z > \dfrac{0 - 0.007233}{0.04135/\sqrt{12}}\right)$$

$$= P(Z > -0.61)$$

$$= 1 - P(Z \leq -0.61)$$

$$= 1 - 0.2709$$

$$= 0.7291 \quad \text{[Tech: 0.7277]}$$

If we take 100 simple random samples of size $n = 12$ months, then about 73 of the samples will result in a mean monthly rate that is positive.

(c) $\mu_{\bar{x}} = \mu = 0.007233$;

$$\sigma_{\bar{x}} = \dfrac{\sigma}{\sqrt{n}} = \dfrac{0.04135}{\sqrt{24}}$$

$$P(\overline{x} > 0) = P\left(Z > \frac{0 - 0.007233}{0.04135 / \sqrt{24}}\right)$$

$$= P(Z > -0.86)$$

$$= 1 - P(Z \le -0.86)$$

$$= 1 - 0.1949$$

$$= 0.8051 \quad [\text{Tech: } 0.8043]$$

If we take 100 simple random samples of size $n = 24$ months, then about 81 of the samples will result in a mean monthly rate that is positive.

(d) $\mu_{\overline{x}} = \mu = 0.007233$;

$$\sigma_{\overline{x}} = \frac{\sigma}{\sqrt{n}} = \frac{0.04135}{\sqrt{36}}$$

$$P(\overline{x} > 0) = P\left(Z > \frac{0 - 0.007233}{0.04135 / \sqrt{36}}\right)$$

$$= P(Z > -1.05)$$

$$= 1 - P(Z \le -1.05)$$

$$= 1 - 0.1469$$

$$= 0.8531 \quad [\text{Tech: } 0.8530]$$

If we take 100 simple random samples of size $n = 36$ months, then about 85 of the samples will result in a mean monthly rate that is positive.

(e) Answers will vary. Based on the results of parts (b)–(d), the likelihood of earning a positive rate of return increases as the investment time horizon increases.

25. (a) Without knowing the shape of the distribution, we would need a sample size of at least 30 so we could apply the Central Limit Theorem.

(b) $\mu_{\overline{x}} = \mu = 11.4$; $\sigma_{\overline{x}} = \frac{\sigma}{\sqrt{n}} = \frac{3.2}{\sqrt{40}}$

$$P(\overline{x} < 10) = P\left(Z < \frac{10 - 11.4}{3.2 / \sqrt{40}}\right)$$

$$= P(Z < -2.77)$$

$$= 0.0028$$

If we take 1000 simple random samples of size $n = 40$ oil changes, then about 3 of the samples will result in a mean time less than 10 minutes.

(c) We need to find c such that the probability that the sample mean will be less than c is 10%. So we have:

$$P(\overline{x} \le c) = P\left(Z \le \frac{c - 11.4}{3.2 / \sqrt{40}}\right) = 0.10$$

The value of Z such that 10% of the area is less than or equal to it is $Z = -1.28$.

So, $\dfrac{c - 11.4}{3.2 / \sqrt{40}} = -1.28$. Solving for Z, we

get: $Z = 11.4 + (-1.28)\dfrac{3.2}{\sqrt{40}} \approx 10.8$

minutes.

27. (a) The sampling distribution of $\overline{x}$ is approximately normal because the sample is large, $n = 50 \ge 30$. From the Central Limit Theorem, as the sample size increases, the sampling distribution of the mean becomes more normal.

(b) Assuming that we are sampling from a population that is exactly at the Food Defect Action Level, $\mu_{\overline{x}} = \mu = 3$ and

$$\sigma_{\overline{x}} = \frac{\sigma}{\sqrt{n}} = \frac{\sqrt{3}}{\sqrt{50}} = \sqrt{\frac{3}{50}} \approx 0.245 \, .$$

(c) $P(\overline{x} \ge 3.6) = P\left(Z \ge \dfrac{3.6 - 3}{\sqrt{3} / \sqrt{50}}\right)$

$$= P(Z \ge 2.45) = 1 - 0.9929$$

$$= 0.0071 \quad [\text{Tech: } 0.0072]$$

If we take 1000 simple random samples of size $n = 50$ ten-gram portions of peanut butter, then about 7 of the samples will result in a mean of at least 3.6 insect fragments. This result is unusual. We might conclude that the sample comes from a population with a mean higher than 3 insect fragments per ten-gram portion.

29. (a) No, the variable "weekly time spent watching television" is not likely normally distributed. It is likely skewed right.

(b) Because the sample is large, $n = 40 \ge 30$, the sampling distribution of $\overline{x}$ is approximately normal with $\mu_{\overline{x}} = \mu = 2.35$ and

$$\sigma_{\overline{x}} = \frac{\sigma}{\sqrt{n}} = \frac{1.93}{\sqrt{40}} \approx 0.305 \, .$$

(c) $P(2 \leq \overline{x} \leq 3)$

$$= P\left(\frac{2-2.35}{1.93/\sqrt{40}} \leq Z \leq \frac{3-2.35}{1.93/\sqrt{40}}\right)$$

$$= P(-1.15 \leq Z \leq 2.13)$$

$$= 0.9834 - 0.1251$$

$$= 0.8583 \quad [\text{Tech: } 0.8577]$$

If we take 100 simple random samples of size $n = 40$ adult Americans, then about 86 of the samples will result in a mean time between 2 and 3 hours watching television on a weekday.

(d) $\mu_{\overline{x}} = \mu = 2.35$; $\sigma_{\overline{x}} = \frac{\sigma}{\sqrt{n}} = \frac{1.93}{\sqrt{35}}$

$$P(\overline{x} \leq 1.89) = P\left(Z \leq \frac{1.89-2.35}{1.93/\sqrt{35}}\right)$$

$$= P(Z \leq -1.41)$$

$$= 0.0793$$

If we take 100 simple random samples of size $n = 35$ adult Americans, then about 8 of the samples will result in a mean time of 1.89 hours or less watching television on a weekday. This result is not unusual, so this evidence is insufficient to conclude that avid Internet users watch less television.

31. (a) $\mu = \frac{\Sigma x}{N} = \frac{278}{6} \approx 46.3$ years

The population mean age is about 46.3 years.

(b) 37, 38;   37, 45;   37, 50;   37, 48;   37, 60;
38, 45;   38, 50;   38, 48;   38, 60;   45, 50;
45, 48;   45, 60;   50, 48;   50, 60;   48, 60

(c) Obtain each sample mean by adding the two ages in a sample and dividing by two.

$$\overline{x} = \frac{37+38}{2} = 37.5 \text{ yr} ; \ \overline{x} = \frac{37+45}{2} = 41 \text{ yr} ;$$

$$\overline{x} = \frac{37+50}{2} = 43.5 \text{ yr} ;$$

$$\overline{x} = \frac{37+48}{2} = 41.5 \text{ yr}, \text{ etc.}$$

$\overline{x}$	$P(\overline{x})$	$\overline{x}$	$P(\overline{x})$
37.5	$1/15 \approx 0.0667$	46.5	$1/15 \approx 0.0667$
41	$1/15 \approx 0.0667$	47.5	$1/15 \approx 0.0667$
41.5	$1/15 \approx 0.0667$	48.5	$1/15 \approx 0.0667$
42.5	$1/15 \approx 0.0667$	49	$2/15 \approx 0.1333$
43	$1/15 \approx 0.0667$	52.5	$1/15 \approx 0.0667$
43.5	$1/15 \approx 0.0667$	54	$1/15 \approx 0.0667$
44	$1/15 \approx 0.0667$	55	$1/15 \approx 0.0667$

(d) $\mu_{\overline{x}} = (37.5)\left(\frac{1}{15}\right) + (41)\left(\frac{1}{15}\right) + \ldots + (55)\left(\frac{1}{15}\right)$

$$\approx 46.3 \text{ years}$$

Notice that this is the same value we obtained in part (a) for the population mean.

(e) $P(43.3 \leq \overline{x} \leq 49.3)$

$$= P(43.5 \leq \overline{x} \leq 49)$$

$$= \frac{1}{15} + \frac{1}{15} + \frac{1}{15} + \frac{1}{15} + \frac{1}{15} + \frac{2}{15} = \frac{7}{15} = 0.467$$

(f) for part (b):
37, 38, 45;   37, 38, 50;   37, 38, 48;
37, 38, 60;   37, 45, 50;   37, 45, 48;
37, 45, 60;   37, 50, 48;   37, 50, 60;
37, 48, 60;   38, 45, 50;   38, 45, 48;
38, 45, 60;   38, 50, 48;   38, 50, 60;
38, 48, 60;   45, 50, 48;   45, 50, 60;
45, 48, 60;   50, 48, 60

for part (c):
Obtain each sample mean by adding the three ages in a sample and dividing by three.

$$\overline{x} = \frac{37+38+45}{3} = 40 \text{ yr} ;$$

$$\overline{x} = \frac{37+38+50}{3} \approx 41.7 \text{ yr} ;$$

$$\overline{x} = \frac{37+38+48}{3} = 41 \text{ yr} ; \text{ etc}$$

$\bar{x}$	$P(\bar{x})$	$\bar{x}$	$P(\bar{x})$
40	$1/20 = 0.05$	47.3	$1/20 = 0.05$
41	$1/20 = 0.05$	47.7	$2/20 = 0.1$
41.7	$1/20 = 0.05$	48.3	$1/20 = 0.05$
43.3	$1/20 = 0.05$	48.7	$1/20 = 0.05$
43.7	$1/20 = 0.05$	49	$1/20 = 0.05$
44	$1/20 = 0.05$	49.3	$1/20 = 0.05$
44.3	$1/20 = 0.05$	51	$1/20 = 0.05$
45	$2/20 = 0.1$	51.7	$1/20 = 0.05$
45.3	$1/20 = 0.05$	52.7	$1/20 = 0.05$

for part (d):

$$\mu_{\bar{x}} = (40)\left(\frac{1}{20}\right) + (41)\left(\frac{1}{20}\right) + \ldots + (52.7)\left(\frac{1}{20}\right)$$
$$\approx 46.3 \text{ years}$$

Notice that this is the same value we obtained previously.

for part (e):

$$P(43.3 \le \bar{x} \le 49.3) = \frac{14}{20} = \frac{7}{10} = 0.7$$

With the larger sample size, the probability of obtaining a sample mean within 3 years of the population mean increases.

33. (a) Assuming that only one number is selected, the probability distribution will be as follows:

x	$P(x)$
35	$1/38 \approx 0.0263$
-1	$37/38 \approx 0.9737$

(b)

$$\mu = (35)(0.0263) + (-1)(0.9737) \approx -\$0.05$$

$$\sigma = \sqrt{(35 - (-.05))^2 (0.0263) + (-1 - (-.05))^2 (0.9737)}$$
$$\approx \$5.76$$

(c) Because the sample size is large, $n = 100 > 30$, the sampling distribution of $\bar{x}$ is approximately normal with
$$\mu_{\bar{x}} = \mu = -\$0.05 \text{ and}$$
$$\sigma_{\bar{x}} = \frac{\sigma}{\sqrt{n}} = \frac{5.76}{\sqrt{100}} \approx \$0.576 .$$

(d) $P(\bar{x} > 0) = P\left(Z > \dfrac{0 - (-0.05)}{5.76/\sqrt{100}} \right)$

$\qquad = P(Z > 0.09)$

$\qquad = 1 - P(Z \le 0.09)$

$\qquad = 1 - 0.5359$

$\qquad = 0.4641$ [Tech: 0.4654]

(e) $\mu_{\bar{x}} = -\$0.05;\ \sigma_{\bar{x}} = \dfrac{5.76}{\sqrt{200}} \approx \$0.407$

$\qquad P(\bar{x} > 0) = P\left(Z > \dfrac{0 - (-0.05)}{5.76/\sqrt{200}} \right)$

$\qquad = P(Z > 0.12)$

$\qquad = 1 - P(Z \le 0.12)$

$\qquad = 1 - 0.5478$

$\qquad = 0.4522$ [Tech: 0.4511]

(f) $\mu_{\bar{x}} = -\$0.05;\ \sigma_{\bar{x}} = \dfrac{5.76}{\sqrt{1000}} \approx \$0.275$

$\qquad P(\bar{x} > 0) = P\left(Z > \dfrac{0 - (-0.05)}{5.76/\sqrt{1000}} \right)$

$\qquad = P(Z > 0.27)$

$\qquad = 1 - P(Z \le 0.27)$

$\qquad = 1 - 0.6064$

$\qquad = 0.3936$ [Tech: 0.3918]

(g) The probability of being ahead decreases as the number of games played increases.

35. The Central Limit Theorem states that, regardless of the distribution of the population, the sampling distribution of the sample mean becomes approximately normal as the sample size, n, increases.

37. Probability (b) will be larger. The standard deviation of the sampling distribution of $\bar{x}$ is decreased as the sample size, n, gets larger.

39. (a) We would expect that Jack's distribution would be skewed left, but not as much as the original distribution. Diane's distribution should be bell-shaped and symmetric, i.e., approximately normal.

(b) We would expect both distributions to have a mean of 50. The mean of the distribution of the sample mean is the same as the mean of the distribution from which the data are drawn.

(c) We expect Jack's distribution to have a standard deviation of $\sigma_{\bar{x}} = \dfrac{10}{\sqrt{3}} \approx 5.8$. We expect Diane's distribution to have a standard deviation of $\sigma_{\bar{x}} = \dfrac{10}{\sqrt{30}} \approx 1.8$.

41. (a) The population of interest is all full time students at your university. The sample is the ten students selected in the simple random sample.

(b) The number of hours of sleep each night is a quantitative random variable because it varies from student to student.

(c) The mean for ten randomly selected students would be a sample mean and therefore a statistic.

(d) The sample mean from part (c) is a random variable because its value will change depending on the ten students in the sample. There are two sources of variation in this problem. One source of variation is that people do not sleep the same number of hours each night, therefore, there will be person-to-person variability. The other source of variation is the variability of sleeping habits of each individual day-to-day.

Section 8.2

1. 0.44; $p = \dfrac{220}{500} = 0.44$

3. False; while it is possible for the sample proportion to have the same value as the population proportion, it will not *always* have the same value.

5. The sampling distribution of $\hat{p}$ is approximately normal when $n \le 0.05N$ and $np(1-p) \ge 10$.

7. $25,000(0.05) = 1250$; the sample size, $n = 500$, is less than 5% of the population size and $np(1-p) = 500(0.4)(0.6) = 120 \ge 10$. The distribution of $\hat{p}$ is approximately normal, with mean $\mu_{\hat{p}} = p = 0.4$ and standard deviation

$$\sigma_{\hat{p}} = \sqrt{\dfrac{p(1-p)}{n}} = \sqrt{\dfrac{0.4(1-0.4)}{500}} \approx 0.022.$$

9. $25,000(0.05) = 1250$; the sample size, $n = 1000$, is less than 5% of the population size and $np(1-p) = 1000(0.103)(0.897) = 92.391 \ge 10$. The distribution of $\hat{p}$ is approximately normal, with mean $\mu_{\hat{p}} = p = 0.103$ and standard deviation

$$\sigma_{\hat{p}} = \sqrt{\dfrac{p(1-p)}{n}} = \sqrt{\dfrac{0.103(1-0.103)}{1000}} \approx 0.010.$$

11. (a) $10,000(0.05) = 500$; the sample size, $n = 75$, is less than 5% of the population size and $np(1-p) = 75(0.8)(0.2) = 12 \ge 10$. The distribution of $\hat{p}$ is approximately normal, with mean $\mu_{\hat{p}} = p = 0.8$ and standard deviation

$$\sigma_{\hat{p}} = \sqrt{\dfrac{p(1-p)}{n}} = \sqrt{\dfrac{0.8(1-0.8)}{75}} \approx 0.046.$$

(b) $P(\hat{p} \ge 0.84) = P\left(Z \ge \dfrac{0.84-0.8}{\sqrt{0.8(0.2)/75}}\right)$
$= P(Z \ge 0.87)$
$= 1 - P(Z < 0.87)$
$= 1 - 0.8078$
$= 0.1922$ [Tech: 0.1932]

About 19 out of 100 random samples of size $n = 75$ will result in 63 or more individuals (that is, 84% or more) with the characteristic.

(c) $P(\hat{p} \le 0.68) = P\left(Z \le \dfrac{0.68-0.8}{\sqrt{0.8(0.2)/75}}\right)$
$= P(Z \le -2.60) = 0.0047$

About 5 out of 1000 random samples of size $n = 75$ will result in 51 or fewer individuals (that is, 68% or less) with the characteristic.

13. **(a)** $1,000,000(0.05) = 50,000$; the sample size, $n = 1000$, is less than 5% of the population size and
$np(1-p) = 1000(0.35)(0.65) = 227.5 \geq 10$.
The distribution of $\hat{p}$ is approximately normal, with mean $\mu_{\hat{p}} = p = 0.35$ and standard deviation
$$\sigma_{\hat{p}} = \sqrt{\frac{p(1-p)}{n}} = \sqrt{\frac{0.35(1-0.35)}{1000}} \approx 0.015.$$

(b) $\hat{p} = \dfrac{x}{n} = \dfrac{390}{1000} = 0.39$

$P(X \geq 390) = P(\hat{p} \geq 0.39)$

$$= P\left(Z \geq \frac{0.39 - 0.35}{\sqrt{0.53(0.65)/1000}}\right)$$

$= P(Z \geq 2.65)$

$= 1 - P(Z < 2.65)$

$= 1 - 0.9960 = 0.0040$

About 4 out of 1000 random samples of size $n = 1000$ will result in 390 or more individuals (that is, 39% or more) with the characteristic.

(c) $\hat{p} = \dfrac{x}{n} = \dfrac{320}{1000} = 0.32$

$P(X \leq 320) = P(\hat{p} \leq 0.32)$

$$= P\left(Z \geq \frac{0.32 - 0.35}{\sqrt{0.53(0.65)/1000}}\right)$$

$= P(Z \leq -1.99)$

$= 0.0233$ [Tech: 0.0234]

About 2 out of 100 random samples of size $n = 1000$ will result in 320 or fewer individuals (that is, 32% or less) with the characteristic.

15. **(a)** The response to the question is qualitative with two possible outcomes, can order a meal in a foreign language or not.

(b) The sample proportion is a random variable because it varies from sample to sample. The source of the variability is the individuals in the sample and their ability to order a meal in a foreign language.

(c) The sample size, $n = 200$, is less than 5% of the population size and
$np(1-p) = 200(0.47)(0.53) = 49.82 \geq 10$.
The distribution of $\hat{p}$ is approximately

normal, with mean $\mu_{\hat{p}} = p = 0.47$ and standard deviation
$$\sigma_{\hat{p}} = \sqrt{\frac{p(1-p)}{n}} = \sqrt{\frac{0.47(1-0.47)}{200}} \approx 0.035.$$

(d) $\hat{p} = \dfrac{x}{n} = \dfrac{100}{200} = 0.5$

$$P(\hat{p} > 0.5) = P\left(Z > \frac{0.5 - 0.47}{\sqrt{0.47(0.53)/200}}\right)$$

$= P(Z > 0.85) = 1 - P(Z \leq 0.85)$

$= 1 - 0.8023$

$= 0.1977$ [Tech: 0.1976]

About 20 out of 100 random samples of size $n = 200$ Americans will result in a sample where more than half can order a meal in a foreign language.

(e) $\hat{p} = \dfrac{x}{n} = \dfrac{80}{200} = 0.4$

$P(X \leq 80) = P(\hat{p} \leq 0.4)$

$$= P\left(Z \leq \frac{0.4 - 0.47}{\sqrt{0.47(0.53)/200}}\right)$$

$= P(Z \leq -1.98)$

$= 0.0239$ [Tech: 0.0237]

About 2 out of 100 random samples of size $n = 200$ Americans will result in a sample where 80 or fewer can order a meal in a foreign language. This result is unusual.

17. **(a)** Our sample size, $n = 500$, is less than 5% of the population size and
$np(1-p) = 500(0.39)(0.61) = 118.95 \geq 10$.
The distribution of $\hat{p}$ is approximately normal, with mean $\mu_{\hat{p}} = p = 0.39$ and standard deviation
$$\sigma_{\hat{p}} = \sqrt{\frac{p(1-p)}{n}} = \sqrt{\frac{0.39(1-0.39)}{500}} \approx 0.022.$$

(b) $P(\hat{p} < 0.38) = P\left(Z < \dfrac{0.38 - 0.39}{\sqrt{0.39(0.61)/500}}\right)$

$= P(Z < -0.46)$

$= 0.3228$ [Tech: 0.3233]

About 32 out of 100 random samples of size $n = 500$ adults will result in less than 38% who believe marriage is obsolete.

(c)

$$P(0.40 < \hat{p} < 0.45)$$

$$= P\left(\frac{0.40 - 0.39}{\sqrt{\dfrac{0.39(0.61)}{500}}} < Z < \frac{0.45 - 0.39}{\sqrt{\dfrac{0.39(0.61)}{500}}} \right)$$

$$= P(0.46 < Z < 2.75)$$

$$= 0.9970 - 0.6772$$

$$= 0.3198 \quad [\text{Tech: } 0.3203]$$

About 32 out of 100 random samples of size $n = 500$ adults will have between 40% and 45% of the respondents who say that marriage is obsolete.

(d)

$$P(X \geq 210)$$

$$= P(\hat{p} \geq 0.42)$$

$$= 1 - P(\hat{p} < 0.42)$$

$$= 1 - P\left(Z < \frac{0.42 - 0.39}{\sqrt{\dfrac{0.39(0.61)}{500}}} \right)$$

$$= 1 - P(Z < 1.38)$$

$$= 1 - 0.9162$$

$$= 0.0838 \quad [\text{Tech: } 0.0845]$$

About 8 out of 100 random samples of size $n = 500$ adults will have at least 210 of the respondents who say that marriage is obsolete.

19. $\hat{p} = \dfrac{x}{n} = \dfrac{121}{1100} = 0.11$

$$P(X \geq 121) = P(\hat{p} \geq 0.11)$$

$$= P\left(Z \geq \frac{0.11 - 0.10}{\sqrt{0.1(0.9)/1100}} \right)$$

$$= P(Z \geq 1.11)$$

$$= 1 - P(Z < 1.11)$$

$$= 1 - 0.8665$$

$$= 0.1335 \quad [\text{Tech: } 0.1345]$$

This result is not unusual, so this evidence is insufficient to conclude that the proportion of Americans who are afraid to fly has increased above 0.10.

21. $\hat{p} = \dfrac{x}{n} = \dfrac{164}{310} = 0.529$

$$P(X \geq 164) = P(\hat{p} \geq 0.529)$$

$$= P\left(Z \geq \frac{0.529 - 0.49}{\sqrt{0.49(0.51)/310}} \right)$$

$$= P(Z \geq 1.37) = 1 - P(Z < 1.37)$$

$$= 1 - 0.9147$$

$$= 0.0853 \quad [\text{Tech: } 0.0846]$$

This result is not unusual, if the true proportion of voters in favor of the referendum is 0.49. So, it is not surprising that we would get a result as extreme as this result. This illustrates the dangers of using exit polling to call elections. Notice that most people in the town do not favor the referendum, however the exit poll result showed a majority favor the referendum.

23. (a) To say the sampling distribution of $\hat{p}$ is approximately normal, we need $np(1-p) \geq 10$ and $n \leq 0.05N$.

With $p = 0.1$, we need

$$n(0.1)(1 - 0.1) \geq 10$$

$$n(0.1)(0.9) \geq 10$$

$$n(0.09) \geq 10$$

$$n \geq 111.11$$

Therefore, we need a sample size of 112, or 62 more adult Americans.

(b) With $p = 0.2$, we need

$$n(0.2)(1 - 0.2) \geq 10$$

$$n(0.2)(0.8) \geq 10$$

$$n(0.16) \geq 10$$

$$n \geq 62.5$$

Therefore, we need a sample size of 63, or 13 more, adult Americans.

25. (a) The response to the question is qualitative with two possible outcomes, believe in reincarnation or not.

(b) The sample proportion will vary because of the variability of the individuals in the sample and their belief in reincarnation. Thus, Bob could get 18 respondents who believe in reincarnation while Alicia could get 22 who believe in reincarnation.

(c) In sampling it is always important to select a random sample of individuals in order to increase the chance the sample is representative of the population. This will allow the results of the sample to be generalized to the population.

(d) In a sample of 100 American teens, we would expect $np = (100)(0.21) = 21$ of the teens to believe in reincarnation.

(e) The normal model should not be used to describe the distribution of the sample proportion in this problem because the sample is not large enough. Specifically, $np(1-p) = 20(0.21)(1-0.21) = 3.318 < 10$

(f) In order to determine the minimum sample size we need to solve the following inequality for n.
$$n(0.21)(1-0.21) > 10$$
After solving for n we get $n > 60.28$ which means the minimum sample size is 61 in order to use the normal model to describe the distribution of the sample proportion.

27. (a) Let X be the number of passengers who do not show up and let $\hat{p}$ be the proportion of passengers who do not show up for the flight. The sampling distribution of $\hat{p}$ is approximately normal with a mean of $\mu_{\hat{p}} = 0.0995$ and a standard deviation of
$$\sigma_{\hat{p}} = \sqrt{\frac{(0.0995)(1-0.0995)}{290}} \approx 0.01758$$

If a flight has 290 reservations, then the probability that 25 or more passengers miss the flight is the probability that the proportion is $25/290 = 0.0862$ or more. Thus,
$$P(X \geq 25) = P(\hat{p} \geq 0.0862)$$
$$= P\left(Z \geq \frac{0.0862-0.0995}{0.01758}\right)$$
$$= P(Z \geq -0.7565)$$
$$= 0.7764 \ [\text{Tech: } 0.7753]$$

(b) Let X be the number of passengers who show up for the flight and let $\hat{p}$ be the proportion of passengers who show up for the flight. The sampling distribution of $\hat{p}$ is approximately normal with a mean of $\mu_{\hat{p}} = 0.9005$ and a standard deviation of

$$\sigma_{\hat{p}} = \sqrt{\frac{(0.9005)(1-0.9005)}{320}} \approx 0.01673$$
$$P(X \leq 300) = P\left(\hat{p} \leq \frac{300}{320}\right)$$
$$= P(\hat{p} \leq 0.9375)$$
$$= P\left(Z \leq \frac{0.9375-0.9005}{0.01673}\right)$$
$$= P(Z \leq 2.21)$$
$$= 0.9864 \ [\text{Tech: } 0.9865]$$

(c) Let X be the number who do not show up for the flight and let $\hat{p}$ be the proportion of passengers who do not show up for the flight. The sampling distribution of $\hat{p}$ is approximately normal with a mean of $\mu_{\hat{p}} = 0.04$ and a standard deviation of
$$\sigma_{\hat{p}} = \sqrt{\frac{(0.04)(1-0.04)}{300}} \approx 0.01131.$$
In order to get on the flight you need 15 passengers to not show up because there are 14 passengers ahead of you.
$$P(X \geq 15) = P\left(\hat{p} \geq \frac{15}{300}\right)$$
$$= P(\hat{p} \geq 0.05)$$
Thus,
$$= P\left(Z \geq \frac{0.05-0.04}{0.01131}\right)$$
$$= P(Z \geq 0.88)$$
$$= 0.1894 \ [\text{Tech: } 0.1884]$$

29. (a)

Plays per Fumble

The distribution is slightly skewed to the right with three outliers (Atlanta Falcons, New Orleans Saints, and New England Patriots).

(b)   **Plays per Fumble**

The New England Patriots is only one outlier.

Chapter 8 Review Exercises

1. Answers will vary. The sampling distribution of a statistic (such as the sample mean) is a probability distribution for all possible values of the statistic computed from a sample of size n.

2. The sampling distribution of $\bar{x}$ is exactly normal if the population distribution is normal. If the population distribution is not normal, we apply the Central Limit Theorem and say that the distribution of $\bar{x}$ is approximately normal for a sufficiently large n (for finite σ). For our purposes, $n \geq 30$ is considered large enough to apply the theorem.

3. The sampling distribution of $\hat{p}$ is approximately normal if $np(1-p) \geq 10$ and $n \leq 0.05N$.

4. For $\bar{x}$: $\mu_{\bar{x}} = \mu$ and $\sigma_{\bar{x}} = \dfrac{\sigma}{\sqrt{n}}$

    For $\hat{p}$: $\mu_{\hat{p}} = p$ and $\sigma_{\hat{p}} = \sqrt{\dfrac{p(1-p)}{n}}$

5. **(a)** The total energy need during pregnancy is quantitative.

    (b) $P(X > 2625) = P\left(Z > \dfrac{2625 - 2600}{50}\right)$

    $= P(Z > 0.5)$

    $= 1 - P(Z \leq 0.5)$

    $= 1 - 0.6915$

    $= 0.3085$

    This result is not unusual.

    (c) Since the population is normally distributed, the sampling distribution of

$\bar{x}$ will be normal, regardless of the sample size. The mean of the distribution is $\mu_{\bar{x}} = \mu = 2600$ kcal, and the standard deviation is

$$\sigma_{\bar{x}} = \frac{\sigma}{\sqrt{n}} = \frac{50}{\sqrt{20}} \approx 11.180 \, \text{kcal.}$$

Answers may vary about the source of the variability in the sample mean.

(d) $P(\bar{x} > 2625) = P\left(Z > \dfrac{2625 - 2600}{50/\sqrt{20}}\right)$

$= P(Z > 2.24)$

$= 1 - P(Z \leq 2.24)$

$= 1 - 0.9875$

$= 0.0125$ [Tech: 0.0127]

If we take 100 simple random samples of size $n = 20$ pregnant women, then about 1 of the samples will result in a mean energy need of more than 2625 kcal/day. This result is unusual.

6. **(a)** Since we have a large sample ($n = 30$), we can use the Central Limit Theorem to say that the sampling distribution of $\bar{x}$ is approximately normal. The mean of the sampling distribution is $\mu_{\bar{x}} = \mu = 0.75$ inch and the standard deviation is

    $$\sigma_{\bar{x}} = \frac{\sigma}{\sqrt{n}} = \frac{0.004}{\sqrt{30}} \approx 0.001 \text{ inch.}$$

    (b) The quality control inspector will determine the machine needs an adjustment if the sample mean is either less than 0.748 inch or greater than 0.752 inch.

    $P(\text{needs adjustment})$

    $= P(\bar{x} < 0.748) + P(\bar{x} > 0.752)$

    $= P\left(Z < \dfrac{0.748 - 0.75}{0.004/\sqrt{30}}\right) + P\left(Z > \dfrac{0.752 - 0.75}{0.004/\sqrt{30}}\right)$

    $= P(Z < -2.74) + P(Z > 2.74)$

    $= P(Z < -2.74) + \left[1 - P(Z \leq 2.74)\right]$

    $= 0.0031 + (1 - 0.9969)$

    $= 0.0062$

    There is a 0.0062 probability that the quality control inspector will conclude the machine needs an adjustment when, in fact, the machine is correctly calibrated.

7. (a) No, the variable "number of television sets" is not likely normally distributed. It is likely skewed right.

(b) $\bar{x} = \dfrac{102}{40} = 2.55$ televisions per household

(c) Because the sample is large, $n = 40 > 30$, the sampling distribution of $\bar{x}$ is approximately normal with
$\mu_{\bar{x}} = \mu = 2.24$ and
$\sigma_{\bar{x}} = \dfrac{\sigma}{\sqrt{n}} = \dfrac{1.38}{\sqrt{40}} \approx 0.218$.

$$P(\bar{x} \geq 2.55) = P\left(Z \geq \dfrac{2.55 - 2.24}{1.38/\sqrt{40}}\right)$$
$$= P(Z \geq 1.42)$$
$$= 1 - P(Z < 1.42)$$
$$= 1 - 0.9222$$
$$= 0.0778 \quad [\text{Tech: } 0.0777]$$

If we take 100 simple random samples of size $n = 40$ households, then about 8 of the samples will result in a mean of 2.55 or more televisions. This result is not unusual, so it does not contradict the results reported by A.C. Nielsen.

8. (a) The variable start your own business versus work for someone else is a qualitative variable with two outcomes.

(b) The sample size, $n = 600$, is less than 5% of the population size and
$np(1 - p) = 600(0.72)(0.28) = 120.96 \geq 10$.
The sampling distribution of $\hat{p}$ is approximately normal, with mean
$\mu_{\hat{p}} = p = 0.72$ and standard deviation
$\sigma_{\hat{p}} = \sqrt{\dfrac{p(1 - p)}{n}} = \sqrt{\dfrac{0.72(1 - 0.72)}{600}} \approx 0.018$.

(b) $P(\hat{p} \leq 0.70) = P\left(Z \leq \dfrac{0.70 - 0.72}{\sqrt{0.72(0.28)/600}}\right)$
$$= P(Z \leq -1.09)$$
$$= 0.1379 \quad [\text{Tech: } 0.1376]$$

About 14 out of 100 random samples of size $n = 600$ 18- to 29-year-olds will result in no more than 70% who would prefer to start their own business.

(c) $\hat{p} = \dfrac{x}{n} = \dfrac{450}{600} = 0.75$
$$P(X \geq 450) = P(\hat{p} \geq 0.75)$$
$$= P\left(Z \geq \dfrac{0.75 - 0.72}{\sqrt{0.72(0.28)/600}}\right)$$
$$= P(Z \geq 1.64)$$
$$= 1 - P(Z < 1.64)$$
$$= 1 - 0.9495$$
$$= 0.0505 \quad [\text{Tech: } 0.0509]$$

About 5 out of 100 random samples of size $n = 600$ 18- to 29-year-olds will result in 450 or more who would prefer to start their own business. Since the probability is approximately 0.05, this result is considered unusual.

9. $\hat{p} = \dfrac{60}{500} = 0.12$; the sample size, $n = 500$, is less than 5% of the population size and
$np(1 - p) = 500(0.1)(0.9) = 45 \geq 10$, so the sampling distribution of $\hat{p}$ is approximately normal with mean $\mu_{\hat{p}} = p = 0.1$ and standard deviation
$\sigma_{\hat{p}} = \sqrt{\dfrac{p(1 - p)}{n}} = \sqrt{\dfrac{0.1(1 - 0.1)}{500}} \approx 0.013$.

$$P(x \geq 60) = P(\hat{p} \geq 0.12)$$
$$= P\left(Z \geq \dfrac{0.12 - 0.1}{\sqrt{0.1(0.9)/500}}\right)$$
$$= P(Z \geq 1.49)$$
$$= 1 - P(Z < 1.49)$$
$$= 1 - 0.9319$$
$$= 0.0681 \quad [\text{Tech: } 0.0680]$$

This result is not unusual, so this evidence is insufficient to conclude that the proportion of adults 25 years of age or older with advanced degrees increased.

10. (a) $np(1 - p) = 500(0.28)(0.72) = 100.8 \geq 10$.
The distribution of $\hat{p}$ is approximately normal, with mean $\mu_{\hat{p}} = p = 0.280$ and standard deviation
$\sigma_{\hat{p}} = \sqrt{\dfrac{p(1 - p)}{n}} = \sqrt{\dfrac{0.28(1 - 0.28)}{500}} \approx 0.020$.

(b)

$$P\left(\hat{p} \geq 0.310\right) = P\left(Z \geq \frac{0.310 - 0.280}{\sqrt{0.280(0.720)/500}}\right)$$

$$= P(Z \geq 1.49)$$

$$= 1 - P(Z < 1.49)$$

$$= 1 - 0.9319$$

$$= 0.0681 \quad [\text{Tech: } 0.0676]$$

If we take 100 simple random samples of size $n = 500$ at-bats, then about 7 of the samples will result in a batting average of 0.310 or greater. This result is not unusual.

(c)

$$P\left(\hat{p} \leq 0.255\right) = P\left(Z \leq \frac{0.255 - 0.280}{\sqrt{0.280(0.720)/500}}\right)$$

$$= P(Z \leq -1.25)$$

$$= 0.1056 \quad [\text{Tech: } 0.1066]$$

If we take 100 simple random samples of size $n = 500$ at-bats, then about 11 of the samples will result in a batting average of 0.255 or lower. This result is not unusual.

(d) $P\left(0.260 \leq \hat{p} \leq 0.300\right)$

$$= P\left(\frac{0.260 - 0.280}{\sqrt{\dfrac{0.280(0.720)}{500}}} Z \leq \frac{0.300 - 0.280}{\sqrt{\dfrac{0.280(0.720)}{500}}}\right)$$

$$= P(-1.00 \leq Z \leq 1.00)$$

$$= 0.8413 - 0.1587$$

$$= 0.6827 \quad [\text{Tech: } 0.6808]$$

Batting averages between 0.260 and 0.300 lie within 1 standard deviation of the mean of the sampling distribution. We expect that about 68% of the time, the sample proportion will be within these bounds.

(e) It is unlikely that a career 0.280 hitter hits exactly 0.280 each season. He will probably have seasons where he bats below 0.280 and other seasons where he bats above 0.280.

$P\left(\hat{p} \leq 0.260\right) \approx 0.16$ and

$P\left(\hat{p} \geq 0.300\right) \approx 0.16$.

Batting averages as low as 0.260 and as high as 0.300 would not be unusual. Based on a single season, we cannot conclude that a player who hit 0.260 is worse than a player who hit 0.300 because neither result would be unusual for players who had identical career batting averages of 0.280.

Chapter 8 Test

1. The Central Limit Theorem states that, regardless of the shape of the population, the sampling distribution of $\bar{x}$ becomes approximately normal as the sample size n increases.

2. $\mu_{\bar{x}} = \mu = 50$; $\sigma_{\bar{x}} = \dfrac{\sigma}{\sqrt{n}} = \dfrac{24}{\sqrt{36}} = 4$

3. **(a)** $P(X > 100) = P\left(Z > \dfrac{100 - 90}{35}\right)$

$$= P(Z > 0.29)$$

$$= 1 - P(Z \leq 0.29)$$

$$= 1 - 0.6141$$

$$= 0.3859 \quad [\text{Tech: } 0.3875]$$

If we select a simple random sample of $n = 100$ batteries of this type, then about 39 batteries would last more than 100 minutes. This result is not unusual.

(b) Since the population is normally distributed, the sampling distribution of $\bar{x}$ will be normal, regardless of the sample size. The mean of the distribution is $\mu_{\bar{x}} = \mu = 90$ minutes, and the standard deviation is

$$\sigma_{\bar{x}} = \frac{\sigma}{\sqrt{n}} = \frac{35}{\sqrt{10}} \approx 11.068 \text{ minutes.}$$

(c) $P(\bar{x} > 100) = P\left(Z > \dfrac{100 - 90}{35/\sqrt{10}}\right)$

$$= P(Z > 0.90)$$

$$= 1 - P(Z \leq 0.90)$$

$$= 1 - 0.8159$$

$$= 0.1841 \quad [\text{Tech: } 0.1831]$$

If we take 100 simple random samples of size $n = 10$ batteries of this type, then about 18 of the samples will result in a mean charge life of more than 100 minutes. This result is not unusual.

(d) $\mu_{\bar{x}} = \mu = 90$; $\sigma_{\bar{x}} = \dfrac{\sigma}{\sqrt{n}} = \dfrac{35}{\sqrt{25}} = 7$

$$P(\bar{x} > 100) = P\left(Z > \dfrac{100-90}{7}\right)$$

$$= P(Z > 1.43) = 1 - P(Z \le 1.43)$$

$$= 1 - 0.9263$$

$$= 0.0764 \quad [\text{Tech: } 0.0766]$$

If we take 100 simple random samples of size $n = 25$ batteries of this type, then about 8 of the samples will result in a mean charge life of more than 100 minutes.

(e) The probabilities are different because a change in n causes a change in $\sigma_{\bar{x}}$.

4. (a) Since we have a large sample ($n = 45 \ge 30$), we can use the Central Limit Theorem to say that the sampling distribution of $\bar{x}$ is approximately normal. The mean of the sampling distribution is $\mu_{\bar{x}} = \mu = 2.0$ liters and the standard deviation is

$$\sigma_{\bar{x}} = \dfrac{\sigma}{\sqrt{n}} = \dfrac{0.05}{\sqrt{45}} \approx 0.007 \text{ liter.}$$

(b) The quality control manager will determine the machine needs an adjustment if the sample mean is either less than 1.98 liters or greater than 2.02 liters.

$P(\text{needs adjustment})$

$$= P(\bar{x} < 1.98) + P(\bar{x} > 2.02)$$

$$= P\left(Z < \dfrac{1.98-2.0}{0.05/\sqrt{45}}\right) + P\left(Z > \dfrac{2.02-2.0}{0.05/\sqrt{45}}\right)$$

$$= P(Z < -2.68) + P(Z > 2.68)$$

$$= P(Z < -2.68) + \left[1 - P(Z \le 2.68)\right]$$

$$= 0.0037 + (1 - 0.9963)$$

$$= 0.0074 \quad [\text{Tech: } 0.0073]$$

There is a 0.0074 probability that the quality control manager will conclude the machine needs an adjustment even though the machine is correctly calibrated.

5. (a) Our sample size, $n = 300$, is less than 5% of the population size and

$$np(1-p) = 300(0.224)(0.776) \approx 52 \ge 10.$$

The distribution of $\hat{p}$ is approximately normal, with mean $\mu_{\hat{p}} = p = 0.224$ and standard deviation

$$\sigma_{\hat{p}} = \sqrt{\dfrac{p(1-p)}{n}} = \sqrt{\dfrac{0.224(1-0.224)}{300}} \approx 0.024.$$

(b) $\hat{p} = \dfrac{x}{n} = \dfrac{50}{300} = \dfrac{1}{6} \approx 0.167$

$$P(X \ge 50) = P\left(\hat{p} \ge \dfrac{1}{6}\right)$$

$$= P\left(Z \ge \dfrac{\dfrac{1}{6} - 0.224}{\sqrt{0.224(0.776)/300}}\right)$$

$$= P(Z \ge -2.38) = 1 - P(Z < -2.38)$$

$$= 1 - 0.0087$$

$$= 0.9913 \quad [\text{Tech: } 0.9914]$$

About 99 out of 100 random samples of size $n = 300$ adults will result in at least 50 adults (that is, at least 16.7%) who are smokers.

(c) $P(\hat{p} \le 0.18) = P\left(Z \le \dfrac{0.18 - 0.224}{\sqrt{0.224(0.776)/300}}\right)$

$$= P(Z \le -1.83)$$

$$= 0.0336 \quad [\text{Tech: } 0.0338]$$

About 3 out of 100 random samples of size $n = 300$ adults will result 18% or less who are smokers. This result would be unusual.

6. (a) For the sample proportion to be normal, the sample size must be large enough to meet the condition $np(1-p) \ge 10$. Since $p = 0.01$, we have:

$$n(0.01)(1 - 0.01) \ge 10$$

$$0.0099n \ge 10$$

$$n \ge \dfrac{10}{0.0099} \approx 1010.1$$

Thus, the sample size must be at least 1011 in order to satisfy the condition.

(b) $\hat{p} = \dfrac{x}{n} = \dfrac{9}{1500} = \dfrac{3}{500} = 0.006$

$P(X < 10) = P(X \leq 9)$

$\qquad = P(\hat{p} \leq 0.006)$

$\qquad = P\left(Z \leq \dfrac{0.006 - 0.01}{\sqrt{0.01(0.99)/1500}} \right)$

$\qquad = P(Z \leq -1.56)$

$\qquad = 0.0594$ [Tech: 0.0597]

About 6 out of 100 random samples of size $n = 1500$ Americans will result in fewer than 10 with peanut or tree nut allergies. This result is not considered unusual.

7. $\hat{p} = \dfrac{82}{1000} = 0.082$; the sample size, $n = 1000$, is less than 5% of the population size and $np(1-p) = 1000(0.07)(0.93) = 65.1 \geq 10$, so the sampling distribution of $\hat{p}$ is approximately normal with mean $\mu_{\hat{p}} = p = 0.07$ and standard deviation

$\sigma_{\hat{p}} = \sqrt{\dfrac{p(1-p)}{n}} = \sqrt{\dfrac{0.07(1-0.07)}{1000}} \approx 0.008$.

$P(x \geq 82) = P(\hat{p} \geq 0.082)$

$\qquad = P\left(Z \geq \dfrac{0.082 - 0.07}{\sqrt{0.07(0.93)/1000}} \right)$

$\qquad = P(Z \geq 1.49) = 1 - P(Z < 1.49)$

$\qquad = 1 - 0.9319$

$\qquad = 0.0681$ [Tech: 0.0685]

This result is not unusual, so this evidence is insufficient to conclude that the proportion of households with a net worth in excess of $1 million has increased above 7%.

Case Study: Sampling Distribution of the Median

1. Open a new MINITAB worksheet, and click on the **Calc** menu. Highlight the **Random Data** option and then **Normal…**. Fill in the information as follows:

    Number of rows of data to generate: 250
    Store in Column(s): c1–c80
    Mean: 50
    Standard deviation: 10
    Select the OK button to generate the samples.

2. Select the **Calc** menu and click on the **Row Statistics…** option. Fill in the information as follows:

    Click the Mean radio.
    Input variables: c1–c10
    Store results in: c81
    Select the OK button to produce the means.

    Note: the values in the left dialogue box fill in automatically as you complete the other information.

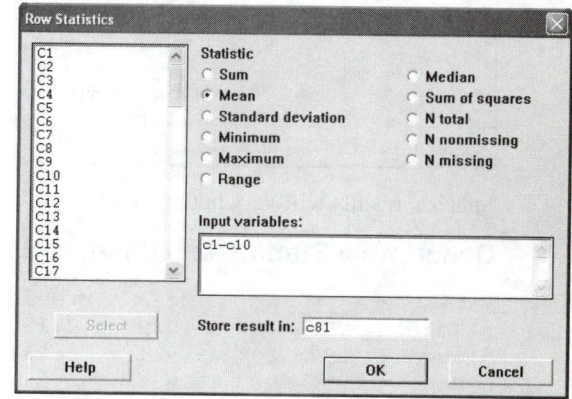

3. Select the **Calc** menu and click on the **Row Statistics…** option. Fill in the information as follows:

    Click the Median radio.
    Input variables: c1–c10
    Store results in: c82
    Select the OK button to produce the means.

    Note: the values in the left dialogue box fill in automatically as you complete the other information.

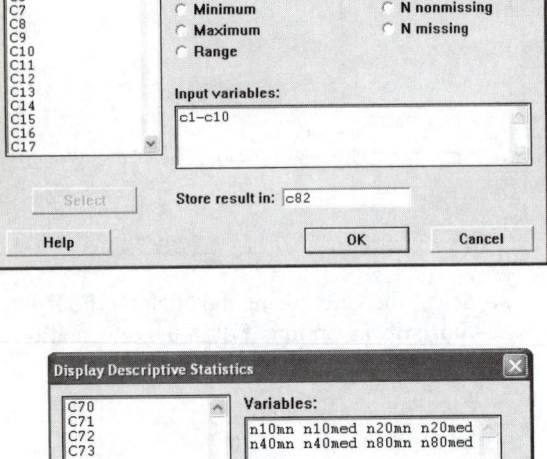

4. Repeat (2) and (3) for the remaining sample sizes: $n = 20$ (c1–c20), $n = 40$ (c1–c40), and $n = 80$ (c1–c80).

5. Select the **Stat** menu. Highlight the **Basic Statistics** option and then **Display Descriptive Statistics…**. Scroll to the bottom of the left dialogue box. Double click on "c81 n10mn" to select it as a variable. It should appear in the Variables: dialogue box. Similarly, select "c82 n10med", "c83 n20mn", and so on through "c88 n80med" as variables. Select the OK button to compute the summary statistics.

Specific results will vary but should be similar to the following.

Descriptive Statistics: n10mn, n10med, n20mn, n20med, n40mn, n40med, n80mn, ...

Variable	N	N*	Mean	SE Mean	StDev	Minimum	Q1	Median	Q3
n10mn	250	0	50.089	0.189	2.989	42.337	48.151	50.155	51.873
n10med	250	0	49.911	0.234	3.698	40.146	47.315	49.839	52.413
n20mn	250	0	49.936	0.133	2.096	44.011	48.550	49.811	51.355
n20med	250	0	50.006	0.164	2.591	42.746	48.073	49.941	51.761
n40mn	250	0	49.949	0.0930	1.471	46.207	49.055	49.914	50.900
n40med	250	0	49.999	0.119	1.877	45.734	48.794	49.933	51.197
n80mn	250	0	49.947	0.0663	1.048	47.065	49.149	49.936	50.804
n80med	250	0	49.977	0.0845	1.337	45.844	49.047	49.919	50.929

Variable	Maximum
n10mn	58.094
n10med	60.061
n20mn	55.999
n20med	56.150
n40mn	54.141
n40med	54.981
n80mn	52.438
n80med	54.033

6. The averages of the sample means are all very close to the actual value of $\mu = 50$. There does not appear to be much of a difference for the various sample sizes.

   The averages of the sample medians are also very close to the actual value of $M = 50$. For smaller sample sizes, the average seems to be slightly more than the true value, while for larger samples the average of the sample medians tends to be slightly less than the actual value.

7. In both cases, the standard deviations decrease as the sample size increases. However, the standard deviation for the sample means is always less than the standard deviation for the sample medians.

8. Answers will vary. It seems that the mean may be a better measure of center since the sampling distribution of the sample mean has less variability (i.e. a smaller standard deviation).

9. We will draw the histograms for the sample means first, and then draw the histograms for the sample medians. We will use a feature in MINITAB that will allow us to draw multiple graphs on the same panel, using the same scale. Select the **Graph** menu and click **Histogram….** Select the **Simple histogram** type and click on the OK button.

Scroll to the bottom of the left dialogue box. Because we want to draw histograms of the sample means, double click on "c81 n10mn" to select it as a variable. It should appear in the **Graph variables:** dialogue box. Similarly, select "c83 n20mn", "c85 n40mn", and "c87 n80mn" as graph variables.

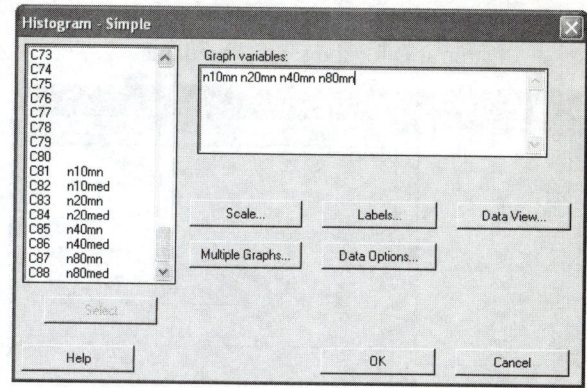

Click on the "Multiple Graphs…" button. Select the "In Separate panels on the same graph" radio button. Place check marks in the boxes next to "Same Y" and "Same X, including same bins." Select the OK button to close the Multiple Graphs window. Select OK again to draw the graphs.

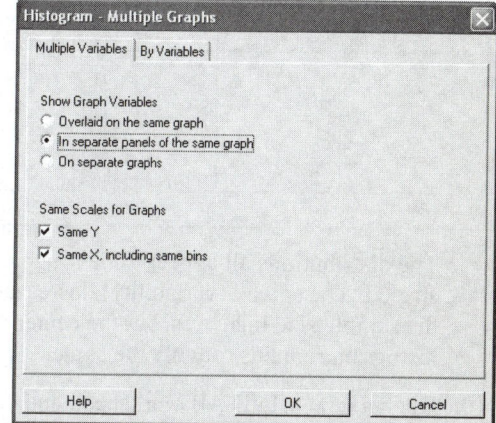

The histograms for the sample means follow:

The distributions all appear to be roughly bell-shaped. The spread (variability) decreases as the sample size increases, but the center of the histogram remains roughly the same.

Next, we repeat the process to draw histograms for the sample medians:

The distributions all appear to be roughly bell-shaped. The spread (variability) decreases as the sample size increases, but the center of the histogram remains roughly the same.

10. Open a new MINITAB worksheet, and click on the **Calc** menu. Highlight the **Random Data** option and then **Exponential…**. Fill in the information as follows:

    Number of rows of data to generate: 250
    Sore in Column(s): c100–c180
    Mean: 50
    Threshold: 0.0
    Select the OK button to generate the samples.

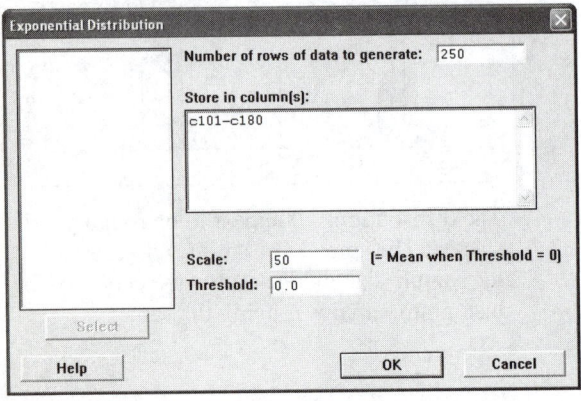

11. Select the **Calc** menu and click on the **Row Statistics…** option. Fill in the information as follows:

    Click the Mean radio.
    Input variables: c100–c110
    Store results in: c181
    Select the OK button to produce the means.

12. Select the **Calc** menu and click on the **Row Statistics…** option. Fill in the information as follows:

    Click the Median radio.
    Input variables: c101–c110
    Store results in: c182
    Select the OK button to produce the means.

13. Repeat (11) and (12) for the remaining sample sizes: $n = 20$ (c101–c120), $n = 40$ (c101–c140), and $n = 80$ (c101–c180).

14. Select the **Stat** menu. Highlight the **Basic Statistics** option and then **Display Descriptive Statistics…**. Scroll to the bottom of the left dialogue box. Double click on "c181 e10mn" to select it as a variable. It should appear in the **Variables:** dialogue box. Similarly, select "c182 e10med", "c183 e20mn", and so on through "c188 e80med" as variables. Select the OK button to compute the summary statistics.

Specific results will vary but should be similar to those shown next.

Descriptive Statistics: e10mn, e10med, e20mn, e20med, e40mn, e40med, e80mn, ...

Variable	N	N*	Mean	SE Mean	StDev	Minimum	Q1	Median	Q3
e10mn	250	0	49.454	0.978	15.462	17.347	38.799	48.674	58.552
e10med	250	0	37.66	1.00	15.81	8.81	26.47	35.64	44.79
e20mn	250	0	49.092	0.669	10.572	27.176	41.450	48.512	56.381
e20med	250	0	35.625	0.724	11.443	14.206	27.759	33.494	41.713
e40mn	250	0	49.786	0.482	7.627	32.322	44.789	48.862	54.227
e40med	250	0	35.365	0.494	7.819	19.930	29.199	34.744	40.131
e80mn	250	0	50.081	0.344	5.445	36.059	46.104	49.952	53.103
e80med	250	0	34.891	0.345	5.454	21.648	30.792	34.859	38.212

Variable	Maximum
e10mn	105.964
e10med	98.75
e20mn	82.229
e20med	84.592
e40mn	73.698
e40med	61.172
e80mn	70.860
e80med	58.424

15. The averages of the sample means are all close to the actual value of $\mu = 50$. The sample mean appears to slightly overestimate the true mean for small samples. As the sample size increases, the estimate is much closer.

$$\lambda = \frac{1}{\mu} = \frac{1}{50} = 0.02 \; ; \quad M = \frac{\ln 2}{\lambda} = \frac{\ln 2}{0.02} = 34.657$$

The averages of the sample medians are also close to the actual value of $M = 34.657$. For smaller sample sizes, the average of the sample medians was not as close as for larger sample sizes.

16. In both cases, the standard deviations decrease as the sample size increases. However, the standard deviation for the sample means is always less than the standard deviation for the sample medians.

17. Answers will vary. It seems that the mean may be a better measure of center since the sampling distribution of the sample mean has less variability (i.e. a smaller standard deviation).

18. We will draw the histograms for the sample means first, and then draw the histograms for the sample medians. We will again draw multiple graphs on the same panel, using the same scale.

Select the **Graph** menu and click **Histogram…**.
Select the **Simple histogram** type and click on
the OK button.

Scroll to the bottom of the left dialogue box.
Because we want to draw histograms of the
sample means, double click on "c181 e10mn"
to select it as a variable. It should appear in
the **Graph variables:** dialogue box.
Similarly, select "c183 e20mn",
"c185 e40mn", and "c187 e80mn" as graph
variables.

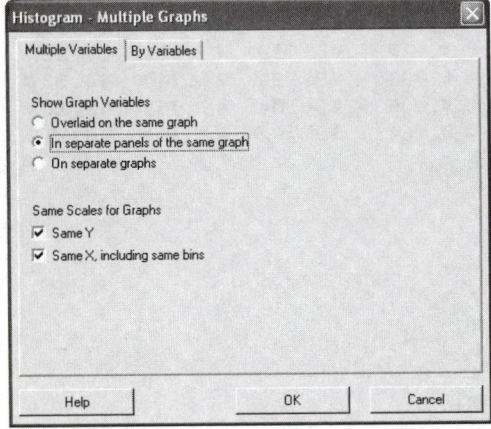

Click on the "Multiple Graphs…" button. Select
the "In Separate panels on the same graph" radio
button. Place check marks in the boxes next to
"Same Y" and "Same X, including same bins."
Select the OK button to close the Multiple
Graphs window. Select OK again to draw the
graphs.

The histograms for the sample means follow:

The distributions begin skewed right but become
more bell-shaped as the sample size increases.
The spread (variability) decreases as the sample
size increases, but the center of the histogram
remains roughly the same.

Next, we repeat the process to draw
histograms for the sample medians:

The distributions begin skewed right but become more bell-shaped as the sample size increases. The distributions appear to stay in the same general location, but the variability decreases as the sample size increases.

19. The Central Limit Theorem can be used to explain why the distributions of sample means always become roughly normal as the sample size increases. However, the Central Limit Theorem is for sample *means* only. It *cannot* be used to explain why the sampling distribution of the sample medians appears to approach a normal distribution as the sample size increases.

Chapter 9
Estimating the Value of a Parameter

Section 9.1

1. Point estimate

3. False. A 95% confidence interval means 95% of intervals will contain the parameter if a large number of samples is obtained.

5. Increases

7. $z_{\frac{\alpha}{2}} = z_{0.05} = 1.645$

9. $z_{\frac{\alpha}{2}} = z_{0.01} = 2.326$

11. $\hat{p} = \dfrac{0.201 + 0.249}{2}$

$= 0.225$

$E = \dfrac{0.249 - 0.201}{2}$

$= 0.024$

$x = 0.225 \cdot 1200$

$= 270$

13. $\hat{p} = \dfrac{0.462 + 0.509}{2}$

$= 0.4855$

$E = \dfrac{0.509 - 0.462}{2}$

$= 0.0235$

$x = 0.4855 \cdot 1680 = 815.64 \approx 816$

15. $\hat{p} = \dfrac{30}{150}$

$= 0.2$

Lower bound:

$0.2 - 1.645 \cdot \sqrt{\dfrac{0.2(1 - 0.2)}{150}} = 0.146$

Upper bound:

$0.2 + 1.645 \cdot \sqrt{\dfrac{0.2(1 - 0.2)}{150}} = 0.254$

17. $\hat{p} = \dfrac{120}{500}$

$= 0.24$

Lower bound:

$0.24 - 2.575 \cdot \sqrt{\dfrac{0.24(1 - 0.24)}{500}} = 0.191$

Upper bound:

$0.24 + 2.575 \cdot \sqrt{\dfrac{0.24(1 - 0.24)}{500}} = 0.289$

19. $\hat{p} = \dfrac{860}{1100}$

$= 0.782$

$z_{0.03} = 1.88$

Lower bound:

$0.782 - 1.88 \cdot \sqrt{\dfrac{0.782(1 - 0.782)}{1100}} = 0.759$

[Tech: 0.758]

Upper bound:

$0.782 + 1.88 \cdot \sqrt{\dfrac{0.782(1 - 0.782)}{1100}} = 0.805$

21. (a) The statement is flawed because no interval has been provided about the population proportion.

(b) The statement is flawed because this interpretation has a varying level of confidence.

(c) The statement is correct because an interval has been provided for the population proportion.

(d) The statement is flawed because this interpretation suggests that this interval sets the standard for all the other intervals, which is not true.

23. We are 95% confident that the population proportion of adult Americans who dread Valentine's day is between $0.18 - 0.045 = 0.135$ and $0.18 + 0.045 = 0.225$.

25. (a) $\hat{p} = \dfrac{417}{2306}$

$\qquad = 0.181$

(b) $n\hat{p}(1-\hat{p}) = 2306 \cdot 0.181(1-0.181)$; The

$\qquad = 341.84 \geq 10$

sample is less than 5% of the population.

(c) Lower bound:

$\qquad 0.181 - 1.645 \cdot \sqrt{\dfrac{0.181(1-0.181)}{2306}} = 0.168$

Upper bound:

$\qquad 0.181 + 1.645 \cdot \sqrt{\dfrac{0.181(1-0.181)}{2306}} = 0.194$

(d) We are 90% confident that the population proportion of adult Americans 18 years and older who have donated blood in the past two years is between 0.168 and 0.194.

27. (a) $\hat{p} = \dfrac{521}{1003}$

$\qquad = 0.519$

(b) $n\hat{p}(1-\hat{p}) = 1003 \cdot 0.519(1-0.519)$; The

$\qquad = 250.39 \geq 10$

sample is less than 5% of the population.

(c) Lower bound:

$\qquad 0.519 - 1.96 \cdot \sqrt{\dfrac{0.519(1-0.519)}{1003}} = 0.488$

[Tech: 0.489]

$\qquad 0.519 + 1.96 \cdot \sqrt{\dfrac{0.519(1-0.519)}{1003}} = 0.550$

(d) It is possible that the population proportion is more than 60%, because it is possible that the true proportion is not captured in the confidence interval. However, it is not likely because 0.6 is outside of the confidence interval.

(e) Lower bound: $1 - 0.550 = 0.450$

Upper bound: $1 - 0.488 = 0.512$
[Tech: 0.511]

29. (a) $\hat{p} = \dfrac{26}{234}$

$\qquad = 0.111$

Lower bound:

$\qquad 0.111 - 1.96 \cdot \sqrt{\dfrac{0.111(1-0.111)}{234}} = 0.071$

Upper bound:

$\qquad 0.111 + 1.96 \cdot \sqrt{\dfrac{0.111(1-0.111)}{234}} = 0.151$

(b) Lower bound:

$\qquad 0.111 - 2.575 \cdot \sqrt{\dfrac{0.111(1-0.111)}{234}} = 0.058$

Upper bound:

$\qquad 0.111 + 2.575 \cdot \sqrt{\dfrac{0.111(1-0.111)}{234}} = 0.164$

(c) Increasing the confidence level increases the margin of error.

31. (a) The sample is the 1000 adults aged 19 and older. The population is adults aged 19 and older.

(b) The variable of interest is whether the individual brings and uses his or her cell phone every trip to the bathroom. It is qualitative with two possible outcomes.

(c) $\hat{p} = \dfrac{241}{1000}$

$\qquad = 0.241$

(d) The point estimate $\hat{p}$ is a statistic because its value is based on the sample. The point estimate is a random variable because its value may change depending on the individuals in the survey. The main source of variability is the individuals selected to be in the study.

(e) Lower bound:

$\qquad 0.241 - 1.96 \cdot \sqrt{\dfrac{0.241(1-0.241)}{1000}} = 0.214$

Upper bound:

$\qquad 0.241 + 1.96 \cdot \sqrt{\dfrac{0.241(1-0.241)}{1000}} = 0.268$

(f) The sample must be representative of the population. Random sampling is required to ensure the individuals in the sample are representative of the population.

33. $\hat{p} = \dfrac{105}{202}$

$= 0.520$

Lower bound:

$0.520 - 1.96 \cdot \sqrt{\dfrac{0.520(1-0.520)}{202}} = 0.451$

Upper bound:

$0.520 + 1.96 \cdot \sqrt{\dfrac{0.520(1-0.520)}{202}} = 0.589$

We are 95% confident that the proportion of days JNJ stock will increase is between 0.451 and 0.589.

35. (a) $n = 0.635(1-0.635)\left(\dfrac{2.575}{0.03}\right)^2$

$= 1708$

(b) $n = 0.25\left(\dfrac{2.575}{0.03}\right)^2$

$= 1842$

37. (a) $n = 0.15(1-0.15)\left(\dfrac{2.33}{0.02}\right)^2$

$= 1731$

(b) $n = 0.25\left(\dfrac{2.33}{0.02}\right)^2$

$= 3394$

39. (a) $n = 0.53(1-0.53)\left(\dfrac{1.96}{0.03}\right)^2$

$= 1064$

(b) $n = 0.25\left(\dfrac{1.96}{0.03}\right)^2$

$= 1067$

(c) The results are close because $0.53(1-0.53) = 0.2491$ is close to 0.25.

41. $n = 0.64(1-0.64)\left(\dfrac{1.96}{0.03}\right)^2$

$= 984$

43. The difference between the two point estimates is within the margin of error.

45. $n = 20 + 1.96^2$

$= 23.84$

$\tilde{p} = \left(\dfrac{1}{23.84}\right)\left(2 + \dfrac{1}{2}1.96^2\right)$

$= 0.164$

Lower bound:

$0.164 - 1.96 \cdot \sqrt{\dfrac{1}{23.84}0.164(1-0.164)} = 0.015$

[Tech: 0.016]

Upper bound:

$0.164 + 1.96 \cdot \sqrt{\dfrac{1}{23.84}0.164(1-0.164)} = 0.313$

We are 95% confident that the proportion of students on Jane's campus who eat cauliflower is between 0.015 and 0.313.

47. (a) The variable of interest in the survey is whether the individual says they wash their hands in a public rest room, or not. It is qualitative with two possible outcomes.

(b) The sample is the 1001 adults interviewed. The population is all adults.

(c) $n\hat{p}(1-\hat{p}) = 1001\left(\dfrac{921}{1001}\right)\left(1-\dfrac{921}{1001}\right)$;

$= 73.6 \geq 10$

The sample is less than 5% of the population.

(d) $\hat{p} = \dfrac{921}{1001}$

$= 0.920$

Lower bound:

$0.920 - 1.96 \cdot \sqrt{\dfrac{0.920(1-0.920)}{1001}} = 0.903$

Upper bound:

$$0.920 + 1.96 \cdot \sqrt{\frac{0.920(1-0.920)}{1001}} = 0.937$$

(e) The variable of interest is whether the individual is observed washing their hands in a public restroom or not. It is qualitative with two outcomes.

(f) Randomness could be achieved by using systematic sampling. For example, select every 10th individual who enters the bathroom. Also, to be able to generalize the results, we would want about half the individuals to be female and half male.

(g) $n\hat{p}(1-\hat{p}) = 6076\left(\frac{4679}{6076}\right)\left(1-\frac{4679}{6076}\right)$;

$$= 1075.80 \geq 10$$

The sample is less than 5% of the population.

(h) $\hat{p} = \dfrac{4679}{6076}$

$= 0.770$

Lower bound:

$$0.770 - 1.96\sqrt{\frac{0.770(1-0.770)}{6076}} = 0.759$$

Upper bound:

$$0.770 + 1.96\sqrt{\frac{0.770(1-0.770)}{6076}} = 0.781$$

(i) The proportion who say that they wash their hands is greater than the proportion who actually do. Explanations as to why may include people lying about their hand-washing habits out of embarrassment.

(j) There is person-to-person variability in the telephone survey. That is, different samples will result in different individuals surveyed, and therefore, yield different results. There is person-to-person variability and individual variability in the observational study, for example, an individual who does not always wash his or her hands in the study, but during this observation does wash his or her hands.

49. Data must be qualitative with two possible outcomes to construct confidence intervals for a proportion.

51. The polling companies use the margin of error formula with $\hat{p} = 0.5$ and

$$0.5(1-0.5)\left(\frac{1.96}{0.03}\right)^2 = 1068.$$

53. Mariya's interval is incorrect because the distance from the upper bound to the point estimate is $0.173 - 0.13 = 0.043$ and the distance from the lower bound to the point estimate is $0.13 - 0.117 = 0.013$.

Section 9.2

1. Decreases

3. α

5. False. The sample size can be large.

7. (a) $t_{0.10} = 1.316$

    (b) $t_{0.05} = 1.697$

    (c) $t_{0.01} = 2.552$

    (d) $t_{0.05} = 1.725$

9. Yes, because $0.987 > 0.928$ and there are no outliers.

11. No, because there is one outlier.

13. $\bar{x} = \dfrac{18+24}{2}$

$= 21$

$E = \dfrac{24-18}{2}$

$= 3$

15. $\bar{x} = \dfrac{23+5}{2}$

$= 14$

$E = \dfrac{23-5}{2}$

$= 9$

17. (a) $t_{0.02} = 2.172$

Lower bound:

$$108 - 2.172 \cdot \frac{10}{\sqrt{25}} = 103.66$$

Upper bound:

$$108 + 2.172 \cdot \frac{10}{\sqrt{25}} = 112.34$$

(b) $t_{0.02} = 2.398$

Lower bound:

$$108 - 2.398 \cdot \frac{10}{\sqrt{10}} = 100.42$$

Upper bound:

$$108 + 2.398 \cdot \frac{10}{\sqrt{10}} = 115.58$$

Decreasing the sample size increases the margin of error.

(c) $t_{0.05} = 1.711$

Lower bound:

$$108 - 1.711 \cdot \frac{10}{\sqrt{25}} = 104.58$$

Upper bound:

$$108 + 1.711 \cdot \frac{10}{\sqrt{25}} = 111.42$$

Decreasing the confidence level decreases the margin of error.

(d) No because the sample sizes are too small.

(b) $t_{0.02} = 2.624$

Lower bound: $50 - 2.624 \cdot \dfrac{8}{\sqrt{15}} = 44.58$

Upper bound: $50 + 2.624 \cdot \dfrac{8}{\sqrt{15}} = 55.42$

Decreasing the sample size increases the margin of error.

(c) $t_{0.025} = 2.093$

Lower bound: $50 - 2.093 \cdot \dfrac{8}{\sqrt{20}} = 46.26$

Upper bound: $50 + 2.093 \cdot \dfrac{8}{\sqrt{20}} = 53.74$

Decreasing the confidence level decreases the margin of error.

(d) No because the sample sizes are too small.

19. **(a)** $t_{0.025} = 2.032$

Lower bound:

$$18.4 - 2.032 \cdot \frac{4.5}{\sqrt{35}} = 16.85$$

Upper bound:

$$18.4 + 2.032 \cdot \frac{4.5}{\sqrt{35}} = 19.95$$

(b) $t_{0.025} = 2.009$

Lower bound:

$$18.4 - 2.009 \cdot \frac{4.5}{\sqrt{50}} = 17.12$$

Upper bound:

$$18.4 + 2.009 \cdot \frac{4.5}{\sqrt{50}} = 19.68$$

Increasing the sample size decreases the margin of error.

(c) $t_{0.005} = 2.728$

Lower bound:

$$18.4 - 2.728 \cdot \frac{4.5}{\sqrt{35}} = 16.32$$

Upper bound:

$$18.4 + 2.728 \cdot \frac{4.5}{\sqrt{35}} = 20.48$$

Increasing the confidence level increases the margin of error.

(d) If $n = 15$, the population must be normal.

21. **(a)** This statement is flawed because this interpretation implies that the population mean varies rather than the interval.

(b) This statement is correct because an interval has been provided for the population mean.

(c) This statement is flawed because this interpretation makes an implication about the individuals rather than the mean.

(d) This statement is flawed because the interpretation should be about the mean number of hours worked by adult Americans, not about adults in Idaho.

23. We are 90% confident that the mean drive-through service time of Taco Bell restaurants is between 161.5 and 164.7 seconds.

25. Increase the sample size and decrease the confidence level to narrow the confidence interval.

27. (a) Since the distribution of blood alcohol concentrations is not normally distributed, the sample size must be large so that the distribution of the sample mean will be approximately normal.

(b) The sample size is less than 5% of the population.

(c) $t_{0.05} = 1.676$

Lower bound:

$$0.167 - 1.676 \cdot \frac{0.010}{\sqrt{51}} = 0.1647$$

Upper bound:

$$0.167 + 1.676 \cdot \frac{0.010}{\sqrt{51}} = 0.1693$$

We are 90% confident that the mean BAC in fatal crashes where the driver had a positive BAC is between 0.1647 and 0.1693 g / dL .

(d) It is possible that the mean BAC is less than 0.08 g / dL because it is possible that the true mean is not captured in the confidence interval, but it is not likely.

29. $t_{0.025} = 1.987$

Lower bound: $356.1 - 1.987 \cdot \dfrac{185.7}{\sqrt{92}} = 317.63$

[Tech: 317.64]

Upper bound: $356.1 + 1.987 \cdot \dfrac{185.7}{\sqrt{92}} = 394.57$

[Tech: 394.56]

We are 95% confident that the mean number of licks to the center of a Tootsie pop is between 317.63 and 394.57.

31. (a) $\bar{x} = \dfrac{58.71}{12}$

$= 4.893$

(b) $s = 0.319$

$t_{0.025} = 2.201$

Lower bound:

$$4.893 - 2.201 \cdot \frac{0.319}{\sqrt{12}} = 4.690$$

Upper bound:

$$4.893 + 2.201 \cdot \frac{0.319}{\sqrt{12}} = 5.096$$

[Tech: 5.095]

We are 95% confident that the mean pH of rain water in Tucker County, West Virginia is between 4.690 and 5.096.

(c) $t_{0.005} = 3.106$

Lower bound:

$$4.893 - 3.106 \cdot \frac{0.319}{\sqrt{12}} = 4.607$$

[Tech: 4.606]

Upper bound:

$$4.893 + 3.106 \cdot \frac{0.319}{\sqrt{12}} = 5.179$$

We are 99% confident that the mean pH of rain water in Tucker County, West Virginia is between 4.607 and 5.179.

(d) The margin of error increases as the confidence level increases.

33. (a) Since 0.961 [Tech: 0.966] > 0.918 , it is reasonable to conclude the data come from a normal population.

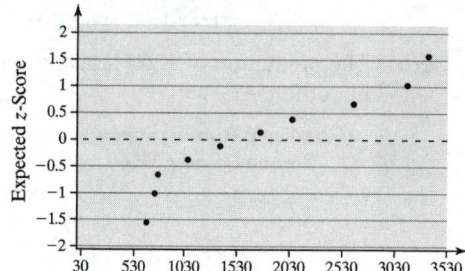

Normal Probabilty Plot of Repair Cost

(b) There are no outliers.

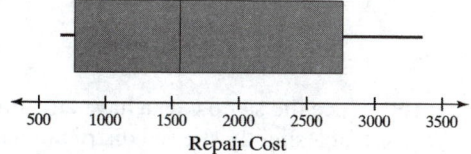

Repair Cost for Low-Impact Collision

(c)
$$\bar{x} = \frac{17565}{10}$$

$$= 1756.5$$

$$s = 1007.5$$

$$t_{0.025} = 2.262$$

Lower bound:

$$1756.50 - 2.262 \cdot \frac{1007.5}{\sqrt{10}} = 1035.8$$

Upper bound:

$$1756.50 + 2.262 \cdot \frac{1007.5}{\sqrt{10}} = 2477.2$$

We are 95% confident the mean repair cost for a low-impact collision involving mini- and micro-vehicles is between $1035.8 and $2477.2.

(d) The confidence interval would likely be narrower because there is less variability in the data because variability associated with the make of the vehicle has been removed.

35. (a) The histogram is slightly skewed to the right.

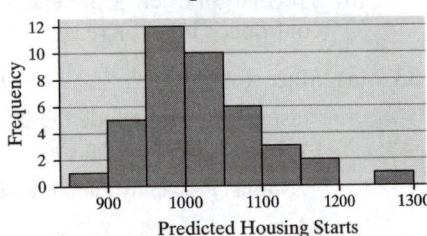

Predicted Housing Starts, Second Quarter 2014

(b) The prediction of 1260 housing starts is in outlier.

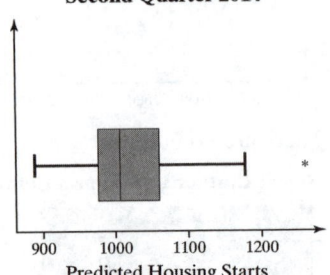

Predicted Housing Starts, Second Quarter 2014

(c) Since the sample data have an outlier and a slightly skewed distribution, it is necessary to have a large sample size to invoke the Central Limit Theorem so

that the sampling distribution of the sample mean is approximately normal.

(d)
$$\bar{x} = \frac{40911}{40}$$

$$= 1022.8$$

$$s = 75.1$$

$$t_{0.025} = 2.023$$

Lower bound:

$$1022.8 - 2.023 \cdot \frac{75.1}{\sqrt{40}} = 998.8$$

[Tech: 998.7]

Upper bound:

$$1022.8 + 2.023 \cdot \frac{75.1}{\sqrt{40}} = 1046.8$$

We are 95% confident the mean number of predicted housing starts in the second quarter of 2014 is between 998.8 and 1046.8.

37. (a) The data are skewed to the right.

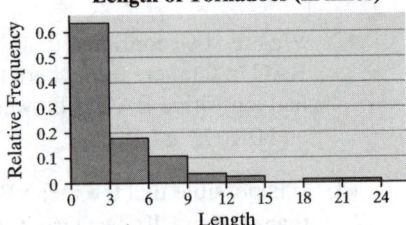

Length of Tornadoes (in miles)

(b) There are 4 outliers in the data.

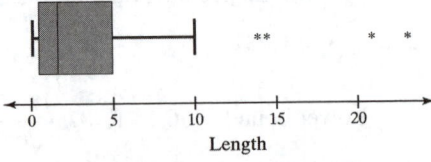

(c) Since the sample data have outliers and a skewed distribution, it is necessary to have a large sample size to invoke the Central Limit Theorem so that the sampling distribution of the sample mean is approximately normal.

(d) Using statistical software, the 95% confidence interval is $(2.400, 4.800)$.

We are 95% confident the mean length of a tornado in Oklahoma is between 2.400 miles and 4.280 miles.

39. (a) $\bar{x} = 22.150$

$$E = \frac{22.209 - 22.091}{2} = 0.059$$

(b) We are 95% confident that the mean age of people when first married is between 22.091 and 22.209 years.

(c) $s = 4.885$

$t_{0.025} = 1.960$

Lower bound:

$$22.150 - 1.960 \cdot \frac{4.885}{\sqrt{26,540}} = 22.091$$

Upper bound:

$$22.150 + 1.960 \cdot \frac{4.885}{\sqrt{26,540}} = 22.209$$

(d) The margin of error is very small because the sample size is very large.

41. For 99% confidence: $n = \left(\dfrac{2.575 \cdot 13.4}{2} \right)^2$

$$= 298$$

For 95% confidence: $n = \left(\dfrac{1.96 \cdot 13.4}{2} \right)^2$

$$= 173$$

Decreasing the confidence level decreases the sample size needed.

43. (a) $n = \left(\dfrac{1.96 \cdot 16.6}{4} \right)^2$

$$= 67$$

(b) $n = \left(\dfrac{1.96 \cdot 16.6}{2} \right)^2$

$$= 265$$

(c) The sample size approximately quadruples.

(d) $n = \left(\dfrac{2.575 \cdot 16.6}{4} \right)^2$

$$= 115$$

Increasing the level of confidence increases the sample size. For a fixed margin of error, greater confidence can be achieved with a larger sample size.

45. (a) set I: $\bar{x} = \dfrac{793}{8} = 99.1$

set II: $\bar{x} = \dfrac{1982}{20} = 99.1$

set III: $\bar{x} = \dfrac{2971}{30} = 99.0$

(b) set I: $s = 19.8$, $t_{0.025} = 2.365$

Lower bound: $99.1 - 2.365 \cdot \dfrac{19.8}{\sqrt{8}} = 82.5$

[Tech: 82.6]

Upper bound: $99.1 + 2.365 \cdot \dfrac{19.8}{\sqrt{8}} = 115.7$

set II: $s = 15.8$, $t_{0.025} = 2.093$

Lower bound: $99.1 - 2.093 \cdot \dfrac{15.8}{\sqrt{20}} = 91.7$

Upper bound: $99.1 + 2.093 \cdot \dfrac{15.8}{\sqrt{20}} = 106.5$

set III: $s = 14.8$, $t_{0.025} = 2.045$

Lower bound: $99.0 - 2.045 \cdot \dfrac{14.8}{\sqrt{30}} = 93.5$

Upper bound:

$$99.0 + 2.045 \cdot \frac{14.8}{\sqrt{30}} = 104.5$$

(c) As the size of the sample increases, the width of the confidence interval decreases.

(d) set I: $\bar{x} = \dfrac{703}{8}$, $s = 35.0$

$$= 87.9$$

Lower bound: $87.9 - 2.365 \cdot \dfrac{35.0}{\sqrt{8}} = 58.6$

Upper bound:

$$87.9 - 2.365 \cdot \frac{35.0}{\sqrt{8}} = 117.2$$

set II: $\bar{x} = \dfrac{1892}{20}$, $s = 24.3$

$\qquad = 94.6$

Lower bound: $94.6 - 2.093 \cdot \dfrac{24.3}{\sqrt{20}} = 83.2$

Upper bound:

$94.6 + 2.093 \cdot \dfrac{24.3}{\sqrt{20}} = 106.0$

set III: $\bar{x} = \dfrac{2881}{30}$, $s = 21.1$

$\qquad = 96.0$

Lower bound: $96.0 - 2.045 \cdot \dfrac{21.1}{\sqrt{30}} = 88.1$

Upper bound:

$96.0 + 2.045 \cdot \dfrac{21.1}{\sqrt{30}} = 103.9$

(e) Each interval contains the population mean. The procedure for constructing the confidence interval is robust. This also illustrates the Law of Large Numbers and the Central Limit Theorem.

47. (a) Answers will vary.

(b) Answers will vary.

(c) Answers will vary.

(d) Answers will vary. Expect 95% of the intervals to contain the population mean.

49. (a) Completely randomized design

(b) The treatment is the smoking cessation program. There are 2 levels.

(c) The response variable is whether or not the smoker had 'even a puff' from a cigarette in the past 7 days.

(d) The statistics reported are 22.3% of participants in the experimental group reported abstience and 13.1% of participants in the control group reported abstinence.

(e) $\dfrac{p(1-q)}{q(1-p)} = \dfrac{0.223(1-0.131)}{0.131(1-0.223)}$

$\qquad \approx 1.90$

This means that reported abstinence is almost twice as likely in the experimental group than in the control group.

(f) The authors are 95% confident that the population odds ratio is between 1.12 and 3.26.

(g) Answers will vary. One possibility is smoking cessation is more likely when the Happy Ending Intervention program is used rather than the control method.

51. The t-distribution has less spread as the degrees of freedom increase because as n increases, s becomes closer to σ by the Law of Large Numbers.

53. The degrees of freedom are the number of data values that are allowed to vary.

55. We expect that the margin of error for population A will be smaller since there should be less variation in the ages of college students than in the ages of the residents of a town, resulting in a smaller standard deviation for population A.

Section 9.3

1. False. The chi-square distribution is skewed to the right.

3. False. To construct a confidence interval about a population variance or standard deviation, the sample must come from a normally distributed population.

5. $\chi^2_{0.95} = 10.117$, $\chi^2_{0.05} = 30.144$

7. $\chi^2_{0.99} = 9.542$, $\chi^2_{0.01} = 40.289$

9. (a) $\chi^2_{0.95} = 10.117$, $\chi^2_{0.05} = 30.144$

Lower bound: $\dfrac{(20-1)(12.6)}{30.144} = 7.94$

Upper bound: $\dfrac{(20-1)(12.6)}{10.117} = 23.66$

(b) $\chi^2_{0.95} = 17.708$, $\chi^2_{0.05} = 42.557$

Lower bound: $\dfrac{(30-1)(12.6)}{42.557} = 8.59$

Upper bound: $\dfrac{(30-1)(12.6)}{17.708} = 20.63$

The width of the interval decreases as the sample size increases.

(c) $\chi^2_{0.99} = 7.633$, $\chi^2_{0.01} = 36.191$

Lower bound: $\dfrac{(20-1)(12.6)}{36.191} = 6.61$

Upper bound: $\dfrac{(20-1)(12.6)}{7.633} = 31.36$

The width of the interval increases as the confidence level increases.

11. $s^2 = (0.319)^2 = 0.102$, $\chi^2_{0.975} = 3.816$, $\chi^2_{0.025} = 21.920$

Lower bound: $\sqrt{\dfrac{(12-1)(0.102)}{21.920}} = 0.226$

Upper bound: $\sqrt{\dfrac{(12-1)(0.102)}{3.816}} = 0.542$

We can be 95% confident that the population standard deviation of the pH of rainwater in Tucker County, West Virginia is between 0.226 and 0.542.

13. $s^2 = (1007.4542)^2 = 1,014,963.9651$, $\chi^2_{0.95} = 3.325$, $\chi^2_{0.05} = 16.919$

Lower bound:

$\sqrt{\dfrac{(10-1)(1,014,963.9651)}{16.919}} = 734.8$

Upper bound:

$\sqrt{\dfrac{(10-1)(1,014,963.9651)}{3.325}} = 1657.5$

We are 90% confident that the population standard deviation of repair costs of a low-impact bumper crash on a mini- or micro-car is between \$734.8 and \$1657.5.

15. (a) From the probability plot shown below, the data appear to be from a population that is normally distributed.

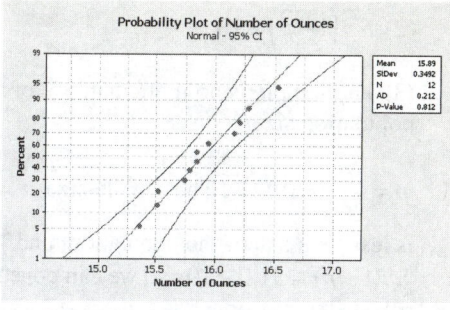

(b) $s = 0.349$ ounce

(c) $s^2 = (0.349)^2 = 0.122$, $\chi^2_{0.95} = 4.575$, $\chi^2_{0.05} = 19.675$

Lower bound: $\sqrt{\dfrac{(12-1)(0.122)}{19.675}} = 0.261$

Upper bound: $\sqrt{\dfrac{(12-1)(0.122)}{4.575}} = 0.542$

[Tech 0.541]

The quality control manager can be 90% confident that the population standard deviation of the number of ounces of peanuts is between 0.261 and 0.542 ounce.

17. Fisher's approximation:

$\chi^2_{0.975} = \dfrac{\left(-1.96 + \sqrt{2(100)-1}\right)^2}{2}$

$= 73.772$

$\chi^2_{0.025} = \dfrac{\left(1.96 + \sqrt{2(100)-1}\right)^2}{2}$

$= 129.070$

Actual values:

$\chi^2_{0.975} = 74.222$

$\chi^2_{0.025} = 129.561$

The values calculated by Fisher's approximation are close to the actual values.

Section 9.4

1. Confidence intervals for a population proportion are constructed on qualitative variables for which there are two possible outcomes.

3. (1) The data must be from a simple random sample or the result of a randomized experiment.

(2) $n\hat{p}\left(1-\hat{p}\right) \geq 10$

(3) The sample is no more than 5% of the population size.

5. $\hat{p} = \dfrac{35}{300} \approx 0.117$. The sample size $n = 300$ is less than 5% of the population, and $n\hat{p}(1-\hat{p}) \approx 31.0 \geq 10$, so we can construct a Z-interval. For 99% confidence the critical value is $z_{0.005} = 2.575$. Then:

Lower bound

$$= \hat{p} - z_{0.005} \cdot \sqrt{\frac{\hat{p}(1-\hat{p})}{n}}$$

$$= 0.117 - 2.575 \cdot \sqrt{\frac{0.117(1-0.117)}{300}}$$

$$\approx 0.117 - 0.048 = 0.069$$

Upper bound

$$= \hat{p} + z_{0.005} \cdot \sqrt{\frac{\hat{p}(1-\hat{p})}{n}}$$

$$= 0.117 + 2.575 \cdot \sqrt{\frac{0.117(1-0.117)}{300}}$$

$$\approx 0.117 + 0.048 = 0.165 \quad [\text{Tech: } 0.164]$$

7. We construct a *t*-interval because we are estimating the population mean, we do not know the population standard deviation, and the underlying population is normally distributed. For 90% confidence, $\alpha/2 = 0.05$. Since $n = 12$, df $= 11$ and $t_{0.05} = 1.796$. Then:

Lower bound $= \bar{x} - t_{0.05} \cdot \dfrac{s}{\sqrt{n}}$

$$= 45 - 1.796 \cdot \frac{14}{\sqrt{12}}$$

$$\approx 45 - 7.26 = 37.74$$

Upper bound $= \bar{x} + t_{0.05} \cdot \dfrac{s}{\sqrt{n}}$

$$= 45 + 1.796 \cdot \frac{14}{\sqrt{12}}$$

$$\approx 45 + 7.26 = 52.26$$

9. We construct a *t*-interval because we are estimating the population mean, we do not know the population standard deviation, and the underlying population is normally distributed. For 99% confidence, $\alpha/2 = 0.005$. Since $n = 40$, df $= 39$ and $t_{0.050} = 2.708$. Then:

Lower bound $= \bar{x} - t_{0.005} \cdot \dfrac{s}{\sqrt{n}}$

$$= 120.5 - 2.708 \cdot \frac{12.9}{\sqrt{40}}$$

$$\approx 120.5 - 5.52 = 114.98$$

Upper bound $= \bar{x} + t_{0.005} \cdot \dfrac{s}{\sqrt{n}}$

$$= 120.5 + 2.708 \cdot \frac{12.9}{\sqrt{40}}$$

$$\approx 120.5 + 5.52 = 126.02$$

11. df $= n - 1 = 12 - 1 = 11$ and, for 90% confidence, $\alpha/2 = 0.05$. From the table, $\chi^2_{0.05} = 19.675$ and $\chi^2_{1-0.05} = \chi^2_{0.95} = 4.575$.

Lower bound $= \dfrac{(n-1)s^2}{\chi^2_{\alpha/2}} = \dfrac{11 \cdot 23.7}{19.675} \approx 13.25$

Upper bound $= \dfrac{(n-1)s^2}{\chi^2_{1-\alpha/2}} = \dfrac{11 \cdot 23.7}{4.575} \approx 56.98$

13. We construct a *t*-interval because we are estimating the population mean, we do not know the population standard deviation, and the underlying population is normally distributed. For 95% confidence, $\alpha/2 = 0.025$. Since $n = 40$, df $= 39$ and $t_{0.025} = 2.023$. Then:

Lower bound $= \bar{x} - t_{0.025} \cdot \dfrac{s}{\sqrt{n}}$

$$= 54 - 2.023 \cdot \frac{8}{\sqrt{40}}$$

$$\approx 54 - 2.6 = 51.4 \text{ months}$$

Upper bound $= \bar{x} + t_{0.025} \cdot \dfrac{s}{\sqrt{n}}$

$$= 54 + 2.023 \cdot \frac{8}{\sqrt{40}}$$

$$\approx 54 + 2.6 = 56.6 \text{ months}$$

We can be 95% confident that the population of felons convicted of aggravated assault serve a mean sentence between 51.4 and 56.6 months.

15. We construct a *t*-interval because we are estimating the population mean, we do not know the population standard deviation, and the underlying population is normally distributed. For 90% confidence, $\alpha/2 = 0.05$. Since $n = 100$, then df $= 99$. There is no row in the table for 99 degrees of freedom, so we use df $= 100$ instead. The

critical value is $t_{0.05} = 1.660$.

$$\text{Lower bound} = \overline{x} - t_{0.05} \cdot \frac{s}{\sqrt{n}}$$

$$= 3421 - 1.660 \cdot \frac{2583}{\sqrt{100}}$$

$$\approx 3421 - 428.8$$

$$= 2992.2 \ [\text{Tech: } 2992.1]$$

$$\text{Upper bound} = \overline{x} + t_{0.05} \cdot \frac{s}{\sqrt{n}}$$

$$= 3421 + 1.660 \cdot \frac{2583}{\sqrt{100}}$$

$$\approx 3421 + 428.8$$

$$= 3849.8 \ [\text{Tech: } 3849.9]$$

The Internal Revenue Service can be 90% confident that the true mean additional tax owed for estate tax returns is between $2,992.2 and $3,849.8.

17. $\hat{p} = \dfrac{526}{1008} \approx 0.522$. The sample size

$n = 1008$ is less than 5% of the population, and $n\hat{p}(1-\hat{p}) \approx 251.5 \geq 10$, so we can construct a Z-interval.

For 90% confidence the critical value is $z_{0.05} = 1.645$. Then:

Lower bound

$$= \hat{p} - z_{0.05} \cdot \sqrt{\frac{\hat{p}(1-\hat{p})}{n}}$$

$$= 0.522 - 1.645 \cdot \sqrt{\frac{0.522(1-0.522)}{1008}}$$

$$\approx 0.522 - 0.026 = 0.496$$

Upper bound

$$= \hat{p} + z_{0.05} \cdot \sqrt{\frac{\hat{p}(1-\hat{p})}{n}}$$

$$= 0.522 + 1.645 \cdot \sqrt{\frac{0.522(1-0.522)}{1008}}$$

$$\approx 0.522 + 0.026 = 0.548$$

The Gallup Organization can be 90% confident that the population proportion of adult Americans who are worried about having enough money to live comfortably in retirement is between 0.496 and 0.548 (i.e., between 49.6% and 54.8%).

19. (a) Pitch speed is a quantitative variable. It is important to know because confidence intervals for a mean are constructed on quantitative data, while confidence intervals for a proportion are constructed on qualitative data with two possible outcomes.

(b) The correlation between raw data and normal score is 0.989 [Tech: 0.992], which is greater than 0.946, the critical value for $n = 18$. Therefore, it is reasonable to conclude that pitch speed comes from a population that is normally distributed.

Normal Probability Plot of Pitch Speed

(c) The data set has no outliers.

Pitch Speed of Kershaw Four-Seam Fastball

(d) Since it is reasonable to conclude the data come from a population that is normally distributed, there are no outliers, and the data is the result of a random sample of pitches, it is reasonable to construct a confidence interval for mean pitch speed.

(e) We construct a t-interval because we are estimating the population mean and we do not know the population standard deviation. For 95% confidence, $\alpha/2 = 0.025$. Since $n = 18$, then df $= 17$. The critical value is $t_{0.025} = 2.110$. Using technology, we find $\overline{x} \approx 94.747$ and $s \approx 0.783$.

$$\text{Lower bound} = \overline{x} - t_{0.025} \cdot \frac{s}{\sqrt{n}}$$

$$= 94.747 - 2.110 \cdot \frac{0.783}{\sqrt{18}}$$

$$\approx 94.747 - 0.389$$

$$= 94.358$$

Upper bound $= \bar{x} + t_{0.025} \cdot \dfrac{s}{\sqrt{n}}$

$$= 94.747 + 2.110 \cdot \dfrac{0.783}{\sqrt{18}}$$

$$\approx 94.747 + 0.389$$

$$= 95.136$$

We can be 95% confident that the mean pitch speed of a Clayton Kershaw four-seam fastball is between 94.358 mph and 95.136 mph.

(f) A 95% confidence interval for the mean pitch speed of all major league pitchers' four-seam fastballs would be wider because of the pitcher-to-pitcher variability in pitch speed is now part of the analysis.

21. (a) A cross-sectional study is a study that has its data obtained at a specific point in time (or over a very short period of time). It also means the study is an observational study.

(b) The variable of interest is whether an individual with sleep apnea has gum disease, or not. This is a qualitative variable because it categorizes the individual with two outcomes ("Has gum disease" and "Does not have gum disease").

(c) $\hat{p} = \dfrac{192}{320} = 0.6$. The sample size $n = 320$ is less than 5% of the population and $n\hat{p}(1-\hat{p}) = 76.8 \geq 10$, so we can construct a z-interval. For 95% confidence, the critical value is $z_{0.025} = 1.960$. Then:

Lower bound

$$= \hat{p} - z_{0.025} \cdot \sqrt{\dfrac{\hat{p}(1-\hat{p})}{n}}$$

$$= 0.6 - 1.960 \cdot \sqrt{\dfrac{0.6(1-0.6)}{320}}$$

$$\approx 0.6 - 0.054 = 0.546$$

Upper bound

$$= \hat{p} + z_{0.025} \cdot \sqrt{\dfrac{\hat{p}(1-\hat{p})}{n}}$$

$$= 0.6 + 1.960 \cdot \sqrt{\dfrac{0.6(1-0.6)}{320}}$$

$$\approx 0.6 + 0.054 = 0.654$$

We are 95% confident the proportion of individuals with sleep apnea who have gum disease is between 0.546 and 0.654 (i.e., between 54.6% and 65.4%).

23. (a) A cohort study is an observational study in which a group of individuals are followed over a period of time and information about the individual is obtained over this time period.

(b) The variable weight gain is quantitative. The variable is measured by determining the change in weight of the individual from a baseline measure to a second point in time, which is 4 years later in this case.

(c) The variable "increase in consumption of french fries" is qualitative because an individual is either categorized as increasing french fry consumption, or not.

(d) Margin of error:
$4.42 - 3.35 = 1.07$ pounds

(e) Individuals whose smoking status changed from former to current smoker lost 2.47 pounds on average over the 4-year period.

(f) We are 95% confident the mean weight loss over a 4-year period of individuals whose smoking status changed from former to current smoker was between 1.12 pounds and 3.82 pounds.

(g) The population applies to males and females who are healthy and not obese.

25. A confidence interval for a proportion should be constructed since the variable of interest is qualitative with two possible outcomes–either morals are getting better, or not.

27. A confidence interval for a mean should be constructed since the variable of interest is quantitative and we would like to estimate the typical reduction in caloric intake.

Section 9.5

1. Bootstrapping is a method of statistical inference that uses a set of sample data as the population. Many samples of size equal to the original sample size are chosen. The samples are chosen with replacement so that the samples are not necessarily the original sample. The statistic of interest is determined for each sample. The distribution of these statistics is used to make a judgment about the population parameter.

3. (a) Yes. The sample size is the same as the original sample, and all of the values in the bootstrap sample appear in the original sample.

   (b) No. The sample size is too small.

   (c) No. The sample size is too large.

   (d) No. The value 63 is not in the original sample.

   (e) Yes. The sample size is the same as the original sample, and all of the values in the bootstrap sample appear in the original sample.

5. Sample 1: $\dfrac{68+72+68+73+72+75}{6} \approx 71.3$

   Sample 2: $\dfrac{71+72+73+72+72+71}{6} \approx 71.8$

   Sample 3: $\dfrac{72+73+68+68+73+71}{6} \approx 70.8$

7. Answers will vary. A sample answer generated by StatCrunch is given.

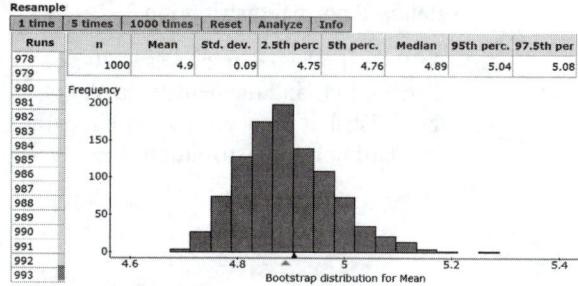

   (a) In Exercise 31 of Section 9.2, we found the 95% t-interval to be $(4.690, 5.076)$.

      The confidence interval constructed using the bootstrapping method is narrower.

   (b) The histogram is skewed right.

9. Answers will vary. A sample answer generated by StatCrunch is given.

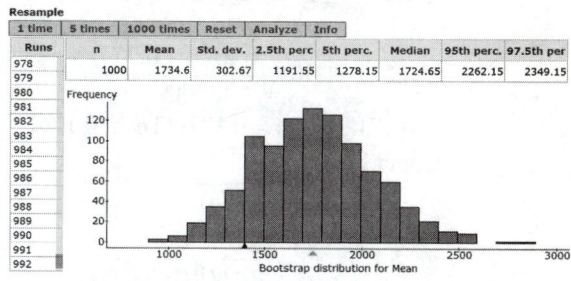

   In Exercise 33 of Section 9.2, we found the 95% t-interval to be $(1035.8, 2477.2)$.

   The confidence interval constructed using the bootstrapping method is narrower.

11. (a) A t-interval cannot be constructed because there are two outliers.

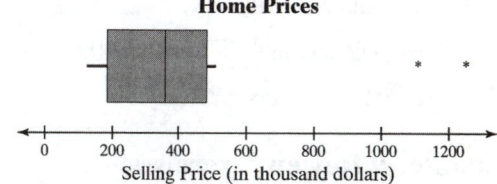

Home Prices

   (b) Answers will vary. A sample answer generated by StatCrunch is given.

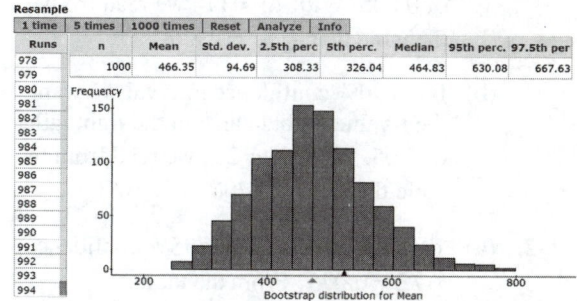

   (c) Lower bound: $308.3 thousand
      Upper bound: $667.6 thousand.

13. (a) $\hat{p} = \dfrac{23}{85}$

      ≈ 0.271

   (b) Answers will vary.

   (c) The sample size $n = 85$ is less than 5% of the population, and $n\hat{p}(1-\hat{p}) \approx 16.8 \geq 10$, so we can construct a 95% confidence interval using the normal model.
      For 95% confidence the critical value is $z_{0.025} = 1.96$. Then:

Lower bound

$$= \hat{p} - z_{0.025} \cdot \sqrt{\frac{\hat{p}(1-\hat{p})}{n}}$$

$$= 0.271 - 1.96 \cdot \sqrt{\frac{0.271(1-0.271)}{85}}$$

$$\approx 0.271 - 0.094 = 0.177 \quad [\text{Tech: } 0.176]$$

Upper bound

$$= \hat{p} + z_{0.025} \cdot \sqrt{\frac{\hat{p}(1-\hat{p})}{n}}$$

$$= 0.271 + 1.96 \cdot \sqrt{\frac{0.271(1-0.271)}{85}}$$

$$\approx 0.271 + 0.094 = 0.365$$

15. Answers will vary.

17. Answers will vary.

19. (a)-(e) Answers will vary.

21. (a) Answers will vary.

    (b) The median IQ score is 100.

    (c)–(f) Answers will vary.

Chapter 9 Review Exercises

1. (a) For a 99% confidence interval we want the t-value with an area in the right tail of 0.005. With $df = 17$, we read from the table that $t_{0.005} = 2.898$.

    (b) For a 90% confidence interval we want the t-value with an area in the right tail of 0.05. With $df = 26$, we read from the table that $t_{0.05} = 1.706$.

2. (a) $df = n - 1 = 21$ and for 95% confidence, $\alpha / 2 = 0.025$. From the table,
$\chi^2_{1-0.025} = \chi^2_{0.975} = 10.283$ and
$\chi^2_{0.025} = 35.479$.

    (b) $df = n - 1 = 11$ and for 99% confidence, $\alpha / 2 = 0.005$. From the table,
$\chi^2_{1-0.005} = \chi^2_{0.995} = 2.603$ and
$\chi^2_{0.005} = 26.757$.

3. In 100 samples, we would expect 95 of the 100 intervals to include the true population mean, 100. Random chance in sampling causes a particular interval to not include the true population mean.

4. In a 95% confidence interval, the 95% represents the proportion of intervals that would contain the parameter of interest (e.g. the population mean, population proportion, or population standard deviation) if a large number of different samples is obtained.

5. If a large number of different samples (of the same size) is obtained, a 90% confidence interval for the population mean will not capture the true value of the population mean 10% of the time.

6. The area to the left of $t = -1.56$ is also 0.0681, because the t-distribution is symmetric about 0.

7. The area under the t-distribution to the right of $t = 2.32$ is greater than the area under the standard normal distribution to the right of $z = 2.32$ because the t-distribution uses s to approximate σ, making it more dispersed than the z-distribution.

8. The properties of Student's t-distribution:

    (1) It is symmetric around $t = 0$.

    (2) It is different for different sample sizes.

    (3) The area under the curve is 1; half the area is to the right of 0 and half the area is to the left of 0.

    (4) As t gets extremely large, the graph approaches, but never equals, zero. Similarly, as t gets extremely small (negative), the graph approaches, but never equals, zero.

    (5) The area in the tails of the t-distribution is greater than the area in the tails of the standard normal distribution.

    (6) As the sample size n increases, the distribution (and the density curve) of the t-distribution becomes more like the standard normal distribution.

9. (a) For 90% confidence, $\alpha/2 = 0.05$. For a sample size of $n = 30$, $df = n-1 = 29$, and the critical value is $t_{\alpha/2} = t_{0.05} = 1.699$. Then:

$$\text{Lower bound} = \bar{x} - t_{0.05} \cdot \frac{s}{\sqrt{n}}$$

$$= 54.8 - 1.699 \cdot \frac{10.5}{\sqrt{30}}$$

$$\approx 54.8 - 3.26 = 51.54$$

$$\text{Upper bound} = \bar{x} + t_{0.05} \cdot \frac{s}{\sqrt{n}}$$

$$= 54.8 + 1.699 \cdot \frac{10.5}{\sqrt{30}}$$

$$\approx 54.8 + 3.26 = 58.06$$

(b) For a sample size of $n = 51$, $df = n-1 = 50$, and the critical value is $t_{\alpha/2} = t_{0.05} = 1.676$. Then:

$$\text{Lower bound} = \bar{x} - t_{0.05} \cdot \frac{s}{\sqrt{n}}$$

$$= 54.8 - 1.676 \cdot \frac{10.5}{\sqrt{51}}$$

$$\approx 54.8 - 2.49 = 52.34$$

$$\text{Upper bound} = \bar{x} + t_{0.05} \cdot \frac{s}{\sqrt{n}}$$

$$= 54.8 + 1.676 \cdot \frac{10.5}{\sqrt{51}}$$

$$\approx 54.8 + 2.49 = 57.26$$

Increasing the sample size decreases the width of the confidence interval.

(c) For 99% confidence, $\alpha/2 = 0.005$. For a sample size of $n = 30$, $df = n-1 = 29$, and the critical value is $t_{\alpha/2} = t_{0.05} = 2.756$. Then:

$$\text{Lower bound} = \bar{x} - t_{0.005} \cdot \frac{s}{\sqrt{n}}$$

$$= 54.8 - 2.756 \cdot \frac{10.5}{\sqrt{30}}$$

$$\approx 54.8 - 5.28 = 49.52$$

$$\text{Upper bound} = \bar{x} + t_{0.005} \cdot \frac{s}{\sqrt{n}}$$

$$= 54.8 + 2.756 \cdot \frac{10.5}{\sqrt{30}}$$

$$\approx 54.8 + 5.28 = 60.08$$

Increasing the level of confidence increases the width of the confidence interval.

10. (a) For 90% confidence, $\alpha/2 = 0.05$. If $n = 15$, then $df = 14$. The critical value is $t_{0.05} = 1.761$. Then:

$$\text{Lower bound} = \bar{x} - t_{0.05} \cdot \frac{s}{\sqrt{n}}$$

$$= 104.3 - 1.761 \cdot \frac{15.9}{\sqrt{15}}$$

$$\approx 104.3 - 7.23 = 97.07$$

$$\text{Upper bound} = \bar{x} + t_{0.05} \cdot \frac{s}{\sqrt{n}}$$

$$= 104.3 + 1.761 \cdot \frac{15.9}{\sqrt{15}}$$

$$\approx 104.3 + 7.23 = 111.53$$

(b) If $n = 25$, then $df = 24$. The critical value is $t_{0.05} = 1.711$. Then:

$$\text{Lower bound} = \bar{x} - t_{0.05} \cdot \frac{s}{\sqrt{n}}$$

$$= 104.3 - 1.711 \cdot \frac{15.9}{\sqrt{25}}$$

$$\approx 104.3 - 5.44 = 98.86$$

$$\text{Upper bound} = \bar{x} + t_{0.05} \cdot \frac{s}{\sqrt{n}}$$

$$= 104.3 + 1.711 \cdot \frac{15.9}{\sqrt{25}}$$

$$\approx 104.3 + 5.44 = 109.74$$

Increasing the sample size decreases the width of the confidence interval.

(c) For 95% confidence, $\alpha/2 = 0.025$. With $df = 14$, $t_{0.025} = 2.145$. Then:

$$\text{Lower bound} = \bar{x} - t_{0.025} \cdot \frac{s}{\sqrt{n}}$$

$$= 104.3 - 2.145 \cdot \frac{15.9}{\sqrt{15}}$$

$$\approx 104.3 - 8.81 = 95.49$$

$$\text{Upper bound} = \bar{x} + t_{0.025} \cdot \frac{s}{\sqrt{n}}$$

$$= 104.3 + 2.145 \cdot \frac{15.9}{\sqrt{15}}$$

$$\approx 104.3 + 8.81 = 113.11$$

Increasing the level of confidence increases the width of the confidence interval.

(d) $df = n-1 = 14$ and for 90% confidence, $\alpha/2 = 0.05$. From the table, $\chi^2_{1-0.05} = \chi^2_{0.95} = 6.571$ and $\chi^2_{0.05} = 23.685$.

$$\text{Lower bound} = \sqrt{\frac{(n-1)s^2}{\chi^2_{\alpha/2}}}$$

$$= \sqrt{\frac{14(15.9)^2}{23.685}} = 12.22$$

$$\text{Upper bound} = \sqrt{\frac{(n-1)s^2}{\chi^2_{\alpha/2}}}$$

$$= \sqrt{\frac{14(15.9)^2}{6.571}} = 23.21$$

11. **(a)** The size of the sample ($n = 35$) is sufficiently large to apply the Central Limit Theorem and conclude that the sampling distribution of $\overline{x}$ is approximately normal.

(b) For 95% confidence, $\alpha/2 = 0.025$. If $n = 35$, then $df = 34$. The critical value is $t_{0.025} = 2.032$. Then:

$$\text{Lower bound} = \overline{x} - t_{0.025} \cdot \frac{s}{\sqrt{n}}$$

$$= 87.9 - 2.032 \cdot \frac{15.5}{\sqrt{35}}$$

$$\approx 87.9 - 5.32 = 82.58$$

$$\text{Upper bound} = \overline{x} + t_{0.025} \cdot \frac{s}{\sqrt{n}}$$

$$= 87.9 + 2.032 \cdot \frac{15.5}{\sqrt{35}}$$

$$\approx 87.9 + 5.32 = 93.22$$

We can be 95% confident that the population mean age people would live to is between 82.58 and 93.22 years.

(c) For 95% confidence, $z_{\alpha/2} = 1.96$ and

$$n = \left(\frac{z_{\alpha/2} \cdot s}{E}\right)^2 = \left(\frac{1.96 \cdot 15.5}{2}\right)^2 \approx 230.7$$

which we must increase to 231. A total of 231 subjects are needed.

12. **(a)** Since the number of emails sent in a day cannot be negative, and one standard deviation to the left of the mean results in a negative number, we expect that the distribution is skewed right.

(b) For 90% confidence, $\alpha/2 = 0.05$. We have $df = n - 1 = 927$. Since there is no row in the table for 927 degrees of freedom, we use $df = 1000$ instead. Thus, $t_{0.05} = 1.646$. Then:

$$\text{Lower bound} = \overline{x} - t_{0.05} \cdot \frac{s}{\sqrt{n}}$$

$$= 10.4 - 1.646 \cdot \frac{28.5}{\sqrt{928}}$$

$$\approx 10.4 - 1.54$$

$$= 8.86 \text{ e-mails}$$

$$\text{Upper bound} = \overline{x} + t_{0.05} \cdot \frac{s}{\sqrt{n}}$$

$$= 10.4 + 1.646 \cdot \frac{28.5}{\sqrt{928}}$$

$$\approx 10.4 + 1.54$$

$$= 11.94 \text{ e-mails}$$

We are 90% confident that the population mean number of e-mails sent per day is between 8.86 and 11.94.

13. **(a)** The sample size is probably small due to the difficulty and expense of locating highly trained cyclists and gathering the data.

(b) For 95% confidence, $\alpha/2 = 0.025$. We have $df = n - 1 = 15$, and $t_{0.025} = 2.131$. Then:

$$\text{Lower bound} = \overline{x} - t_{0.025} \cdot \frac{s}{\sqrt{n}}$$

$$= 218 - 2.131 \cdot \frac{31}{\sqrt{16}}$$

$$\approx 218 - 16.5$$

$$= 201.5 \text{ kilojoules}$$

$$\text{Upper bound} = \overline{x} + t_{0.025} \cdot \frac{s}{\sqrt{n}}$$

$$= 218 + 2.131 \cdot \frac{31}{\sqrt{16}}$$

$$\approx 218 + 16.5$$

$$= 234.5 \text{ kilojoules}$$

The researchers can be 95% confident that the population mean total work performed for the sports-drink treatment group is between 201.5 and 234.5 kilojoules.

(c) Yes; it is possible that the mean total work performed is less than 198 kilojoules since it is possible that the true mean is not captured in the confidence interval. However, it is not very likely since we are 95% confident the true mean total work performed is between 201.5 and 234.5 kilojoules.

(d) For 95% confidence, $\alpha/2 = 0.025$. We have $df = n - 1 = 15$, and $t_{0.025} = 2.131$. Then:

$$\text{Lower bound} = \bar{x} - t_{0.025} \cdot \frac{s}{\sqrt{n}}$$
$$= 178 - 2.131 \cdot \frac{31}{\sqrt{16}}$$
$$\approx 178 - 16.5$$
$$= 161.5 \text{ kilojoules}$$

$$\text{Upper bound} = \bar{x} + t_{0.025} \cdot \frac{s}{\sqrt{n}}$$
$$= 178 + 2.131 \cdot \frac{31}{\sqrt{16}}$$
$$\approx 178 + 16.5$$
$$= 194.5 \text{ kilojoules}$$

The researchers can be 95% confident that the population mean total work performed for the placebo treatment group is between 161.5 and 194.5 kilojoules.

(e) Yes; it is possible that the mean total work performed is more than 198 kilojoules since it is possible that the true mean is not captured in the confidence interval. However, it is not very likely since we are 95% confident the true mean total work performed is between 161.5 and 194.5 kilojoules.

(f) Yes; our findings support the researchers' conclusion. The confidence intervals do not overlap, so we are confident that the mean total work performed for the sports-drink treatment is greater than the mean total work performed for the placebo treatment.

14. (a) Since the sample size is large ($n \geq 30$), $\bar{x}$ has an approximately normal distribution.

(b) We construct a *t*-interval because we are estimating a population mean and we do not know the population standard deviation. For 95% confidence, $\alpha/2 = 0.025$. Since $n = 60$, then $df = 59$. There is no row in the table for 59 degrees of freedom, so we use $df = 60$ instead. The critical value is $t_{0.025} = 2.000$. Then:

$$\text{Lower bound} = \bar{x} - t_{0.025} \cdot \frac{s}{\sqrt{n}}$$
$$= 2.27 - 2.000 \cdot \frac{1.22}{\sqrt{60}}$$
$$\approx 2.27 - 0.32$$
$$= 1.95 \text{ children}$$

$$\text{Upper bound} = \bar{x} + t_{0.025} \cdot \frac{s}{\sqrt{n}}$$
$$= 2.27 + 2.000 \cdot \frac{1.22}{\sqrt{60}}$$
$$\approx 2.27 + 0.32$$
$$= 2.59 \text{ children}$$
[Tech: 2.58]

We are 95% confident that couples who have been married for 7 years have a population mean number of children between 1.95 and 2.59.

(c) For 99% confidence, $\alpha/2 = 0.005$. The critical value is $t_{0.005} = 2.660$. Then:

$$\text{Lower bound} = \bar{x} - t_{0.025} \cdot \frac{s}{\sqrt{n}}$$
$$= 2.27 - 2.660 \cdot \frac{1.22}{\sqrt{60}}$$
$$\approx 2.27 - 0.42$$
$$= 1.85 \text{ children}$$

$$\text{Upper bound} = \bar{x} + t_{0.025} \cdot \frac{s}{\sqrt{n}}$$
$$= 2.27 + 2.660 \cdot \frac{1.22}{\sqrt{60}}$$
$$\approx 2.27 + 0.42$$
$$= 2.69 \text{ children}$$

We are 99% confident that couples who have been married for 7 years have a population mean number of children between 1.85 and 2.69.

15. (a) Using technology, we obtain $\bar{x} \approx 147.3$ cm and $s \approx 28.8$ cm.

(b) Yes. All the data values lie within the bounds on the normal probability plot, indicating that the data should come from a population that is normal. The boxplot does not show any outliers.

(c) We construct a *t*-interval because we are estimating a population mean and we do not know the population standard deviation. For 95% confidence, $\alpha/2 = 0.025$. Since $n = 12$, then $df = 11$. The critical value is

$t_{0.025} = 2.201$. Then:

Lower bound $= \bar{x} - t_{0.025} \cdot \dfrac{s}{\sqrt{n}}$

$= 147.3 - 2.201 \cdot \dfrac{28.8}{\sqrt{12}}$

$\approx 147.3 - 18.3$

$= 129.0$ cm

Upper bound $= \bar{x} + t_{0.025} \cdot \dfrac{s}{\sqrt{n}}$

$= 147.3 + 2.201 \cdot \dfrac{28.8}{\sqrt{12}}$

$\approx 147.3 + 18.3$

$= 165.6$ cm

We are 95% confident that the population mean diameter of a Douglas fir tree in the western Washington Cascades is between 129.0 and 165.6 centimeters.

(d) From the table, with df $= 11$ and $\alpha/2 = 0.025$, we find

$\chi^2_{1-0.025} = \chi^2_{0.975} = 3.816$ and

$\chi^2_{0.025} = 21.920$. Then:

Lower bound $= \sqrt{\dfrac{(n-1)s^2}{\chi^2_{\alpha/2}}}$

$= \sqrt{\dfrac{11(28.8)^2}{21.920}} \approx 20.4$ cm

Upper bound $= \sqrt{\dfrac{(n-1)s^2}{\chi^2_{\alpha/2}}}$

$= \sqrt{\dfrac{11(28.8)^2}{3.816}} \approx 48.9$ cm

We are 95% confident that the population standard deviation diameter of a Douglas fir tree in the western Washington Cascades is between 20.4 and 48.9 centimeters.

16. (a) $\hat{p} = \dfrac{x}{n} = \dfrac{58}{678} \approx 0.086$

(b) For 95% confidence, $z_{\alpha/2} = z_{.025} = 1.96$.

Lower bound

$= \hat{p} - z_{.025} \cdot \sqrt{\dfrac{\hat{p}(1-\hat{p})}{n}}$

$= 0.086 - 1.96 \cdot \sqrt{\dfrac{0.086(1-0.086)}{678}}$

$\approx 0.086 - 0.021 = 0.065$

Upper bound

$= \hat{p} + z_{.025} \cdot \sqrt{\dfrac{\hat{p}(1-\hat{p})}{n}}$

$= 0.086 + 1.96 \cdot \sqrt{\dfrac{0.086(1-0.086)}{678}}$

$\approx 0.086 + 0.021 = 0.107$

The Centers for Disease Control can be 95% confident that the population proportion of adult males aged 20–34 years who have hypertension is between 0.065 and 0.107 (i.e. between 6.5% and 10.7%).

(c) $n = \hat{p}(1-\hat{p}) \left(\dfrac{z_{\alpha/2}}{E} \right)^2$

$= 0.086(1-0.086) \left(\dfrac{1.96}{.03} \right)^2$

≈ 335.5

which we must increase to 336. You would need a sample of size 336 for your estimate to be within 3 percentage points of the true proportion, with 95% confidence.

(d) $n = 0.25 \left(\dfrac{z_{\alpha/2}}{E} \right)^2 = 0.25 \left(\dfrac{1.96}{0.03} \right)^2$

≈ 1067.1

which we must increase to 1068. Without the prior estimate, you would need a sample size of 1068.

17. Answers will vary. A sample answer generated by StatCrunch is given.

(a) Summary Statistics

Column	Mean	2.5th Percentile	97.5th Percentile
Mean	147.25333	132.625	162.29167

Lower bound: 132.625, upper bound: 162.292;
The confidence interval constructed using the bootstrapping method is narrower.

(b) The histogram resembles that of a normal distribution, which is not surprising, since the distribution of the original data was approximately normal.

Histogram of Means

(c) Summary statistics:

Column	Median	2.5th Percentile	97.5th Percentile
Median	151.5	121.5	167.5

Lower bound: 121.5, upper bound: 167.5

Chapter 9 Test

1. **(a)** For a 96% confidence interval we want the t-value with an area in the right tail of 0.02. With df $= 25$, we read from the table that $t_{0.02} = 2.167$.

   (b) For a 98% confidence interval we want the t-value with an area in the right tail of 0.01. With df df $= 17$, we read from the table that $t_{0.01} = 2.567$.

2. **(a)** df $= n - 1 = 15$ and for 98% confidence, $\alpha / 2 = 0.01$. From the table,
   $\chi^2_{1-0.01} = \chi^2_{0.99} = 5.229$ and $\chi^2_{0.01} = 30.578$.

   (b) df $= n - 1 = 29$ and for 90% confidence, $\alpha / 2 = 0.05$. From the table,
   $\chi^2_{1-0.05} = \chi^2_{0.95} = 17.708$ and $\chi^2_{0.05} = 42.557$.

3. $\overline{x} = \dfrac{125.8 + 152.6}{2} = \dfrac{278.4}{2} = 139.2$

   $E = \dfrac{152.6 - 125.8}{2} = \dfrac{26.8}{2} = 13.4$

4. **(a)** We would expect the distribution to be skewed right. We expect most respondents to have relatively few family members in prison, but there will be some with several (possibly many) family members in prison.

   (b) We construct a t-interval because we are estimating a population mean and we do not know the population standard deviation. For 99% confidence,

$\alpha / 2 = 0.005$. Since $n = 499$, then df $= 498$ and $t_{0.005} = 2.586$. Then:

Lower bound $= \overline{x} - t_{0.005} \cdot \dfrac{s}{\sqrt{n}}$

$= 1.22 - 2.586 \cdot \dfrac{0.59}{\sqrt{499}}$

$\approx 1.22 - 0.068 = 1.152$

Upper bound $= \overline{x} + t_{0.005} \cdot \dfrac{s}{\sqrt{n}}$

$= 1.22 + 2.586 \cdot \dfrac{0.59}{\sqrt{499}}$

$\approx 1.22 + 0.068 = 1.288$

We are 99% confident that the population mean number of family members in jail is between 1.152 and 1.288.

5. **(a)** We construct a t-interval because we are estimating a population mean and we do not know the population standard deviation. For 90% confidence, $\alpha / 2 = 0.05$. Since $n = 50$, then df $= 49$. There is no row in the table for 49 degrees of freedom, so we use df $= 50$ instead. Thus, $t_{0.05} = 1.676$, and
Lower bound

$= \overline{x} - t_{0.05} \cdot \dfrac{s}{\sqrt{n}}$

$= 4.58 - 1.676 \cdot \dfrac{1.10}{\sqrt{50}}$

$\approx 4.58 - 0.261 = 4.319$ yrs

Upper bound $= \overline{x} + t_{0.05} \cdot \dfrac{s}{\sqrt{n}}$

$= 4.58 + 1.676 \cdot \dfrac{1.10}{\sqrt{50}}$

$\approx 4.58 + 0.261 = 4.841$ yrs

We are 90% confident that the population mean time to graduate is between 4.319 years and 4.841 years.

   (b) Yes; we are 90% confident that the mean time to graduate is between 4.319 years and 4.841 years. Since the entire interval is above 4 years, our evidence contradicts the belief that it takes 4 years to complete a bachelor's degree.

6. (a) Using technology, we obtain $\bar{x} = 57.75$ inches and $s \approx 15.45$ inches.

(b) Yes. The plotted points are generally linear and stay within the bounds of the normal probability plot. The boxplot shows that there are no outliers.

(c) We construct a t-interval because we are estimating a population mean and we do not know the population standard deviation. For 95% confidence, $\alpha/2 = 0.025$. Since $n = 12$, then df $= 11$ and $t_{0.025} = 2.201$. Then:

$$\text{Lower bound} = \bar{x} - t_{0.025} \cdot \frac{s}{\sqrt{n}}$$

$$= 57.75 - 2.201 \cdot \frac{15.45}{\sqrt{12}}$$

$$\approx 57.75 - 9.82 = 47.93 \text{ in.}$$
[Tech: 47.94 in.]

$$\text{Upper bound} = \bar{x} + t_{0.025} \cdot \frac{s}{\sqrt{n}}$$

$$= 57.75 + 2.201 \cdot \frac{15.4}{\sqrt{12}}$$

$$\approx 57.75 + 9.82 = 67.57 \text{ in.}$$
[Tech: 67.56 in.]

The researcher can be 95% confident that the population mean depth of visibility of the Secchi disk is between 47.93 and 67.57 inches.

(d) For 99% confidence, $\alpha/2 = 0.005$. For df $= 11$, $t_{0.005} = 3.106$. Then:

Lower bound

$$= \bar{x} - t_{0.005} \cdot \frac{s}{\sqrt{n}}$$

$$= 57.75 - 3.106 \cdot \frac{15.45}{\sqrt{12}}$$

$$\approx 57.75 - 13.85 = 43.90 \text{ in.}$$

Upper bound

$$= \bar{x} + t_{0.005} \cdot \frac{s}{\sqrt{n}}$$

$$= 57.75 + 3.106 \cdot \frac{15.45}{\sqrt{12}}$$

$$\approx 57.75 + 13.85 = 71.60 \text{ in.}$$

The researcher can be 99% confident that the population mean depth of visibility of the Secchi disk is between 43.90 and 71.60 inches.

(e) From the table, with df $= 11$ and $\alpha/2 = 0.025$, we find

$$\chi^2_{1-0.025} = \chi^2_{0.975} = 3.816 \text{ and}$$

$$\chi^2_{0.025} = 21.920. \text{ Then:}$$

$$\text{Lower bound} = \sqrt{\frac{(n-1)s^2}{\chi^2_{\alpha/2}}} = \sqrt{\frac{11(15.45)^2}{21.920}}$$

$$\approx 10.9 \text{ in.}$$

$$\text{Upper bound} = \sqrt{\frac{(n-1)s^2}{\chi^2_{\alpha/2}}} = \sqrt{\frac{11(15.45)^2}{3.816}}$$

$$\approx 26.2 \text{ in.}$$

The researcher can be 99% confident that the population standard deviation of the depth of visibility of the Secchi disk is between 10.9 and 26.2 inches.

7. (a) $\hat{p} = \frac{x}{n} = \frac{1139}{1201} \approx 0.948$

(b) For 99% confidence, $z_{\alpha/2} = z_{.005} = 2.575$.

Lower bound

$$= \hat{p} - z_{.005} \cdot \sqrt{\frac{\hat{p}(1-\hat{p})}{n}}$$

$$= 0.948 - 2.575 \cdot \sqrt{\frac{0.948(1-0.948)}{1201}}$$

$$\approx 0.948 - 0.016 = 0.932$$

Upper bound

$$= \hat{p} + z_{.005} \cdot \sqrt{\frac{\hat{p}(1-\hat{p})}{n}}$$

$$= 0.948 + 2.575 \cdot \sqrt{\frac{0.948(1-0.948)}{1201}}$$

$$\approx 0.948 + 0.016 = 0.964 \quad [\text{Tech: } 0.965]$$

The EPA can be 99% confident that the population proportion of Americans who live in neighborhoods with acceptable levels of carbon monoxide is between 0.932 and 0.964 (i.e., between 93.2% and 96.4%).

(c) $n = \hat{p}(1-\hat{p})\left(\frac{z_{\alpha/2}}{E}\right)^2$

$$= 0.948(1-0.948)\left(\frac{1.645}{.015}\right)^2$$

$$= 592.9$$

which we must increase to 593. A sample size of 593 would be needed for the estimate to be within 1.5 percentage points with 90% confidence.

(d) $n = 0.25 \left(\dfrac{z_{\alpha/2}}{E} \right)^2 = 0.25 \left(\dfrac{1.645}{.015} \right)^2$

≈ 3006.7

which we must increase to 3007. Without the prior estimate, a sample size of 3007 would be needed for the estimate to be within 1.5 percentage points with 90% confidence.

8. (a) $\bar{x} = \dfrac{\sum x}{40} = \dfrac{5336}{40} = 133.4$ minutes

(b) Because the population is not normally distributed, a large sample is needed to apply the Central Limit Theorem and say that the $\bar{x}$ distribution is approximately normal.

(c) For 99% confidence, $\alpha / 2 = 0.05$. Since $n = 40$, then df $= 39$ and $t_{0.005} = 2.708$. Then:

Lower bound

$= \bar{x} - t_{0.005} \cdot \dfrac{s}{\sqrt{n}}$

$= 133.4 - 2.708 \cdot \dfrac{43.3}{\sqrt{40}}$

$\approx 133.4 - 18.54 = 114.86$ minutes

[Tech: 114.87 minutes]

Upper bound

$= \bar{x} + t_{0.005} \cdot \dfrac{s}{\sqrt{n}}$

$= 133.4 + 2.708 \cdot \dfrac{43.3}{\sqrt{40}}$

$\approx 133.4 + 18.54 = 151.94$ minutes

[Tech: 151.93 minutes]

The tennis enthusiast is 99% confident that the population mean length of men's singles matches during Wimbledon is between 114.86 and 151.94 minutes.

(d) For 95% confidence the critical value is $t_{0.025} = 2.023$. Then:

Lower bound $= \bar{x} - t_{0.025} \cdot \dfrac{s}{\sqrt{n}}$

$= 133.4 - 2.023 \cdot \dfrac{43.3}{\sqrt{40}}$

$\approx 133.4 - 13.84$

$= 119.6$ minutes

Upper bound $= \bar{x} + t_{0.025} \cdot \dfrac{s}{\sqrt{n}}$

$= 133.4 + 2.023 \cdot \dfrac{43.3}{\sqrt{40}}$

$\approx 133.4 + 13.84$

$= 147.2$ minutes

The tennis enthusiast is 95% confident that the population mean length of men's singles matches during Wimbledon is between 119.6 and 147.2 minutes.

(e) Increasing the level of confidence increases the width of the interval.

(f) No; the tennis enthusiast only sampled Wimbledon matches. Therefore, his results cannot be generalized to all professional tournaments.

9. (a) The distribution is skewed right and includes an outlier, and the sample size is small.

(b) The bootstrapping method uses the given set of sample data as the population. Many samples of size equal to the original sample size are chosen. The samples are chosen with replacement so that the samples are not necessarily the original sample. The mean is determined for each sample. The distribution of these sample means is used to construct the 95% confidence interval.

(c) Answers will vary. A sample answer generated by StatCrunch is given. Summary Statistics

Column	Mean	2.5th Percentile	97.5th Percentile
Mean	45.220856	25.857143	67.67857

Lower bound: 25.857, upper bound: 67.679

Case Study: Fire-Safe Cigarettes

Proportion of self-extinguish cigarettes:

Let $\hat{p}$ be the proportion of cigarettes that self-extinguish.

Brand A: $\hat{p} = \dfrac{27}{40} = 0.675$. The sample size $n = 40$ is less than 5% of the population, but $n\hat{p}(1 - \hat{p}) \approx 8.8 < 10$, so we cannot construct a confidence interval using the normal model. Construct a 95% confidence interval using the Agresti-Coull model.

$$n = 40 + 1.96^2$$

$$\approx 43.84$$

$$\widetilde{p} = \left(\frac{1}{43.84}\right)\left(27 + \frac{1}{2}\cdot 1.96^2\right)$$

$$\approx 0.660$$

Lower bound:

$$0.660 - 1.96\cdot\sqrt{\frac{1}{43.84}\cdot 0.660\left(1 - 0.660\right)} = 0.520$$

[Tech: 0.519]

Upper bound:

$$0.660 + 1.96\cdot\sqrt{\frac{1}{43.84}\cdot 0.660\left(1 - 0.660\right)} = 0.800$$

We are 95% confident that the proportion of Brand A cigarettes that self extinguish is between 0.520 and 0.800 (i.e., between 52.0% and 80.0%.

Brand B: $\hat{p} = \frac{1}{40} = 0.025$. The sample size $n = 40$ is less than 5% of the population, but $n\hat{p}(1 - \hat{p}) \approx 1.0 < 10$, so we cannot construct a confidence interval using the normal model. Construct a 95% confidence interval using the Agresti-Coull model.

$$\widetilde{p} = \left(\frac{1}{43.84}\right)\left(1 + \frac{1}{2}\cdot 1.96^2\right)$$

$$\approx 0.067$$

Lower bound:

$$0.067 - 1.96\cdot\sqrt{\frac{1}{43.84}\cdot 0.067\left(1 - 0.067\right)} = -0.007$$

Upper bound:

$$0.067 + 1.96\cdot\sqrt{\frac{1}{43.84}\cdot 0.067\left(1 - 0.067\right)} = 0.141$$

[Tech: 0.140]

We are 95% confident that the proportion of Brand B cigarettes that self extinguish is between 0 and 0.141 (i.e., between 0% and 14.1%).

Mean of nicotine in cigarettes:

Brand A: Since the normal probability plot shows that the data are not approximately normal, a confidence interval cannot be constructed using the Student t-distribution. Use bootstrapping to find the confidence interval.

Using technology, the 95% confidence interval is $(1.144, 1.281)$ (Answers will vary). We are 95% confident that the population mean amount of nicotine in a Brand A cigarette is between 1.144 and 1.281.

Brand B: We can reasonably assume that the sample size is small relative to the population size. All the data lie within the bounds of the normal probability plot. The boxplot does not reveal any outliers.

Using technology, we find $\bar{x} = 1.234$ and $s \approx 0.079$. For 95% confidence, $\alpha/2 = 0.025$. Since $n = 15$, then df $= 14$ and $t_{0.025} = 2.145$. Thus:

$$\text{Lower bound} = \bar{x} - t_{0.025}\cdot\frac{s}{\sqrt{n}}$$

$$= 1.234 - 2.145\cdot\frac{0.079}{\sqrt{15}}$$

$$\approx 1.234 - 0.044 = 1.190$$

$$\text{Upper bound} = \bar{x} + t_{0.025}\cdot\frac{s}{\sqrt{n}}$$

$$= 1.234 + 2.145\cdot\frac{0.079}{\sqrt{15}}$$

$$\approx 1.234 + 0.044 = 1.278$$

We are 95% confident that the population mean amount of nicotine in a Brand B cigarette is between 1.190 and 1.278.

Chapter 10
Hypothesis Tests Regarding a Parameter

Section 10.1

1. Hypothesis

3. Null hypothesis

5. I

7. Level of significance

9. Right-tailed, μ

11. Two-tailed, σ

13. Left-tailed, μ

15. (a) $H_0: p = 0.399$
$H_1: p > 0.399$

(b) Making a Type I error would mean concluding that the proportion of students who enroll in Joliet Junior College and earn a bachelor's degree within six years is greater than 0.399 when, in fact, the proportion of students who enroll in Joliet Junior College and earn a bachelor's degree within six years is not greater than 0.399.

(c) Making a Type II error would mean concluding that the proportion of students who enroll in Joliet Junior College and earn a bachelor's degree within six years is not greater than 0.399 when, in fact, the proportion of students who enroll in Joliet Junior College and earn a bachelor's degree within six years is greater than 0.399.

17. (a) $H_0: \mu = \$245,700$
$H_1: \mu < \$245,700$

(b) Making a Type I error would mean concluding that the existing home prices in the broker's neighborhood are lower than \$245,700 when, in fact, the existing home prices in the broker's neighborhood are not lower than \$245,700.

(c) Making a Type II error would mean concluding that the existing home prices in the broker's neighborhood are not lower than \$245,700 when, in fact, the existing home prices in the broker's neighborhood are lower than \$245,700.

19. (a) $H_0: \sigma = 0.7$ psi
$H_1: \sigma < 0.7$ psi

(b) Making a Type I error would mean concluding that the pressure variability has been reduced below 0.7 psi when, in fact, the pressure variability has not been reduced below 0.7 psi.

(c) Making a Type II error would mean concluding that the pressure variability has not been reduced below 0.7 psi when, in fact, the pressure variability has been reduced below 0.7 psi.

21. (a) $H_0: \mu = \$48.79$
$H_1: \mu \neq \$48.79$

(b) Making a Type I error would mean concluding that the mean monthly revenue per cell phone is not \$48.79 when, in fact, the mean monthly revenue per cell phone is \$48.79.

(c) Making a Type II error would mean concluding that the mean monthly revenue per cell phone is \$48.79 when, in fact, the mean monthly revenue per cell phone is not \$48.79.

23. There is sufficient evidence to conclude that the proportion of students who enroll in Joliet Junior College and earn a bachelor's degree within six years is greater than 0.399.

25. There is not sufficient evidence to conclude that the existing home prices in the broker's neighborhood are lower than \$245,700.

27. There is not sufficient evidence to conclude that the pressure variability has been reduced below 0.7 psi.

29. There is sufficient evidence to conclude that the mean monthly revenue per cell phone is different from $48.79.

31. There is not sufficient evidence to conclude that the proportion of students who enroll in Joliet Junior College and earn a bachelor's degree within six years is greater than 0.399.

33. There is sufficient evidence to conclude that the existing home prices in the broker's neighborhood are lower than $245,700.

35. (a) $H_0: \mu = 12$ ounces
    $H_1: \mu \neq 12$ ounces

    (b) There is sufficient evidence to conclude that the filling machine is either over-filling or under-filling the cans.

    (c) Type I error, because the null hypothesis was rejected when, in fact, the null hypothesis was true.

    (d) Answers will vary. Sample answer: 0.01, because a low level of significance will make it less likely that the null hypothesis will be rejected.

37. (a) $H_0: p = 0.028$
    $H_1: p > 0.028$

    (b) There is not sufficient evidence to conclude that more than 2.8% of the students at the counselor's school use e-cigs.

    (c) Type II error, because the null hypothesis was not rejected when, in fact, the alternative hypothesis was true.

39. (a) $H_0: \mu = 4$ hours
    $H_1: \mu < 4$ hours

    (b) The manufacturer's claim is not true.

41. (a) $\dfrac{\$750,000}{30} = \$25,000$

    (b) This is a binomial experiment because the trials (asking adults their opinion) are independent and there are only two possible mutually exclusive results ("Yes" and "No"); $n = 20$, $p = 0.16$

    (c) $P(8) = {}_{20}C_8 \cdot (0.16)^8 (0.84)^{12} \approx 0.0067$

(d) Using technology,
    $$P(X < 8) = P(X \leq 7) \approx 0.9912.$$

(e) The normal model can be used to approximate the sampling distribution of this sample proportion because
    $np(1 - p) = 500(0.16)(0.84) = 67.2,$
    which is greater than 10. The distribution is bell-shaped, the center is
    $\mu_X = np = (500)(0.16) = 80,$ and the spread is
    $$\sigma_X = \sqrt{np(1-p)} = \sqrt{67.2} \approx 8.20.$$

(f) $P(X \geq 100) \approx P(X \geq 99.5)$
    $$= P\left(Z \geq \frac{99.5 - 80}{\sqrt{67.2}} \right)$$
    $$= P(Z \geq 2.38)$$
    $$\approx 1 - 0.9913$$
    $$= 0.0087 \quad [\text{Tech: } 0.0087]$$

    This result is fairly unusual; the probability of 0.0087 means that this type of result would be expected in about 8 or 9 out of 1000 samples.

43. It increases, because a Type II error occurs if the null hypothesis is not rejected when, in fact, the alternative hypothesis is true. Lower levels of significance make it harder to reject the null hypothesis, even if it should be rejected.

45. Answers will vary.

Section 10.2

1. Statistically significant

3. False. If the P-value is small, we reject the null hypothesis.

5. -1.28

7. $np_0(1 - p_0) = 200(0.3)(1 - 0.3) = 42 > 10$

    The sample proportion is $\hat{p} = \dfrac{75}{200} = 0.375.$

    The test statistic is
    $$z_0 = \frac{0.375 - 0.3}{\sqrt{\dfrac{0.3(1 - 0.3)}{200}}} \approx 2.31.$$

(a) Because this is a right-tailed test, we determine the critical value at the $\alpha = 0.05$ level of significance to be $z_{0.05} = 1.645$. Because the test statistic is greater than the critical value, reject the null hypothesis.

(b) Because this is a right-tailed test, the P-value is the area under the standard normal distribution to the right of the test statistic, $z_0 = 2.31$. So,

P-value $= P(z > 2.31) \approx 0.0104$.

Because the P-value is less than the level of significance, reject the null hypothesis.

9. $np_0(1-p_0) = 150(0.55)(0.45) = 37.125 > 10$

The sample proportion is $\hat{p} = \dfrac{78}{150} = 0.52$.

The test statistic is

$$z_0 = \dfrac{0.52 - 0.55}{\sqrt{\dfrac{0.55(1-0.55)}{150}}} \approx -0.74$$

(a) Because this is a left-tailed test, we determine the critical value at the $\alpha = 0.1$ level of significance to be $-z_{0.1} = -1.28$. Because the test statistic is greater than the critical value, do not reject the null hypothesis.

(b) Because this is a left-tailed test, the P-value is the area under the standard normal distribution to the right of the test statistic, $z_0 = -0.74$. So,

P-value $= P(z < -0.74) \approx 0.2296$.

Because the P-value is greater than the level of significance, do not reject the null hypothesis.

11. $np_0(1-p_0) = 500(0.9)(1-0.9) = 45 > 10$

The sample proportion is $\hat{p} = \dfrac{440}{500} = 0.88$.

The test statistic is

$$z_0 = \dfrac{0.88 - 0.9}{\sqrt{\dfrac{0.9(1-0.9)}{500}}} \approx -1.49.$$

(a) Because this is a two-tailed test, we determine the critical values at the $\alpha = 0.05$ level of significance to be

$-z_{0.05/2} = -z_{0.025} = -1.96$ and $z_{0.05/2} = z_{0.025} = 1.96$. Because the test statistic does not lie in the critical region, do not reject the null hypothesis.

(b) Because this is a two-tailed test, the P-value is the area under the standard normal distribution to the left of $-|z_0| = -1.49$ and to the right of $|z_0| = 1.49$. So,

P-value $= 2P(Z < -1.49) \approx 2(0.0681) = 0.1362$.

Because the P-value is greater than the level of significance, do not reject the null hypothesis.

13. The P-value of 0.2743 means that if the null hypothesis is true, this type of result would be expected in about 27 or 28 out of 100 samples. The observed results are not unusual. Because the P-value is large, do not reject the null hypothesis.

15. (a) $\dfrac{320}{678} \approx 0.472$

(b) $H_0: p = 0.5$
$H_1: p < 0.5$

(c) While is it not possible for an individual person's predictions to be independent from one another, it is at least reasonable to assume that the sample is random. Since $np_0(1-p_0) = 678(0.5)(1-0.5) = 169.5 > 10$, the normal model may be used for this test.

(d)

(e) The test statistic is

$$z_0 = \dfrac{0.472 - 0.5}{\sqrt{\dfrac{0.5(1-0.5)}{678}}} \approx -1.45.$$ The P-value

is $P(z < -1.45) \approx 0.0735$. (Tech: 0.0722)

(f) The *P*-value of 0.0735 means that if the null hypothesis is true, this type of result would be expected in about 7 or 8 out of 100 samples.

(g) Because $0.0735 > 0.05$, do not reject the null hypothesis. There is not sufficient evidence to conclude that Jim Cramer's predictions are correct less than half of the time.

17. Hypotheses: $H_0: p = 0.019$
$\qquad\qquad\;\; H_1: p > 0.019$

It is reasonable to assume the patients in the trials were randomly selected and independent. Since
$np_0(1 - p_0) = 863(0.019)(1 - 0.019) \approx 16.1 > 10$,
the normal model may be used for this test.

Sample proportion: $\dfrac{19}{863} \approx 0.022$

Test statistic: $z_0 = \dfrac{0.022 - 0.019}{\sqrt{\dfrac{0.019(1 - 0.019)}{863}}} \approx 0.65$

Critical value: $z_{0.01} = 2.33$

P-value: $P(z > 0.65) \approx 0.2578$ (Tech: 0.2582)

Since $0.65 < 2.33$ and $0.2578 > 0.01$, do not reject the null hypothesis. There is not sufficient evidence to conclude that more than 1.9% of Lipitor users experience flulike symptoms as a side effect.

19. Hypotheses: $H_0: p = 0.36$
$\qquad\qquad\;\; H_1: p > 0.36$

It is reasonable to assume the traffic fatalities in the sample are independent, and it is given that the sample is random. Since
$np_0(1 - p_0) = 105(0.36)(1 - 0.36) = 24.192 > 10$,
the normal model may be used for this test.

Sample proportion: $\dfrac{51}{105} \approx 0.486$

Test statistic: $z_0 = \dfrac{0.486 - 0.36}{\sqrt{\dfrac{0.36(1 - 0.36)}{105}}} \approx 2.69$

Critical value: $z_{0.05} = 1.645$

P-value: $P(z > 2.69) \approx 0.0036$

Since $2.69 > 1.645$ and $0.0036 < 0.05$, reject the null hypothesis. There is sufficient evidence to conclude that Hawaii has a higher proportion of traffic fatalities involving a positive BAC than the United States as a whole.

21. Hypotheses: $H_0: p = 0.52$
$\qquad\qquad\;\; H_1: p \neq 0.52$

It is reasonable to assume the adults in the survey were randomly selected and independent. Since
$np_0(1 - p_0) = 800(0.52)(1 - 0.52) = 199.68 > 10$,
the normal model may be used for this test.

Sample proportion: $\dfrac{256}{800} = 0.32$

Test statistic: $z_0 = \dfrac{0.32 - 0.52}{\sqrt{\dfrac{0.52(1 - 0.52)}{800}}} \approx -11.32$

Critical values:
$-z_{0.05/2} = -z_{0.025} = -1.96$,
$z_{0.05/2} = z_{0.025} = 1.96$

P-value: $2P(z < -11.32) \approx 0.0000$

Since $-11.32 < -1.96$ and $0.0000 < 0.05$, reject the null hypothesis. There is sufficient evidence to conclude that parents feel differently today about high school students not being taught enough math and science being a serious problem.

23. The 95% confidence interval for *p* based on the Gallup poll has a lower bound of 0.401 and an upper bound of 0.462. Because 0.47 is not within the bounds of the confidence interval, there is sufficient evidence to conclude that the proportion of parents with children in grades K–12 that are satisfied with the quality of education the students receive has changed from 2002.

25. In order to make this claim, the *P*-value for a right-tailed test claiming that the competing fast food restaurant's drive thru accuracy is greater than 96.4% must be less than 0.1. That is, the area under the normal curve to the right of the corresponding test statistic must be less than 0.1.

$P(Z > z_0) < 0.1 \rightarrow z_0 > 1.28$

$$\frac{\hat{p}-0.964}{\sqrt{\dfrac{0.964(0.036)}{350}}} > 1.28$$

$$\hat{p} > 1.28\sqrt{\frac{0.964(0.036)}{350}} + 0.964$$

$$\hat{p} > 0.977$$

$$\frac{x}{350} > 0.977$$

$$x > 341.95$$

The drive thru must fill out at least 342 out of 350 accurate orders in order for the manager to be able to claim her drive thru has a statistically significantly better accuracy record than Chick-fil-A.

27. (a) $H_0: p = 0.5$
    $H_1: p > 0.5$

    (b) It is reasonable to assume that the sample is random and that the predictions are independent. However, since
    $np_0(1-p_0) = 16(0.5)(1-0.5) = 4 < 10$,
    the normal model is not appropriate for this test.

    (c) This is a binomial experiment because the trials (students taking the course) are independent and there are only two possible mutually exclusive results (completing the course with a letter grade of A, B, or C, or not completing the course with a letter grade of A, B, or C).

    (d) Using technology, $P(X \geq 11) \approx 0.1051$. Since $0.1051 > 0.05$, do not reject the null hypothesis. There is not sufficient evidence to conclude that the course is effective.

    (e) It is reasonable to assume that the sample is random and that the predictions are independent. Since
    $np_0(1-p_0) = 48(0.5)(1-0.5) = 12 > 10$,
    the normal model may be used for this test.

    (f) Using technology, $P(X \geq 33) \approx 0.0047$. Since $0.0047 < 0.05$, reject the null hypothesis. There is sufficient evidence to conclude that the course is effective.

    (g) Answers will vary.

29. (a) $H_0: p = 0.5$
    $H_1: p > 0.5$

    (b) Using technology,
    $P(X \geq 28) \approx 0.00000232$.

    (c) Answers will vary, but should make some reference to how correlation does not imply causation.

31. Hypotheses: $H_0: p = 0.6$
    $H_1: p > 0.6$

    It is reasonable to assume the adult Americans in the sample are independent, and it is given that the sample is random. Since
    $np_0(1-p_0) = 134(0.6)(1-0.6) = 32.16 > 10$,

    the normal model may be used for this test.

    Sample proportion: $\dfrac{89}{134} \approx 0.664$

    Test statistic: $z_0 = \dfrac{0.664 - 0.6}{\sqrt{\dfrac{0.6(1-0.6)}{134}}} \approx 1.51$

    Critical value: $z_{0.05} = 1.645$

    P-value: $P(z > 1.51) \approx 0.0655$ (Tech: 0.0647)

    Since $1.51 < 1.645$ and $0.0655 > 0.05$, do not reject the null hypothesis. There is not sufficient evidence to conclude that a supermajority of adult Americans believe there is income inequality among males and females with the same experience and education.

33. Hypotheses: $H_0: p = 0.5$
    $H_1: p \neq 0.5$

    It is reasonable to assume the games in the sample are independent, and it is given that the sample is random. Since
    $np_0(1-p_0) = 45(0.5)(1-0.5) = 11.25 > 10$,
    the normal model may be used for this test.

    Sample proportion: $\dfrac{19}{45} \approx 0.422$

    Test statistic: $z_0 = \dfrac{0.422 - 0.5}{\sqrt{\dfrac{0.5(1-0.5)}{45}}} \approx -1.05$

    P-value: $2P(z < -1.05) \approx 0.2938$ (Tech: 0.2967)

Since 0.2938 is greater than all the usual levels of confidence, do not reject the null hypothesis. Yes, the data suggest sports books establish accurate spreads.

35. (a) The P-values and conclusions for the various hypothesis tests are shown in the table below.

Value of p_0	P-value	Reject H_0?
0.42	0.0258	Yes
0.43	0.0434	Yes
0.44	0.0698	No
0.45	0.1078	No
0.46	0.1602	No
0.47	0.2293	No
0.48	0.3169	No
0.49	0.4236	No
0.50	0.5485	No
0.51	0.6891	No
0.52	0.8414	No
0.53	0.9999	No
0.54	0.8410	No
0.55	0.6877	No
0.56	0.5456	No
0.57	0.4191	No
0.58	0.3110	No
0.59	0.2225	No
0.60	0.1530	No
0.61	0.1010	No
0.62	0.0637	No
0.63	0.0383	Yes
0.64	0.0219	Yes

The values of p_0 represent the population proportion in each test.

(b) Using technology, the interval is $(0.432, 0.628)$.

(c) The range of values of p_0 for which the null hypothesis would not be rejected

would increase. This makes sense because α represents the probability of incorrectly rejecting the null hypothesis.

37. (a) The statement of no change or no effect is that the individual is randomly guessing at the color of the card. The statement we would be looking to demonstrate is that the individual has the ability to predict card color. The hypotheses would then be $H_0: p = 0.5$ and $H_1: p > 0.5$.

(b) If the null is 0.50 and someone got 24 out of 40 guesses, the P-value is 0.2059 and the null hypothesis would not be rejected. Flipping a coin is an action with two possible results that are mutually exclusive, collectively exhaustive, and equally likely. It thus represents the act of guessing one of two colors that are equally likely to be on a randomly selected card.

(c) Answers will vary.

(d) This is a binomial experiment because the trials (guessing the card's color) are independent and there are only two possible mutually exclusive results (guessing correctly, guessing incorrectly).

(e) Using technology, $P(X \geq 24) \approx 0.1341$.

(f) Answers will vary.

(g) Answers will vary, but are likely to be that the savant is probably only guessing.

39. (a) Randomizing which toy the baby sees first avoids bias. If the toys are shown in the same order to all of the babies, they may end up predisposed to one or the other.

(b) $H_0: p = 0.5$
$H_1: p > 0.5$

(c) $P(X \geq 14) = P(14) + P(15) + P(16)$
$\approx 0.0018 + 0.0002 + 0.0000$
$= 0.0020$

(d) The P-value of 0.0002 means that if the null hypothesis is true, all 12 babies would prefer the helper toy in about 2 out of 1000 samples.

41. The *P*-value of 0.23 means that if the null hypothesis is true, this type of result would be expected in about 23 out of 100 samples. The observed results are not unusual. Because the *P*-value is large, do not reject the null hypothesis.

43. Answers will vary.

45. Answers will vary.

Section 10.3

1. (a) 2.602

   (b) −1.729

   (c) ±2.179

3. (a) $t_0 = \dfrac{\overline{x} - \mu_0}{\frac{s}{\sqrt{n}}}$

   $= \dfrac{47.1 - 50}{\frac{10.3}{\sqrt{24}}}$

   ≈ -1.379

   (b) With $24 - 1 = 23$ degrees of freedom, the critical value is $-t_{0.05} = -1.714$.

   (c)

   (d) The researcher will not reject the null hypothesis because the test statistic is not in the critical region.

5. (a) $t_0 = \dfrac{\overline{x} - \mu_0}{\frac{s}{\sqrt{n}}}$

   $= \dfrac{104.8 - 100}{\frac{9.2}{\sqrt{23}}}$

   ≈ 2.502

   (b) With $23 - 1 = 22$ degrees of freedom, the critical values are
   $-t_{0.01/2} = -t_{0.005} = -2.819$ and
   $t_{0.01/2} = t_{0.005} = 2.819$.

   (c)

   (d) The researcher will not reject the null hypothesis because the test statistic is not in the critical region.

   (e) Using technology, the 99% confidence interval is $(99.393, 110.207)$.

7. (a) $t_0 = \dfrac{\overline{x} - \mu_0}{\frac{s}{\sqrt{n}}}$

   $= \dfrac{18.3 - 20}{\frac{4.3}{\sqrt{18}}}$

   ≈ -1.677

   (b)

   (c) With $18 - 1 = 17$ degrees of freedom, the *P*-value is $P(t_0 < -1.677) \approx 0.0559$.

   (d) The researcher will not reject the null hypothesis because the *P*-value is not less than the level of significance.

9. (a) No, because the sample size is at least 30.

   (b) $t_0 = \dfrac{\overline{x} - \mu_0}{\frac{s}{\sqrt{n}}}$

   $= \dfrac{101.9 - 105}{\frac{5.9}{\sqrt{35}}}$

   ≈ -3.108

   (c)

   (d) With $35 - 1 = 34$ degrees of freedom, the *P*-value is $2P(t_0 < -3.108) \approx 0.0038$.

   (e) The researcher will reject the null hypothesis because the *P*-value is less than the level of significance.

11. (a) $H_0: \mu = \$67$
$H_1: \mu > \$67$

(b) The *P*-value of 0.02 means that if the null hypothesis is true, this type of result would be expected in about 2 out of 100 samples.

(c) Reject the null hypothesis because the *P*-value is less than the level of significance. There is sufficient evidence that people withdraw more money from a PayEase ATM.

13. (a) $H_0: \mu = 22$
$H_1: \mu > 22$

(b) It is given that the sample is random. The sample size, 200, is at least 30. It is reasonable to assume the sampled values are independent.

(c) $t_0 = \dfrac{22.6 - 22}{\frac{3.9}{\sqrt{200}}}$
≈ 2.176

This test statistic follows a t-distribution with $200 - 1 = 199$ degrees of freedom.

Classical approach: The critical value is $t_{0.05} = 1.653$. Since $2.176 > 1.653$, reject the null hypothesis.

P-value approach: The *P*-value is $P(t_0 > 2.176) \approx 0.0154$. Since $0.0154 < 0.05$, reject the null hypothesis.

(d) There is sufficient evidence to conclude that students who complete the core curriculum are ready for college-level mathematics.

15. Hypotheses: $H_0: \mu = 9.02$ cm^3
$H_1: \mu < 9.02$ cm^3

It is given that the sample is random and that the hippocampal volume is approximately normally distributed. It is reasonable to assume the sampled values are independent.

Test statistic: $t_0 = \dfrac{8.10 - 9.02}{\frac{0.7}{\sqrt{12}}}$
≈ -4.553

This test statistic follows a t-distribution with $12 - 1 = 11$ degrees of freedom.

Critical value: $-t_{0.01} = -2.718$

P-value: $P(t_0 < -4.553) \approx 0.0004$

Since $-4.553 < -2.718$ and $0.0004 < 0.01$, reject the null hypothesis. There is sufficient evidence to conclude that hippocampal volumes in alcoholic adolescents are less than the normal volume of 9.02 cm^3.

17. Hypotheses: $H_0: \mu = 703.5$
$H_1: \mu > 703.5$

It is given that the sample is random. The sample size, 40, is at least 30. It is reasonable to assume the sampled values are independent.

Test statistic: $t_0 = \dfrac{714.2 - 703.5}{\frac{83.2}{\sqrt{40}}}$
≈ 0.813

This test statistic follows a t-distribution with $40 - 1 = 39$ degrees of freedom.

Critical value: $t_{0.05} = 1.685$

P-value: $P(t_0 > 0.813) \approx 0.2106$

Since $0.813 < 1.685$ and $0.2106 < 0.05$, do not reject the null hypothesis. There is not sufficient evidence to conclude that high income individuals have higher credit scores.

19. Using technology, the 95% confidence interval is $(35.439, 42.361)$. The interval contains 40.7, which implies that the mean age of a death-row inmate has not changed from 2002.

21. (a) Yes, because the normal probability plot looks approximately linear, and the boxplot is roughly symmetric with no outliers. According to Table VI, the critical value for the correlation between waiting time and expected z-scores for $n = 10$ is 0.918. Since $0.971 > 0.918$, it is reasonable to assume the sample has been drawn from an approximately normal population.

(b) Hypotheses: $H_0: \mu = 84.3$ min
$H_1: \mu < 84.3$ min

Sample mean: $\bar{x} = 78$

Sample standard deviation: $s = 15.2$

Test statistic: $t_0 = \dfrac{78 - 84.3}{\dfrac{15.2}{\sqrt{10}}}$

≈ -1.311

(Tech: -1.310)

This test statistic follows a t-distribution with $10 - 1 = 9$ degrees of freedom.

Critical value: $-t_{0.1} = -1.383$

P-value: $P(t_0 < -1.311) \approx 0.1112$
(Tech: 0.1113)

Since $-1.311 > -1.383$ and $0.1112 > 0.1$, do not reject the null hypothesis. There is not sufficient evidence to conclude that the new system has decreased the wait time.

23. Hypotheses: $H_0: \mu = 0.11$ mg/L
    $H_1: \mu \neq 0.11$ mg/L

It is given that the sample is random. The results from the normal probability plot and boxplot suggest the sample is drawn from a population that is approximately normal. It is reasonable to assume the sampled values are independent.

Sample mean: $\bar{x} = 0.1568$

Sample standard deviation: $s = 0.087$

Test statistic: $t_0 = \dfrac{0.1568 - 0.11}{\dfrac{0.087}{\sqrt{10}}}$

≈ 1.701

(Tech: 1.707)

This test statistic follows a t-distribution with $10 - 1 = 9$ degrees of freedom.

Critical values: $-t_{0.05/2} = -t_{0.025} = -2.262$ and $t_{0.05/2} = t_{0.025} = 2.262$

P-value: $2P(t_0 > 1.701) \approx 0.1232$
(Tech: 0.1220)

Since $-2.262 < 1.701 < 2.262$ and $0.1232 > 0.05$, do not reject the null hypothesis. There is not sufficient evidence to conclude that calcium concentrations have changed.

25. (a)

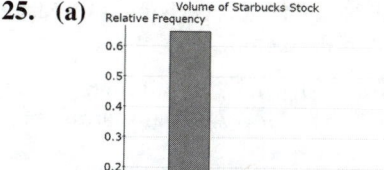

This histogram is unimodal and a bit skewed right.

(b)

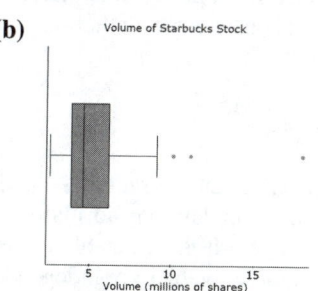

Yes, there are several outliers to the right of the box.

(c) One of the conditions for this test is that "the sample has no outliers and the population from which the sample is drawn is normally distributed, or the sample size is large." The histogram and boxplot do not suggest a normal population, and the boxplot shows several outliers, so the only way to satisfy this condition is with a large sample size.

(d) Hypotheses: $H_0: \mu = 7.52$ mil shares
    $H_1: \mu \neq 7.52$ mil shares

Use technology to conduct the test on this data set.

Test statistic: $t_0 \approx -4.202$

P-value: $P(t_0 < -4.202) \approx 0.0001$

Since $0.0001 > 0.05$, reject the null hypothesis. There is sufficient evidence to conclude that the volume of Starbucks stock has changed from 2014.

27. Hypotheses: $H_0: \mu = 0.11$ mg/L
    $H_1: \mu \neq 0.11$ mg/L

Using technology, the 95% confidence interval is $(0.095, 0.219)$. Since 0.11 falls within this interval, do not reject the null hypothesis. There is not sufficient evidence to conclude that calcium concentrations have changed. Note that this decision is consistent with that of Problem 23.

29. Hypotheses: H_0: $\mu = 7.52$ mil shares
    H_1: $\mu \neq 7.52$ mil shares

    Using technology, the 95% confidence interval is $(4.6676, 6.5214)$. Since 7.52 falls outside this interval, reject the null hypothesis. There is sufficient evidence to conclude that the volume of Starbucks stock has changed from 2014. Note that this decision is consistent with that of Problem 25.

31. (a) H_0: $\mu = 515$
    H_1: $\mu > 515$

    (b) It is given that the sample is random and that the population is normally distributed. It is reasonable to assume the sampled values are independent.

    Test statistic: $t_0 = \dfrac{519 - 515}{\dfrac{111}{\sqrt{1800}}}$
    ≈ 1.529

    This test statistic follows a t-distribution with $1800 - 1 = 1799$ degrees of freedom.

    Critical value: $t_{0.1} = 1.282$

    P-value: $P(t_0 > 1.529) \approx 0.0632$

    Since $1.529 > 1.282$ and $0.0632 < 0.1$, reject the null hypothesis. There is sufficient evidence to conclude that the review course increases the scores of students on the math portion of the SAT exam.

    (c) This score increase likely does not have any practical significance. School admissions administrators are likely not going to care very much about a four point difference on a test where scores are routinely in the 500s.

    (d) Test statistic: $t_0 = \dfrac{519 - 515}{\dfrac{111}{\sqrt{400}}}$
    ≈ 0.721

    This test statistic follows a t-distribution with $400 - 1 = 399$ degrees of freedom.

    Critical value: $t_{0.1} = 1.284$

P-value: $P(t_0 > 0.721) \approx 0.2357$

(Tech: 0.2358)

Since $0.721 < 1.284$ and $0.2357 > 0.1$, do not reject the null hypothesis. There is not sufficient evidence to conclude that the review course increases the scores of students on the math portion of the SAT exam. Larger samples will make it more likely that the null hypothesis will be rejected, assuming the sample statistics stay the same.

33. (a) The researcher is testing to see if the mean IQ of JJC students is greater than 100, the mean IQ score of all humans.

    It is given that the sample is random. The sample size, 40, is at least 30. It is reasonable to assume the students in the sample are independent

    Test statistic: $t_0 = \dfrac{103.4 - 100}{\dfrac{13.2}{\sqrt{40}}}$
    ≈ 1.629

    This test statistic follows a t-distribution with $40 - 1 = 39$ degrees of freedom.

    Critical value: $t_{0.05} = 1.685$

    P-value: $P(t_0 > 1.629) \approx 0.0557$

    Since $1.629 < 1.685$ and $0.0557 > 0.05$, do not reject the null hypothesis. There is not sufficient evidence to conclude that the mean IQ of JJC students is greater than 100.

    (b) The researcher is testing to see if the mean IQ of JJC students is greater than 101.

    It is given that the sample is random. The sample size, 40, is at least 30. It is reasonable to assume the students in the sample are independent

    Test statistic: $t_0 = \dfrac{103.4 - 101}{\dfrac{13.2}{\sqrt{40}}}$
    ≈ 1.150

    This test statistic follows a t-distribution with $40 - 1 = 39$ degrees of freedom.

    Critical value: $t_{0.05} = 1.685$

P-value: $P(t_0 > 1.150) \approx 0.1286$

Since $1.150 < 1.685$ and $0.1286 > 0.05$, do not reject the null hypothesis. There is not sufficient evidence to conclude that the mean IQ of JJC students is greater than 101.

(c) The researcher is testing to see if the mean IQ of JJC students is greater than 102.

It is given that the sample is random. The sample size, 40, is at least 30. It is reasonable to assume the students in the sample are independent

Test statistic: $t_0 = \dfrac{103.4 - 102}{\dfrac{13.2}{\sqrt{40}}}$
≈ 0.671

This test statistic follows a t-distribution with $40 - 1 = 39$ degrees of freedom.

Critical value: $t_{0.05} = 1.685$

P-value: $P(t_0 > 0.671) \approx 0.2531$
(Tech: 0.2532)

Since $0.671 < 1.685$ and $0.2531 > 0.05$, do not reject the null hypothesis. There is not sufficient evidence to conclude that the mean IQ of JJC students is greater than 102.

(d) Answers will vary.

35. (a) Answers will vary.

(b) Answers will vary.

(c) Since $\alpha = 0.05$ is the probability of making a Type I error, we would expect $100(0.05) = 5$ samples to result in a Type I error.

(d) Answers will vary.

37. Hypothesis tests are meant to be conducted on samples. The researcher found the mean and standard deviation of the entire population, so there is no need to conduct a hypothesis test.

39. Answers will vary.

Section 10.4

1. (a) 28.869

(b) 14.041

(c) 16.047, 45.722

3. (a) $\chi_0^2 = \dfrac{(n-1)s^2}{\sigma_0^2}$
$= \dfrac{(24-1)(47.2^2)}{50^2}$
≈ 20.496

(b) With $24 - 1 = 23$ degrees of freedom, the critical value is $\chi_{0.95}^2 \approx 13.091$.

(c)

(d) The researcher will not reject the null hypothesis because the test statistic is not in the critical region.

5. (a) $\chi_0^2 = \dfrac{(n-1)s^2}{\sigma_0^2}$
$= \dfrac{(18-1)(2.4^2)}{1.8^2}$
≈ 30.222

(b) With $18 - 1 = 17$ degrees of freedom, the critical value is $\chi_{0.1}^2 \approx 24.769$.

(c)

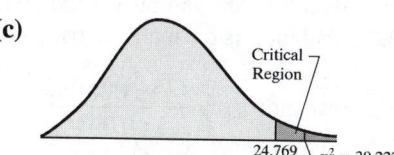

(d) The researcher will reject the null hypothesis because the test statistic is in the critical region.

7. (a) $\chi_0^2 = \dfrac{(n-1)s^2}{\sigma_0^2}$
$= \dfrac{(12-1)(4.8^2)}{4.3^2}$
≈ 13.706

(b) With $12 - 1 = 11$ degrees of freedom, the critical values are $\chi_{0.975}^2 \approx 3.816$ and $\chi_{0.025}^2 \approx 21.920$.

(c)

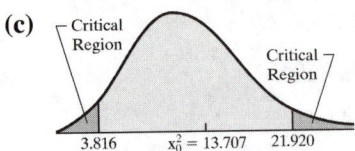

(d) The researcher will not reject the null hypothesis because the test statistic is not in the critical region.

9. Hypotheses: $H_0: \sigma = 0.04$
$H_1: \sigma < 0.04$

It is given that the sample is random and that the population is normally distributed.

Test statistic: $\chi_0^2 = \dfrac{(25-1)(0.0301^2)}{0.04^2}$
≈ 13.590

This test statistic follows a χ^2-distribution with $25-1 = 24$ degrees of freedom.

Critical value: $\chi_{0.95}^2 \approx 13.848$

P-value: $P(\chi^2 < 13.590) \approx 0.0446$

Since $13.590 < 13.848$ and $0.0446 < 0.05$, reject the null hypothesis. There is sufficient evidence to conclude that the fund has moderate risk.

11. Hypotheses: $H_0: \sigma = 0.004$
$H_1: \sigma < 0.004$

It is given that the sample is random and that the population is normally distributed.

Test statistic: $\chi_0^2 = \dfrac{(25-1)(0.0025^2)}{0.004^2}$
$= 9.375$

This test statistic follows a χ^2-distribution with $25-1 = 24$ degrees of freedom.

Critical value: $\chi_{0.99}^2 \approx 10.856$

P-value: $P(\chi^2 < 9.375) \approx 0.0033$

Since $9.375 < 10.856$ and $0.0033 < 0.01$, reject the null hypothesis. There is sufficient evidence to conclude that the recalibration was effective.

13. Hypotheses: $H_0: \sigma = 18.0$ sec
$H_1: \sigma < 18.0$ sec

It is given that the sample is random and that the population is normally distributed.

Sample standard deviation: $s \approx 15.2$ sec

Test statistic: $\chi_0^2 = \dfrac{(10-1)(15.2^2)}{18.0^2}$
≈ 6.418

(Tech: 6.422)

This test statistic follows a χ^2-distribution with $10-1 = 9$ degrees of freedom.

Critical value: $\chi_{0.95}^2 \approx 3.325$

P-value: $P(\chi^2 < 6.418) \approx 0.3025$
(Tech: 0.3030)

Since $6.418 > 3.325$ and $0.3025 > 0.05$, do not reject the null hypothesis. There is not sufficient evidence to conclude that the standard deviation of the wait time is less than 18.0 seconds.

15. Hypotheses: $H_0: \sigma = 8.3$ points
$H_1: \sigma < 8.3$ points

It is given that the sample is random and that the population is approximately normally distributed.

Test statistic: $\chi_0^2 = \dfrac{(25-1)(6.7^2)}{8.3^2}$
≈ 15.639

This test statistic follows a χ^2-distribution with $25-1 = 24$ degrees of freedom.

Critical value: $\chi_{0.9}^2 \approx 15.658$

P-value: $P(\chi^2 < 15.639) \approx 0.0993$

Since $15.639 < 15.658$ and $0.0993 < 0.1$, reject the null hypothesis. There is sufficient evidence to conclude that Derrick Rose is more consistent than other shooting guards.

17. (a)

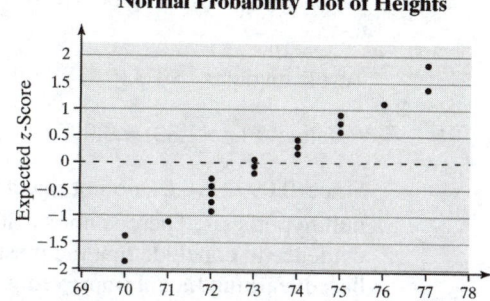

Normal Probability Plot of Heights

The points roughly form a straight line, so it is reasonable to assume the population is approximately normally distributed.

(b) $s \approx 2.059$

(c) Hypotheses: H_0: $\sigma = 2.9$ in
H_1: $\sigma < 2.9$ in

It is given that the sample is random. It has been shown that the population is approximately normally distributed.

Test statistic: $\chi_0^2 = \dfrac{(20-1)(2.059^2)}{2.9^2}$
≈ 9.578

This test statistic follows a χ^2-distribution with $20-1=19$ degrees of freedom.

Critical value: $\chi_{0.99}^2 \approx 7.633$

P-value: $P(\chi^2 < 9.578) \approx 0.0374$

Since $9.578 < 7.633$ and $0.0374 < 0.01$, do not reject the null hypothesis. There is not sufficient evidence to conclude that the standard deviation of heights for major-league baseball players is less than 2.9 inches.

Section 10.5

1. Hypotheses: H_0: $\mu = 1.0$ min
H_1: $\mu < 1.0$ min

It is given that the sample is random and that the population is normally distributed. It is reasonable to assume the sampled values are independent.

Test statistic: $t_0 \approx -2.179$

This test statistic follows a t-distribution with $19-1=18$ degrees of freedom.

P-value: $P(t_0 < -2.179) \approx 0.0214$

Since $0.0214 > 0.01$, do not reject the null hypothesis. There is not sufficient evidence to conclude that the population mean is less than 1.0.

3. Hypotheses: H_0: $\sigma^2 = 95$
H_1: $\sigma^2 < 95$

It is given that the sample is random and that the population is normally distributed.

Test statistic: $\chi_0^2 = 9.86$

This test statistic follows a χ^2-distribution with $20-1=19$ degrees of freedom.

P-value: $P(\chi^2 < 9.86) \approx 0.0437$

Since $0.0437 < 0.1$, reject the null hypothesis. There is sufficient evidence to conclude that the population variance is less than 95.

5. Hypotheses: H_0: $\sigma^2 = 10$
H_1: $\sigma^2 > 10$

It is given that the sample is random and that the population is normally distributed.

Test statistic: $\chi_0^2 = 20.55$

This test statistic follows a χ^2-distribution with $16-1=15$ degrees of freedom.

P-value: $P(\chi^2 > 20.55) \approx 0.1518$

Since $0.1518 > 0.05$, do not reject the null hypothesis. There is not sufficient evidence to conclude that the population variance is greater than 10.

7. Hypotheses: H_0: $\mu = 100$
H_1: $\mu > 100$

It is given that the sample is random. The sample size, 40, is at least 30. It is reasonable to assume the sampled values are independent.

Test statistic: $t_0 \approx 3.003$

This test statistic follows a t-distribution with $40-1=39$ degrees of freedom.

P-value: $P(t_0 > 3.003) \approx 0.0023$

Since $0.0023 < 0.05$, reject the null hypothesis. There is sufficient evidence to conclude that the population mean is greater than 100.

9. Hypotheses: $H_0: \mu = 100$
$H_1: \mu > 100$

It is given that the sample is random and that the population is normally distributed. It is reasonable to assume the sampled values are independent.

Test statistic: $t_0 \approx 1.278$

This test statistic follows a t-distribution with $20 - 1 = 19$ degrees of freedom.

P-value: $P(t_0 > 1.278) \approx 0.1084$

Since $0.1084 > 0.05$, do not reject the null hypothesis. There is not sufficient evidence to conclude that children whose mothers play Mozart in the house at least 30 minutes each day until they give birth have higher IQs.

11. Hypotheses: $H_0: \sigma = 7000$ psi
$H_1: \sigma > 7000$ psi

It is reasonable to assume that the sample is random. It is given that the population is normally distributed.

Test statistic: $\chi_0^2 \approx 21.811$

This test statistic follows a χ^2-distribution with $20 - 1 = 19$ degrees of freedom.

P-value: $P(\chi^2 > 21.811) \approx 0.2938$

Since $0.2938 > 0.01$, do not reject the null hypothesis. There is not sufficient evidence to conclude that the standard deviation of yield strength exceeds 7000 psi.

13. (a) The variable of interest is whether or not the student passed. This variable is qualitative because it allows for classification of the students based on whether or not the students passed.

   (b) Hypotheses: $H_0: p = 0.526$
   $H_1: p > 0.526$

   It is reasonable to assume the students in the sample are independent. While the sample is not random, it is not random in a fashion that does not affect the results much, if at all. Since

$np_0(1 - p_0) = 480(0.526)(0.474) \approx 119.7 > 10$, the normal model may be used for this test.

Test statistic: $z_0 \approx 1.33$

P-value: $P(z > 1.33) \approx 0.0922$

Since $0.0922 > 0.01$, do not reject the null hypothesis. There is not sufficient evidence to conclude that the mastery-based learning model improved pass rates.

   (c) Answers will vary.

15. Hypotheses: $H_0: \mu = 28$
$H_1: \mu \neq 28$

It is given that the sample is random. It is reasonable to assume the sampled values are independent. The sample size, 10, is not at least 30, so construct a normal probability plot.

Normal Probability Plot of Gas Mileage

The plot looks roughly linear. The value of the correlation is $r \approx 0.960$, which is greater than 0.918, the critical value from Table VI for a sample size of 10. Thus, it can be assumed that the population is approximately normally distributed.

Test statistic: $t_0 \approx -2.963$

This test statistic follows a t-distribution with $10 - 1 = 9$ degrees of freedom.

P-value: $P(t_0 < -2.963) \approx 0.0159$

The rejection decision depends on the level of significance chosen. If $\alpha = 0.1$ or $\alpha = 0.05$ were chosen, then reject the null hypothesis; there is sufficient evidence to conclude that individuals are getting different gas mileage than the EPA states should be attained. If $\alpha = 0.01$ was chosen, then do not reject the null hypothesis; there is not sufficient evidence to conclude that individuals are getting different gas mileage than the EPA states should be attained.

17. Hypotheses: $H_0: p = 0.26$
$H_1: p \neq 0.26$

It is reasonable to assume the individuals in the sample are independent, and it is given that the sample is random. Since $np_0(1-p_0) = 60(0.26)(1-0.26) = 11.544 > 10$, the normal model may be used for this test.

Test statistic: $z_0 \approx 1.59$

P-value: $P(z > 1.59) \approx 0.1120$

Since $0.1120 > 0.05$, do not reject the null hypothesis. There is not sufficient evidence to conclude that the data contradict the results of Toluna.

19. A small level of significance would be best, as the congresswoman does not want to conclude that the population favors a tax increase when, in fact, they do not.
$\alpha = 0.05$ or $\alpha = 0.01$ would be reasonable choices.

Hypotheses: $H_0: p = 0.5$
$H_1: p > 0.5$

It is reasonable to assume the sample is random and the individuals in the sample are independent. Since $np_0(1-p_0) = 8250(0.5)(1-0.5) = 2062.5 > 10$, the normal model may be used for this test.

Test statistic: $z_0 \approx 1.76$

P-value: $P(z > 1.79) \approx 0.0391$

The conclusion to this test depends on the level of significance chosen. If $\alpha = 0.05$ was chosen, then reject the null hypothesis; there is sufficient evidence to conclude that a majority of the district favor the tax increase. If $\alpha = 0.01$ was chosen, then do not reject the null hypothesis; there is not sufficient evidence to conclude that a majority of the district favor the tax increase.

21. (a)

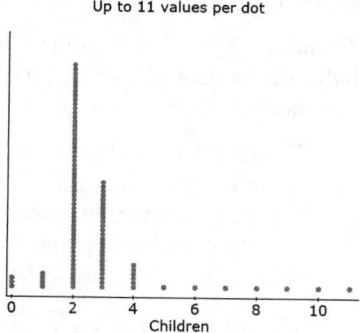

The dot plot appears to be skewed right.

(b) 2

(c) Mean: 2.418

Median: 2

Standard deviation: 1.064

Interquartile range: 1

(d) The sample is pretty strongly skewed right. A large sample size is necessary to satisfy the second condition for the hypothesis test for a population mean.

(e) Hypotheses: $H_0: \mu = 2.64$ children
$H_1: \mu \neq 2.64$ children

Test statistic: $t_0 \approx -6.597$

This test statistic follows a t-distribution with $1000 - 1 = 999$ degrees of freedom.

P-value: $P(t_0 < -6.597) \approx 0.0000$

Since $0.0000 < 0.05$, reject the null hypothesis. There is sufficient evidence to conclude that people's beliefs as to the ideal number of children have changed.

23. Hypothesis test; $H_0: \mu = 16$ oz
$H_1: \mu < 16$ oz

25. Confidence interval for the population proportion

27. Hypothesis test; $H_0: p = 0.14$
$H_1: p > 0.14$

29. Hypothesis test; $H_0: \sigma = 15$
$H_1: \sigma < 15$

Section 10.6

1. To make a Type II error is to not reject the null hypothesis when, in fact, the alternative hypothesis is true.

3. (a) To make a Type II error on this test is to not reject the null hypothesis that the population proportion is 0.30 when, in fact, the population proportion is less than 0.30.

(b) $\hat{p} = p_0 - z_\alpha \sqrt{\dfrac{p_0(1-p_0)}{n}}$

$= 0.30 - 1.645\sqrt{\dfrac{0.30(1-0.30)}{300}}$

≈ 0.256

$\beta = P(\hat{p} > 0.256 \text{ given that } p = 0.28)$

$= P\left(z > \dfrac{0.256 - 0.28}{\sqrt{\dfrac{0.28(1-0.28)}{300}}}\right)$

$\approx P(z > -0.93)$

≈ 0.8238

The power of the test is
$1 - 0.8238 = 0.1762$.
(StatCrunch: 0.1821)

(c) $\beta = P(\hat{p} > 0.256 \text{ given that } p = 0.25)$

$= P\left(z > \dfrac{0.256 - 0.25}{\sqrt{\dfrac{0.25(1-0.25)}{300}}}\right)$

$= P(z > 0.24)$

≈ 0.4052

The power of the test is
$1 - 0.4052 = 0.5948$.
(StatCrunch: 0.6023)

5. (a) To make a Type II error on this test is to not reject the null hypothesis that the population proportion is 0.65 when, in fact, the population proportion is greater than 0.65.

(b) $\hat{p} = p_0 + z_\alpha \sqrt{\dfrac{p_0(1-p_0)}{n}}$

$= 0.65 + 2.33\sqrt{\dfrac{0.65(1-0.65)}{100}}$

≈ 0.761

$\beta = P(\hat{p} < 0.761 \text{ given that } p = 0.70)$

$= P\left(z < \dfrac{0.761 - 0.70}{\sqrt{\dfrac{0.70(1-0.70)}{100}}}\right)$

$\approx P(z < 1.33)$

≈ 0.9082

The power of the test is
$1 - 0.9082 = 0.0918$.
(StatCrunch: 0.0917)

(c) $\beta = P(\hat{p} < 0.761 \text{ given that } p = 0.72)$

$= P\left(z < \dfrac{0.761 - 0.72}{\sqrt{\dfrac{0.72(1-0.72)}{100}}}\right)$

$\approx P(z < 0.91)$

≈ 0.8186

The power of the test is
$1 - 0.8186 = 0.1814$.
(StatCrunch: 0.1808)

7. (a) To make a Type II error on this test is to not reject the null hypothesis that the population proportion is 0.45 when, in fact, the population proportion is not 0.45.

(b) $\hat{p}_L = p_0 - z_{\alpha/2} \sqrt{\dfrac{p_0(1-p_0)}{n}}$

$= 0.45 - 1.645\sqrt{\dfrac{0.45(1-0.45)}{500}}$

≈ 0.413

$\hat{p}_U = p_0 + z_{\alpha/2} \sqrt{\dfrac{p_0(1-p_0)}{n}}$

$= 0.45 + 1.645\sqrt{\dfrac{0.45(1-0.45)}{500}}$

≈ 0.487

$\beta = P(0.413 < \hat{p} < 0.487 \text{ given that } p = 0.49)$

$= P\left(\dfrac{0.413 - 0.49}{\sqrt{\dfrac{0.49(1-0.49)}{500}}} < z < \dfrac{0.487 - 0.49}{\sqrt{\dfrac{0.49(1-0.49)}{500}}}\right)$

$\approx P(-3.44 < z < -0.13)$

≈ 0.4480

The power of the test is
$1 - 0.4480 = 0.5520$.
(StatCrunch: 0.5608)

(c)

$$\beta = P(0.413 < \hat{p} < 0.487 \text{ given that } p = 0.47)$$

$$= P\left(\frac{0.413 - 0.47}{\sqrt{\frac{0.47(1-0.47)}{500}}} < z < \frac{0.487 - 0.47}{\sqrt{\frac{0.47(1-0.47)}{500}}} \right)$$

$$\approx P(-2.55 < z < 0.76)$$

$$\approx 0.7710$$

The power of the test is
$1 - 0.7710 = 0.2290$.
(StatCrunch: 0.2342)

9. (a) To make a Type II error on this test is to not reject the null hypothesis that an equal number of adults in the United States believe and do not believe they will not have enough money for retirement when, in fact, the a majority of adults in the United States believe they will not have enough money for retirement.

(b) $\hat{p} = p_0 + z_\alpha \sqrt{\dfrac{p_0(1-p_0)}{n}}$

$$= 0.5 + 1.645 \sqrt{\frac{0.5(1-0.5)}{676}}$$

$$\approx 0.532$$

$$\beta = P(\hat{p} < 0.532 \text{ given that } p = 0.53)$$

$$= P\left(z < \frac{0.532 - 0.53}{\sqrt{\frac{0.53(1-0.53)}{676}}} \right)$$

$$\approx P(z < 0.10)$$

$$\approx 0.5398$$

The power of the test is
$1 - 0.5398 = 0.4602$.
(StatCrunch: 0.4661)

(c) $\beta = P(\hat{p} < 0.532 \text{ given that } p = 0.55)$

$$= P\left(z < \frac{0.532 - 0.55}{\sqrt{\frac{0.55(1-0.55)}{676}}} \right)$$

$$\approx P(z < -0.94)$$

$$\approx 0.1736$$

The power of the test is
$1 - 0.1736 = 0.8264$.
(StatCrunch: 0.8315)

11. (a) To make a Type II error on this test is to not reject the null hypothesis that the population proportion of parents of children in grades K–12 that are satisfied with the quality of education the students receive has not changed from 2002 when, in fact, the population proportion of parents of children in grades K–12 that are satisfied with the quality of education the students receive has decreased since 2002.

(b) $\hat{p} = p_0 - z_\alpha \sqrt{\dfrac{p_0(1-p_0)}{n}}$

$$= 0.47 - 1.28 \sqrt{\frac{0.47(1-0.47)}{1013}}$$

$$\approx 0.450$$

$$\beta = P(\hat{p} > 0.450 \text{ given that } p = 0.42)$$

$$= P\left(z > \frac{0.450 - 0.42}{\sqrt{\frac{0.42(1-0.42)}{1013}}} \right)$$

$$\approx P(z > 1.93)$$

$$\approx 0.0268$$

The power of the test is
$1 - 0.0268 = 0.9732$.
(StatCrunch: 0.9731)

(c) $\beta = P(\hat{p} > 0.450 \text{ given that } p = 0.46)$

$$= P\left(z > \frac{0.450 - 0.46}{\sqrt{\frac{0.46(1-0.46)}{1013}}} \right)$$

$$\approx P(z > -0.64)$$

$$\approx 0.7389$$

The power of the test is
$1 - 0.7389 = 0.2611$.
(StatCrunch: 0.2595)

13. (a) To make a Type II error on this test is to not reject the null hypothesis that parents feel the same about it being a serious problem that high school students were not being taught enough math and science as in 1994 when, in fact, parents feel differently about it being a serious problem that high school students were not being taught enough math and science than they did in 1994.

(b) $\hat{p}_L = p_0 - z_{\alpha/2} \sqrt{\dfrac{p_0(1-p_0)}{n}}$

$$= 0.52 - 1.96 \sqrt{\frac{0.52(1-0.52)}{800}}$$

$$\approx 0.485$$

$$\hat{p}_U = p_0 + z_{\alpha/2}\sqrt{\frac{p_0(1-p_0)}{n}}$$

$$= 0.52 + 1.96\sqrt{\frac{0.52(1-0.52)}{800}}$$

$$\approx 0.555$$

$$\beta = P(0.485 < \hat{p} < 0.555 \text{ given that } p = 0.50)$$

$$= P\left(\frac{0.485 - 0.50}{\sqrt{\dfrac{0.50(1-0.50)}{800}}} < z < \frac{0.555 - 0.50}{\sqrt{\dfrac{0.50(1-0.50)}{800}}}\right)$$

$$\approx P(-0.85 < z < 3.11)$$

$$\approx 0.8014$$

The power of the test is
$1 - 0.8014 = 0.1986$.
(StatCrunch: 0.2051)

(c)

$$\beta = P(0.485 < \hat{p} < 0.555 \text{ given that } p = 0.48)$$

$$= P\left(\frac{0.485 - 0.48}{\sqrt{\dfrac{0.48(1-0.48)}{800}}} < z < \frac{0.555 - 0.48}{\sqrt{\dfrac{0.48(1-0.48)}{800}}}\right)$$

$$\approx P(0.28 < z < 4.25)$$

$$\approx 0.3897$$

The power of the test is
$1 - 0.3897 = 0.6103$.
(StatCrunch: 0.6197)

15. $\hat{p} = p_0 - z_{\alpha}\sqrt{\dfrac{p_0(1-p_0)}{n}}$

$$= 0.30 - 2.33\sqrt{\frac{0.30(1-0.30)}{300}}$$

$$\approx 0.238$$

$$\beta = P(\hat{p} > 0.238 \text{ given that } p = 0.28)$$

$$= P\left(z > \frac{0.238 - 0.28}{\sqrt{\dfrac{0.28(1-0.28)}{300}}}\right)$$

$$\approx P(z > -1.62)$$

$$\approx 0.9474$$

The power of the test is $1 - 0.9474 = 0.0526$.
(StatCrunch: 0.0545)

The power of the test has decreased from
0.1762 to 0.0526. This makes sense because
α and β are inversely related, so when
α decreases, β increases, leading to a
decrease in $1 - \beta$.

17.

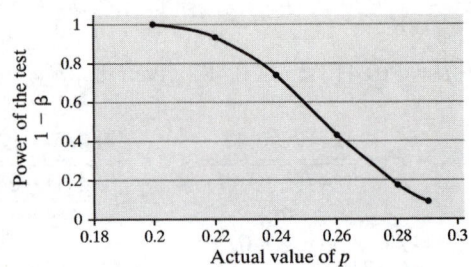

19. (a) $\bar{x} = \mu_0 - t_{\alpha}\sqrt{\dfrac{s^2}{n}}$

$$= 50 - 1.714\sqrt{\frac{6^2}{24}}$$

$$\approx 47.9$$

(b) $\beta = P(\bar{x} > 47.9 \text{ given that } \mu = 48.9)$

$$= P\left(t > \frac{47.9 - 48.9}{\sqrt{\dfrac{6^2}{24}}}\right)$$

$$\approx P(t > -0.816)$$

$$\approx 0.7886$$

The power of the test is
$1 - 0.7886 = 0.2114$.

21. Increasing the sample size increases the
power of the test.

Chapter 10 Review Exercises

1. (a) $H_0: \mu = \$3173$
   $H_1: \mu < \$3173$

(b) Making a Type I error would mean
concluding that the mean outstanding
credit-card debt of college graduates has
decreased since 2010 when, in fact, the
mean outstanding credit-card debt of
college graduates has not decreased
since 2010.

(c) Making a Type II error would mean
concluding that the mean outstanding
credit-card debt of college graduates has
not decreased since 2010 when, in fact,
the mean outstanding credit-card debt of
college graduates has decreased since
2010.

(d) The mean outstanding credit-card debt
of college graduates has not decreased
since 2010.

(e) The mean outstanding credit-card debt
of college graduates has decreased since
2010.

2. (a) $H_0: p = 0.13$
$H_1: p \neq 0.13$

(b) Making a Type I error would mean concluding that the proportion of issued credit cards that result in default is different from 0.13 when, in fact, the proportion of issued credit cards that result in default is not different from 0.13.

(c) Making a Type II error would mean concluding that the proportion of issued credit cards that result in default is not different from 0.13 when, in fact, the proportion of issued credit cards that result in default is different from 0.13.

(d) The proportion of issued credit cards that result in default is not different from 0.13.

(e) The proportion of issued credit cards that result in default is different from 0.13.

3. 0.05.

4. 0.113; 0.887; the probability that the null hypothesis will be correctly rejected is 0.887.

5. (a) The shape of the population distribution is not known to be normal, and there is no information indicating it can be assumed to be at least approximately normal.

(b) Test statistic: $t_0 = \dfrac{104.3 - 100}{\dfrac{12.4}{\sqrt{35}}}$
≈ 2.052

This test statistic follows a t-distribution with $35 - 1 = 34$ degrees of freedom.

Critical value: $t_{0.05} = 1.691$

P-value: $P(t_0 > 2.052) \approx 0.0240$

Since $2.052 > 1.691$ and $0.0240 < 0.05$, reject the null hypothesis. There is sufficient evidence to conclude that $\mu > 100$.

6. (a) The sample size, 15, is not at least 30.

(b) Test statistic: $t_0 = \dfrac{48.1 - 50}{\dfrac{4.1}{\sqrt{15}}}$
≈ -1.795

This test statistic follows a t-distribution with $15 - 1 = 14$ degrees of freedom.

Critical values: $-t_{0.025} = -2.145$ and $t_{0.025} = 2.145$

P-value: $2P(t_0 < -1.795) \approx 0.0943$

Since $-2.145 < -1.795 < 2.145$ and $0.0943 > 0.05$, do not reject the null hypothesis. There is not sufficient evidence to conclude that $\mu \neq 50$.

7. $np_0(1 - p_0) = 250(0.6)(1 - 0.6) = 60 > 10$

The sample proportion is $\hat{p} = \dfrac{165}{250} = 0.66$.

The test statistic is $z_0 = \dfrac{0.66 - 0.6}{\sqrt{\dfrac{0.6(1 - 0.6)}{250}}} \approx 1.94$.

(a) Because this is a right-tailed test, we determine the critical value at the $\alpha = 0.05$ level of significance to be $z_{0.05} = 1.645$. Because the test statistic is greater than the critical value, reject the null hypothesis.

(b) Because this is a right-tailed test, the P-value is the area under the standard normal distribution to the right of the test statistic, $z_0 = 1.94$. So,
P-value $= P(z > 1.94) \approx 0.0262$.
(Tech: 0.0264) Because the P-value is less than the level of significance, reject the null hypothesis.

8.

$np_0(1 - p_0) = 420(0.35)(1 - 0.35) = 95.55 > 10$

The sample proportion is $\hat{p} = \dfrac{138}{420} \approx 0.329$.

The test statistic is
$z_0 = \dfrac{0.329 - 0.35}{\sqrt{\dfrac{0.35(1 - 0.35)}{420}}} \approx -0.90$.

(Tech: -0.92)

(a) Because this is a two-tailed test, we determine the critical values at the $\alpha = 0.05$ level of significance to be $-z_{0.05/2} = -z_{0.025} = -1.96$ and $z_{0.05/2} = z_{0.025} = 1.96$. Because the test statistic does not lie in the critical region, do not reject the null hypothesis.

(b) Because this is a two-tailed test, the P-value is the area under the standard normal distribution to the left of $-|z_0| = -0.90$ and to the right of $|z_0| = 0.90$. So,

P-value $= 2P(Z < -0.90) \approx 2(0.1841) = 0.3682$.

(Tech: 0.3572) Because the P-value is greater than the level of significance, do not reject the null hypothesis.

9. (a) $\chi_0^2 = \dfrac{(n-1)s^2}{\sigma_0^2}$

$= \dfrac{(18-1)(4.9^2)}{5.2^2}$

≈ 15.095

(b) With $18 - 1 = 17$ degrees of freedom, the critical values are $\chi_{0.025}^2 \approx 7.564$ and $\chi_{0.975}^2 \approx 30.191$.

P-value: $2P(\chi_0^2 < 15.095) \approx 0.8227$

Since $7.564 < 15.095 < 30.191$ and $0.8227 > 0.05$, do not reject the null hypothesis. There is not sufficient evidence to conclude that $\sigma \neq 5.2$.

10. (a) $\chi_0^2 = \dfrac{(n-1)s^2}{\sigma_0^2}$

$= \dfrac{(25-1)(16.5^2)}{15.7^2}$

≈ 26.508

(b) With $25 - 1 = 24$ degrees of freedom, the critical value is $\chi_{0.95}^2 \approx 33.196$

P-value: $P(\chi_0^2 > 26.508) \approx 0.3279$

Since $26.508 < 33.196$ and $0.3279 > 0.1$, do not reject the null hypothesis. There is not sufficient evidence to conclude that $\sigma > 15.7$.

11. (a) $H_0: p = 0.733$
$H_1: p \neq 0.733$

(b) It is reasonable to assume that the observations are independent, and it is given that the sample is random. Since $np_0(1 - p_0) = 100(0.733)(0.267) \approx 19.6 < 10$, the normal model is appropriate for this test.

(c) Sample proportion: $\dfrac{78}{100} = 0.78$

Test statistic: $z_0 = \dfrac{0.78 - 0.733}{\sqrt{\dfrac{0.733(0.267)}{100}}}$

≈ 1.06

P-value: $2P(z > 1.06) \approx 0.2892$

(Tech: 0.2881)

Since 0.2892 is fairly high, do not reject the null hypothesis. There is not sufficient evidence to conclude that the proportion of individuals who cover their mouth when sneezing is not 0.733.

12. Hypotheses: $H_0: p = 0.05$
$H_1: p > 0.05$

It is reasonable to assume the patients in the sample are independent, and it is given that the sample is random. Since $np_0(1 - p_0) = 250(0.05)(1 - 0.05) = 11.875 > 10$, the normal model may be used for this test.

Sample proportion: $\dfrac{17}{250} \approx 0.068$

Test statistic: $z_0 = \dfrac{0.068 - 0.05}{\sqrt{\dfrac{0.05(1 - 0.05)}{250}}}$

≈ 1.31

P-value: $P(z > 1.31) \approx 0.0951$

(Tech: 0.0958)

The administrator should be a little concerned. The rejection decision and conclusion depends on the chosen level of significance. If the $\alpha = 0.1$ level of significance was chosen, the administrator would reject the null hypothesis and conclude there is strong evidence that more ER patients for this hospital die within a year than normal. If either the $\alpha = 0.05$ or $\alpha = 0.01$ level of significance were chosen, the administrator would not reject the null hypothesis and conclude there is not strong evidence that more ER patients for this hospital die within a year than normal.

13. **(a)** Yes. It is given that the sample is random. The sample size, 36, is at least 30. It is reasonable to assume the retaining rings are independent.

(b) Hypotheses: $H_0: \mu = 0.875$ in
$H_1: \mu > 0.875$ in

(c) Making a Type I error would be very costly for the company, because they would shut down the machine and recalibrate it.

(d) Test statistic: $t_0 = \dfrac{0.876 - 0.875}{\dfrac{0.005}{\sqrt{36}}}$

$= 1.2$

This test statistic follows a t-distribution with $36 - 1 = 35$ degrees of freedom.

Critical value: $t_{0.01} \approx 2.438$

P-value: $P(t_0 > 1.2) \approx 0.1191$

Since $1.2 < 2.438$ and $0.1191 > 0.01$, do not reject the null hypothesis. There is not sufficient evidence to conclude that the machine needs to be recalibrated.

(e) Making a Type I error means shutting down and recalibrating the machine when it is working as intended. Making a Type II error means letting the machine run when it needs to be shut down and recalibrated.

14. Hypotheses: $H_0: \mu = 98.6°F$
$H_1: \mu < 98.6°F$

It is reasonable to assume that the sample is random and that the sampled values are independent.. The sample size, 148, is at least 30.

Test statistic: $t_0 = \dfrac{98.2 - 98.6}{\dfrac{0.7}{\sqrt{148}}}$

≈ -6.9517

This test statistic follows a t-distribution with $148 - 1 = 147$ degrees of freedom.

Critical value: $-t_{0.01} \approx -2.352$

P-value: $P(t_0 < -6.9517) \approx 0.0000$

Since $-6.9517 < -2.352$ and $0.0000 < 0.01$, reject the null hypothesis. There is sufficient evidence to conclude that the mean temperature of humans is less than 98.6°F.

15. **(a)** Yes, because the normal probability plot looks approximately linear, and the boxplot has no outliers. According to Table VI, the critical value for the correlation between waiting time and expected z-scores for $n = 12$ is 0.928. Since $0.951 > 0.928$, it is reasonable to assume the sample has been drawn from an approximately normal population.

(b) Hypotheses: $H_0: \mu = 1.68$ in
$H_1: \mu \neq 1.68$ in

Using technology, the 95% confidence interval is $(1.6782, 1.6838)$. Since the hypothesized mean of 1.68 lies within the confidence interval, do not reject the null hypothesis. There is not sufficient evidence to conclude that the golf balls do not conform to USGA standards.

16. Hypotheses: $H_0: \mu = 480$ min
$H_1: \mu < 480$ min

It is given that the sample is random and that the time spent studying is approximately normally distributed. It is reasonable to assume the sampled values are independent.

Test statistic: $t_0 \approx -1.577$

This test statistic follows a t-distribution with $11 - 1 = 10$ degrees of freedom.

P-value: $P(t_0 < -1.577) \approx 0.0730$

Since $0.0730 > 0.05$, do not reject the null hypothesis. There is not sufficient evidence to conclude that the students in this College Algebra class are studying less than 480 minutes each week.

17. Hypotheses: $H_0: p = 0.5$
$H_1: p > 0.5$

It is reasonable to assume the women in the sample are independent, and it is given that the sample is random. Since $np_0(1 - p_0) = 150(0.50)(1 - 0.50) = 37.5 > 10$, the normal model may be used for this test.

Sample proportion: $\dfrac{81}{150} = 0.54$

Test statistic: $z_0 = \dfrac{0.54 - 0.50}{\sqrt{\dfrac{0.50(1-0.50)}{150}}}$

≈ 0.98

P-value: $P(z > 0.98) \approx 0.1635$

(Tech: 0.1636)

Since $0.1635 > 0.05$, do not reject the null hypothesis. There is not sufficient evidence to conclude that a majority of pregnant women nap at least twice a week.

18. (a) To make a Type II error on this test is to not reject the null hypothesis that an equal amount of pregnant women do and do not nap twice a week when, in fact, the majority of pregnant women nap at least twice a week.

    (b) $\hat{p} = p_0 + z_\alpha \sqrt{\dfrac{p_0(1-p_0)}{n}}$

    $= 0.50 + 1.645\sqrt{\dfrac{0.50(1-0.50)}{150}}$

    ≈ 0.567

    $\beta = P(\hat{p} < 0.567 \text{ given that } p = 0.53)$

    $= P\left(z < \dfrac{0.567 - 0.53}{\sqrt{\dfrac{0.53(1-0.53)}{150}}}\right)$

    $\approx P(z < 0.91)$
    ≈ 0.8186

    The power of the test is
    $1 - 0.8186 = 0.1814$.
    (StatCrunch: 0.1810)

19. Hypotheses: $H_0: \sigma = 505.6$ g
    $H_1: \sigma > 505.6$ g

    It is given that the sample is random and the population is normally distributed.

    Test statistic: $\chi_0^2 = \dfrac{(41-1)(840^2)}{505.6^2}$

    ≈ 110.409

    This test statistic follows a χ^2-distribution with $41 - 1 = 40$ degrees of freedom.

    P-value: $P(\chi^2 > 110.409) \approx 0.0000$

    Since $0.0000 < 0.05$, reject the null hypothesis. There is sufficient evidence to conclude that the variability in the birth

weight of preterm babies is more than the variability in birth weight of full-term babies.

20. Hypotheses: $H_0: p = 0.52$
    $H_1: p > 0.52$

    It is reasonable to assume the students in the sample are independent, and it is given that the sample is random. Since
    $np_0(1 - p_0) = 2843(0.52)(0.48) \approx 710 > 10$,
    the normal model may be used for this test.

    Sample proportion: $\dfrac{1516}{2843} \approx 0.533$

    Test statistic: $z_0 = \dfrac{0.533 - 0.52}{\sqrt{\dfrac{0.52(1-0.52)}{2843}}}$

    ≈ 1.39

    (Tech: 1.41)

    P-value: $P(z > 1.39) \approx 0.0823$

    (Tech: 0.0788)

    Since $0.0788 < 0.1$, reject the null hypothesis. There is sufficient evidence to conclude that the proportion of students that return for their second year under the new policy is greater than 0.52. However, the results do not seem to be very practically significant, as the difference in proportions is very small.

21. Hypotheses: $H_0: p = 0.4$
    $H_1: p > 0.4$

    It is reasonable to assume that the sample is random and that the adolescents in the sample are independent. However, since
    $np_0(1 - p_0) = 40(0.4)(1-0.4) = 9.6 < 10$, the
    normal model is not appropriate for this test. This is a binomial experiment because the trials (adolescents) are independent and there are only two possible mutually exclusive results (praying daily or not praying daily), so use the binomial model instead.

    Using technology, $P(X \geq 18) \approx 0.3115$.

    Since $0.3115 > 0.05$, do not reject the null hypothesis. There is not sufficient evidence to conclude that the proportion of adolescents who pray daily has increased.

22. Hypotheses: $H_0: \mu = 73.2$
$H_1: \mu < 73.2$

It is given that the sample is random. The sample size, 3851, is at least 30. It is reasonable to assume that the sampled values are independent.

Test statistic: $t_0 = \dfrac{72.8 - 73.2}{\dfrac{12.3}{\sqrt{3851}}}$
≈ -2.018

This test statistic follows a t-distribution with $3851 - 1 = 3850$ degrees of freedom.

Critical value: $-t_{0.05} \approx -1.645$

P-value: $P(t_0 < -2.018) \approx 0.0218$

Since $-2.018 < -1.645$ and $0.0218 < 0.05$, reject the null hypothesis. There is sufficient evidence to conclude that the mean final exam score for students that use the self-study format is less than the regular mean final exam score. However, the results do not seem to be very practically significant, as the difference in scores is very small.

23. Accepting a null hypothesis is to believe the null hypothesis is true. Not rejecting the null hypothesis is to believe there is not enough evidence to claim the null hypothesis is false.

24. $H_0: \mu = 1.22$ hr and $H_1: \mu < 1.22$ hr; the P-value of 0.0329 means that if the null hypothesis is true, this type of result would be expected in about 3 or 4 out of 100 samples.

25. To test a hypothesis using the Classical Approach, state the null and alternative hypothesis, verify that the conditions for the test have been met, calculate the test statistics, and then find the critical value for the level of significance. If the test statistic is in the critical region, reject the null hypothesis; otherwise, do not reject the null hypothesis.

26. To test a hypothesis using the P-value Approach, state the null and alternative hypothesis, verify that the conditions for the test have been met, calculate the test statistics, and then calculate the P-value for the level of significance. If the P-value is less than the level of significance, reject the null

hypothesis; otherwise, do not reject the null hypothesis.

Chapter 10 Test

1. **(a)** $H_0: \mu = 42.6$ minutes
$H_1: \mu > 42.6$ minutes

(b) There is sufficient evidence to conclude that the amount of daily time spent on phone calls and answering or writing email has increased.

(c) Making a Type I error would mean concluding that the amount of daily time spent on phone calls and answering or writing email has increased when, in fact, the amount of daily time spent on phone calls and answering or writing email has not increased.

(d) Making a Type II error would mean concluding that the amount of daily time spent on phone calls and answering or writing email has not increased when, in fact, the amount of daily time spent on phone calls and answering or writing email has increased.

2. **(a)** $H_0: \mu = 167.1$ sec
$H_1: \mu < 167.1$ sec

(b) Choosing a small level of significance reduces the probability of making a Type I error. In this case, making a Type I error would be very costly, because McDonald's would likely spend a lot of money implementing a process that is ineffective.

(c) Test statistic: $t_0 = \dfrac{163.9 - 167.1}{\dfrac{15.3}{\sqrt{70}}}$
$= -1.750$

This test statistic follows a t-distribution with $70 - 1 = 69$ degrees of freedom.

Critical value: $-t_{0.01} \approx -2.382$

P-value: $P(t_0 < -1.750) \approx 0.0423$

Since $-1.750 > -2.382$ and $0.0423 > 0.01$, do not reject the null hypothesis. There is not sufficient evidence to conclude that the policy is effective.

3. Hypotheses: $H_0: \mu = 8$ hrs
$H_1: \mu < 8$ hrs

It is reasonable to assume that the sample is random and that the sampled values are independent. The sample size, 151, is at least 30.

Test statistic: $t_0 = \dfrac{7.8 - 8}{\dfrac{1.4}{\sqrt{151}}}$
≈ -1.755

This test statistic follows a t-distribution with $151 - 1 = 150$ degrees of freedom.

Critical value: $-t_{0.05} \approx -1.655$

P-value: $P(t_0 < -1.755) \approx 0.0407$
(Tech: 0.0406)

Since $-1.755 < -1.655$ and $0.0407 < 0.05$, reject the null hypothesis. There is sufficient evidence to conclude that postpartum women do not get enough sleep.

4. Hypotheses: $H_0: \mu = 1.3825$ in
$H_1: \mu \neq 1.3825$ in

Using technology, the 95% confidence interval is $(1.38236, 1.38284)$. Since the hypothesized mean is within the bounds of the interval, do not reject the null hypothesis. There is not sufficient evidence to conclude that the part is not being manufactured to specifications.

5. Hypotheses: $H_0: p = 0.6$
$H_1: p > 0.6$

It is reasonable to assume that the sample is random and that the respondents in the sample are independent. Since
$np_0(1 - p_0) = 1561(0.6)(1 - 0.6) = 374.64 > 10$,
the normal model may be used for this test.

Sample proportion: $\dfrac{954}{1561} \approx 0.611$

Test statistic: $z_0 = \dfrac{0.611 - 0.6}{\sqrt{\dfrac{0.6(1 - 0.6)}{1561}}}$
≈ 0.89
(Tech: 0.90)

Critical value: $z_{0.05} = 1.645$

P-value: $P(z > 0.89) \approx 0.1867$
(Tech: 0.1843)

Since $0.89 < 1.645$ and $0.1867 > 0.05$, do not reject the null hypothesis. There is not sufficient evidence to conclude that a supermajority of Americans felt the United States would have to fight Japan within their lifetimes.

6. Hypotheses: $H_0: \mu = 0$ lb
$H_1: \mu > 0$ lb

It is reasonable to assume that the sample is random and that the sampled values are independent. The sample size, 79, is at least 30.

Test statistic: $t_0 = \dfrac{1.6 - 0}{\dfrac{5.4}{\sqrt{79}}}$
$= 2.634$

This test statistic follows a t-distribution with $79 - 1 = 78$ degrees of freedom.

Critical value: $t_{0.05} \approx 1.645$

P-value: $P(t_0 > 2.634) \approx 0.0051$

Since $2.634 > 1.665$ and $0.0051 < 0.05$, reject the null hypothesis. There is sufficient evidence to conclude that the new policies were effective. There is not much practical significance, however; most people attempting to lose weight are unlikely to care about a weight loss of less than 2 pounds after 12 months.

7. Hypotheses: $H_0: p = 0.37$
$H_1: p > 0.37$

It is reasonable to assume that the sample is random and that the respondents in the sample are independent. However, since
$np_0(1 - p_0) = 30(0.37)(1 - 0.37) = 6.993 < 10$,
the normal model is not appropriate for this test. This is a binomial experiment because the trials (adolescents) are independent and there are only two possible mutually exclusive results (using only a cellular telephone or not using only a cellular telephone), so use the binomial model instead.

Using technology, $P(X \geq 16) \approx 0.0501$.

Since $0.0501 < 0.1$, reject the null hypothesis. There is sufficient evidence to conclude that the proportion of 20- to 24-year-olds who live on their own and don't have a landline is greater than 0.37.

8. Hypotheses: $H_0: \sigma = 0.18$
$H_1: \sigma < 0.18$

It is given that the sample is random. It is reasonable to assume that the population is normally distributed.

Test statistic: $\chi_0^2 = \dfrac{(10-1)(0.16^2)}{0.18^2}$
≈ 7.111

This test statistic follows a χ^2-distribution with $10-1=9$ degrees of freedom.

Critical value: $\chi_{0.05}^2 \approx 3.325$

P-value: $P(\chi^2 < 7.111) \approx 0.3744$

Since $7.111 > 3.325$ and $0.3744 > 0.05$, do not reject the null hypothesis. There is not sufficient evidence to conclude that investment manager's portfolio has less risk than the general market.

9. $\hat{p} = p_0 + z_\alpha \sqrt{\dfrac{p_0(1-p_0)}{n}}$

$= 0.60 + 1.645 \sqrt{\dfrac{0.60(1-0.60)}{1561}}$

≈ 0.620

$\beta = P(\hat{p} < 0.620 \text{ given that } p = 0.63)$

$= P\left(z < \dfrac{0.620 - 0.63}{\sqrt{\dfrac{0.63(1-0.63)}{1561}}} \right)$

$\approx P(z < -0.82)$
≈ 0.2061

The power of the test is $1 - 0.2061 = 0.7939$. (StatCrunch: 0.7841)

Case Study: How Old is Stonehenge?

Reports will vary. The reports should include the following components:

Antler samples from the ditch:

Hypotheses: $H_0: \mu = 2950$
$H_1: \mu \neq 2950$

It is reasonable to assume that the sample is random and that the sampled values are independent. It is assumed that population is approximately normal.

Test statistic: $t_0 = \dfrac{3033.1 - 2950}{\dfrac{66.9}{\sqrt{9}}}$
≈ 3.726

This test statistic follows a t-distribution with $9-1=8$ degrees of freedom.

P-value: $2P(t_0 > 3.726) \approx 0.0058$

Since $0.0058 < 0.05$, reject the null hypothesis. There is sufficient evidence to conclude that Corbin's claim is incorrect.

Animal bone samples in the ditch terminals:

Hypotheses: $H_0: \mu = 2950$
$H_1: \mu \neq 2950$

It is reasonable to assume that the sample is random and that the sampled values are independent. It is assumed that population is approximately normal.

Test statistic: $t_0 = \dfrac{3187.5 - 2950}{\dfrac{67.4}{\sqrt{4}}}$
≈ 7.047

This test statistic follows a t-distribution with $4-1=3$ degrees of freedom.

P-value: $2P(t_0 > 7.047) \approx 0.0059$

Since $0.0059 < 0.05$, reject the null hypothesis. There is sufficient evidence to conclude that the mean age of the site is different from 2950 B.C.

Animal artifacts from the Bluestones:

Hypotheses: $H_0: \mu = 2950$
$H_1: \mu \neq 2950$

While the sample may not be random, it would be not random in a way that may not influence the test very much. It is assumed that population is approximately normal. It is reasonable to assume that the sampled values are independent.

Test statistic: $t_0 = \dfrac{2193.3 - 2950}{\dfrac{104.1}{\sqrt{3}}}$
≈ -12.590

This test statistic follows a t-distribution with $3-1=2$ degrees of freedom.

P-value: $2P(t_0 < -12.590) \approx 0.0062$

Since $0.0062 < 0.05$, reject the null hypothesis. There is sufficient evidence to conclude that the mean age of the artifacts is different from 2950 B.C.

Additional antler samples from Y and Z holes:

Hypotheses: $H_0: \mu = 2950$
$\quad\quad\quad\quad H_1: \mu \neq 2950$

While the sample may not be random, it would be not random in a way that may not influence the test very much. It is assumed that population is approximately normal. It is reasonable to assume that the sampled values are independent.

Test statistic: $t_0 = \dfrac{1671.7 - 2950}{\dfrac{99.7}{\sqrt{3}}}$
$\quad\quad\quad\quad\quad \approx -22.207$

This test statistic follows a t-distribution with $3-1=2$ degrees of freedom.

P-value: $2P(t_0 < -22.207) \approx 0.0020$

Since $0.0020 < 0.05$, reject the null hypothesis. There is sufficient evidence to conclude that the population mean age of the Y and Z holes dated by Dillion is different from 2950 B.C.

Hypothesis test analysis:

It appears that none of the mean ages of the artifacts from the ditch, the ditch terminals, the Bluestones, or the Y and Z holes are consistent with Corbin's claimed mean age of 2950 B.C. for the construction of the ditch. Since hypothesis tests only investigate a specific possible value of a population parameter, it is not a sure thing to judge the likely construction order of the four structures. However, the most likely order, from oldest to youngest, seems to be the ditch terminals, the ditch, the Bluestones, and the Y and Z holes.

Confidence intervals:

Technology is used to construct all of the 95% confidence intervals below.

Ditch: $(2981.68, 3084.52)$

Ditch terminals: $(3080.25, 3294.75)$

Bluestones: $(1934.70, 2451.90)$

Y and Z holes: $(1424.03, 1919.37)$

None of these confidence intervals contain 2950, so none of them support Corbin's claim. In analyzing the overlap (or lack thereof) of the four intervals, it seems the most likely construction order, from oldest to youngest, is the ditch terminals, the ditch, the Bluestones, and the Y and Z holes.

Hypothesis testing vs. confidence intervals:

The confidence intervals are more useful in assessing the age and likely construction order of the Stonehenge structures. They provide a range of likely values, as opposed to rejecting or not rejecting the possibility of a single value. Also, the overlap between the ditch interval and the ditch terminals interval, which leaves open the possibility that the ditch is the oldest, could not be noticed from the hypothesis tests.

Limitations and assumptions:

The two major limitations and assumptions for these results are the assumption of random sampling and the assumption of approximately normal populations. Since these artifacts were found via archaeological digs, we cannot be sure that they are a random, representative sample of all of the artifacts. Also, the extremely low sample sizes for the ditch terminals, the Bluestones, and the X and Y holes make the normality assumption a dangerous one. At the very least, the actual ages of the artifacts involved would be useful to know, so that normal probability plots can be constructed and correlation values can be calculated. If there was another statistical procedure that did not require the normality assumption, it may be better to apply that one instead.

Chapter 11
Inferences on Two Samples

Section 11.1

1. Independent

3. Since the members of the two samples are married to each other, the sampling is dependent. The data are quantitative, since a numeric response was given by each subject.

5. Because the two samples were drawn at separate times with two different groups of people, the sampling is independent. The data are qualitative, since each respondent gave a "Yes" or "No" response.

7. Two independent populations are being studied: students who receive the new curriculum and students who receive the traditional curriculum. The people in the two groups are completely independent, so the sampling is independent. The data are quantitative, since it is the test score for each respondent.

9. (a) $H_0 : p_1 = p_2$
   $H_1 : p_1 > p_2$

   (b) The two sample estimates are
   $$\hat{p}_1 = \frac{x_1}{n_1} = \frac{368}{541} \approx 0.6802 \text{ and}$$
   $$\hat{p}_2 = \frac{x_2}{n_2} = \frac{351}{593} \approx 0.5919.$$
   The pooled estimate is
   $$\hat{p} = \frac{x_1 + x_2}{n_1 + n_2} = \frac{368 + 351}{541 + 593} \approx 0.6340.$$
   The test statistic is
   $$z_0 = \frac{\hat{p}_1 - \hat{p}_2}{\sqrt{\hat{p}(1-\hat{p})}\sqrt{\frac{1}{n_1} + \frac{1}{n_2}}}$$
   $$= \frac{0.6802 - 0.5919}{\sqrt{0.6340(1-0.6340)}\sqrt{\frac{1}{541} + \frac{1}{593}}}$$
   $$\approx 3.08$$

   (c) This is a right-tailed test, so the critical value is $z_\alpha = z_{0.05} = 1.645.$

(d) $P\text{-value} = P(z_0 \geq 3.08)$
   $$= 1 - 0.9990 = 0.0010$$

   Since $z_0 = 3.08 > z_{0.05} = 1.645$ and $P\text{-value} = 0.0010 < \alpha = 0.05,$ we reject H_0. There is sufficient evidence to conclude that $p_1 > p_2$.

11. (a) $H_0 : p_1 = p_2$
    $H_1 : p_1 \neq p_2$

    (b) The two sample estimates are
    $$\hat{p}_1 = \frac{x_1}{n_1} = \frac{28}{254} \approx 0.1102 \text{ and}$$
    $$\hat{p}_2 = \frac{x_2}{n_2} = \frac{36}{301} \approx 0.1196.$$
    The pooled estimate is
    $$\hat{p} = \frac{x_1 + x_2}{n_1 + n_2} = \frac{28 + 36}{254 + 301} \approx 0.1153.$$
    The test statistic is
    $$z_0 = \frac{\hat{p}_1 - \hat{p}_2}{\sqrt{\hat{p}(1-\hat{p})}\sqrt{\frac{1}{n_1} + \frac{1}{n_2}}}$$
    $$= \frac{0.1102 - 0.1196}{\sqrt{0.1153(1-0.1153)}\sqrt{\frac{1}{254} + \frac{1}{301}}}$$
    $$\approx -0.35 \quad [\text{Tech}: -0.34]$$

    (c) This is a two-tailed test, so the critical values are $\pm z_{\alpha/2} = \pm z_{0.025} = \pm 1.96.$

    (d) $P\text{-value} = 2 \cdot P(z_0 \leq -0.35)$
    $$= 2 \cdot 0.3632$$
    $$= 0.7264 \quad [\text{Tech}: 0.7307]$$

    Since $z_0 = -0.35$ falls between $-z_{0.025} = -1.96$ and $z_{0.025} = 1.96$ and $P\text{-value} = 0.7264 > \alpha = 0.05,$ we do not reject H_0. There is not sufficient evidence to conclude that $p_1 \neq p_2$.

13. We have $\hat{p}_1 = \dfrac{x_1}{n_1} = \dfrac{368}{541} \approx 0.6802$ and $\hat{p}_2 = \dfrac{x_2}{n_2} = \dfrac{421}{593} \approx 0.7099$. For a 90% confidence level, we use

$\pm z_{0.05} = \pm 1.645$. Then:

Lower Bound: $(\hat{p}_1 - \hat{p}_2) - z_{\alpha/2} \cdot \sqrt{\dfrac{\hat{p}_1(1-\hat{p}_1)}{n_1} + \dfrac{\hat{p}_2(1-\hat{p}_2)}{n_2}}$

$= (0.6802 - 0.7099) - 1.645 \cdot \sqrt{\dfrac{0.6802(1-0.6802)}{541} + \dfrac{0.7099(1-0.7099)}{593}}$

≈ -0.075

Upper Bound: $(\hat{p}_1 - \hat{p}_2) + z_{\alpha/2} \cdot \sqrt{\dfrac{\hat{p}_1(1-\hat{p}_1)}{n_1} + \dfrac{\hat{p}_2(1-\hat{p}_2)}{n_2}}$

$= (0.6802 - 0.7099) + 1.645 \cdot \sqrt{\dfrac{0.6802(1-0.6802)}{541} + \dfrac{0.7099(1-0.7099)}{593}}$

≈ 0.015

15. We have $\hat{p}_1 = \dfrac{x_1}{n_1} = \dfrac{28}{254} \approx 0.1102$ and $\hat{p}_2 = \dfrac{x_2}{n_2} = \dfrac{36}{301} \approx 0.1196$. For a 95% confidence interval we use

$\pm z_{0.025} = \pm 1.96$. Then:

Lower Bound: $(\hat{p}_1 - \hat{p}_2) - z_{\alpha/2} \cdot \sqrt{\dfrac{\hat{p}_1(1-\hat{p}_1)}{n_1} + \dfrac{\hat{p}_2(1-\hat{p}_2)}{n_2}}$

$= (0.1102 - 0.1196) - 1.96 \cdot \sqrt{\dfrac{0.1102(1-0.1102)}{254} + \dfrac{0.1196(1-0.1196)}{301}}$

≈ -0.063

Upper Bound: $(\hat{p}_1 - \hat{p}_2) + z_{\alpha/2} \cdot \sqrt{\dfrac{\hat{p}_1(1-\hat{p}_1)}{n_1} + \dfrac{\hat{p}_2(1-\hat{p}_2)}{n_2}}$

$= (0.1102 - 0.1196) + 1.96 \cdot \sqrt{\dfrac{0.1102(1-0.1102)}{254} + \dfrac{0.1196(1-0.1196)}{301}}$

≈ 0.044

---

17. We first verify the requirements to perform the hypothesis test: (1) Each sample can be thought of as a simple random sample; (2) We have $x_1 = 107$, $n_1 = 710$, $x_2 = 67$, and

$n_2 = 611$, so $\hat{p}_1 = \dfrac{x_1}{n_1} = \dfrac{107}{710} \approx 0.1507$ and

$\hat{p}_2 = \dfrac{x_2}{n_2} = \dfrac{67}{611} \approx 0.1097$. Thus, $n_1 \hat{p}_1 (1 - \hat{p}_1)$

$= 710(0.1507)(1-0.1507) \approx 91 \geq 10$ and

$n_2 \hat{p}_2 (1 - \hat{p}_2) = 611(0.1097)(1-0.1097) \approx 60$

≥ 10; and (3) Each sample is less than 5% of the population. Thus, the requirements are met, and we can conduct the test.

$H_0 : p_1 = p_2$

$H_1 : p_1 > p_2$

From before, the two sample estimates are $\hat{p}_1 \approx 0.1507$ and $\hat{p}_2 \approx 0.1097$. The pooled estimate is

$\hat{p} = \dfrac{x_1 + x_2}{n_1 + n_2} = \dfrac{107 + 67}{710 + 611} \approx 0.1317$.

The test statistic is

$z_0 = \dfrac{\hat{p}_1 - \hat{p}_2}{\sqrt{\hat{p}(1-\hat{p})}\sqrt{1/n_1 + 1/n_2}}$

$= \dfrac{0.1507 - 0.1097}{\sqrt{0.1317(1-0.1317)}\sqrt{1/710 + 1/611}}$

≈ 2.20

Classical approach: This is a right-tailed test, so the critical value is $z_\alpha = z_{0.05} = 1.645$. Since $z_0 > z_{0.05}$ (the test statistic lies within the critical region), we reject H_0.

P-value approach:
P-value $= P(z_0 \geq 2.20) = 1 - 0.9861 = 0.0139$. Since P-value $< \alpha = 0.05$, we reject H_0.

Conclusion: There is sufficient evidence at the $\alpha = 0.05$ level of significance to conclude that a higher proportion of subjects in the treatment group (taking Prevnar) experienced fever as a side effect than in the control (placebo) group.

19. We first verify the requirements to perform the hypothesis test: (1) Each sample is a simple random sample; (2) we have $x_{1947} = 407$, $n_{1947} = 1100$, $x_{recent} = 333$, and $n_{recent} = 1100$, so $\hat{p}_{1947} = \dfrac{x_{1947}}{n_{1947}} = \dfrac{407}{1100} = 0.37$ and

$\hat{p}_{recent} = \dfrac{x_{recent}}{n_{recent}} = \dfrac{333}{1100} = 0.30$. Thus,

$n_{1947}\hat{p}_{1947}\left(1 - \hat{p}_{1947}\right) = 1100(0.37)(1 - 0.37) \approx 256.4 \geq 10$ and $n_{recent}\hat{p}_{recent}\left(1 - \hat{p}_{recent}\right) = 1100(0.30)(1 - 0.30) \approx 232.2 \geq 10$; and (3) each sample is less than 5% of the population. Thus, the requirements are met, so we can conduct the test.

$H_0 : p_{1947} = p_{recent}$
$H_1 : p_{1947} \neq p_{recent}$

From before, the two sample estimates are $\hat{p}_{1947} = 0.37$ and $\hat{p}_{recent} = 0.30$. The pooled estimate is

$$\hat{p} = \frac{x_{1947} + x_{recent}}{n_{1947} + n_{recent}} = \frac{407 + 333}{1100 + 1100} = 0.336.$$

The test statistic is

$$z_0 = \frac{\hat{p}_{1947} - \hat{p}_{2010}}{\sqrt{\hat{p}(1 - \hat{p})}\sqrt{1/n_{1947} + 1/n_{2010}}}$$

$$= \frac{0.37 - 0.30}{\sqrt{0.336(1 - 0.336)}\sqrt{1/1100 + 1/1100}}$$

$$\approx 3.34$$

Classical approach: This is a two-tailed test, so the critical values are $\pm z_{\alpha/2} = \pm z_{0.025} = \pm 1.96$. Since the test statistic $z_0 = 3.34$ does not lie between $-z_{0.025} = -1.96$ and $z_{0.025} = 1.96$ (the test statistic falls in the critical region), we reject H_0.

P-value approach: P-value $= 2 \cdot P(z_0 \geq 3.34) = 2 \cdot 0.0004 = 0.0008$ [Tech: 0.0008] Since P-value $< \alpha = 0.05$, we reject H_0.

Conclusion: There is sufficient evidence at the $\alpha = 0.05$ level of significance to conclude that the proportion of adult Americans who were abstainers in 1947 is different from the proportion of recent abstainers.

---

21. $H_0 : p_m = p_f$
$H_1 : p_m \neq p_f$

We have $x_m = 181$, $n_m = 1205$, $x_f = 143$, and $n_f = 1097$, so $\hat{p}_m = \dfrac{x_m}{n_m} = \dfrac{181}{1205} \approx 0.1502$ and

$\hat{p}_f = \dfrac{x_f}{n_f} = \dfrac{143}{1097} \approx 0.1304$. For a 95% confidence interval we use $\pm z_{0.025} = \pm 1.96$. Then:

Lower Bound: $(\hat{p}_m - \hat{p}_f) - z_{\alpha/2} \cdot \sqrt{\dfrac{\hat{p}_m(1 - \hat{p}_m)}{n_m} + \dfrac{\hat{p}_f(1 - \hat{p}_f)}{n_f}}$

$$= (0.1502 - 0.1304) - 1.96 \cdot \sqrt{\frac{0.1502(1 - 0.1502)}{1205} + \frac{0.1304(1 - 0.1304)}{1097}}$$

$$\approx -0.009$$

Upper Bound: $(\hat{p}_m - \hat{p}_f) + z_{\alpha/2} \cdot \sqrt{\dfrac{\hat{p}_m(1-\hat{p}_m)}{n_m} + \dfrac{\hat{p}_f(1-\hat{p}_f)}{n_f}}$

$= (0.1502 - 0.1304) + 1.96 \cdot \sqrt{\dfrac{0.1502(1-0.1502)}{1205} + \dfrac{0.1304(1-0.1304)}{1097}}$

≈ 0.048

We are 95% confident that the difference in the proportion of males and females that have at least one tattoo is between -0.009 and 0.048. Because the interval includes zero, we do not reject the null hypothesis. There is no significant difference in the proportion of males and females that have tattoos.

---

23. (a) Each sample is less than 5% of the population size. The data were obtained from two independent simple random samples.

$$p_m = \frac{178}{540} \approx 0.330$$

$$p_f = \frac{206}{560} \approx 0.368$$

$$n_m \hat{p}_m (1-\hat{p}_m) = 119 \geq 10$$

$$n_f \hat{p}_f (1-\hat{p}_f) = 130 \geq 10$$

(b) $H_0 : p_f = p_m$

$H_1 : p_f > p_m$

(c) $\hat{p}_f - \hat{p}_m$ is approximately normal with

$\mu_{\hat{p}_f - \hat{p}_m} = 0$ and

$$\sigma_{\hat{p}_f - \hat{p}_m} = \sqrt{\frac{0.330(1-0.330)}{540} + \frac{0.368(1-0.368)}{560}} = 0.029$$

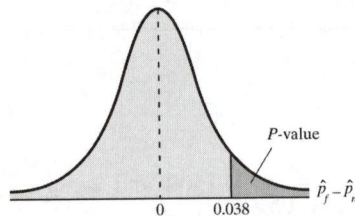

(d) $\hat{p} = \dfrac{178+206}{540+560} \approx 0.349$

Test statistic:

$$z_0 = \frac{0.368 - 0.330}{\sqrt{0.349(1-0.349)}\sqrt{\dfrac{1}{540} + \dfrac{1}{560}}}$$

$= 1.32$ [Tech: 1.33]

P-value $= 0.0934$ [Tech: 0.0918]

(e) If we obtain 1000 different simple random samples as described in the original problem, we would expect about 93 to have a difference in sample proportion of 0.038 or higher if the population proportion difference were zero.

(f) Since P-value $> \alpha$, do not reject the null hypothesis. There is not sufficient evidence to conclude the proportion of females annoyed by people who repeatedly check their mobile phones while having an in-person conversation is greater than the proportion of males.

25. (a) This is a completely randomized design.

(b) The response variable is whether the subject experiences dry mouth, or not. It is a qualitative variable with two possible outcomes.

(c) The explanatory variable is the type of drug. It has two levels—Clarinex or placebo.

(d) Double-blind means that neither the subject nor the individual monitoring the subject knows which treatment the subject is receiving.

(e) It is important to have a placebo group for two reasons. (1) So there is a baseline group against which to judge the Clarinex group, and (2) to eliminate any effect due to psychosomatic behavior.

(f) We first verify the requirements to perform the hypothesis test: (1) Each sample can be thought of as a simple random sample; (2) we have $x_C = 50$, $n_C = 1655$, $x_p = 31$, and

$n_p = 1652$, so $\hat{p}_C = \dfrac{50}{1655} \approx 0.0302$ and

$\hat{p}_p = \dfrac{31}{1652} \approx 0.0188$. Thus,

$n_C \hat{p}_C (1 - \hat{p}_C) =$

$1655(0.0302)(1 - 0.0302) \approx 48 \geq 10$ and

$n_p \hat{p}_p (1 - \hat{p}_p) = 1652(0.0188)(1 - 0.0188) \approx$

$30 \geq 10$; and (3) each sample is less than 5% of the population. So, the requirements are met, so we can conduct the test.

$H_0 : p_C = p_p$

$H_1 : p_C > p_p$

From before, the two sample estimates are $\hat{p}_C \approx 0.0302$ and $\hat{p}_p \approx 0.0188$. The pooled estimate is

$\hat{p} = \dfrac{x_C + x_p}{n_C + n_p} = \dfrac{50 + 31}{1655 + 1652} \approx 0.0245.$

Test statistic:

$z_0 = \dfrac{\hat{p}_C - \hat{p}_p}{\sqrt{\hat{p}(1 - \hat{p})}\sqrt{1/n_C + 1/n_p}}$

$= \dfrac{0.0302 - 0.0188}{\sqrt{0.0245(1 - 0.0245)}\sqrt{\dfrac{1}{1655} + \dfrac{1}{1652}}}$

≈ 2.12 [Tech: 2.13]

Classical approach: This is a right-tailed test, so the critical value is $z_{0.05} = 1.645$. Since $z_0 > z_{0.05} = 1.645$ (the test statistic falls in the critical region), we reject H_0.

P-value approach: P-value $= P(z_0 \geq 2.12)$ $= 1 - 0.9830 = 0.0170$ [Tech: 0.0166]. Since this P-value is less than the $\alpha = 0.05$ level of significance, we reject H_0.

Conclusion: There is sufficient evidence at the $\alpha = 0.05$ level of significance to conclude that the proportion of individuals taking Clarinex and experiencing dry mouth is greater than that of those taking a placebo.

(g) No, the difference between the experimental group and the control group is not practically significant. Both Clarinex and the placebo have fairly low proportions of individuals that experience dry mouth.

27. (a) The choices were randomly rotated in order to remove any potential nonsampling error due to the respondent hearing the word "right" or "wrong" first.

(b) We have $x_{2003} = 1086$, $n_{2003} = 1508$, $x_{2010} = 618$, and $n_{2008} = 1508$, so $\hat{p}_{2003} = \dfrac{x_{2003}}{n_{2003}} = \dfrac{1086}{1508} \approx 0.7202$

and $\hat{p}_{2010} = \dfrac{x_{2010}}{n_{2010}} = \dfrac{618}{1508} = 0.4098$. For a 90% confidence interval we use $\pm z_{0.05} = \pm 1.645$. Then:

Lower Bound: $(\hat{p}_{2003} - \hat{p}_{2010}) - z_{\alpha/2} \cdot \sqrt{\dfrac{\hat{p}_{2003}(1 - \hat{p}_{2003})}{n_{2003}} + \dfrac{\hat{p}_{2010}(1 - \hat{p}_{2010})}{n_{2010}}}$

$= (0.7202 - 0.4098) - 1.645 \cdot \sqrt{\dfrac{0.7202(1 - 0.7202)}{1508} + \dfrac{0.4098(1 - 0.4098)}{1508}} \approx 0.282$

Upper Bound: $(\hat{p}_{2003} - \hat{p}_{2010}) + z_{\alpha/2} \cdot \sqrt{\dfrac{\hat{p}_{2003}(1 - \hat{p}_{2003})}{n_{2003}} + \dfrac{\hat{p}_{2010}(1 - \hat{p}_{2010})}{n_{2010}}}$

$= (0.7202 - 0.4098) + 1.645 \cdot \sqrt{\dfrac{0.7202(1 - 0.7202)}{1508} + \dfrac{0.4098(1 - 0.4098)}{1508}} \approx 0.338$

[Tech: 0.339]

We are 90% confident that the difference in the proportion of adult Americans who believe the United States made the right decision to use military force in Iraq from 2003 to 2008 is between 0.282 and 0.338. The attitude regarding the decision to go to war changed substantially.

---

29. (a) We have $x_{males} = 26$, $n_{males} = 82$, $x_{females} = 25$, and $n_{females} = 117$, so

$\hat{p}_{males} = \dfrac{x_{males}}{n_{males}} = \dfrac{26}{82} = 0.317$ and

$\hat{p}_{females} = \dfrac{x_{females}}{n_{females}} = \dfrac{25}{117} = 0.214$.

(b) We first verify the requirements to perform the hypothesis test: (1) Each sample is a simple random sample; (2)

$n_{males} \hat{p}_{males} (1 - \hat{p}_{males}) =$
$82(0.317)(1 - 0.317) \approx 17.8 \geq 10$ and

$n_{females} \hat{p}_{females} (1 - \hat{p}_{females}) =$
$117(0.214)(1 - 0.214) \approx 19.7 \geq 10$; and (3)

each sample is less than 5% of the population. Thus, the requirements are met, so we can conduct the test.

$H_0 : p_{males} = p_{females}$
$H_1 : p_{males} \neq p_{females}$
The pooled estimate of p is

$\hat{p} = \dfrac{x_{males} + x_{females}}{n_{males} + n_{females}} = \dfrac{26 + 25}{82 + 117} = 0.256$.

Test statistic:

$z_0 = \dfrac{\hat{p}_{males} - \hat{p}_{females}}{\sqrt{\hat{p}(1 - \hat{p})}\sqrt{1/n_{males} + 1/n_{females}}}$

$= \dfrac{0.317 - 0.214}{\sqrt{0.256(1 - 0.256)}\sqrt{1/82 + 1/117}}$

≈ 1.64

Classical approach: This is a two-tailed test, so the critical values are: $\pm z_{\alpha/2} = \pm z_{0.025} = \pm 1.96$. Since the test statistic $z_0 = 1.61$ lies between $-z_{0.025} = -1.96$ and $z_{0.025} = 1.96$ (the test

statistic does not fall in the critical region), we do not reject H_0.

P-value approach: P-value=$2 \cdot P(z_0 \geq 1.64)$
$= 2 \cdot 0.0505 = 0.1010$ [Tech: 0.1001] Since P-value $> \alpha = 0.05$, we do not reject H_0.

Conclusion: There is not sufficient evidence at the $\alpha = 0.05$ level of significance to conclude that the proportion of men and women who are willing to pay higher taxes to reduce the deficit differs.

31. (a) In Sentence A, the verbs are "was having" and "was taking." In sentence B, the verbs are "had" and "took."

(b) We first verify the requirements to perform the hypothesis test: (1) Each sample is a simple random sample; (2) we have $x_A = 71$, $n_A = 98$, $x_B = 49$, and $n_B = 98$,

so $\hat{p}_A = \dfrac{x_A}{n_A} = \dfrac{71}{98} = 0.724$ and

$\hat{p}_B = \dfrac{x_B}{n_B} = \dfrac{49}{98} = 0.500$. Therefore,

$n_A \hat{p}_A (1 - \hat{p}_A) = 98(0.724)(1 - 0.724) \approx$
$19.6 \geq 10$ and $n_B \hat{p}_B (1 - \hat{p}_B) =$
$98(0.500)(1 - 0.500) = 24.5 \geq 10$; and (3)

each sample is less than 5% of the population. Thus, the requirements are met, so we can conduct the test.

$H_0 : p_A = p_B$
$H_1 : p_A \neq p_B$
The pooled estimate of p is

$\hat{p} = \dfrac{x_A + x_B}{n_A + n_B} = \dfrac{71 + 49}{98 + 98} = 0.612$.

Test statistic:

$$z_0 = \frac{\hat{p}_A - \hat{p}_B}{\sqrt{\hat{p}(1-\hat{p})}\sqrt{1/n_A + 1/n_B}}$$

$$= \frac{0.724 - 0.500}{\sqrt{0.612(1-0.612)}\sqrt{1/98 + 1/98}}$$

$$\approx 3.22 \quad [\text{Tech: } 3.23]$$

Classical approach: This is a two-tailed test, so the critical values are: $\pm z_{\alpha/2} = \pm z_{0.025} = 1.96.$ Since the test statistic $z_0 = 3.22$ does not lie between $-z_{0.025} = -1.96$ and $z_{0.025} = 1.96$ (the test statistic falls in the critical region), we reject H_0.

P-value approach:
$$P\text{-value} = 2 \cdot P(z_0 \geq 3.22)$$
$$= 2 \cdot 0.0006 = 0.0012 \quad [\text{Tech: } 0.0013] \text{ Since}$$
P-value $< \alpha = 0.05$, we reject H_0.

Conclusion: There is sufficient evidence at the $\alpha = 0.05$ level of significance to conclude that the sentence structure makes a difference.

(c) Answers will vary. The wording in Sentence A suggests that the actions were taking place over a period of time and suggests habitual behavior that may be continuing in the present or might be expected to continue at some time in the future. The wording in Sentence B suggests events that are concluded and were possibly brief in duration or one-time occurrences.

33. (a)

$$n = n_1 = n_2$$

$$= \left[\hat{p}_1(1-\hat{p}_1) + \hat{p}_2(1-\hat{p}_2)\right]\left(\frac{z_{\alpha/2}}{E}\right)^2$$

$$= \left[0.219(1-0.219) + 0.197(1-0.197)\right]\left(\frac{1.96}{0.03}\right)^2$$

$$\approx 1405.3$$
We increase this result to 1406.

(b) $n = n_1 = n_2$

$$= 0.5\left(\frac{z_{\alpha/2}}{E}\right)^2 = 0.5\left(\frac{1.96}{0.03}\right)^2 = 2134.2$$

We increase this result to 2135.

35. (a) The response variable is whether the fund manager outperforms the market, or not. The explanatory variable is whether it is a high-dispersion year or a low-dispersion year.

(b) The individuals are the fund managers.

(c) This study applies to the population of fund managers.

(d) $H_0 : p_{\text{high}} = p_{\text{low}}$

$H_1 : p_{\text{high}} > p_{\text{low}}$

(e) If the null hypothesis were true and we conducted the study 100 times, we would expect to observe the results as extreme or more extreme than the results observed in about eight of the studies. There is some evidence to suggest that the proportion of fund managers who outperform the market in high-dispersion years is greater than the proportion of fund managers who outperform the market in low-dispersion years.

Conclusion: There is sufficient evidence at the $\alpha = 0.01$ level of significance to conclude that the proportion of children in the experimental group who contracted polio is less than the proportion of children in the control group who contracted polio.

37. A pooled estimate of p is the best point estimate of the common population proportion, p. However, when finding a confidence interval, the sample proportions are not pooled because no assumption about their equality is made.

Section 11.2

1. $<$

3. (a) We measure differences as $d_i = X_i - Y_i$.

Obs	1	2	3	4	5	6	7
X_i	7.6	7.6	7.4	5.7	8.3	6.6	5.6
Y_i	8.1	6.6	10.7	9.4	7.8	9.0	8.5
d_i	−0.5	1.0	−3.3	−3.7	0.5	−2.4	−2.9

(b) $\Sigma d_i = (-0.5) + 1.0 + (-3.3) + (-3.7) + 0.5$
$$+ (-2.4) + (-2.9) = -11.3$$

so $\bar{d} = \dfrac{\Sigma d_i}{n} = \dfrac{-11.3}{7} \approx -1.614$

$\Sigma d_i^2 = (-0.5)^2 + (1.0)^2 + (-3.3)^2 + (-3.7)^2$
$$+ (0.5)^2 + (-2.4)^2 + (-2.9)^2 = 40.25$$

So,

$$s_d = \sqrt{\dfrac{\Sigma d_i^2 - \dfrac{(\Sigma d_i)^2}{n}}{n-1}}$$

$$= \sqrt{\dfrac{40.25 - \dfrac{(-11.3)^2}{7}}{7-1}} \approx 1.915$$

(c) $H_0 : \mu_d = 0$
$H_1 : \mu_d < 0$.

The level of significance is $\alpha = 0.05$. The test statistic is

$$t_0 = \dfrac{\bar{d}}{\dfrac{s_d}{\sqrt{n}}} = \dfrac{-1.614}{\dfrac{1.915}{\sqrt{7}}} \approx -2.230.$$

Classical approach: Since this is a left-tailed test with 6 degrees of freedom, the critical value is $-t_{0.05} = -1.943$. Since the test statistic $t_0 \approx -2.230$ is less than the critical value $-t_{0.05} = -1.943$ (i.e., since the test statistic falls within the critical region), we reject H_0.

P-value approach: The P-value for this left-tailed test is the area under the t-distribution with 6 degrees of freedom to the left of the test statistic $t_0 = -2.230$, which by symmetry is equal to the area to the right of $t_0 = 2.230$. From the t-distribution table (Table VI) in the row corresponding to 6 degrees of freedom, 2.230 falls between 1.943 and 2.447 whose right-tail areas are 0.05 and 0.025, respectively. So, $0.025 < P\text{-value} < 0.05$. [Tech: P-value = 0.0336.] Because the P-value is less than the level of significance $\alpha = 0.05$, we reject H_0.

Conclusion: There is sufficient evidence at the $\alpha = 0.05$ level of significance to

reject the null hypothesis that $\mu_d = 0$ and conclude that $\mu_d < 0$.

(d) For $\alpha = 0.05$ and df = 6,
$t_{\alpha/2} = t_{0.025} = 2.447$. Then:

Lower bound:
$$\bar{d} - t_{0.025} \cdot \dfrac{s_d}{\sqrt{n}} = -1.614 - 2.447 \cdot \dfrac{1.915}{\sqrt{7}}$$
$$\approx -3.39$$

Upper bound:
$$\bar{d} + t_{0.025} \cdot \dfrac{s_d}{\sqrt{n}} = -1.614 + 2.447 \cdot \dfrac{1.915}{\sqrt{7}}$$
$$\approx 0.16$$

We can be 95% confident that the mean difference is between -3.39 and 0.16.

5. (a) $H_0 : \mu_d = 0$
$H_1 : \mu_d > 0$

(b) The level of significance is $\alpha = 0.05$. The test statistic is $t_0 = \dfrac{\bar{d}}{\dfrac{s_d}{\sqrt{n}}} = \dfrac{1.14}{\dfrac{1.75}{\sqrt{16}}} \approx 2.606$.

Classical approach: Since this is a right-tailed test with 15 degrees of freedom, the critical value is $t_{0.05} = 1.753$. Since the test statistic $t_0 \approx 2.606$ falls beyond the critical value $t_{0.05} = 1.753$ (i.e., since the test statistic falls within the critical region), we reject H_0.

P-value approach: The P-value for this one-tailed test is the area under the t-distribution with 15 degrees of freedom to the right of the test statistic. From the t-distribution table in the row corresponding to 15 degrees of freedom, 2.606 falls between 2.602 and 2.947 whose right-tail areas are 0.01 and 0.005, respectively. So, $0.005 < P\text{-value} < 0.01$. [Tech: P-value = 0.0099.] Because the P-value is less than the level of significance $\alpha = 0.05$, we reject H_0.

Conclusion: There is sufficient evidence at the $\alpha = 0.05$ level of significance to support the claim that $\mu_d > 0$, suggesting that a baby will watch the climber approach the hinderer toy for a longer time than the baby will watch the climber approach the helper toy.

(c) Answers will vary. The fact that the babies watch the surprising behavior for a longer period of time suggests that they are curious about it.

7. (a) These are matched-pairs data because two measurements (A and B) are taken on the same round.

(b) We measure differences as $d_i = A_i - B_i$.

Obs	1	2	3	4	5	6
A	793.8	793.1	792.4	794.0	791.4	792.4
B	793.2	793.3	792.6	793.8	791.6	791.6
d_i	0.6	−0.2	−0.2	0.2	−0.2	0.8

Obs	7	8	9	10	11	12
A	791.4	792.3	789.6	794.4	790.9	793.5
B	791.6	792.4	788.5	794.7	791.3	793.5
d_i	0.1	−0.1	1.1	−0.3	−0.4	0

We compute the mean and standard deviation of the differences and obtain $\bar{d} \approx 0.1167$ feet per second, rounded to four decimal places, and $s_d \approx 0.4745$ feet per second, rounded to four decimal places. The hypotheses are $H_0 : \mu_d = 0$ versus $H_1 : \mu_d \neq 0$. The level of significance is $\alpha = 0.01$. The test statistic is

$$t_0 = \frac{\bar{d}}{\frac{s_d}{\sqrt{n}}} = \frac{0.1167}{\frac{0.4745}{\sqrt{12}}} \approx 0.852.$$

Classical approach: Since this is a two-tailed test with 11 degrees of freedom, the critical values are $\pm t_{0.005} = \pm 3.106$. Since the test statistic $t_0 \approx 0.852$ falls between the critical values $\pm t_{0.005} = \pm 3.106$ (i.e., since the test statistic does not fall within the critical regions), we do not reject H_0.

P-value approach: The P-value for this two-tailed test is the area under the t-distribution with 11 degrees of freedom to the right of the test statistic $t_0 = 0.853$ plus the area to the left of $t_0 = −0.853$. From the t-distribution table, in the row corresponding to 11 degrees of freedom, 0.852 falls between 0.697 and 0.876 whose right-tail areas are 0.25 and 0.20, respectively.

We must double these values in order to get the total area in both tails: 0.50 and 0.40.

So, $0.50 > P\text{-value} > 0.40$. [Tech: P-value = 0.4125.] Because the P-value is greater than the level of significance $\alpha = 0.01$, we do not reject H_0.

Conclusion: There is not sufficient evidence at the $\alpha = 0.01$ level of significance to conclude that there is a difference in the measurements of velocity between device A and device B.

(c) For $\alpha = 0.01$ and df = 11, $t_{\alpha/2} = t_{0.005} = 3.106$. Then:

Lower bound:

$$\bar{d} - t_{0.005} \cdot \frac{s_d}{\sqrt{n}} = 0.1167 - 3.106 \cdot \frac{0.4745}{\sqrt{12}}$$
$$\approx -0.31$$

Upper bound:

$$\bar{d} + t_{0.005} \cdot \frac{s_d}{\sqrt{n}} = 0.1167 + 3.106 \cdot \frac{0.4745}{\sqrt{12}}$$
$$\approx 0.54$$

We are 99% confident that the population mean difference in measurement is between −0.31 and 0.54 feet per second.

(d)

Yes. Since a difference of 0 is located in the middle 50%, the boxplot supports that there is no difference in measurements.

9. (a) These are matched-pairs data because both cars are involved in the same collision.

(b) We measure differences as $d_i = SUV_i - Car_i$.

Obs	1	2	3	4	5	6	7
SUV	1721	1434	850	2329	1415	1470	2884
Car	1274	2327	3223	2058	3095	3386	4560
d_i	447	−893	−2372	271	−1680	−1916	−1676

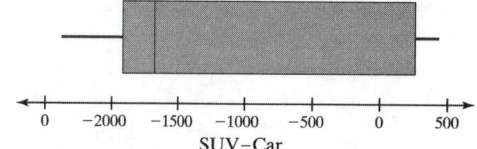

The median of the differences is well to the left of 0, suggesting that SUVs do have a lower repair cost.

(c) We compute the mean and standard deviation of the differences and obtain
$\overline{d} = -\$1117.1$ and $s_d \approx \$1100.6$.
$$H_0 : \mu_d = 0$$
$$H_1 : \mu_d < 0$$
The level of significance is $\alpha = 0.05$. The test statistic is
$$t_0 = \frac{\overline{d}}{\frac{s_d}{\sqrt{n}}} = \frac{-1117.1}{\frac{1100.6}{\sqrt{6}}} \approx -2.685.$$

Classical approach: Since this is a left-tailed test with 6 degrees of freedom, the critical value is $-t_{0.05} = -1.943$. Since the test statistic $t_0 \approx -2.685$ falls beyond the critical value $-t_{0.05} = -1.943$ (i.e., since the test statistic falls within the critical regions), we reject H_0.

P-value approach: The P-value for this two-tailed test is the area under the t-distribution with 6 degrees of freedom to the left of the test statistic $t_0 = -2.685$. From the t-distribution table, in the row corresponding to 6 degrees of freedom, 2.685 falls between 2.612 and 3.143 whose right-tail areas are 0.02 and 0.01, respectively. So, $0.01 > P\text{-value} > 0.02$. [Tech: P-value = 0.0181]. Because the P-value is less than the level of significance $\alpha = 0.05$, we reject H_0.

Conclusion: There is sufficient evidence at the $\alpha = 0.01$ level of significance to support the claim that the repair cost for the SUV is lower, or that the repair cost for the car is higher.

11. We measure differences as $d_i = Y_i - X_i$.

Obs	1	2	3	4	5	6	7
X_i	70.3	67.1	70.9	66.8	72.8	70.4	71.8
Y_i	74.1	69.2	66.9	69.2	68.9	70.2	70.4
d_i	3.8	2.1	−4.0	2.4	−3.9	−0.2	−1.4

Obs	8	9	10	11	12	13
X_i	70.1	69.9	70.8	70.2	70.4	72.4
Y_i	69.3	75.8	72.3	69.2	68.6	73.9
d_i	−0.8	5.9	1.5	−1.0	−1.8	1.5

We compute the mean and standard deviation of the differences and obtain $\overline{d} \approx 0.3154$ inches, rounded to four decimal places, and $s_d \approx 2.8971$ inches, rounded to four decimal places.
$$H_0 : \mu_d = 0$$
$$H_1 : \mu_d > 0$$
The level of significance is $\alpha = 0.1$. The test statistic is $t_0 = \dfrac{\overline{d}}{\frac{s_d}{\sqrt{n}}} = \dfrac{0.3154}{\frac{2.8971}{\sqrt{13}}} \approx 0.393.$

Classical approach: Since this is a right-tailed test with 12 degrees of freedom, the critical value is $t_{0.10} = 1.356$. Since the test statistic $t_0 \approx 0.393$ does not fall to the right of the critical value $t_{0.10} = 1.356$ (i.e., since the test statistic falls outside the critical region), we do not reject H_0.

P-value approach: The P-value for this right-tailed test is the area under the t-distribution with 12 degrees of freedom to the right of the test statistic $t_0 = 0.393$. From the t-distribution table, in the row corresponding to 12 degrees of freedom, 0.393 falls to the left of 0.695, whose right-tail area is 0.25. So, $P\text{-value} > 0.25$ [Tech: P-value = 0.3508]. Because the P-value is greater than the level of significance $\alpha = 0.10$, we do not reject H_0.

Conclusion: No, there is not sufficient evidence at the $\alpha = 0.10$ level of significance to conclude that sons are taller than their fathers.

13. We measure differences as
$d_i = \text{diamond} - \text{steel}$.

Specimen	1	2	3	4	5	6	7	8	9
Steel ball	50	57	61	71	68	54	65	51	53
Diamond	52	56	61	74	69	55	68	51	56
d_i	2	–1	0	3	1	1	3	0	3

We compute the mean and standard deviation of the differences and obtain $\bar{d} \approx 1.3333$, rounded to four decimal places, and $s_d = 1.5$.

$H_0 : \mu_d = 0$

$H_1 : \mu_d \neq 0$

For $\alpha = 0.05$ and df $= 8$, $t_{\alpha/2} = t_{0.025} = 2.306$. Then:

Lower bound:

$$\bar{d} - t_{0.025} \cdot \frac{s_d}{\sqrt{n}} = 1.3333 - 2.306 \cdot \frac{1.5}{\sqrt{9}} \approx 0.2$$

Upper bound:

$$\bar{d} + t_{0.025} \cdot \frac{s_d}{\sqrt{n}} = 1.3333 + 2.306 \cdot \frac{1.5}{\sqrt{9}} \approx 2.5$$

We can be 95% confident that the population mean difference in hardness reading is between 0.2 and 2.5. This interval does not include 0, so we reject H_0.

Conclusion: There is sufficient evidence to conclude that the two indenters produce different hardness readings.

15. (a) It is important to randomly select whether the student would first be tested with normal or impaired vision in order to control for any "learning" that may occur in using the simulator.

(b) We measure differences as $d_i = Y_i - X_i$.

Subject	1	2	3	4	5
Normal, X_i	4.47	4.24	4.58	4.65	4.31
Impaired, Y_i	5.77	5.67	5.51	5.32	5.83
d_i	1.30	1.43	0.93	0.67	1.52

Subject	6	7	8	9
Normal, X_i	4.80	4.55	5.00	4.79
Impaired, Y_i	5.49	5.23	5.61	5.63
d_i	0.69	0.68	0.61	0.84

We compute the mean and standard deviation of the differences and obtain

$\bar{d} \approx 0.9633$ seconds, rounded to four decimal places, and $s_d \approx 0.3576$ seconds,

rounded to four decimal places. For $\alpha = 0.05$ and df $= 8$, $t_{\alpha/2} = t_{0.025} = 2.306$.

Lower bound:

$$\bar{d} - t_{0.025} \cdot \frac{s_d}{\sqrt{n}} = 0.9633 - 2.306 \cdot \frac{0.3576}{\sqrt{9}}$$
$$\approx 0.688$$

Upper bound:

$$\bar{d} + t_{0.025} \cdot \frac{s_d}{\sqrt{n}} = 0.9633 + 2.306 \cdot \frac{0.3576}{\sqrt{9}}$$
$$\approx 1.238$$

We can be 95% confident that the population mean difference in reaction time when teenagers are driving impaired from when driving normally is between 0.688 and 1.238 seconds. This interval does not include 0, so we reject H_0.

Conclusion: There is sufficient evidence to conclude that there is a difference in braking time with impaired vision and normal vision.

17. (a) Matching by driver and car for the two different octane levels is important as a means of controlling the experiment. Drivers and cars behave differently, so this matching reduces variability in miles per gallon that is attributable to the driver's driving style.

(b) Conducting the experiment on a closed track allows for control so that all cars and drivers can be put through the exact same driving conditions.

(c) No, neither variable is normally distributed. Both have at least one point outside the bounds of the normal probability plot.

(d) Yes, the differences in mileages appear to be approximately normally distributed since all of the points fall within the boundaries of the probability plot.

(e) From the MINITAB printout, $\bar{d} \approx 5.091$ miles and $s_d \approx 14.876$ miles, where the differences are 92 octane minus 87 octane.

$H_0 : \mu_d = 0$

$H_1 : \mu_d > 0$, where $d_i = 92 \text{ Oct} - 87 \text{ Oct}$. The level of significance is $\alpha = 0.05$. The test statistic is $t_0 \approx 1.14$.

Classical approach: Since this is a right-tailed test with 10 degrees of freedom, the critical value is $t_{0.05} = 1.812$. Since the test statistic $t_0 \approx 1.14$ does not fall to the right of the critical value $t_{0.05} = 1.812$ (i.e., since the test statistic does not fall within the critical region), we fail to reject H_0.

P-value approach: From the MINITAB printout, we find that P-value ≈ 0.141. Because the P-value is greater than the level of significance $\alpha = 0.05$, we do not reject H_0.

Conclusion: We would expect to get the results we obtained in about 14 out of 100 samples if the statement in the null hypothesis were true. Our result is not unusual. Thus, there is not sufficient evidence at the $\alpha = 0.05$ level of significance to conclude that cars get better mileage when using 92-octane gasoline than when using 87-octane gasoline.

Section 11.3

1. (a) $H_0 : \mu_1 = \mu_2$
$H_1 : \mu_1 \neq \mu_2$
The level of significance is $\alpha = 0.05$. Since the sample size of both groups is 15, we use $n_1 - 1 = 14$ degrees of freedom. The test statistic is

$$t_0 = \frac{(\bar{x}_1 - \bar{x}_2) - (\mu_1 - \mu_2)}{\sqrt{\frac{s_1^2}{n_1} + \frac{s_2^2}{n_2}}} = \frac{(15.3 - 14.2) - 0}{\sqrt{\frac{3.2^2}{15} + \frac{3.5^2}{15}}}$$

$$\approx 0.898$$

Classical approach: Since this is a two-tailed test with 14 degrees of freedom, the critical values are $\pm t_{0.025} = \pm 2.145$. Since the test statistic $t_0 \approx 0.898$ is between the critical values $-t_{0.025} = -2.145$ and $t_{0.025} = 2.145$ (i.e., since the test statistic does not fall within the critical region), we do not reject H_0.

P-value approach: The P-value for this two-tailed test is the area under the t-distribution with 14 degrees of freedom to the right of $t_0 = 0.898$ plus the area to the left of -0.898. From the t-distribution

table in the row corresponding to 14 degrees of freedom, 0.898 falls between 0.868 and 1.076 whose right-tail areas are 0.20 and 0.15, respectively. We must double these values in order to get the total area in both tails: 0.40 and 0.30. So, $0.30 < P\text{-value} < 0.40$ [Tech: P-value = 0.3767]. Because the P-value is greater than the level of significance $\alpha = 0.05$, we do not reject H_0.

Conclusion: There is not sufficient evidence at the $\alpha = 0.05$ level of significance to conclude that the population means are different.

(b) For a 95% confidence interval with df = 14, we use $t_{\alpha/2} = t_{0.025} = 2.145$. Then:

Lower bound:
$$(\bar{x}_1 - \bar{x}_2) - t_{\alpha/2} \cdot \sqrt{\frac{s_1^2}{n_1} + \frac{s_2^2}{n_2}}$$
$$= (15.3 - 14.2) - 2.145 \cdot \sqrt{\frac{3.2^2}{15} + \frac{3.5^2}{15}}$$
$$= -1.53 \text{ [Tech: } -1.41]$$

Upper bound:
$$(\bar{x}_1 - \bar{x}_2) + t_{\alpha/2} \cdot \sqrt{\frac{s_1^2}{n_1} + \frac{s_2^2}{n_2}}$$
$$= (15.3 - 14.2) + 2.145 \cdot \sqrt{\frac{3.2^2}{15} + \frac{3.5^2}{15}}$$
$$\approx 3.73 \text{ [Tech: } 3.61]$$

We can be 95% confident that the mean difference is between -1.53 and 3.73 [Tech: between -1.41 and 3.61].

3. (a) $H_0 : \mu_1 = \mu_2$
$H_1 : \mu_1 > \mu_2$
The level of significance is $\alpha = 0.10$. Since the smaller sample size is $n_2 = 18$, we use $n_2 - 1 = 18 - 1 = 17$ degrees of freedom. The test statistic is

$$t_0 = \frac{(\bar{x}_1 - \bar{x}_2) - (\mu_1 - \mu_2)}{\sqrt{\frac{s_1^2}{n_1} + \frac{s_2^2}{n_2}}} = \frac{(50.2 - 42.0) - 0}{\sqrt{\frac{6.4^2}{25} + \frac{9.9^2}{18}}}$$

$$\approx 3.081$$

Classical approach: Since this is a right-tailed test with 17 degrees of freedom, the critical value is $t_{0.10} = 1.333$.

Since the test statistic $t_0 \approx 3.081$ is to the right of the critical value (i.e., since the test statistic falls within the critical region), we reject H_0.

P-value approach: The P-value for this right-tailed test is the area under the t-distribution with 17 degrees of freedom to the right of $t_0 = 3.081$. From the t-distribution table, in the row corresponding to 17 degrees of freedom, 3.081 falls between 2.898 and 3.222 whose right-tail areas are 0.005 and 0.0025, respectively. Thus, $0.0025 < P\text{-value} < 0.005$ [Tech: P-value = 0.0024]. Because the P-value is less than the level of significance $\alpha = 0.10$, we reject H_0.

Conclusion: There is sufficient evidence at the $\alpha = 0.01$ level of significance to conclude that $\mu_1 > \mu_2$.

(b) For a 90% confidence interval with df = 17, we use $t_{\alpha/2} = t_{0.05} = 1.740$. Then:

Lower bound:

$$(\overline{x}_1 - \overline{x}_2) - t_{\alpha/2} \cdot \sqrt{\frac{s_1^2}{n_1} + \frac{s_2^2}{n_2}}$$

$$= (50.2 - 42.0) - 1.740 \cdot \sqrt{\frac{6.4^2}{25} + \frac{9.9^2}{18}}$$

$$\approx 3.57 \;\; [\text{Tech: } 3.67]$$

Upper bound:

$$(\overline{x}_1 - \overline{x}_2) + t_{\alpha/2} \cdot \sqrt{\frac{s_1^2}{n_1} + \frac{s_2^2}{n_2}}$$

$$= (50.2 - 42.0) + 1.740 \cdot \sqrt{\frac{6.4^2}{25} + \frac{9.9^2}{18}}$$

$$\approx 12.83 \;\; [\text{Tech: } 12.73]$$

We can be 90% confident that the mean difference is between 3.57 and 12.83 [Tech: between 3.67 and 12.73].

5. $H_0 : \mu_1 = \mu_2$
$H_1 : \mu_1 < \mu_2$
The level of significance is $\alpha = 0.02$. Since the smaller sample size is $n_2 = 25$, we use $n_2 - 1 = 25 - 1 = 24$ degrees of freedom. The

test statistic is

$$t_0 = \frac{(\overline{x}_1 - \overline{x}_2) - (\mu_1 - \mu_2)}{\sqrt{\dfrac{s_1^2}{n_1} + \dfrac{s_2^2}{n_2}}} = \frac{(103.4 - 114.2) - 0}{\sqrt{\dfrac{12.3^2}{32} + \dfrac{13.2^2}{25}}}$$

$$\approx -3.158$$

Classical approach: Since this is a left-tailed test with 24 degrees of freedom, the critical value is $-t_{0.02} = -2.172$. Since the test statistic $t_0 \approx -3.158$ is to the left of the critical value (i.e., since the test statistic falls within the critical region), we reject H_0.

P-value approach: The P-value for this left-tailed test is the area under the t-distribution with 24 degrees of freedom to the left of $t_0 \approx -3.158$, which is equivalent to the area to the right of $t_0 \approx 3.158$. From the t-distribution table in the row corresponding to 24 degrees of freedom, 3.158 falls between 3.091 and 3.467 whose right-tail areas are 0.0025 and 0.001, respectively. Thus, $0.001 < P\text{-value} < 0.0025$ [Tech: P-value = 0.0013]. Because the P-value is less than the level of significance $\alpha = 0.02$, we reject H_0.

Conclusion: There is sufficient evidence at the $\alpha = 0.02$ level of significance to conclude that $\mu_1 < \mu_2$.

7. (a) The response variable is time to graduate. The explanatory variable is whether the student went to community college or not.

(b) There are two independent groups: those who enroll directly in four-year institutions, and those who first enroll in community college and then transfer. The response variable, time to graduate, is quantitative. Treat each sample as a simple random sample. Each sample size is large, so each sample mean is approximately normal. Each population is small relative to its population size.

(c) $H_0 : \mu_{\text{CC}} = \mu_{\text{NT}}$
$H_1 : \mu_{\text{CC}} > \mu_{\text{NT}}$
The level of significance is $\alpha = 0.01$. Since the smaller sample size is $n_2 = 268$, we use $n_2 - 1 = 268 - 1 = 267$ degrees of

freedom. The test statistic is

$$t_0 = \frac{(\bar{x}_1 - \bar{x}_2) - (\mu_1 - \mu_2)}{\sqrt{\frac{s_1^2}{n_1} + \frac{s_2^2}{n_2}}} = \frac{(5.43 - 4.43) - 0}{\sqrt{\frac{1.162^2}{268} + \frac{1.015^2}{1145}}}$$

$$\approx 12.977$$

Classical approach: This is a right-tailed test with 267 degrees of freedom. However, since our t-distribution table does not contain a row for 267, we use df $= 100$ Thus, the critical value is $t_{0.01} = 2.364$ [Tech: $t_{0.01} = 2.340$]. Since the test statistic $t_0 \approx 12.977$ is to the right of the critical value (i.e., since the test statistic falls within the critical region), we reject H_0.

P-value approach: The P-value for this right-tailed test is the area under the t-distribution with 267 degrees of freedom to the right of $t_0 = 12.977$. From the t-distribution table, in the row corresponding to 100 degrees of freedom (since the table does not contain a row for df = 267), 12.977 falls to the right of 3.390 whose right-tail area is 0.0005.

Thus, P-value < 0.0005 [Tech: P-value < 0.0001]. Because the P-value is less than the level of significance $\alpha = 0.01$, we reject H_0.

Conclusion: There is sufficient evidence at the $\alpha = 0.01$ level of significance to conclude that $\mu_{CC} > \mu_{NT}$. That is, the evidence suggests that the mean time to graduate for students who first start in community college is longer than the mean time to graduate for those who do not transfer.

(d) For a 95% confidence interval with 100 degrees of freedom (since the table does not contain a row for df = 267), we use $t_{\alpha/2} = t_{0.025} = 1.984$ [Tech: 1.969]. Then:

Lower bound:

$$(\bar{x}_1 - \bar{x}_2) - t_{\alpha/2} \cdot \sqrt{\frac{s_1^2}{n_1} + \frac{s_2^2}{n_2}}$$

$$= (5.43 - 4.43) - 1.984 \cdot \sqrt{\frac{1.162^2}{268} + \frac{1.015^2}{1145}}$$

$$\approx 0.847 \text{ [Tech: 0.848]}$$

Upper bound:

$$(\bar{x}_1 - \bar{x}_2) + t_{\alpha/2} \cdot \sqrt{\frac{s_1^2}{n_1} + \frac{s_2^2}{n_2}}$$

$$= (5.43 - 4.43) + 1.984 \cdot \sqrt{\frac{1.162^2}{268} + \frac{1.015^2}{1145}}$$

$$\approx 1.153 \text{ [Tech: 1.152]}$$

We can be 95% confident that the mean additional time to graduate for students who start in community college is between 0.847 and 1.153 years [Tech: between 0.848 and 1.152 years].

(e) No, the results from parts (c) and (d) do not imply that community college *causes* one to take extra time to earn a bachelor's degree. This is observational data. Community college students may be working more hours, which does not allow them to take additional classes.

9. (a) This is an observational study, since no treatment is imposed. The researcher did not influence the data.

(b) Though we do not know if the population is normally distributed, the samples are independent with sizes that are sufficiently large ($n_{arrival} = n_{departure} = 35$).

(c) $H_0 : \mu_{arrival} = \mu_{departure}$

$H_1 : \mu_{arrival} \neq \mu_{departure}$

The level of significance is $\alpha = 0.05$. Since the sample size of both groups is 35, we use $n_{arrival} - 1 = 34$ degrees of freedom. The test statistic is

$$t_0 = \frac{(\bar{x}_{arrival} - \bar{x}_{departure}) - (\mu_{arrival} - \mu_{departure})}{\sqrt{\frac{s_{arrival}^2}{n_{arrival}} + \frac{s_{departure}^2}{n_{departure}}}}$$

$$= \frac{(269 - 260) - 0}{\sqrt{\frac{53^2}{35} + \frac{34^2}{35}}} \approx 0.846$$

Classical approach: Since this is a two-tailed test with 34 degrees of freedom, the critical values are $\pm t_{0.025} = \pm 2.032$. Since the test statistic $t_0 \approx 0.846$ is between the critical values (i.e., since the test statistic does not fall within the critical regions), we do not reject H_0.

P-value approach: The *P*-value for this two-tailed test is the area under the *t*-distribution with 34 degrees of freedom to the right of $t_0 \approx 0.846$ plus the area to the left of $t_0 \approx -0.846$. From the *t*-distribution table in the row corresponding to 34 degrees of freedom, 0.846 falls between 0.682 and 0.852 whose right-tail areas are 0.25 and 0.20, respectively. We must double these values in order to get the total area in both tails: 0.50 and 0.40. So, $0.40 < P\text{-value} < 0.50$ [Tech: $P\text{-value} \approx 0.4013$]. Since the *P*-value is greater than the level of significance $\alpha = 0.05$, we do not reject H_0.

Conclusion: There is not sufficient evidence at the $\alpha = 0.05$ level of significance to conclude that travelers walk at different speeds depending on whether they are arriving or departing an airport.

11. (a) Both samples include 200 observations, so there are 199 degrees of freedom. For a 95% confidence interval with df $= 199$, we use row 100 of Table VI to get $t_{\alpha/2} = t_{0.025} = 1.984$. Then:

Lower bound:

$$(\bar{x}_1 - \bar{x}_2) - t_{\alpha/2} \cdot \sqrt{\frac{s_1^2}{n_1} + \frac{s_2^2}{n_2}}$$

$$= (23.4 - 17.9) - 1.984 \cdot \sqrt{\frac{4.1^2}{200} + \frac{3.9^2}{200}}$$

$$\approx 4.71$$

Upper bound:

$$(\bar{x}_1 - \bar{x}_2) + t_{\alpha/2} \cdot \sqrt{\frac{s_1^2}{n_1} + \frac{s_2^2}{n_2}}$$

$$= (23.4 - 17.9) + 1.984 \cdot \sqrt{\frac{4.1^2}{200} + \frac{3.9^2}{200}}$$

$$\approx 6.29$$

We can be 95% confident that the mean difference in the scores is between students who think about being a professor and students who think about soccer hooligans is between 4.71 and 6.29.

(b) Since the 95% confidence interval does not contain 0, the results suggest that priming does have an effect on scores.

13. (a) The five number summaries follow:

Meters Off:
17, 26, 37, 41, 52

Meters On:
25, 31, 42, 48, 56

Based on the boxplots, there does appear to be a difference in the speeds.

(b) $H_0 : \mu_{\text{On}} = \mu_{\text{Off}}$

$H_1 : \mu_{\text{On}} > \mu_{\text{Off}}$

The level of significance is $\alpha = 0.10$. Both samples involve 15 observations, so we use $n_1 - 1 = 15 - 1 = 14$ degrees of freedom. The test statistic is

$$t_0 = \frac{(\bar{x}_{\text{On}} - \bar{x}_{\text{Off}}) - (\mu_{\text{On}} - \mu_{\text{Off}})}{\sqrt{\dfrac{s_{\text{On}}^2}{n_{\text{On}}} + \dfrac{s_{\text{Off}}^2}{n_{\text{Off}}}}}$$

$$= \frac{(40.67 - 34.53) - 0}{\sqrt{\dfrac{10.04^2}{15} + \dfrac{9.56^2}{15}}}$$

$$\approx 1.713$$

Classical approach: This is a right-tailed test with 14 degrees of freedom. Thus, the critical value is $t_{0.10} = 1.345$. Since the test statistic $t_0 \approx 1.713$ is to the right of the critical value (i.e., since the test statistic falls within the critical region), we reject H_0.

P-value approach: The *P*-value for this right-tailed test is the area under the *t*-distribution with 14 degrees of freedom to the right of $t_0 = 1.713$. From the *t*-distribution table in the row corresponding to 14 degrees of freedom, 1.713 falls between 1.345 and 1.761 whose right-tail areas are 0.10 and 0.05, respectively.

Thus, $0.05 < P\text{-value} < 0.10$ [Tech: $P\text{-value} \approx 0.0489$]. Because the P-value is less than the level of significance $\alpha = 0.10$, we reject H_0.

Conclusion: There is sufficient evidence at the $\alpha = 0.10$ level of significance to conclude that the ramp meters are effective in maintaining higher speed on the freeway.

15. $H_0 : \mu_{\text{carpeted}} = \mu_{\text{uncarpeted}}$

$H_1 : \mu_{\text{carpeted}} > \mu_{\text{uncarpeted}}$

The level of significance is $\alpha = 0.05$. The sample statistics for the data are

$\overline{x}_{\text{carpeted}} = 11.2$, $s_{\text{carpeted}} \approx 2.6774$, $n_{\text{carpeted}} = 8$,

$\overline{x}_{\text{uncarpeted}} = 9.7875$, $s_{\text{uncarpeted}} \approx 3.2100$, and

$n_{\text{uncarpeted}} = 8$. Since the sample size of both groups is 8, we use $n_{\text{carpeted}} - 1 = 7$ degrees of freedom.

The test statistic is

$$t_0 = \frac{(\overline{x}_{\text{carpeted}} - \overline{x}_{\text{uncarpeted}}) - (\mu_{\text{carpeted}} - \mu_{\text{uncarpeted}})}{\sqrt{\dfrac{s_{\text{carpeted}}^2}{n_{\text{carpeted}}} + \dfrac{s_{\text{uncarpeted}}^2}{n_{\text{uncarpeted}}}}}$$

$$= \frac{(11.2 - 9.7875) - 0}{\sqrt{\dfrac{2.6774^2}{8} + \dfrac{3.2100^2}{8}}} \approx 0.956$$

Classical approach: Since this is a right-tailed test with 7 degrees of freedom, the critical value is $t_{0.05} = 1.895$. Since the test statistic $t_0 \approx 0.954$ is not to the right of the critical value (i.e., since the test statistic does not fall within the critical region), we do not reject H_0.

P-value approach: The P-value for this right-tailed test is the area under the t-distribution with 7 degrees of freedom to the right of $t_0 \approx 0.956$. From the t-distribution table in the row corresponding to 7 degrees of freedom, 0.956 falls between 0.896 and 1.119 whose right-tail areas are 0.20 and 0.15, respectively. So, $0.15 < P\text{-value} < 0.20$ [Tech: $P\text{-value} \approx 0.1780$]. Since the P-value is greater than the level of significance $\alpha = 0.05$, we do not reject H_0.

Conclusion: There is not sufficient evidence at the $\alpha = 0.05$ level of significance to

conclude that carpeted rooms have more bacteria than uncarpeted rooms.

17. (a)

(b) (1) Treat each sample as a simple random sample. (2) Each sample is obtained independently of the other. (3) Each sample size is large ($n_{\text{cc}} = 96$ and $n_{\text{I}} = 98$). (4) Each sample size is small relative to the size of the population. (5) The response variable is quantitative and there are two groups to compare.

(c) $H_0 : \mu_{\text{cc}} = \mu_{\text{I}}$

$H_1 : \mu_{\text{cc}} \neq \mu_{\text{I}}$

The level of significance is $\alpha = 0.05$. The sample statistics for the data are

$\overline{x}_{\text{cc}} = 6.595$, $s_{\text{cc}} = 19.078$, and

$\overline{x}_{\text{I}} = 14.425$, $s_{\text{I}} = 23.851$. Since the smaller sample size is 96, we use $n_{\text{cc}} - 1 = 95$ degrees of freedom. The test statistic is

$$t_0 = \frac{(\overline{x}_{\text{cc}} - \overline{x}_{\text{I}}) - (\mu_{\text{cc}} - \mu_{\text{I}})}{\sqrt{\dfrac{s_{\text{cc}}^2}{n_{\text{cc}}} + \dfrac{s_{\text{I}}^2}{n_{\text{I}}}}}$$

$$= \frac{(6.595 - 14.425) - 0}{\sqrt{\dfrac{19.078^2}{96} + \dfrac{23.851^2}{98}}} \approx -2.528$$

Classical approach: We are not provided with 95 degrees of freedom in the Student t-distribution table, so use 100 degrees of freedom. Since this is a two-tailed test with 100 degrees of freedom, the critical values are $\pm t_{0.025} = \pm 1.984$. Since the test statistic $t_0 \approx -2.528$ falls to the left of the critical value -1.984 (i.e., since the test statistic falls within the critical region), we reject H_0.

P-value approach: The *P*-value for this two-tailed test is the area under the *t*-distribution with 95 degrees of freedom to the left of $t_0 \approx -2.528$ plus the area to the right of $-t_0 \approx 2.528$. From the *t*-distribution table in the row corresponding to 100 degrees of freedom, 2.528 falls between 2.364 and 2.626 whose right-tail areas are 0.01 and 0.005, respectively. We must double these values in order to get the total area in both tails: 0.02 and 0.01. So, $0.01 < P\text{-value} < 0.02$ [Tech: *P*-value = 0.0123]. Since the *P*-value is less than the level of significance $\alpha = 0.05$, we reject H_0.

Conclusion: There is sufficient evidence at the $\alpha = 0.05$ level of significance to conclude that the mean rate of return on consumer cyclical stocks differs from the mean rate of return on industrial stocks.

(d) We are not provided with 95 degrees of freedom in the Student *t*-distribution table, so use 100 degrees of freedom. So, $t_{0.025} = 1.984$.

Lower bound:

$$\left(\overline{x}_I - \overline{x}_{cc}\right) - t_{\alpha/2} \cdot \sqrt{\frac{s_I^2}{n_I} + \frac{s_{cc}^2}{n_{cc}}}$$

$$= (14.425 - 6.595) - 1.984\sqrt{\frac{23.851^2}{98} + \frac{19.078^2}{96}}$$

$$= 1.684 \text{ [Tech: } 1.719]$$

Upper bound:

$$\left(\overline{x}_I - \overline{x}_{cc}\right) + t_{\alpha/2} \cdot \sqrt{\frac{s_I^2}{n_I} + \frac{s_{cc}^2}{n_{cc}}}$$

$$= (14.425 - 6.595) + 1.984\sqrt{\frac{23.851^2}{98} + \frac{19.078^2}{96}}$$

$$= 13.976 \text{ [Tech: } 13.942]$$

We are 95% confident that the mean difference in rate of return of industrial stocks versus consumer cyclical stocks is between 1.684 and 13.976. This suggests that the one-year rate of return on industrial stocks was higher than consumer cyclical stocks by somewhere

between 1.684% and 13.976% for this time period.

19. For this 90% confidence level with df $= 40 - 1$ $= 39$, we use $t_{\alpha/2} = t_{0.05} = 1.685$. Then:

Lower bound:

$$(\overline{x}_{\text{no children}} - \overline{x}_{\text{children}}) - t_{\alpha/2} \cdot \sqrt{\frac{s_{\text{no children}}^2}{n_{\text{no children}}} + \frac{s_{\text{children}}^2}{n_{\text{children}}}}$$

$$= (5.62 - 4.10) - 1.685 \cdot \sqrt{\frac{2.43^2}{40} + \frac{1.82^2}{40}}$$

$$\approx 0.71 \text{ [Tech: } 0.72]$$

Upper bound:

$$(\overline{x}_{\text{no children}} - \overline{x}_{\text{children}}) + t_{\alpha/2} \cdot \sqrt{\frac{s_{\text{no children}}^2}{n_{\text{no children}}} + \frac{s_{\text{children}}^2}{n_{\text{children}}}}$$

$$= (5.62 - 4.10) + 1.685 \cdot \sqrt{\frac{2.43^2}{40} + \frac{1.82^2}{40}}$$

$$\approx 2.33 \text{ [Tech: } 2.32]$$

We can be 90% confident that the mean difference in daily leisure time between adults without children and those with children is between 0.71 and 2.33 hours [Tech: between 0.72 and 2.33 hours]. Since the confidence interval does not include zero, we can conclude that there is a significant difference in the leisure time of adults without children and those with children.

21. (a) $H_0 : \mu_{\text{men}} = \mu_{\text{women}}$

$H_1 : \mu_{\text{men}} < \mu_{\text{women}}$

(b) From the MINITAB printout, *P*-value = 0.0051. Because the *P*-value is less than the level of significance $\alpha = 0.01$, the researcher will reject H_0. Thus, there is sufficient evidence at the $\alpha = 0.01$ level of significance to conclude that the mean step pulse of men is lower than the mean step pulse of women.

(c) From the MINITAB printout, the 95% confidence interval is:
Lower bound: -10.7; Upper bound: -1.5.
We are 95% confident that the mean step pulse of men is between 1.5 and 10.7 beats per minute lower than the mean step pulse of women.

23. (a) This is an experiment using a completely randomized design.

(b) The response variable is scores on the final exam. The treatments are online homework system versus old-fashioned paper-and-pencil homework.

(c) Factors that are controlled in the experiment are the teacher, the text, the syllabus, the tests, the meeting time, the meeting location.

(d) In this case, the assumption is that the students "randomly" enrolled in the course.

(e) $H_0: \mu_{fall} = \mu_{spring}$

$H_1: \mu_{fall} > \mu_{spring}$

The level of significance is $\alpha = 0.05$. We are given the summary statistics $\bar{x}_{fall} = 73.6$, $s_{fall} = 10.3$, and $n_{fall} = 27$, and $\bar{x}_{spring} = 67.9$, $s_{spring} = 12.4$, and $n_{spring} = 25$. Since the smaller sample size is $n_{spring} = 25$, we use $n_{spring} - 1 = 24$ degrees of freedom.

Test statistic:

$$t_0 = \frac{(\bar{x}_{fall} - \bar{x}_{spring}) - (\mu_{fall} - \mu_{spring})}{\sqrt{\dfrac{s_{fall}^2}{n_{fall}} + \dfrac{s_{spring}^2}{n_{spring}}}}$$

$$= \frac{(73.6 - 67.9) - 0}{\sqrt{\dfrac{10.3^2}{27} + \dfrac{12.4^2}{25}}} \approx 1.795$$

Classical approach: Since this is a right-tailed test with 24 degrees of freedom, the critical values are $t_{0.05} = 1.711$. Since the test statistic $t_0 \approx 1.795$ falls to the right of the critical value 1.711 (i.e., since the test statistic falls within the critical region), we reject H_0.

P-value approach: The P-value for this right-tailed test is the area under the t-distribution with 24 degrees of freedom to the right of $t_0 \approx 1.795$.

From the t-distribution table, in the row corresponding to 24 degrees of freedom, 1.795 falls between 1.711 and 2.064 whose right-tail areas are 0.05 and 0.025, respectively. So, $0.025 < P\text{-value} < 0.05$ [Tech: P-value = 0.0395]. Since the

P-value is less than the level of significance $\alpha = 0.05$, we reject H_0.

Conclusion: There is sufficient evidence at the $\alpha = 0.05$ level of significance to conclude that final exam scores in the fall semester were higher than final exam scores in the spring semester. It would appear to be the case that the online homework system helped in raising final exam scores.

(f) Answers will vary. One possibility follows: One factor that may confound the results is that the weather is pretty lousy at the end of the fall semester, but pretty nice at the end of the spring semester. If "spring fever" kicked in for the spring semester students, then they probably studied less for the final exam.

25. The sampling method is independent (the freshman cannot be matched to the corresponding seniors.) Therefore, the inferential method that may be applied is a two-sample t-test. This comparison, however, has major shortcomings. The goal of the CLA+ is to measure gains in critical thinking, analytical reasoning, and so on, as a result of four years of college. The logical design is to measure this as a matched-pairs design where the exam is administered before and after college to the same student.

27. The degrees of freedom obtained from Formula (2) are larger than the smaller of $n_1 - 1$ or $n_2 - 1$ and t_α increases as the degrees of freedom increase. The larger the critical value is, the harder it is to reject the null hypothesis.

Section 11.4

1. $F_{0.05, 9, 10} = 3.02$

3. $\alpha = 0.05$, so $\dfrac{\alpha}{2} = 0.025$. Thus, $F_{0.025, 6, 8} = 4.65$

and $F_{0.975, 6, 8} = \dfrac{1}{F_{0.025, 8, 6}} = \dfrac{1}{5.60} \approx 0.18$.

5. $F_{0.90, 25, 20} = \dfrac{1}{F_{0.10, 20, 25}} = \dfrac{1}{1.72} \approx 0.58$

7. $F_{0.05,45,15} = \dfrac{F_{0.05,40,15} + F_{0.05,50,15}}{2}$

$= \dfrac{2.20 + 2.18}{2}$

$= 2.19$

9. $H_0 : \sigma_1 = \sigma_2$
$H_1 : \sigma_1 \neq \sigma_2$

The test statistic is $F_0 = \dfrac{s_1^2}{s_2^2} = \dfrac{3.2^2}{3.5^2} \approx 0.84$.

Classical approach: This is a two-tailed test with $n_1 = n_2 = 16$ and $\alpha = 0.05$. So, $\dfrac{\alpha}{2} = 0.025$, and the critical values are $F_{0.025,15,15} = 2.86$ and

$F_{0.975,15,15} = \dfrac{1}{F_{0.025,15,15}} = \dfrac{1}{2.86} \approx 0.35$. Since

$F_0 = 0.84$ is between $F_{0.975,15,15} = 0.35$ and $F_{0.025,15,15} = 2.86$ (the test statistic does not fall in the critical regions), we do not reject H_0.

P-value approach: Using technology, we find P-value ≈ 0.7330. Since this P-value is greater than the $\alpha = 0.05$ level of significance, we do not reject H_0.

Conclusion: There is not sufficient evidence at the $\alpha = 0.05$ level of significance to conclude that $\sigma_1 \neq \sigma_2$.

11. $H_0 : \sigma_1 = \sigma_2$
$H_1 : \sigma_1 > \sigma_2$

The test statistic is $F_0 = \dfrac{s_1^2}{s_2^2} = \dfrac{9.9^2}{6.4^2} \approx 2.39$.

Classical approach: This is a right-tailed test with $n_1 = 26$, $n_2 = 19$, and $\alpha = 0.01$. So, the critical value is $F_{0.01,25,18}$, which we approximate by $F_{0.01,25,20} = 2.84$. Since

$F_0 = 2.39 < F_{0.01,25,20} = 2.84$ the test statistic does not fall in the critical region, we do not reject H_0.

P-value approach: Using technology, we find P-value ≈ 0.0303. Since this P-value is greater than the $\alpha = 0.01$ level of significance, we do not reject H_0.

Conclusion: There is not sufficient evidence at the $\alpha = 0.01$ level of significance to conclude that $\sigma_1 > \sigma_2$.

13. $H_0 : \sigma_1 = \sigma_2$
$H_1 : \sigma_1 < \sigma_2$

The test statistic is $F_0 = \dfrac{s_1^2}{s_2^2} = \dfrac{8.3^2}{13.2^2} \approx 0.40$.

Classical approach: This is a left-tailed test with $n_1 = 51$, $n_2 = 26$, and $\alpha = 0.10$. So, the critical value is $F_{0.90,50,25} = \dfrac{1}{F_{0.10,25,50}} = \dfrac{1}{1.53} \approx 0.65$.

Since $F_0 = 0.40 < F_{0.90,50,25} = 0.65$ (the test statistic falls in a critical region), we reject H_0.

P-value approach: Using technology, we find P-value ≈ 0.0226. Since this P-value is less than the $\alpha = 0.10$ level of significance, we reject H_0.

Conclusion: There is sufficient evidence at the $\alpha = 0.1$ level of significance to conclude that $\sigma_1 < \sigma_2$.

15. $H_0 : \sigma_1 = \sigma_2$
$H_1 : \sigma_1 \neq \sigma_2$

The test statistic is $F_0 = \dfrac{s_1^2}{s_2^2} = \dfrac{1.162^2}{1.015^2} \approx 1.311$.

Classical approach: This is a two-tailed test with $n_1 = 268$, $n_2 = 1145$, and $\alpha = 0.05$. The critical values are $F_{0.025,267,1144}$ and $F_{0.975,267,1144}$, which we approximate by $F_{0.025,267,1144} \approx F_{0.025,120,1000} = 1.29$ and

$F_{0.975,267,1144} = \dfrac{1}{F_{0.025,1144,267}} \approx \dfrac{1}{F_{0.025,1000,200}} = \dfrac{1}{1.25} \approx 0.80$. Since $F_0 = 1.311$ does not fall

between $F_{0.975,267,1144} \approx 0.80$ and $F_{0.025,267,1144} \approx 1.29$ (the test statistic falls in the critical region), we reject H_0.

P-value approach: Using technology, we find P-value ≈ 0.0036. Since this P-value is less than the $\alpha = 0.05$ level of significance, we reject H_0.

Conclusion: There is sufficient evidence at the $\alpha = 0.05$ level of significance to conclude that the standard deviation in time to earn a bachelor's degree is different for students who attend a community college than for students who immediately attend a 4-year institution.

17. $H_0 : \sigma_1 = \sigma_2$

$H_1 : \sigma_1 > \sigma_2$

The test statistic is $F_0 = \dfrac{s_1^2}{s_2^2} = \dfrac{17.07^2}{6.48^2} \approx 6.94$.

Classical approach: This is a right-tailed test with $n_1 = n_2 = 65$ and $\alpha = 0.01$. So, the critical value is $F_{0.01,64,64}$, which we approximate by $F_{0.01,60,50} = 1.91$. Since $F_0 \approx 6.94 > F_{0.01,60,50} = 1.91$ (the test statistic falls in the critical region), we reject H_0.

P-value approach: Using technology, we find P-value P-style < 0.0001. Since this P-value is less than the $\alpha = 0.01$ level of significance, we reject H_0.

Conclusion: There is sufficient evidence at the $\alpha = 0.01$ level of significance to conclude that the treatment group had a higher standard deviation for serum retinal concentration than did the control group.

19. **(a)** The hypotheses are $H_0 : \sigma_1 = \sigma_2$ versus $H_1 : \sigma_1 < \sigma_2$, where $\sigma_1 =$ the standard deviation of waiting time for a single line. We compute the sample standard deviations to be $s_1 \approx 0.574$ minutes and $s_2 \approx 1.012$ minutes. The test statistic is

$$F_0 = \frac{s_1^2}{s_2^2} = \frac{0.574^2}{1.012^2} \approx 0.32.$$

Classical approach: This is a left-tailed test with $n_1 = n_2 = 20$ and $\alpha = 0.05$. The critical value is $F_{0.95,19,19}$. We approximate this by $F_{0.95,20,20} = \dfrac{1}{F_{0.05,20,20}} = \dfrac{1}{2.12} \approx 0.47$.

Since $F_0 = 0.32 < F_{0.95,20,20} \approx 0.47$ (the test statistic falls in the critical region), we reject H_0.

P-value approach: Using technology, we find P-value ≈ 0.0087 . Since this P-value is less than the $\alpha = 0.05$ level of significance, we reject H_0.

Conclusion: There is sufficient evidence at the $\alpha = 0.05$ level of significance to conclude that the standard deviation of waiting time in a single line is less than the standard deviation for wait time in the multiple lines.

(b) We compute the five-number summary for each variable:

Single: 1.1, 1.9, 2.2, 2.8, 3.1

Multiple: 0.9, 1.8, 2.45, 2.9, 4.6

[Minitab: 0.9, 1.7, 2.45, 2.9, 4.6]

The boxplot indicates that there is much greater variability in the waiting times for multiple lines than for a single line.

21. The normal probability plots suggest that the data are approximately normally distributed. Thus, this requirement of the F-test is satisfied.

$H_0 : \sigma_1 = \sigma_{cc}$

$H_1 : \sigma_1 > \sigma_{cc}$

We compute the sample standard deviations to be $s_1 \approx 23.851$ and $s_{cc} \approx 19.078$. The test statistic is $F_0 = \dfrac{s_1^2}{s_{cc}^2} = \dfrac{23.851^2}{19.078^2} \approx 1.56$.

Classical approach: This is a left-tailed test with $n_1 = 98$ and $n_{cc} = 96$. Let $\alpha = 0.05$. The critical value is $F_{0.05,97,95}$, which we approximate by $F_{0.05,120,100} = 1.38$. Since $F_0 = 1.56 > F_{0.05,97,95} \approx F_{0.05,120,100} \approx 1.38$ (the test statistic falls in the critical region), we reject H_0.

P-value approach: Using technology, we find
P-value ≈ 0.0153. Since this *P*-value is less
than the $\alpha = 0.05$ level of significance, we
reject H_0.

Conclusion: There is sufficient evidence at the
$\alpha = 0.05$ level of significance to conclude that
industrial stocks are riskier than consumer
cyclical stocks

Section 11.5

1. We verify the requirements to perform the
hypothesis test: (1) We are told that the
samples are random samples; (2) We have
$x_1 = 43$, $n_1 = 120$, $x_2 = 56$, and $n_2 = 130$, so

$$\hat{p}_1 = \frac{43}{120} \approx 0.3583 \text{ and } \hat{p}_2 = \frac{56}{130} \approx 0.4308.$$

Thus,

$n_1 \hat{p}_1 (1 - \hat{p}_1) = 120(0.3583)(1 - 0.3583) \approx$
$28 \geq 10$ and $n_2 \hat{p}_2 (1 - \hat{p}_2) =$
$130(0.4308)(1 - 0.4308) \approx 32 \geq 10$; and (3)

We assume each sample is less than 5% of the
population. Thus, the requirements are met,
and we can conduct the test.

The hypotheses are $H_0 : p_1 = p_2$ versus
$H_1 : p_1 \neq p_2$. From before, the two sample
estimates are $\hat{p}_1 \approx 0.3583$ and $\hat{p}_2 \approx 0.4308$.

The pooled estimate is $\hat{p} = \dfrac{43 + 56}{120 + 130} = 0.396$.

The test statistic is

$$z_0 = \frac{\hat{p}_1 - \hat{p}_2}{\sqrt{\hat{p}(1 - \hat{p})}\sqrt{\dfrac{1}{n_1} + \dfrac{1}{n_2}}}$$

$$= \frac{0.3583 - 0.4308}{\sqrt{0.396(1 - 0.396)}\sqrt{\dfrac{1}{120} + \dfrac{1}{130}}}$$

$$\approx -1.17$$

Classical approach: This is a two-tailed test,
so the critical values for $\alpha = 0.01$ are
$\pm z_{\alpha/2} = \pm z_{0.005} = \pm 2.575$. Since $z_0 = -1.17$
falls between $-z_{0.005} = -2.575$ and
$z_{0.025} = 2.575$ (the test statistic does not fall in
the critical region), we do not reject H_0.

P-value approach: The *P*-value is two times
the area under the standard normal distribution

to the left of the test statistic, $z_0 = -1.17$.
$$P\text{-value} = 2 \cdot P(z_0 \leq -1.17)$$
$$= 2 \cdot (0.1210) = 0.2420$$

Since the *P*-value $= 0.2420 > \alpha = 0.01$, we do
not reject H_0.

Conclusion: There is not sufficient evidence
at the $\alpha = 0.01$ level of significance to
conclude that $p_1 \neq p_2$.

3. $H_0 : \mu_1 = \mu_2$
   $H_1 : \mu_1 \neq \mu_2$

We assume that all requirements to conduct
the test are satisfied. The level of significance
is $\alpha = 0.05$. The smaller sample size is
$n_1 = 13$, so we use $n_1 - 1 = 13 - 1 = 12$ degrees
of freedom. The test statistic is

$$t_0 = \frac{(\bar{x}_1 - \bar{x}_2) - (\mu_1 - \mu_2)}{\sqrt{\dfrac{s_1^2}{n_1} + \dfrac{s_2^2}{n_2}}} = \frac{(45.3 - 52.1) - 0}{\sqrt{\dfrac{12.4^2}{13} + \dfrac{14.7^2}{18}}}$$

$$\approx -1.393$$

Classical approach: Since this is a two-tailed
test with 12 degrees of freedom, the critical
values are $\pm t_{0.025} = \pm 2.179$. Since the test
statistic $t_0 \approx -1.393$ is between the critical
values $-t_{0.025} = -2.179$ and $t_{0.025} = 2.179$ (the
test statistic does not fall in the critical region),
we do not reject H_0.

P-value approach: The *P*-value for this two-
tailed test is the area under the *t*-distribution
with 12 degrees of freedom to the left of
$t_0 = -1.393$ plus the area to the right of 1.393.
From the *t*-distribution table in the row
corresponding to 12 degrees of freedom, 1.393
falls between 1.356 and 1.782 whose right-tail
areas are 0.10 and 0.05, respectively. We must
double these values in order to get the total area
in both tails: 0.20 and 0.10. So,
$0.10 < P\text{-value} < 0.20$ [Tech:
P-value $= 0.1745$]. Because the *P*-value is
greater than the level of significance $\alpha = 0.05$,
we do not reject H_0.

Conclusion: There is not sufficient evidence at
the $\alpha = 0.05$ level of significance to conclude
that there is a difference in the population
means.

5. $H_0 : \sigma_1 = \sigma_2$

$H_1 : \sigma_1 \neq \sigma_2$

We assume that all requirements to conduct the test are satisfied. The test statistic is

$$F_0 = \frac{s_1^2}{s_2^2} = \frac{18.3^2}{13.5^2} \approx 1.84.$$

Classical approach: This is a two-tailed test with $n_1 = 61$, $n_2 = 57$, and $\alpha = 0.05$. The critical values are $F_{0.025,60,56}$ and $F_{0.975,60,56}$, which we approximate by $F_{0.025,60,50} = 1.72$ and

$$F_{0.975,60,56} = \frac{1}{F_{0.025,56,60}} \approx \frac{1}{F_{0.025,50,50}} = \frac{1}{1.75} \approx$$

0.57. Since $F_0 = 1.84 > F_{0.025,60,56} \approx 1.72$ (the test statistic falls in a critical region), we reject H_0.

P-value approach: Using technology, we find P-value ≈ 0.0228. Since this P-value is less than the $\alpha = 0.05$ level of significance, we reject H_0.

Conclusion: There is sufficient evidence at the $\alpha = 0.05$ level of significance to conclude that $\sigma_1 \neq \sigma_2$.

7. These are matched-pair data because two measurements, X_i and Y_i, are taken on the same individual. We measure differences as $d_i = Y_i - X_i$.

Individual	1	2	3	4	5
X_i	93	102	90	112	107
Y_i	95	100	95	115	107
d_i	2	-2	5	3	0

We compute the mean and standard deviation of the differences and obtain $\bar{d} = 1.6$ and $s_d \approx 2.7019$, rounded to four decimal places.

$H_0 : \mu_d = 0$

$H_1 : \mu_d > 0$

The level of significance is $\alpha = 0.05$. The test statistic is $t_0 = \dfrac{\bar{d}}{\dfrac{s_d}{\sqrt{n}}} = \dfrac{1.6}{\dfrac{2.7019}{\sqrt{5}}} \approx 1.324$.

Classical approach: Since this is a right-tailed test with 4 degrees of freedom, the critical value is $t_{0.05} = 2.132$. Since the test statistic

$t_0 \approx 1.324 < t_{0.05} = 2.132$ (the test statistic does not fall within the critical region), we do not reject H_0.

P-value approach: The P-value for this right-tailed test is the area under the t-distribution with 4 degrees of freedom to the right of the test statistic $t_0 = 1.324$. From the t-distribution table, in the row corresponding to 4 degrees of freedom, 1.324 falls between 1.190 and 1.533 whose right-tail areas are 0.15 and 0.10, respectively. So, $0.10 < P$-value < 0.15 [Tech: P-value = 0.1280]. Because the P-value is greater than the level of significance $\alpha = 0.05$, we do not reject H_0.

Conclusion: There is not sufficient evidence at the $\alpha = 0.05$ level of significance to conclude that there is a difference in the measure of the variable before and after a treatment. That is, there is not sufficient evidence to conclude that the treatment is effective.

9. (a) Collision claims tend to be skewed right because there are a few very large collision claims relative to the majority of claims.

(b) $H_0 : \mu_{30-59} = \mu_{20-24}$

$H_1 : \mu_{30-59} < \mu_{20-24}$

All requirements are satisfied to conduct the test for two independent population means. The level of significance is $\alpha = 0.05$. The sample sizes are both $n = 40$, so we use $n - 1 = 40 - 1 = 39$ degrees of freedom. Test statistic:

$$t_0 = \frac{(\bar{x}_1 - \bar{x}_2) - (\mu_1 - \mu_2)}{\sqrt{\dfrac{s_1^2}{n_1} + \dfrac{s_2^2}{n_2}}}$$

$$= \frac{(3669 - 4586) - 0}{\sqrt{\dfrac{2029^2}{40} + \dfrac{2302^2}{40}}} \approx -1.890$$

Classical approach: Since this is a left-tailed test with 39 degrees of freedom, the critical value is $-t_{0.05} = -1.685$. Since $t_0 = -1.890 < -t_{0.05} = -1.685$ (the test statistic falls in the critical region), we reject H_0.

P-value approach: The P-value for this left-tailed test is the area under the t-distribution with 39 degrees of freedom to the left of $t_0 = -1.890$. From the t-distribution table in the row corresponding to 39 degrees of freedom, 1.890 falls between 1.685 and 2.023 whose right-tail areas are 0.05 and 0.025, respectively. So, $0.025 < P\text{-value} < 0.05$ [Tech: $P\text{-value} = 0.0313$. Because the P-value is less than the level of significance $\alpha = 0.05$, we reject H_0.

Conclusion: There is sufficient evidence at the $\alpha = 0.05$ level of significance to conclude that the mean collision claim of a 30- to 59-year-old is less than the mean claim of a 20- to 24-year-old. Given that 20- to 24-year-olds tend to claim more for each accident, it makes sense to charge them more for coverage.

11. (a) The response variable in this study is age.

(b) The sampling method is dependent because each husband is matched with the wife.

(c) To estimate the mean difference, construct a 95% confidence interval for the mean difference of the data. Use the matched pairs method because each husband is matched with the wife. Compute the differences as $d_i = \text{Husband}_i - \text{Wife}_i$.

Obs	1	2	3	4	5	6	7	8
Husband	47	53	45	50	28	65	25	56
Wife	43	43	45	48	29	61	27	51
d_i	4	10	0	2	-1	4	-2	5

The differences are approximately normal because a normal probability plot shows that the differences are approximately normal and the boxplot reveals that there are no outliers.

We find that $\bar{d} = 2.75$ and $s_d \approx 3.88$. Since there are $n - 1 = 8 - 1 = 7$ degrees of freedom, $t_{0.025} = 2.365$.

Lower bound: $\bar{d} - t_{\alpha/2} \cdot \dfrac{s_d}{\sqrt{n}}$

$$= 2.75 - 2.365 \cdot \frac{3.88}{\sqrt{8}}$$

$$\approx -0.49 \ [\text{Tech } -0.50]$$

Upper bound: $\bar{d} + t_{\alpha/2} \cdot \dfrac{s_d}{\sqrt{n}}$

$$= 2.75 + 2.365 \cdot \frac{3.88}{\sqrt{8}}$$

$$\approx 5.99 \ [\text{Tech: 6.00}]$$

We are 95% confident the mean difference in age between a husband and wife is between -0.49 year and 5.99 years.

13. This hypothesis test requires a single sample t-test for a population mean.

$$H_0 : \mu = 92.0 \text{ mph}$$
$$H_1 : \mu > 92.0 \text{ mph}$$

The sample size $n = 14$ is small. However, a normal probability plot indicates the data could come from a population that is normally distributed. A boxplot indicates there are no outliers.

Using technology, $\bar{x} = 92.25$ and $s = 0.86$.
Test Statistic:

$$t_0 = \frac{\bar{x} - \mu_0}{s / \sqrt{n}} = \frac{92.25 - 92.0}{0.86 / \sqrt{14}}$$
$$\approx 1.088 \ [\text{Tech: 1.087}]$$

Classical approach:
Because this is a right-tailed test with $n - 1 = 13$ degrees of freedom, the critical value for 95% confidence is $t_{0.05} = 1.771$.
Since $t_0 = 1.088 < t_{0.05} = 1.771$, the test statistic does not fall within the critical region. Therefore, we do not reject the null hypothesis.

P-value approach:
This is a right-tailed test with 13 degrees of freedom. The P-value is the area under the t-distribution to the right of the test statistic, $t_0 = 1.088$. From the t-distribution table in the row corresponding to 13 degrees of freedom, 1.088 falls to between 1.079 and 1.350, whose right-tail area are 0.15 and 0.10, respectively. So, $0.10 < P\text{-value} < 0.15$ [Tech: 0.1485]. Since $P\text{-value} > \alpha = 0.05$, we do not reject the null hypothesis.

Conclusion: There is not sufficient evidence to conclude that the pitcher has a fastball that exceeds 92.0 mph.

15. Since an election is won by getting over half the votes, the hypotheses are $H_0 : p = 0.5$ versus $H_1 : p > 0.5$. Since

$$np_0(1 - p_0) = 469(0.5)(1 - 0.5) = 117.25 \geq 10,$$

the requirements of the hypothesis test are satisfied. Use a level of significance of $\alpha = 0.05$ for this hypothesis test. From the survey, $\hat{p} = \dfrac{469}{695} \approx 0.675$. The test statistic is

$$z_0 = \frac{\hat{p} - p_0}{\sqrt{\dfrac{p_0(1 - p_0)}{n}}} = \frac{0.675 - 0.5}{\sqrt{\dfrac{0.5(1 - 0.5)}{695}}} \approx 9.22$$

Classical approach:
This is a right-tailed test, so the critical value is

$z_{0.05} = 1.645$. Since $z_0 = 9.22 > z_{0.05} = 1.645$, the test statistic falls in the critical region, so we reject the null hypothesis.

P-value approach:

$$P\text{-value} = P(Z > 9.22)$$
$$= 1 - P(Z \leq 9.232)$$
$$< 0.0001$$

Since $P\text{-value} < 0.0001 < \alpha = 0.05,$ we reject the null hypothesis.

Conclusion: There is sufficient evidence to conclude that a quick one-second view of a black and white photo represents enough information to judge the winner of an election.

---

17. $H_0 : p_{>100K} = p_{<100K}$

$H_1 : p_{>100K} \neq p_{<100K}$

We have $x_{>100K} = 710$, $n_{>100K} = 1205$, $x_{<100K} = 695$, and $n_{<100K} = 1310$, so $\hat{p}_{>100K} = \dfrac{x_{>100K}}{n_{>100K}} = \dfrac{710}{1205} \approx 0.5892$

and $\hat{p}_{<100K} = \dfrac{x_{<100K}}{n_{<100K}} = \dfrac{695}{1310} = 0.5305$. For a 95% confidence interval we use $\pm z_{0.025} = \pm 1.96$. Then:

Lower Bound: $(\hat{p}_{>100K} - \hat{p}_{<100K}) - z_{\alpha/2} \cdot \sqrt{\dfrac{\hat{p}_{>100K}(1 - \hat{p}_{>100K})}{n_{>100K}} + \dfrac{\hat{p}_{<100K}(1 - \hat{p}_{<100K})}{n_{<100K}}}$

$$= (0.5892 - 0.5305) - 1.96 \cdot \sqrt{\frac{0.5892(1 - 0.5892)}{1205} + \frac{0.5305(1 - 0.5305)}{1310}} \approx 0.020$$

Upper Bound: $(\hat{p}_{>100K} - \hat{p}_{<100K}) + z_{\alpha/2} \cdot \sqrt{\dfrac{\hat{p}_{>100K}(1 - \hat{p}_{>100K})}{n_{>100K}} + \dfrac{\hat{p}_{<100K}(1 - \hat{p}_{<100K})}{n_{<100K}}}$

$$= (0.5892 - 0.5305) + 1.96 \cdot \sqrt{\frac{0.5892(1 - 0.5892)}{1205} + \frac{0.5305(1 - 0.5305)}{1310}} \approx 0.097$$

We are 95% confident that the difference in the proportion of individuals who believe it is morally wrong for unwed women to have children (for individuals who earned more than \$100,000 versus individuals who earned less than \$100,000) is between 0.020 and 0.097. Because the confidence interval does not include 0, there is sufficient evidence at the $\alpha = 0.05$ level of significance to conclude that there is a difference in the proportions. It appears that a higher proportion of individuals who earn over \$100,000 per year feel it is morally wrong for unwed women to have children.

---

19. (a) The response variable is the score on the quiz. The explanatory variable is whether texting was required or the cell phone was turned off (no texting allowed).

(b) Students were randomly given instructions at the beginning of the class. Both groups received the same lecture.

(c) The sampling method is independent because the individuals selected for the texting required group have no relation to the individuals selected for the no texting allowed group.

(d) $H_0 : \mu_{\text{text}} = \mu_{\text{cell off}}$

$H_1 : \mu_{\text{text}} < \mu_{\text{cell off}}$

Since $n_{\text{text}} = n_{\text{cell off}} = 31$, the sample sizes are large enough and the requirements are satisfied to conduct the test for two independent population means. The level of significance is $\alpha = 0.1$. The sample sizes are both $n = 31$, so we use $n - 1 = 31 - 1 = 30$ degrees of freedom.

Test statistic:

$$t_0 = \frac{(\bar{x}_{\text{text}} - \bar{x}_{\text{cell off}}) - (\mu_{\text{text}} - \mu_{\text{cell off}})}{\sqrt{\dfrac{s_{\text{text}}^2}{n_{\text{text}}} + \dfrac{s_{\text{cell off}}^2}{n_{\text{cell off}}}}}$$

$$= \frac{(42.81 - 58.67) - 0}{\sqrt{\dfrac{9.91^2}{31} + \dfrac{10.42^2}{31}}}$$

$$\approx -6.141$$

Classical approach: Since this is a left-tailed test with 30 degrees of freedom, the critical value is $-t_{0.05} = -1.697$. Since $t_0 = -6.141 < -t_{0.05} = -1.697$ (the test statistic falls in the critical region), we reject H_0.

P-value approach: The P-value for this left-tailed test is the area under the t-distribution with 30 degrees of freedom, to the left of $t_0 = -6.141$. From the t-distribution table in the row corresponding to 30 degrees of freedom, 6.141 falls to the right of 3.646 whose right-tail area is 0.0005. So, P-value < 0.0005 [Tech: P-value < 0.0001]. Because the P-value is less than the level of significance $\alpha = 0.05$, we reject H_0.

Conclusion: There is sufficient evidence at the $\alpha = 0.05$ level of significance to suggest the mean score in the texting group is less than the mean score in the cell phone off group. Apparently, students cannot multitask.

(e) Since $n_{\text{text}} = n_{\text{cell off}} = 31$, the sample sizes are large enough to conduct a confidence interval. For a 90% confidence interval with $df = 30$, we use $t_{\alpha/2} = t_{0.05} = 1.697$. Then:

Lower bound:

$$(\bar{x}_{\text{text}} - \bar{x}_{\text{cell off}}) - t_{\alpha/2} \cdot \sqrt{\dfrac{s_{\text{text}}^2}{n_{\text{text}}} + \dfrac{s_{\text{cell off}}^2}{n_{\text{cell off}}}}$$

$$= (42.81 - 58.67) - 1.697 \cdot \sqrt{\dfrac{9.91^2}{31} + \dfrac{10.42^2}{31}}$$

$$\approx -20.243 \ [\text{Tech: } -20.175]$$

Upper bound:

$$(\bar{x}_{\text{text}} - \bar{x}_{\text{cell off}}) + t_{\alpha/2} \cdot \sqrt{\dfrac{s_{\text{text}}^2}{n_{\text{text}}} + \dfrac{s_{\text{cell off}}^2}{n_{\text{cell off}}}}$$

$$= (42.81 - 58.67) + 1.697 \cdot \sqrt{\dfrac{9.91^2}{31} + \dfrac{10.42^2}{31}}$$

$$\approx -11.477 \ [\text{Tech: } -11.545]$$

We can be 90% confident the texting group scored between 11.464 and 20.256 points worse, on average, than the cell-phone off group.

(f) First determine whether the requirements are satisfied. Since the samples are from a completely randomized design and the populations from which the samples are drawn are normally distributed, the requirements are satisfied.

$H_0 : \sigma_{\text{text}}^2 = \sigma_{\text{cell off}}^2$

$H_1 : \sigma_{\text{text}}^2 \neq \sigma_{\text{cell off}}^2$

Use a level of significance of $\alpha = 0.05$. We are given the summary statistics $s_{\text{text}} = 9.91$, $n_{\text{text}} = 31$, $s_{\text{cell off}} = 10.42$, and $n_{\text{cell off}} = 31$. Since both sample sizes are 31, we use $31 - 1 = 30$ degrees of freedom.

Test statistic:

$$F_0 = \frac{s_{\text{cell off}}^2}{s_{\text{text}}^2} = \frac{10.42^2}{9.91^2} \approx 1.106$$

Classical approach: This is a two-tailed test with $n_{text} = n_{cell\,off} = 31$ and $\alpha = 0.05$. So,

$\dfrac{\alpha}{2} = 0.025$, and the critical values are

$F_{0.025,30,25} = 2.18$ and

$F_{0.975,15,15} = \dfrac{1}{F_{0.025,30,25}} = \dfrac{1}{2.18} \approx 0.46$. Since

$F_0 = 1.10$ is between $F_{0.975,30,25} = 0.46$ and

$F_{0.025,30,25} = 2.18$ (the test statistic does not fall in the critical regions), we do not reject H_0.

P-value approach: Using technology, we find P-value ≈ 0.7852. Since this P-value is greater than the $\alpha = 0.05$ level of significance, we do not reject H_0.

Conclusion: There is not sufficient evidence at the $\alpha = 0.05$ level of significance to suggest the variability in scores differs between the two groups.

21. Age: Since $n_R = 637$ and $n_{NR} = 3137$, the sample sizes are large enough and the requirements are satisfied to conduct the test for two independent population means.

$H_0 : \mu_R = \mu_{NR}$

$H_1 : \mu_R > \mu_{NR}$

Let $\alpha = 0.05$.

Test statistic:

$t_0 = \dfrac{(\overline{x}_R - \overline{x}_{NR}) - (\mu_R - \mu_{NR})}{\sqrt{\dfrac{s_R^2}{n_R} + \dfrac{s_{NR}^2}{n_{NR}}}}$

$= \dfrac{(78.0 - 76.0) - 0}{\sqrt{\dfrac{13.7^2}{637} + \dfrac{15.8^2}{3137}}}$

≈ 3.27

Using technology, P-value ≈ 0.0006. Therefore, reject the null hypothesis.

Length of stay: The sample sizes are large enough and the requirements are satisfied to conduct the test for two independent population means.

$H_0 : \mu_R = \mu_{NR}$

$H_1 : \mu_R > \mu_{NR}$

Let $\alpha = 0.05$. The test statistic is:

$t_0 = \dfrac{(\overline{x}_R - \overline{x}_{NR}) - (\mu_R - \mu_{NR})}{\sqrt{\dfrac{s_R^2}{n_R} + \dfrac{s_{NR}^2}{n_{NR}}}}$

$= \dfrac{(4.8 - 3.9) - 0}{\sqrt{\dfrac{4.3^2}{637} + \dfrac{3.1^2}{3137}}}$

≈ 5.02

Using technology, P-value < 0.0001. Therefore, reject the null hypothesis.

Admission in previous calendar year: We first verify the requirements to perform the hypothesis test: (1) Each sample can be thought of as a simple random sample. (2) We have $x_R = 139$, $n_R = 637$, $x_{NR} = 445$, and

$n_{NR} = 3137$, so $\hat{p}_R = \dfrac{139}{637} \approx 0.2182$ and

$\hat{p}_{NR} = \dfrac{445}{3137} \approx 0.1419$. Thus,

$n_R \hat{p}_R (1 - \hat{p}_R) =$
$637(0.2182)(1 - 0.2182) \approx 108.7 \geq 10$ and

$n_{NR} \hat{p}_{NR} (1 - \hat{p}_{NR}) =$
$3137(0.1419)(1 - 0.1419) \approx 381.9 \geq 10$.

(3) Each sample is less than 5% of the population. So, the requirements are met, so we can conduct the test.

$H_0 : p_R = p_{NR}$

$H_1 : p_R > p_{NR}$

Let $\alpha = 0.05$. The pooled estimate is

$\hat{p} = \dfrac{x_R + x_{NR}}{n_R + n_{NR}} = \dfrac{139 + 445}{637 + 3137} \approx 0.1547$

Test statistic:

$z_0 = \dfrac{\hat{p}_R - \hat{p}_{NR}}{\sqrt{\hat{p}(1 - \hat{p})}\sqrt{1/n_R + 1/n_{NR}}}$

$= \dfrac{0.2182 - 0.1419}{\sqrt{0.1547(1 - 0.1547)}\sqrt{\dfrac{1}{637} + \dfrac{1}{3137}}}$

≈ 4.86

Using technology, P-value < 0.0001. Therefore, reject the null hypothesis.

Season: We first verify the requirements to perform the hypothesis test: (1) Each sample can be thought of as a simple random sample. (2) We have $x_R = 199$, $n_R = 637$, $x_{NR} = 765$, and $n_{NR} = 3137$, so

$$\hat{p}_R = \frac{199}{637} \approx 0.3124 \text{ and}$$

$$\hat{p}_{NR} = \frac{765}{3137} \approx 0.2439 \text{ . Thus,}$$

$$n_R \hat{p}_R (1 - \hat{p}_R) =$$
$$637(0.3124)(1 - 0.3124) \approx 136.8 \geq 10 \text{ and}$$

$$n_{NR} \hat{p}_{NR} (1 - \hat{p}_{NR}) =$$
$$3137(0.2439)(1 - 0.2439) \approx 578.5 \geq 10 \text{ .}$$

(3) Each sample is less than 5% of the population. So, the requirements are met, so we can conduct the test.

$$H_0 : p_R = p_{NR}$$
$$H_1 : p_R > p_{NR}$$

Let $\alpha = 0.05$. The pooled estimate is

$$\hat{p} = \frac{x_R + x_{NR}}{n_R + n_{NR}} = \frac{199 + 765}{637 + 3137} \approx 0.2554$$

Test statistic

$$z_0 = \frac{\hat{p}_R - \hat{p}_{NR}}{\sqrt{\hat{p}(1 - \hat{p})}\sqrt{1/n_R + 1/n_{NR}}}$$
$$= \frac{0.3124 - 0.2439}{\sqrt{0.2554(1 - 0.2554)}\sqrt{\dfrac{1}{637} + \dfrac{1}{3137}}}$$

$$\approx 3.61$$

Using technology, P-value ≈ 0.0001. Therefore, reject the null hypothesis.

Floor: We first verify the requirements to perform the hypothesis test: (1) Each sample can be thought of as a simple random sample. (2) We have $x_R = 332$, $n_R = 637$, $x_{NR} = 1617$, and $n_{NR} = 3137$, so

$$\hat{p}_R = \frac{332}{637} \approx 0.5212 \text{ and}$$

$$\hat{p}_{NR} = \frac{1617}{3137} \approx 0.5155 \text{ . Thus,}$$

$$n_R \hat{p}_R (1 - \hat{p}_R) =$$
$$637(0.5212)(1 - 0.5212) \approx 159.0 \geq 10 \text{ and}$$

$$n_{NR} \hat{p}_{NR} (1 - \hat{p}_{NR}) =$$
$$3137(0.5155)(1 - 0.5155) \approx 783.5 \geq 10 \text{ . (3)}$$
Each sample is less than 5% of the population. So, the requirements are met, so

we can conduct the test.

$$H_0 : p_R = p_{NR}$$
$$H_1 : p_R > p_{NR}$$

Let $\alpha = 0.05$. The pooled estimate is

$$\hat{p} = \frac{x_R + x_{NR}}{n_R + n_{NR}} = \frac{332 + 1617}{637 + 3137} \approx 0.5164$$

Test statistic

$$z_0 = \frac{\hat{p}_R - \hat{p}_{NR}}{\sqrt{\hat{p}(1 - \hat{p})}\sqrt{1/n_R + 1/n_{NR}}}$$
$$= \frac{0.5212 - 0.5155}{\sqrt{0.5164(1 - 0.5164)}\sqrt{\dfrac{1}{637} + \dfrac{1}{3137}}}$$

$$\approx 0.26$$

Using technology, P-value ≈ 0.3959. Therefore, reject the null hypothesis.

From the analysis, the readmits were older and had a longer length of stay. A higher proportion of readmits were admitted in the previous calendar year and a higher proportion of readmits were discharged in the winter. The proportion of readmits who were on the cardiac floor is not significantly different from the proportion of non-readmits who were on the cardiac floor.

23. This study is best done using a matched-pairs design. Researchers should match males and females based on the characteristics that determine salary, such as career, experience, level of education, and geographic location. Determine the differences in each pair and see if the mean differences (men salary minus women salary) is significantly greater than 0.

25. Hypothesis test for two proportions—independent sample

27. Confidence interval for a single mean

29. Two sample t-test of independent means

31. Use a matched-pairs design. Analyze the data using t-test for a dependent sample.

33. Use a hypothesis test of two independent proportions.

Chapter 11 Review Exercises

1. This is dependent since the members of the two samples are matched by diagnosis. The response variable, length of stay, is quantitative.

2. This is independent because the subjects are randomly selected from two distinct populations. The response variable, commute times, is quantitative.

3. (a) $F_{0.05,8,9} = 3.23$

(b) The level of significance is $\alpha = 0.05$, so $\dfrac{\alpha}{2} = 0.025$. Thus, $F_{0.025,10,5} = 6.62$ and

$$F_{0.975,10,5} = \frac{1}{F_{0.025,5,10}} = \frac{1}{4.24} \approx 0.24.$$

4. (a) We compute each difference, $d_i = X_i - Y_i$.

Obs	1	2	3	4	5	6
X_i	34.2	32.1	39.5	41.8	45.1	38.4
Y_i	34.9	31.5	39.5	41.9	45.5	38.8
d_i	−0.7	0.6	0	−0.1	−0.4	−0.4

(b) Using technology, $\bar{d} \approx -0.1667$ and $s_d \approx 0.4502$, rounded to four decimal places.

(c) The hypotheses are $H_0 : \mu_d = 0$ versus $H_1 : \mu_d < 0$. The level of significance is $\alpha = 0.05$. The test statistic is

$$t_0 = \frac{\bar{d}}{\frac{s_d}{\sqrt{n}}} = \frac{-0.1667}{\frac{0.4502}{\sqrt{6}}} \approx -0.907.$$

Classical approach: Since this is a left-tailed test with 5 degrees of freedom, the critical value is $-t_{0.05} = -2.015$. Since the test statistic $t_0 = -0.907 > -t_{0.05} = -2.015$ (the test statistic does not fall within the critical region), we do not reject H_0.

P-value approach: The *P*-value for this left-tailed test is the area under the *t*-distribution with 5 degrees of freedom to the left of the test statistic $t_0 \approx -0.907$, which is equivalent to the area to the right of 0.907.

From the *t*-distribution table in the row corresponding to 5 degrees of freedom, 0.907 falls between 0.727 and 0.920 whose right-tail areas are 0.25 and 0.20, respectively. So, $0.20 < P\text{-value} < 0.25$ [Tech: *P*-value = 0.2030]. Because the *P*-value is greater than the $\alpha = 0.05$ level of significance, we do not reject H_0.

Conclusion: There is not sufficient evidence at the $\alpha = 0.05$ level of significance to conclude that the mean difference is less than zero.

(d) For 98% confidence, we use $\alpha = 0.02$. With 5 degrees of freedom, we have $t_{\alpha/2} = t_{0.01} = 3.365$. Then:

$$\text{Lower bound} = \bar{d} - t_{0.01} \cdot \frac{s_d}{\sqrt{n}}$$

$$= -0.1667 - 3.365 \cdot \frac{0.4502}{\sqrt{6}}$$

$$\approx -0.79$$

$$\text{Upper bound} = \bar{d} + t_{0.01} \cdot \frac{s_d}{\sqrt{n}}$$

$$= -0.167 + 3.365 \cdot \frac{0.450}{\sqrt{6}}$$

$$\approx 0.45$$

We can be 98% confident that the mean difference is between −0.79 and 0.45.

5. (a) $H_0 : \mu_1 = \mu_2$
$H_1 : \mu_1 \neq \mu_2$
The level of significance is $\alpha = 0.1$. Since the smaller sample size is $n_2 = 8$, we use $n_2 - 1 = 7$ degrees of freedom. The test statistic is

$$t_0 = \frac{(\bar{x}_1 - \bar{x}_2) - (\mu_1 - \mu_2)}{\sqrt{\frac{s_1^2}{n_1} + \frac{s_2^2}{n_2}}} = \frac{(32.4 - 28.2) - 0}{\sqrt{\frac{4.5^2}{13} + \frac{3.8^2}{8}}}$$

$$\approx 2.290$$

Classical approach: Since this is a two-tailed test with 7 degrees of freedom, the critical values are $\pm t_{0.05} = \pm 1.895$. Since the test statistic $t_0 \approx 2.290$ is to the right of the critical value 1.895 (the test statistic falls within a critical region), we reject H_0.

P-value approach: The *P*-value for this two-tailed test is the area under the *t*-distribution with 7 degrees of freedom to the right of $t_0 = 2.290$ plus the area to the left of $-t_0 = -2.290$. From the *t*-distribution table in the row corresponding to 7 degrees of freedom, 2.290 falls between 1.895 and 2.365 whose right-tail areas are 0.05 and 0.025, respectively. We must double these values in order to get the total area in both tails: 0.10 and 0.05. Thus, $0.05 < P\text{-value}$

< 0.10 [Tech: P-value $= 0.0351$]. Because the P-value is less than the $\alpha = 0.10$ level of significance, we reject H_0.

Conclusion: There is sufficient evidence at the $\alpha = 0.1$ level of significance to conclude that $\mu_1 \neq \mu_2$.

(b) For a 90% confidence interval with df $= 7$, we use $t_{\alpha/2} = t_{0.05} = 1.895$. Then:

Lower bound

$$= (\bar{x}_1 - \bar{x}_2) - t_{\alpha/2} \cdot \sqrt{\frac{s_1^2}{n_1} + \frac{s_2^2}{n_2}}$$

$$= (32.4 - 28.2) - 1.895 \cdot \sqrt{\frac{4.5^2}{13} + \frac{3.8^2}{8}}$$

$$\approx 0.73 \quad [\text{Tech: } 1.01]$$

Upper bound

$$= (\bar{x}_1 - \bar{x}_2) + t_{\alpha/2} \cdot \sqrt{\frac{s_1^2}{n_1} + \frac{s_2^2}{n_2}}$$

$$= (32.4 - 28.2) + 1.895 \cdot \sqrt{\frac{4.5^2}{13} + \frac{3.8^2}{8}}$$

$$\approx 7.67 \quad [\text{Tech: } 7.39]$$

We are 90% confident that the population mean difference is between 0.73 and 7.67. [Tech: between 1.01 and 7.39].

(c) $H_0 : \sigma_1 = \sigma_2$

$H_1 : \sigma_1 \neq \sigma_2$

The test statistic is $F_0 = \dfrac{s_1^2}{s_2^2} = \dfrac{4.5^2}{3.8^2} \approx 1.40$.

Classical approach: This is a two-tailed test with $n_1 = 13$, $n_2 = 8$, and $\alpha = 0.05$. The critical values are $F_{0.025,12,7}$, which we approximate by $F_{0.025,10,7} = 4.76$ and

$$F_{0.975,12,7} = \frac{1}{F_{0.025,7,12}} = \frac{1}{3.61} \approx 0.28. \text{ Since}$$

$F_0 = 1.40$ falls between $F_{0.975,12,7} \approx 0.28$

and $F_{0.025,12,7} \approx 4.76$ (the test statistic does not fall in the critical region), we do not reject H_0.

P-value approach: Using technology, we find P-value ≈ 0.6734. Since this P-value

is greater than the $\alpha = 0.05$ level of significance, we do not reject H_0.

Conclusion: There is not sufficient evidence at the $\alpha = 0.05$ level of significance to conclude that the standard deviation in population 1 is different from the standard deviation in population 2.

6. (a) $H_0 : \mu_1 = \mu_2$

$H_1 : \mu_1 > \mu_2$

The level of significance is $\alpha = 0.01$. Since the smaller sample size is $n_2 = 41$, we use $n_2 - 1 = 40$ degrees of freedom. The test statistic is

$$t_0 = \frac{(\bar{x}_1 - \bar{x}_2) - (\mu_1 - \mu_2)}{\sqrt{\dfrac{s_1^2}{n_1} + \dfrac{s_2^2}{n_2}}}$$

$$= \frac{(48.2 - 45.2) - 0}{\sqrt{\dfrac{8.4^2}{45} + \dfrac{10.3^2}{41}}} \approx 1.472$$

Classical approach: Since this is a right-tailed test with 40 degrees of freedom, the critical value is $t_{0.01} = 2.423$. Since the test statistic $t_0 \approx 1.472 < t_{0.01} = 2.423$ (the test statistic does not fall within the critical region), we do not reject H_0.

P-value approach: The P-value for this right-tailed test is the area under the t-distribution with 40 degrees of freedom to the right of $t_0 \approx 1.472$. From the t-distribution table in the row corresponding to 40 degrees of freedom, 1.472 falls between 1.303 and 1.684 whose right-tail areas are 0.10 and 0.05, respectively.

Thus, $0.10 > P$-value > 0.05 [Tech: P-value $= 0.0726$]. Because the P-value is greater than the $\alpha = 0.01$ level of significance, we do not reject H_0.

Conclusion: There is not sufficient evidence at the $\alpha = 0.01$ level of significance to conclude that the mean of population 1 is larger than the mean of population 2.

(b) $H_0 : \sigma_1 = \sigma_2$

$H_1 : \sigma_1 < \sigma_2$

The test statistic is

$$F_0 = \frac{s_1^2}{s_2^2} = \frac{8.4^2}{10.3^2} \approx 0.67.$$

Classical approach: This is a left-tailed test with $n_1 = 45$, $n_2 = 41$, and $\alpha = 0.01$.

The critical value is $F_{0.99, 44, 40} = \dfrac{1}{F_{0.01, 40, 44}}$

$\approx \dfrac{1}{F_{0.01, 40, 50}} \approx \dfrac{1}{2.01} \approx 0.50$. Since

$F_0 = 0.67 > F_{0.99, 44, 40} \approx 0.50$ (the test statistic does not fall in the critical region), we do not reject H_0.

P-value approach: Using technology, we find P-value ≈ 0.0940. Since this P-value is greater than the $\alpha = 0.01$ level of significance, we do not reject H_0.

Conclusion: There is not sufficient evidence at the $\alpha = 0.01$ level of significance to conclude that the standard deviation in population 1 is less than the standard deviation in population 2.

7. $H_0 : p_1 = p_2$

   $H_1 : p_1 \neq p_2$

   The level of significance is $\alpha = 0.05$. The two sample estimates are $\hat{p}_1 = \dfrac{451}{555} \approx 0.8126$

   and $\hat{p}_2 = \dfrac{510}{600} = 0.85$. The pooled estimate is

   $\hat{p} = \dfrac{451 + 510}{555 + 600} \approx 0.8320.$

   The test statistic is

   $$z_0 = \frac{\hat{p}_1 - \hat{p}_2}{\sqrt{\hat{p}(1-\hat{p})}\sqrt{\dfrac{1}{n_1} + \dfrac{1}{n_2}}}$$

   $$= \frac{0.8126 - 0.85}{\sqrt{0.8320(1-0.8320)}\sqrt{\dfrac{1}{555} + \dfrac{1}{600}}} \approx -1.70$$

   Classical approach: This is a two-tailed test, so the critical values for $\alpha = 0.05$ are

   $\pm z_{\alpha/2} = \pm z_{0.025} = \pm 1.96$. Since $z_0 = -1.70$

   is between $-z_{0.025} = -1.96$ and $z_{0.025} = 1.96$

(the test statistic does not fall in the critical region), we do not reject H_0.

P-value approach: The P-value is two times the area under the standard normal distribution to the left of the test statistic, $z_0 = -1.70$.

P-value $= 2 \cdot P(z \leq -1.70) = 2(0.0446)$
      $= 0.0892$ [Tech: 0.0895]

Since the P-value is greater than the $\alpha = 0.05$ level of significance, we do not reject H_0.

Conclusion: There is not sufficient evidence at the $\alpha = 0.05$ level of significance to conclude that the proportion in population 1 is different from the proportion in population 2.

8. (a) Since the same individual is used for both measurements, the sampling method is dependent.

   (b) We measure differences as $d_i = A_i - H_i$.

Student	1	2	3	4	5
Height, H_i	59.5	69	77	59.5	74.5
Arm span, A_i	62	65.5	76	63	74
$d_i = A_i - H_i$	2.5	−3.5	−1	3.5	−0.5

Student	6	7	8	9	10
Height, H_i	63	61.5	67.5	73	69
Arm span, A_i	66	61	69	70	71
$d_i = A_i - H_i$	3	−0.5	1.5	−3	2

   We compute the mean and standard deviation of the differences and obtain $\bar{d} = 0.4$ inches and $s_d \approx 2.4698$ inches, rounded to four decimal places.

   $H_0 : \mu_d = 0$

   $H_1 : \mu_d \neq 0$

   The level of significance is $\alpha = 0.05$. The test statistic is

   $$t_0 = \frac{\bar{d}}{\dfrac{s_d}{\sqrt{n}}} = \frac{0.4}{\dfrac{2.4698}{\sqrt{10}}} \approx 0.512.$$

   Classical approach: Since this is a two-tailed test with 9 degrees of freedom, the critical values are $\pm t_{0.025} = \pm 2.262$. Since $t_0 = 0.512$ falls between $-t_{0.025} = -2.262$ and $t_{0.025} = 2.262$ (the test statistic does not fall within the critical regions), we do not reject H_0.

_P_-value approach: The _P_-value for this two-tailed test is the area under the _t_-distribution with 9 degrees of freedom to the right of the test statistic $t_0 = 0.512$ plus the area to the left of $-t_0 = -0.512$. From the _t_-distribution table in the row corresponding to 9 degrees of freedom, 0.512 falls to the left of 0.703 whose right-tail area is 0.25. We must double this value in order to get the total area in both tails: 0.50. So, _P_-value > 0.50 [Tech: _P_-value = 0.6209]. Because the _P_-value is greater than the $\alpha = 0.05$ level of significance, we do not reject H_0.

Conclusion: There is not sufficient evidence at the $\alpha = 0.05$ level of significance to conclude that an individual's arm span is different from the individual's height. That is, the sample evidence does not contradict the belief that arm span and height are the same.

9. **(a)** The sampling method is independent since the cars selected for the McDonald's sample has no bearing on the cars chosen for the Wendy's sample.

   (b) The variable of interest is wait time and it is a quantitative variable.

   (c) $H_0 : \mu_{McD} = \mu_W$

   $H_1 : \mu_{McD} \neq \mu_W$

   We compute the means and standard deviations, rounded to four decimal places when necessary, of both samples and obtain $\overline{x}_{McD} = 133.601$, $s_{McD} \approx 39.6110$, $n_{McD} = 30$, and $\overline{x}_W \approx 219.1444$, $s_W \approx 102.8459$, and $n_W = 27$. Since the smaller sample size is $n_W = 27$, we use $n_W - 1 = 26$ degrees of freedom.

   The test statistic is
   $$t_0 = \frac{(\overline{x}_{McD} - \overline{x}_W) - (\mu_{McD} - \mu_W)}{\sqrt{\dfrac{s_{McD}^2}{n_{McD}} + \dfrac{s_W^2}{n_W}}}$$
   $$= \frac{(133.601 - 219.1444) - 0}{\sqrt{\dfrac{39.6110^2}{30} + \dfrac{102.8459^2}{27}}} \approx -4.059$$

   Classical approach: Since this is a two-tailed test with 26 degrees of freedom, the

critical values are $\pm t_{0.05} = \pm 1.706$. Since the test statistic $t_0 \approx -4.059$ is to the left of -1.706 (the test statistic falls within a critical region), we reject H_0.

_P_-value approach: The _P_-value for this two-tailed test is the area under the _t_-distribution with 26 degrees of freedom to the left of $t_0 \approx -4.059$ plus the area to the right of 4.059. From the _t_-distribution table in the row corresponding to 26 degrees of freedom, 4.059 falls to the right of 3.707 whose right-tail area is 0.0005. We must double this value in order to get the total area in both tails: 0.001. So, _P_-value < 0.001 [Tech: _P_-value = 0.0003]. Since the _P_-value is less than $\alpha = 0.1$, we reject H_0.

Conclusion: There is sufficient evidence at the $\alpha = 0.1$ level of significance to conclude that the population mean wait times are different for McDonald's and Wendy's drive-through windows.

(d) We compute the five-number summary for each restaurant:

McDonald's: 71.37, 111.84, 127.13, 147.28, 246.59 [Minitab: 71.37, 109.14, 127.13, 148.23, 246.59]

Wendy's: 71.02, 133.56, 190.91, 281.90, 471.62 [Minitab: 71.0, 133.6, 190.9, 281.9, 471.6]

Yes, the boxplots support the results from part (c). Based on the boxplot, it would appear to be the case that the wait time at McDonald's is less than the wait time at Wendy's drive-through windows.

10. (a) This is a completely randomized design with two treatments. The response variable is whether the subject has a bone fracture over the course of one year, or not. The treatments are 5 mg of Actonel versus the placebo.

(b) A double-blind experiment is one in which neither the subject nor the individual administering the treatment knows to which group (experimental or control) the subject belongs.

(c) We first verify the requirements to perform the hypothesis test: (1) We assume the data represent a simple random sample. (2) We have $x_{exp} = 27$, $n_{exp} = 696$, $x_{control} = 49$, and $n_{control} = 678$, so

$$\hat{p}_{exp} = \frac{27}{696} \approx 0.0388 \text{ and } \hat{p}_{control} = \frac{49}{678} \approx 0.0723 . \text{ So,}$$

$$n_{exp}\hat{p}_{exp}\left(1 - \hat{p}_{exp}\right) = 696(0.0388)(1 - 0.0388) \approx 26 \geq 10 \text{ and}$$

$$n_{control}\hat{p}_{control}\left(1 - \hat{p}_{control}\right) = 678(0.0723)(1 - 0.0723) \approx 45 \geq 10 . \text{ (3) Each sample is less than 5\% of the}$$

population. Thus, the requirements are met, and we can conduct the test.

$$H_0 : p_{exp} = p_{control}$$

$$H_1 : p_{exp} < p_{control}$$

From before, the two sample estimates are $\hat{p}_{exp} \approx 0.0388$ and $\hat{p}_{control} \approx 0.0723$. The pooled estimate is

$$\hat{p} = \frac{x_{exp} + x_{control}}{n_{exp} + n_{control}} = \frac{27 + 49}{696 + 678} \approx 0.0553 . \text{ The test statistic is}$$

$$z_0 = \frac{\hat{p}_{exp} - \hat{p}_{control}}{\sqrt{\hat{p}(1 - \hat{p})}\sqrt{\dfrac{1}{n_{exp}} + \dfrac{1}{n_{control}}}} = \frac{0.0388 - 0.0723}{\sqrt{0.0553(1 - 0.0553)}\sqrt{\dfrac{1}{696} + \dfrac{1}{678}}} \approx -2.71 .$$

Classical approach: This is a left-tailed test, so the critical value is $-z_{0.01} = -2.33$. Since $z_0 = -2.71 < -z_{0.01} = -2.33$ (the test statistic falls in the critical region), we reject H_0.

P-value approach: P-value $= P(z_0 \leq -2.71) = 0.0034$ [Tech: 0.0033]. Since this P-value is less than the $\alpha = 0.01$ level of significance, we reject H_0.

Conclusion: There is sufficient evidence at the $\alpha = 0.01$ level of significance to suggest that the drug is effective in preventing bone fractures.

(d) For a 95% confidence interval we use $\pm z_{0.025} = \pm 1.96$. Then:

$$\text{Lower bound} = (\hat{p}_{exp} - \hat{p}_{control}) - z_{\alpha/2} \cdot \sqrt{\frac{\hat{p}_{exp}(1 - \hat{p}_{exp})}{n_{exp}} + \frac{\hat{p}_{control}(1 - \hat{p}_{control})}{n_{control}}}$$

$$= (0.0388 - 0.0723) - 1.96 \cdot \sqrt{\frac{0.0388(1 - 0.0388)}{696} + \frac{0.0723(1 - 0.0723)}{678}} \approx -0.06$$

$$\text{Upper bound} = (\hat{p}_{exp} - \hat{p}_{control}) + z_{\alpha/2} \cdot \sqrt{\frac{\hat{p}_{exp}(1 - \hat{p}_{exp})}{n_{exp}} + \frac{\hat{p}_{control}(1 - \hat{p}_{control})}{n_{control}}}$$

$$= (0.0388 - 0.0723) + 1.96 \cdot \sqrt{\frac{0.0388(1 - 0.0388)}{696} + \frac{0.0723(1 - 0.0723)}{678}} \approx -0.01$$

We can be 95% confident that the difference in the proportion of women who experienced a bone fracture between the experimental and control group is between −0.06 and −0.01.

11. (a) $n = n_1 = n_2$

$$= \left[\hat{p}_1(1 - \hat{p}_1) + \hat{p}_2(1 - \hat{p}_2) \right] \left(\frac{z_{\alpha/2}}{E} \right)^2$$

$$= \left[0.188(1 - 0.188) + 0.205(1 - 0.205) \right] \left(\frac{1.645}{0.02} \right)^2$$

$$\approx 2135.3, \text{ which we must increase to } 2136.$$

(b) $n = n_1 = n_2 = 0.5 \left(\frac{z_{\alpha/2}}{E} \right)^2 = 0.5 \left(\frac{1.645}{0.02} \right)^2$

$$\approx 3382.5 \text{ which we increase to } 3383.$$

12. $H_0 : \sigma_{\text{McD}} = \sigma_{\text{W}}$

$H_1 : \sigma_{\text{McD}} < \sigma_{\text{W}}$

From problem 9, we know $s_{\text{McD}} \approx 39.6110$
and $s_{\text{W}} \approx 102.8459$. The test statistic is

$$F_0 = \frac{s_{\text{McD}}^2}{s_{\text{W}}^2} = \frac{39.6110^2}{102.8459^2} \approx 0.15 .$$

Classical approach: This is a left-tailed test
with $n_{\text{McD}} = 30$ and $n_{\text{W}} = 27$. The critical value
is $F_{0.95, 29, 26}$, which we approximate by

$$F_{0.95, 25, 25} = \frac{1}{F_{0.05, 25, 25}} = \frac{1}{1.96} \approx 0.51. \text{ Since}$$

$F_0 = 0.15 < F_{0.95, 29, 26} \approx 0.51$ (the test statistic
falls in a critical region), we reject H_0.

P-value approach: Using technology, we find
P-value < 0.0001. Since this P-value is less
than the $\alpha = 0.05$ level of significance, we
reject H_0.

Conclusion: There is sufficient evidence at the
$\alpha = 0.05$ level of significance to conclude that
the standard deviation in wait time at Wendy's
is more than the standard deviation in wait
time at McDonald's.

13. From problem 8, we have $\bar{d} = 0.4$ and
$s_d \approx 2.4698$. For 95% confidence, we use
$\alpha = 0.05$, so with 9 degrees of freedom, we
have $t_{\alpha/2} = t_{0.025} = 2.262$. Then:

$$\text{Lower bound} = \bar{d} - t_{0.025} \cdot \frac{s_d}{\sqrt{n}}$$

$$= 0.4 - 2.262 \cdot \frac{2.4698}{\sqrt{10}} \approx -1.37$$

$$\text{Upper bound} = \bar{d} + t_{0.025} \cdot \frac{s_d}{\sqrt{n}}$$

$$= 0.4 + 2.262 \cdot \frac{2.4698}{\sqrt{10}} \approx 2.17$$

We can be 95% confident that the mean
difference between height and arm span is
between -1.37 and 2.17. The interval
contains zero, so we conclude that there is not
sufficient evidence at the $\alpha = 0.05$ level of
significance to reject the claim that arm span
and height are equal.

14. From problem 9, we have $\bar{x}_{\text{McD}} = 133.601$, $s_{\text{McD}} \approx 39.6110$, $n_{\text{McD}} = 30$, $\bar{x}_{\text{W}} \approx 219.1444$, $s_{\text{W}} \approx 102.8459$,
and $n_{\text{W}} = 27$. For a 95% confidence interval with 26 degrees of freedom, we use $t_{\alpha/2} = t_{0.025} = 2.056$. Then:

$$\text{Lower bound} = (\bar{x}_{\text{McD}} - \bar{x}_{\text{W}}) - t_{\alpha/2} \cdot \sqrt{\frac{s_{\text{McD}}^2}{n_{\text{McD}}} + \frac{s_{\text{W}}^2}{n_{\text{W}}}} = (133.601 - 219.1444) - 2.056 \cdot \sqrt{\frac{39.6110^2}{30} + \frac{102.8459^2}{27}}$$

$$\approx -128.869 \quad [\text{Tech:} -128.422]$$

$$\text{Upper bound} = (\bar{x}_{\text{McD}} - \bar{x}_{\text{W}}) + t_{\alpha/2} \cdot \sqrt{\frac{s_{\text{McD}}^2}{n_{\text{McD}}} + \frac{s_{\text{W}}^2}{n_{\text{W}}}} = (133.601 - 219.1444) + 2.056 \cdot \sqrt{\frac{39.6110^2}{30} + \frac{102.8459^2}{27}}$$

$$\approx -42.218 \quad [\text{Tech:} -42.665]$$

We can be 95% confident that the mean difference in wait times between McDonald's and Wendy's drive-
through windows is between -128.869 and -42.218 seconds [Tech: between -128.4 and -42.67].

Answer will vary. One possibility is that a marketing campaign could be initiated by McDonald's touting the
fact that wait times are up to 2 minutes less at McDonald's.

Chapter 11 Test

1. This is independent because the subjects are randomly selected from two distinct populations.

2. This is dependent since the members of the two samples are matched by type of crime committed.

3. **(a)** We compute $d_i = X_i - Y_i$.

Obs	1	2	3	4	5	6	7
X_i	18.5	21.8	19.4	22.9	18.3	20.2	23.1
Y_i	18.3	22.3	19.2	22.3	18.9	20.7	23.9
d_i	0.2	−0.5	0.2	0.6	−0.6	−0.5	−0.8

(b) Using technology, $\bar{d} = -0.2$ and $s_d \approx 0.5260$, rounded to four decimal places.

(c) $H_0 : \mu_d = 0$
$H_1 : \mu_d \neq 0$
The level of significance is $\alpha = 0.01$.
The test statistic is
$$t_0 = \frac{\bar{d}}{\frac{s_d}{\sqrt{n}}} = \frac{-0.2}{\frac{0.5260}{\sqrt{7}}} \approx -1.006 .$$

Classical approach: Since this is a two-tailed test with 6 degrees of freedom, the critical values are $\pm t_{0.005} = \pm 3.707$. Since $t_0 \approx -1.006$ falls between $-t_{0.005} = -3.707$ and $t_{0.005} = 3.707$ (the test statistic does not fall within the critical regions), we do not reject H_0.

P-value approach: The P-value for this two-tailed test is the area under the t-distribution with 6 degrees of freedom to the left of the test statistic $t_0 = -1.006$ plus the area to the right of $t_0 = 1.006$. From the t-distribution table in the row corresponding to 6 degrees of freedom, 1.006 falls between 0.906 and 1.134 whose right-tail areas are 0.20 and 0.15, respectively. We must double these values in order to get the total area in both tails: 0.40 and 0.30. So, $0.30 < P$-value < 0.40 [Tech: P-value ≈ 0.3532]. Because the P-value is greater than the $\alpha = 0.01$ level of significance, we do not reject H_0.

Conclusion: There is not sufficient evidence at the $\alpha = 0.01$ level of significance to conclude that the population mean difference is different from zero.

(d) For $\alpha = 0.05$ and 7 degrees of freedom, $t_{\alpha/2} = t_{0.025} = 2.447$. Then:

Lower bound
$$= \bar{d} - t_{0.025} \cdot \frac{s_d}{\sqrt{n}}$$
$$= -0.2 - 2.447 \cdot \frac{0.5260}{\sqrt{7}} \approx -0.69$$

Upper bound
$$= \bar{d} - t_{0.025} \cdot \frac{s_d}{\sqrt{n}}$$
$$= -0.2 + 2.447 \cdot \frac{0.5260}{\sqrt{7}} \approx 0.29$$

We are 95% confident that the population mean difference is between −0.69 and 0.29.

4. **(a)** $H_0 : \mu_1 = \mu_2$
$H_1 : \mu_1 \neq \mu_2$
The level of significance is $\alpha = 0.1$.
Since the smaller sample size is $n_1 = 24$, we use $n_1 - 1 = 23$ degrees of freedom.
The test statistic is
$$t_0 = \frac{(\bar{x}_1 - \bar{x}_2) - (\mu_1 - \mu_2)}{\sqrt{\frac{s_1^2}{n_1} + \frac{s_2^2}{n_2}}}$$
$$= \frac{(104.2 - 110.4) - 0}{\sqrt{\frac{12.3^2}{24} + \frac{8.7^2}{27}}} \approx -2.054$$

Classical approach: Since this is a two-tailed test with 23 degrees of freedom, the critical values are $\pm t_{0.05} = \pm 1.714$. Since the test statistic $t_0 \approx -2.054$ is to the left of the critical value −1.714 (the test statistic falls within a critical region), we reject H_0.

P-value approach: The P-value for this two-tailed test is the area under the t-distribution with 23 degrees of freedom to the left of $t_0 = -2.054$ plus the area to the right of $t_0 = 2.054$. From the t-distribution table in the row corresponding to 23 degrees of freedom, 2.054 falls between 1.714 and 2.069, whose right-tail areas are 0.05 and 0.025, respectively. We must double these values in order to get the total area in both tails: 0.10 and 0.05. Thus, $0.05 < P\text{-value} < 0.10$ [Tech: $P\text{-value} \approx 0.0464$]. Because the P-value is less than the $\alpha = 0.1$ level of significance, we reject H_0.

Conclusion: There is sufficient evidence at the $\alpha = 0.1$ level of significance to conclude that the means are different.

(b) For a 95% confidence interval with 23 degrees of freedom, we use $t_{\alpha/2} = t_{0.025} = 2.069$. Then:

Lower bound

$$= (\overline{x}_1 - \overline{x}_2) - t_{\alpha/2} \cdot \sqrt{\frac{s_1^2}{n_1} + \frac{s_2^2}{n_2}}$$

$$= (104.2 - 110.4) - 2.069 \cdot \sqrt{\frac{12.3^2}{24} + \frac{8.7^2}{27}}$$

$$\approx -12.44 \quad \text{[Tech: } -12.3\text{]}$$

Upper bound:

$$= (\overline{x}_1 - \overline{x}_2) + t_{\alpha/2} \cdot \sqrt{\frac{s_1^2}{n_1} + \frac{s_2^2}{n_2}}$$

$$= (104.2 - 110.4) + 2.069 \cdot \sqrt{\frac{12.3^2}{24} + \frac{8.7^2}{27}}$$

$$\approx 0.04 \quad \text{[Tech:} -0.10\text{]}$$

We are 95% confident that the population mean difference is between -12.44 and 0.04 [Tech: between -12.3 and -0.10].

(c) The hypotheses are $H_0 : \sigma_1 = \sigma_2$ versus $H_1 : \sigma_1 > \sigma_2$. The test statistic is

$$F_0 = \frac{s_1^2}{s_2^2} = \frac{12.3^2}{8.7^2} \approx 2.00.$$

Classical approach: This is a right-tailed test with $n_1 = 24$, $n_2 = 27$, and $\alpha = 0.10$. The critical value is $F_{0.10, 23, 26}$, which we approximate by $F_{0.10, 25, 25} = 1.68$. Since

$F_0 = 2.00 > F_{0.10, 23, 26} \approx 1.68$ (the test

statistic falls in the critical region), we reject H_0.

P-value approach: Using technology, we find $P\text{-value} \approx 0.0448$. Since this P-value is less than the $\alpha = 0.05$ level of significance, we reject H_0.

Conclusion: There is sufficient evidence at the $\alpha = 0.1$ level of significance to conclude that the standard deviation in population 1 is greater than the standard deviation in population 2.

5. (a) $H_0 : \mu_1 = \mu_2$

$H_1 : \mu_1 < \mu_2$

The level of significance is $\alpha = 0.05$. Since the smaller sample size is $n_2 = 8$, we use $n_2 - 1 = 7$ degrees of freedom. The test statistic is

$$t_0 = \frac{(\overline{x}_1 - \overline{x}_2) - (\mu_1 - \mu_2)}{\sqrt{\frac{s_1^2}{n_1} + \frac{s_2^2}{n_2}}}$$

$$= \frac{(96.6 - 98.3) - 0}{\sqrt{\frac{3.2^2}{13} + \frac{2.5^2}{8}}} \approx -1.357$$

Classical approach: Since this is a left-tailed test with 7 degrees of freedom, the critical value is $-t_{0.05} = -1.895$. Since $t_0 = -1.375 > -t_{0.05} = -1.895$ (the test statistic does not fall within the critical region), we do not reject H_0.

P-value approach: The P-value for this left-tailed test is the area under the t-distribution with 7 degrees of freedom to the left of $t_0 = -1.375$, which is equivalent to the area to the right of $t_0 = 1.375$. From the t-distribution table in the row corresponding to 7 degrees of freedom, 1.375 falls between 1.119 and 1.415 whose right-tail areas are 0.15 and 0.10, respectively. Thus, $0.10 < P\text{-value} < 0.15$ [Tech: $P\text{-value} \approx 0.0959$]. Because the P-value is greater than the $\alpha = 0.05$ level of significance, we do not reject H_0.

Conclusion: There is not sufficient evidence at the $\alpha = 0.05$ level of significance to conclude that the mean of population 1 is less than the mean of population 2.

(b) $H_0 : \sigma_1 = \sigma_2$

$H_1 : \sigma_1 \neq \sigma_2$

The test statistic is $F_0 = \dfrac{s_1^2}{s_2^2} = \dfrac{3.2^2}{2.5^2} \approx 1.64$.

Classical approach: This is a two-tailed test with $n_1 = 13$, $n_2 = 8$, and $\alpha = 0.05$. The critical values are $F_{0.025,12,7}$, which we approximate by $F_{0.025,10,7} = 4.76$ and

$F_{0.975,12,7} = \dfrac{1}{F_{0.025,7,12}} = \dfrac{1}{3.61} \approx 0.28$. Since

$F_0 = 1.64$ is between $F_{0.025,12,7} \approx 4.76$ and $F_{0.975,12,7} \approx 0.28$ (the test statistic does not fall in the critical region), we do not reject H_0.

P-value approach: Using technology, we find P-value ≈ 0.5243. Since this P-value is greater than the $\alpha = 0.05$ level of significance, we do not reject H_0.

Conclusion: There is not sufficient evidence at the $\alpha = 0.05$ level of significance to conclude that the standard deviation in population 1 is different from the standard deviation in population 2.

6. $H_0 : p_1 = p_2$

$H_1 : p_1 < p_2$

The level of significance is $\alpha = 0.05$. The two sample estimates are $\hat{p}_1 = \dfrac{156}{650} = 0.24$ and

$\hat{p}_2 = \dfrac{143}{550} = 0.26$. The pooled estimate is

$\hat{p} = \dfrac{156 + 143}{650 + 550} \approx 0.2492$.

The test statistic is

$$z_0 = \frac{\hat{p}_1 - \hat{p}_2}{\sqrt{\hat{p}(1-\hat{p})}\sqrt{\dfrac{1}{n_1} + \dfrac{1}{n_2}}}$$

$$= \frac{0.24 - 0.26}{\sqrt{0.2492(1 - 0.2492)}\sqrt{\dfrac{1}{650} + \dfrac{1}{550}}} \approx -0.80$$

Classical approach: This is a left-tailed test, so the critical value is $-z_{0.05} = -1.645$. Since $z_0 = -0.80 > -z_{0.05} = -1.645$ (the test statistic does not fall in the critical region), we do not reject H_0.

P-value approach: The P-value is the area under the standard normal distribution to the left of the test statistic, $z_0 = -0.80$. P-value = $P(z \leq -0.80) = 0.2119$ [Tech: 0.2124]. Since the P-value is greater than the $\alpha = 0.05$ level of significance, we do not reject H_0.

Conclusion: There is not sufficient evidence at the $\alpha = 0.05$ level of significance to conclude that the proportion in population 1 is less than the proportion in population 2.

7. (a) Since the dates selected for the Texas sample have no bearing on the dates chosen for the Illinois sample, the testing method is independent.

(b) Because the sample sizes are small, both samples must come from populations that are normally distributed.

(c) We compute the five-number summary for each restaurant:

Texas: 4.11, 4.555, 4.735, 5.12, 5.22 [Minitab: 4.11, 4.508, 4.735, 5.13, 5.22]

Illinois: 4.22, 4.40, 4.505, 4.64, 4.75 [Minitab: 4.22, 4.39, 4.505, 4.6525, 4.75]

The boxplots that follow indicate that rain in Chicago, Illinois has a lower pH than rain in Houston, Texas.

Illinois

Texas

4.0   4.5   5.0
Acidity of Rain (pH)

(d) $H_0 : \mu_{\text{Texas}} = \mu_{\text{Illinois}}$

$H_1 : \mu_{\text{Texas}} \neq \mu_{\text{Illinois}}$

We compute the summary statistics, rounded to four decimal places when necessary:

$\bar{x}_{\text{Texas}} = 4.77$, $s_{\text{Texas}} \approx 0.3696$, $n_{\text{Texas}} = 12$,

$\bar{x}_{\text{Illinois}} \approx 4.5093$, $s_{\text{Illinois}} \approx 0.1557$, and

$n_{\text{Illinois}} = 14$. Since the smaller sample size is $n_{\text{Texas}} = 12$, we use $n_{\text{Texas}} - 1 = 11$ degrees of freedom. The test statistic is

$$t_0 = \frac{(\bar{x}_{\text{Texas}} - \bar{x}_{\text{Illinois}}) - (\mu_{\text{Texas}} - \mu_{\text{Illinois}})}{\sqrt{\dfrac{s_{\text{Texas}}^2}{n_{\text{Texas}}} + \dfrac{s_{\text{Illinois}}^2}{n_{\text{Illinois}}}}}$$

$$= \frac{(4.77 - 4.5093) - 0}{\sqrt{\dfrac{0.3696^2}{12} + \dfrac{0.1557^2}{14}}} \approx 2.276$$

Classical approach: Since this is a two-tailed test with 11 degrees of freedom, the critical values are $\pm t_{0.025} = \pm 2.201$. Since the test statistic $t_0 \approx 2.276$ is to the right of 2.201 (the test statistic falls within a critical region), we reject H_0.

P-value approach: The P-value for this two-tailed test is the area under the t-distribution with 11 degrees of freedom to the right of $t_0 \approx 2.276$ plus the area to the left of $-t_0 \approx -2.276$. From the t-distribution table in the row corresponding to 11 degrees of freedom, 2.276 falls between 2.201 and 2.328, whose right-tail areas are 0.025 and 0.02, respectively. We must double these values in order to get the total area in both tails: 0.05 and 0.04. So, $0.04 < P\text{-value} < 0.05$ [Tech: $P\text{-value} \approx 0.0387$]. Since the P-value is less than $\alpha = 0.05$, we reject H_0.

Conclusion: There is sufficient evidence at the $\alpha = 0.05$ level of significance to conclude that the acidity of the rain near

Houston, Texas is different from the acidity of rain near Chicago, Illinois.

8. (a) The response variable is the student's GPA. The explanatory variable is whether the student has a sleep disorder or not.

(b) $H_0 : \mu_{\text{Sleep Disorder}} = \mu_{\text{No Sleep Disorder}}$

$H_1 : \mu_{\text{Sleep Disorder}} < \mu_{\text{No Sleep Disorder}}$

The level of significance is $\alpha = 0.05$. Since the smaller sample size is

$n_{\text{Sleep Disorder}} = 503$, we use

$n_{\text{Sleep Disorder}} - 1 = 503 - 1 = 502$ degrees of freedom. The test statistic is

$$t_0 = \frac{(\bar{x}_{\text{Sleep Disorder}} - \bar{x}_{\text{No Sleep Disorder}}) - (0)}{\sqrt{\dfrac{s_{\text{Sleep Disorder}}^2}{n_{\text{Sleep Disorder}}} + \dfrac{s_{\text{No Sleep Disorder}}^2}{n_{\text{No Sleep Disorder}}}}}$$

$$= \frac{(2.65 - 2.82) - 0}{\sqrt{\dfrac{0.87^2}{503} + \dfrac{0.83^2}{1342}}}$$

$$\approx -3.784$$

Classical approach: Since this is a left-tailed test with 502 degrees of freedom, we use the closest table value, 100 degrees of freedom, and the critical value is $t_{0.05} = 1.660$. Since the test statistic $t_0 \approx -3.784$ is to the left of the critical value (i.e., since the test statistic falls within the critical region), we reject H_0.

P-value approach: The P-value for this left-tailed test is the area under the t-distribution with 502 degrees of freedom to the left of $t_0 \approx -3.784$, which is equivalent to the area to the right of $-t_0 \approx 3.784$. From the t-distribution table, we use the row corresponding to 100 degrees of freedom, 3.784 falls to the right of 3.390 whose right-tail area is 0.0005. Thus, $P\text{-value} < 0.0005$. [Tech: $P\text{-value} < 0.0001$]. Because the P-value is less than the level of significance, $\alpha = 0.05$, we reject H_0.

Conclusion: There is sufficient evidence at the $\alpha = 0.05$ level of significance to conclude that sleep disorders adversely affect a student's GPA.

9. We measure differences as $d_i = SUV_i - Car_i$.

SUV_i	2091	1338	1053	1872
Car_i	1510	2559	4921	4555
d_i	581	−1221	−3868	−2683

SUV_i	1428	2208	6015
Car_i	5114	5203	3852
d_i	−3686	−2995	2163

We compute the mean and standard deviation of the differences and obtain $\bar{d} \approx -1672.71$ minutes, rounded to two decimal places, and $s_d \approx 2296.29$ minutes, rounded to two decimal places.

$H_0 : \mu_d = 0$

$H_1 : \mu_d < 0$

The level of significance is $\alpha = 0.05$. The test statistic is $t_0 = \dfrac{\bar{d}}{\dfrac{s_d}{\sqrt{n}}} = \dfrac{-1672.71}{\dfrac{2296.29}{\sqrt{7}}} \approx -1.927$.

Classical approach: Since this is a left-tailed test with $7 - 1 = 6$ degrees of freedom, the critical value is $-t_{0.10} = -1.440$. Since the test statistic $t_0 \approx -1.927$ falls to the left of the critical value $-t_{0.10} = -1.440$ (i.e., since the test statistic falls in the critical region), we reject H_0 .

P-value approach: The P-value for this left-tailed test is the area under the t-distribution with $7 - 1 = 6$ degrees of freedom to the left of the test statistic $t_0 \approx -1.927$, which is equivalent to the area to the right of 1.927. From the t-distribution table, in the row corresponding to 6 degrees of freedom, 1.927 falls between 1.440 and 1.943, whose right-tail areas are 0.10 and 0.05, respectively. So, $0.05 < P\text{-value} < 0.10$ [Tech: P-value = 0.0511]. Because the P-value is less than the level of significance, $\alpha = 0.10$, we reject H_0 .

Conclusion: There is sufficient evidence at the $\alpha = 0.10$ level of significance to conclude that the repair cost for the car is higher.

10. (a) This is a completely randomized design. The subjects were randomly divided into two groups.

(b) The response variable is whether the subjects get dry mouth or not.

(c) We verify the requirements to perform the hypothesis test: (1) We treat the sample as if it was a simple random sample.

(2) We have $x_{exp} = 77$, $n_{exp} = 553$, $x_{control} = 34$, and $n_{control} = 373$, so

$$\hat{p}_{exp} = \frac{x_{exp}}{n_{exp}} = \frac{77}{553} \approx 0.1392 \text{ and}$$

$$\hat{p}_{control} = \frac{x_{control}}{n_{control}} = \frac{34}{373} \approx 0.0912. \text{ So,}$$

$n_{exp}\hat{p}_{exp}\left(1 - \hat{p}_{exp}\right) = 553(0.1392)(1 - 0.1392)$

$\approx 66 \geq 10$ and $n_{control}\hat{p}_{control}\left(1 - \hat{p}_{control}\right) =$

$373(0.0912)(1 - 0.0912) \approx 31 \geq 10$. (3)

Each sample is less than 5% of the population. Thus, the requirements are met, and we can conduct the test.

$H_0 : p_{exp} = p_{control}$

$H_1 : p_{exp} > p_{control}$

From before, the two sample estimates are $\hat{p}_{exp} \approx 0.1392$ and $\hat{p}_{control} \approx 0.0912$. The pooled estimate is

$$\hat{p} = \frac{x_{exp} + x_{control}}{n_{exp} + n_{control}} = \frac{77 + 34}{553 + 373} \approx 0.1199 .$$

The test statistic is

$$z_0 = \frac{\hat{p}_{exp} - \hat{p}_{control}}{\sqrt{\hat{p}(1 - \hat{p})}\sqrt{\dfrac{1}{n_{exp}} + \dfrac{1}{n_{control}}}}$$

$$= \frac{0.1392 - 0.0912}{\sqrt{0.1199(1 - 0.1199)}\sqrt{\dfrac{1}{553} + \dfrac{1}{373}}}$$

≈ 2.21

Classical approach: This is a right-tailed test, so the critical value is $z_{0.05} = 1.645$. Since $z_0 = 2.21 > z_{0.05} = 1.645$ (the test statistic falls in the critical region), we reject H_0 .

P-value approach: The P-value is the area under the standard normal curve to the right of $z_0 = 2.21$.

$P\text{-value} = P(z_0 \geq 2.21)$

$\qquad = 1 - P(z_0 < 2.21)$

$\qquad = 1 - 0.9864$

$\qquad = 0.0136$

Since this P-value is less than the $\alpha = 0.05$ level of significance, we reject H_0.

Conclusion: There is sufficient evidence at the $\alpha = 0.05$ level of significance to conclude that a higher proportion of subjects in the experimental group experienced dry mouth than in the control group.

---

11. We have $x_{\mathrm{F}} = 644$, $n_{\mathrm{F}} = 2800$, $x_{\mathrm{M}} = 840$, and $n_{\mathrm{M}} = 2800$, so $\hat{p}_{\mathrm{F}} = \dfrac{x_{\mathrm{F}}}{n_{\mathrm{F}}} = \dfrac{644}{2800} = 0.23$ and

$\hat{p}_{\mathrm{M}} = \dfrac{x_{\mathrm{M}}}{n_{\mathrm{M}}} = \dfrac{840}{2800} = 0.3$.

$H_0 : p_{\mathrm{M}} = p_{\mathrm{F}}$

$H_1 : p_{\mathrm{M}} \neq p_{\mathrm{F}}$

For a 90% confidence interval we use $\pm z_{0.05} = \pm 1.645$. Then:

Lower bound:

$$(\hat{p}_{\mathrm{M}} - \hat{p}_{\mathrm{F}}) - z_{\alpha/2} \cdot \sqrt{\frac{\hat{p}_{\mathrm{M}}(1-\hat{p}_{\mathrm{M}})}{n_M} + \frac{\hat{p}_{\mathrm{F}}(1-\hat{p}_{\mathrm{F}})}{n_F}} = (0.3 - 0.23) - 1.645 \cdot \sqrt{\frac{0.3(1-0.3)}{2800} + \frac{0.23(1-0.23)}{2800}} \approx 0.051$$

Upper bound:

$$(\hat{p}_{\mathrm{M}} - \hat{p}_{\mathrm{F}}) + z_{\alpha/2} \cdot \sqrt{\frac{\hat{p}_{\mathrm{M}}(1-\hat{p}_{\mathrm{M}})}{n_M} + \frac{\hat{p}_{\mathrm{F}}(1-\hat{p}_{\mathrm{F}})}{n_F}} = (0.3 - 0.23) + 1.645 \cdot \sqrt{\frac{0.3(1-0.3)}{2800} + \frac{0.23(1-0.23)}{2800}} \approx 0.089$$

We are 90% confident that the difference in the proportions for males and females for which hypnosis led to quitting smoking is between 0.051 and 0.089. Because the confidence interval does not include 0, we reject the null hypothesis. There is sufficient evidence at the $\alpha = 0.1$ level of significance to conclude that the proportion of males and females for which hypnosis led to quitting smoking is different.

---

12. (a) $n = n_1 = n_2$

$$= \left[\hat{p}_1(1-\hat{p}_1) + \hat{p}_2(1-\hat{p}_2) \right] \left(\frac{z_{\alpha/2}}{E} \right)^2$$

$$= \left[0.322(1-0.322) + 0.111(1-0.111) \right] \left(\frac{1.96}{0.04} \right)^2$$

$$\approx 761.1, \text{which we must increase to 762.}$$

(b)

$n = n_1 = n_2$

$$= 0.5 \left(\frac{z_{\alpha/2}}{E} \right)^2 = 0.5 \left(\frac{1.96}{0.04} \right)^2$$

$$= 1200.5, \text{ which we must increase to 1201.}$$

13. We have $n_{\mathrm{p}} = 72$, $\bar{x}_{\mathrm{p}} = 1.9$, $s_{\mathrm{p}} = 0.22$, $n_{\mathrm{n}} = 75$, $\bar{x}_{\mathrm{n}} = 0.8$, and $s_{\mathrm{n}} = 0.21$. The smaller sample size is $n_{\mathrm{p}} = 72$, so we have $n_{\mathrm{p}} - 1 = 71$ degrees of freedom. Since our t-distributions table does not have a row for

df $= 71$, we will use df $= 70$. For a 95% confidence interval with 70 degrees of freedom, we use $t_{\alpha/2} = t_{0.025} = 1.994$. Then:

Lower bound

$$= (\bar{x}_{\mathrm{p}} - \bar{x}_{\mathrm{n}}) - t_{\alpha/2} \cdot \sqrt{\frac{s_p^2}{n_p} + \frac{s_n^2}{n_n}}$$

$$= (1.9 - 0.8) - 1.994 \cdot \sqrt{\frac{0.22^2}{72} + \frac{0.21^2}{75}}$$

$$\approx 1.03$$

Upper bound

$$= (\bar{x}_{\mathrm{p}} - \bar{x}_{\mathrm{n}}) + t_{\alpha/2} \cdot \sqrt{\frac{s_p^2}{n_p} + \frac{s_n^2}{n_n}}$$

$$= (1.9 - 0.8) + 1.994 \cdot \sqrt{\frac{0.22^2}{72} + \frac{0.21^2}{75}}$$

$$\approx 1.17$$

We can be 95% confident that the mean difference in weight of the placebo group versus the Naltrexone group is between 1.03 and 1.17 pounds. Because the confidence interval does not contain 0, we reject H_0.
There is sufficient evidence to conclude that Naltrexone is effective in preventing weight gain among individuals who quit smoking. Answers will vary regarding practical significance, but one must ask, "Do I want to take a drug so that I can keep about 1 pound off?" Probably not.

14. $H_0 : \sigma_{\text{Texas}} = \sigma_{\text{Illinois}}$

 $H_1 : \sigma_{\text{Texas}} \neq \sigma_{\text{Illinois}}$

From problem 7, we know $s_{\text{Texas}} \approx 0.3696$, $n_{\text{Texas}} = 12$, $s_{\text{Illinois}} \approx 0.1557$, and $n_{\text{Illinois}} = 14$. The test statistic is

$$F_0 = \frac{s_{\text{Texas}}^2}{s_{\text{Illinois}}^2} = \frac{0.3696^2}{0.1557^2} \approx 5.63 \ [\text{Tech: } 5.64].$$

Classical approach: This is a two-tailed test with $n_{\text{Texas}} = 12$ and $n_{\text{Illinois}} = 14$. The critical values are $F_{0.025,11,13} \approx F_{0.025,12,15} = 2.96$ and

$$F_{0.975,11,13} = \frac{1}{F_{0.025,13,11}} \approx \frac{1}{F_{0.025,15,12}} = \frac{1}{3.18} \approx 0.31.$$

Since $F_0 = 5.63 > F_{0.025,11,13} \approx 2.96$ (the test statistic falls in a critical region), we reject H_0.

P-value approach: Using technology, we find P-value ≈ 0.0044. Since this P-value is less than the $\alpha = 0.05$ level of significance, we reject H_0.

Conclusion: There is sufficient evidence at the $\alpha = 0.05$ level of significance to conclude that the standard deviation in the acidity of rain near Houston, Texas is different from the standard deviation in the acidity of rain near Chicago, Illinois.

Case Study: Control in the Design of an Experiment

1. Since the individuals selected for the ESRD sample have no bearing on the individuals selected for the control group, the testing method is independent.

2. Hypothesis Tests for each clinical and biochemical parameter:

For each test of parameters:

 $H_0 : \mu_{\text{ESRD}} = \mu_{\text{control}}$

 $H_1 : \mu_{\text{ESRD}} \neq \mu_{\text{control}}$

We use technology to find each P-value. The sample sizes are $n_{\text{ESRD}} = 75$ and $n_{\text{control}} = 57$, so we use $n_{\text{control}} - 1 = 56$ degrees of freedom.

Clinical Characteristics

Age: P-value ≈ 0.0024

Sex: Sample statistics indicate no difference.

Body surface area: P-value ≈ 0.00002

Body mass index: P-value ≈ 0.0034

Systolic BP: P-value ≈ 0.3437

Diastolic BP: P-value ≈ 0.0636

Mean BP: P-value ≈ 0.7442

Pulse pressure: P-value ≈ 0.0092

Heart rate: P-value ≈ 0.000006

Biological Findings

Total cholesterol: P-value ≈ 0.0469

HDL cholesterol: P-value ≈ 0.00001

Triglycerides: P-value ≈ 0.0006

Serum albumin: P-value $\approx 2.6 \times 10^{-17}$

Plasma fibrinogen: P-value $\approx 4.0 \times 10^{-17}$

Plasma creatinine: P-value $\approx 2.5 \times 10^{-61}$

Blood urea: P-value $\approx 1.5 \times 10^{-97}$

Calcium: P-value ≈ 0.5675

Phosphates: P-value $\approx 2.5 \times 10^{-32}$

Thus, the parameters for age, body surface area, body mass index, pulse pressure, heart rate, HDL cholesterol, Triglycerides, serum albumin, plasma fibrinogen, plasma creatinine, blood urea, and phosphates all have P-values less than 0.01. In addition, the parameter for total cholesterol has a P-value less than 0.05. For each of these parameters, we would reject the null hypothesis and conclude that

$\mu_{ESRD} \neq \mu_{control}$. In other words, for each of these parameters, the ESRD group and the control group are different.

The parameters for sex, systolic BP, diastolic BP, mean BP, and calcium each had P-values above 0.05. For each of these parameters, we would not reject the null hypothesis and conclude that $\mu_{ESRD} = \mu_{control}$. In other words, for each of these parameters, the ESRD group and the control group have equivalent means.

3. Assumptions will vary. Additional information and statistical procedures desired for analysis will vary.

4. Based on the above findings, it does not appear that the ESRD and control groups have similar initial clinical characteristics and biochemical findings. It does not appear that the authors of the article were successful in reducing the likelihood that a confounding effect might obscure the results.

5. Using the nonrandom process, the authors of the article were able to sufficiently control for sex and blood pressure as they desired, but they were not able to control for the other parameters. In fact, one could argue that the authors, by using the nonrandom process, actually caused potential confounding effects to be more prevalent. Thus, Dr. Nicholls can determine that the effort to control confounding effects by using the nonrandom technique did not work well. She would likely be better off using a random process to assign patients to the treatment and control groups. For example, a stratified random assignment would allow for control of some variables while maintaining randomization for the remaining variables.

6. Reports and recommendations will vary.

Chapter 12
Inference on Categorical Data

Section 12.1

1. True. As the number of degrees of freedom increases, the distribution becomes more symmetric.

3. Expected counts; np_i

5.

$n = 500$				
p_i	0.2	0.1	0.45	0.25
Expected counts	100	50	225	125

7. (a) $\chi^2 = \dfrac{(30-25)^2}{25} + \dfrac{(20-25)^2}{25} + \dfrac{(28-25)^2}{25} + \dfrac{(22-25)^2}{25}$
$= 2.72$

(b) $df = 4-1 = 3$

(c) $\chi^2_{0.05} = 7.815$

(d) Do not reject H_0. There is insufficient evidence to conclude that at least one of the proportions is different from the others.

9. (a)
$\chi^2 = \dfrac{(1-1.6)^2}{1.6} + \dfrac{(38-25.6)^2}{25.6} + \dfrac{(132-153.6)^2}{153.6} + \dfrac{(440-409.6)^2}{409.6} + \dfrac{(389-409.6)^2}{409.6}$
≈ 12.56

(b) $df = 5-1 = 4$

(c) $\chi^2_{0.05} = 9.488$

(d) Reject H_0. There is sufficient evidence to conclude that the random variable X is not binomial with $n = 4$ and $p = 0.8$.

11. H_0: the distribution of colors is as stated
H_1: the distribution is different from that stated

Color	Observed	Expected
Brown	57	52
Yellow	64	56
Red	54	52
Blue	75	96
Orange	86	80
Green	64	64

$\chi^2 = \dfrac{(57-52)^2}{52} + \dfrac{(64-56)^2}{56} + \dfrac{(54-52)^2}{52} + \dfrac{(75-96)^2}{96} + \dfrac{(86-80)^2}{80} + \dfrac{(64-64)^2}{64}$
≈ 6.74

$df = 6-1 = 5$

$\chi^2_{0.05} = 11.07$

P-value ≈ 0.241 [Tech: 0.240]

Do not reject H_0. There is not sufficient evidence to conclude that the distribution is different than that which was stated.

13. (a) In order to increase the likelihood that the person actually is guilty if the test deems so, use $\alpha = 0.01$.

(b) H_0: the digits obey Benford's Law
H_1: the digits do not obey Benford's Law

First digit	Observed	Expected
1	36	60.2
2	32	35.2
3	28	25
4	26	19.4
5	23	15.8
6	17	13.4
7	15	11.6
8	16	10.2
9	7	9.2

$$\chi^2 = \frac{(36-60.2)^2}{60.2} + \frac{(32-35.2)^2}{35.2} + \frac{(28-25)^2}{25} +$$
$$\frac{(26-19.4)^2}{19.4} + \frac{(23-15.8)^2}{15.8} + \frac{(17-13.4)^2}{13.4} +$$
$$\frac{(15-11.6)^2}{11.6} + \frac{(16-10.2)^2}{10.2} + \frac{(7-9.2)^2}{9.2}$$
$$\approx 21.69$$

$df = 9 - 1 = 8$

$\chi^2_{0.01} = 20.09$

P-value ≈ 0.006

Reject H_0. There is sufficient evidence to conclude that the first digits in the checks do not obey Benford's Law.

(c) Yes, it is likely that the employee is guilty of embezzlement because it is likely that the checks do not follow Benford's Law.

15. (a)

H_0: the distribution of fatal injuries are the same
H_1: the distribution of fatal injuries are not the same

Location of injury	Observed	Expected
Multiple Locations	1036	1178.76
Head	864	641.08
Neck	38	62.04
Thorax	83	124.08
Abdomen/Lumbar/Spine	47	62.04

$$\chi^2 = \frac{(1036-1178.76)^2}{1178.76} + \frac{(864-641.08)^2}{641.08} +$$
$$\frac{(38-62.04)^2}{62.04} + \frac{(83-124.08)^2}{124.08} + \frac{(47-62.04)^2}{62.04}$$
$$\approx 121.37$$

$df = 9 - 1 = 8$

$\chi^2_{0.05} = 9.49$

P-value ≈ 0.000

Reject H_0. There is sufficient evidence to conclude that the distribution of fatal injuries for riders not wearing a helmet is different than the distribution for riders wearing a helmet.

(b) The observed count for head injuries is much higher than expected, while the observed counts for all the other categories are lower. This shows that it is likely that fatalities from head injuries occur more frequently for riders not wearing a helmet.

17. (a) In groups 1, 2, and 3, 84 students attended on average. In group 4, 81 students attended on average.

H_0: there is a significant difference in attendance
H_1: there is no significant difference in attendance

$$\chi^2 = \frac{(84-83)^2}{83} + \frac{(84-83)^2}{83} +$$
$$\frac{(84-83)^2}{83} + \frac{(81-83)^2}{83}$$
$$\approx 0.08$$

$df = 4 - 1 = 3$

$\chi^2_{0.05} = 7.81$

P-value ≈ 0.994

Do not reject H_0. There is not sufficient evidence to conclude that there is a difference in attendance patterns between the first and second half of the semester.

(b) In group 1, 84 students attended on average, in group 2, 81 students attended on average, in group 3, 78 students attended on average. In group 4, 76 students attended on average.

H_0: there is a significant difference in attendance
H_1: there is no significant difference in attendance

$$\chi^2 = \frac{(84-80)^2}{83} + \frac{(81-80)^2}{83} +$$
$$\frac{(78-80)^2}{83} + \frac{(76-80)^2}{83}$$
$$\approx 0.46$$

$df = 4 - 1 = 3$

$\chi^2_{0.05} = 7.81$

P-value ≈ 0.928 [Tech: 0.927]

Do not reject H_0. There is not sufficient evidence to conclude that there is a difference in attendance patterns. It is curious that the farther a group's original position is located from the front of the room, the more the attendance rate for the group decreases.

(c) It would be expected that $100 \times 0.20 = 20$ students in each group would be in the top 20% of the class if seat location plays no role in grades.

H_0: there is a significant difference in grades
H_1: there is no significant difference in grades

$$\chi^2 = \frac{(25-20)^2}{20} + \frac{(21-20)^2}{20} +$$
$$\frac{(15-20)^2}{20} + \frac{(19-20)^2}{20}$$
$$= 2.6$$

$df = 4 - 1 = 3$

$\chi^2_{0.05} = 7.81$

P-value ≈ 0.457

Do not reject H_0. There is not sufficient evidence to conclude that there is a significant difference in grades.

(d) Though not statistically significant, the group located in the front had both better attendance and a larger number of students in the top 20%. Therefore, it could be advantageous to choose the front.

19. H_0: the birth dates are uniformly distributed
H_1: the birth dates are not uniformly distributed

Birth Month	Observed	Expected
January–March	278	207.5
April–June	246	207.5
July–September	163	207.5
October–December	143	207.5

$$\chi^2 = \frac{(278-207.5)^2}{207.5} + \frac{(246-207.5)^2}{207.5} +$$
$$\frac{(163-207.5)^2}{207.5} + \frac{(143-207.5)^2}{207.5}$$
$$\approx 60.69$$

$df = 4 - 1 = 3$

$\chi^2_{0.05} = 7.81$

P-value ≈ 0.000

Reject H_0. There is sufficient evidence to conclude that hockey players' birthdates are not evenly distributed throughout the year.

21.

H_0: pedestrian deaths are equally frequent each day
H_1: pedestrian deaths are not equally frequent each day

Day of the week	Observed	Expected
Sunday	39	42.86
Monday	40	42.86
Tuesday	30	42.86
Wednesday	40	42.86
Thursday	41	42.86
Friday	49	42.86
Saturday	61	42.86

$$\chi^2 = \frac{(39-42.86)^2}{42.86} + \frac{(40-42.86)^2}{42.86} +$$
$$\frac{(30-42.86)^2}{42.86} + \frac{(40-42.86)^2}{42.86} +$$
$$\frac{(41-42.86)^2}{42.86} + \frac{(49-42.86)^2}{42.86} +$$
$$\frac{(61-42.86)^2}{42.86}$$
$$\approx 13.23$$

$df = 7 - 1 = 6$

$\chi^2_{0.05} = 12.59$

P-value ≈ 0.040

Reject H_0. There is sufficient evidence to conclude that fatalities involving pedestrians do not occur with equal frequency on every day of the week.

23. (a) There were 64 students enrolled in the MRP program.

Grade	Observed	Expected
A	7	8.512
B	16	12.224
C	10	15.744
D	13	6.656
F	6	7.296
W	12	13.568

(b)

H_0: the grade distributions are the same
H_1: the grade distributions are not the same

$$\chi^2 = \frac{(7-8.512)^2}{8.512} + \frac{(16-12.224)^2}{12.224} +$$
$$\frac{(10-15.744)^2}{15.744} + \frac{(13-6.656)^2}{6.656} +$$
$$\frac{(6-7.296)^2}{7.296} + \frac{(12-13.568)^2}{66.67}$$
$$\approx 9.99$$

$df = 6 - 1 = 5$

$\chi^2_{0.01} = 15.09$

P-value ≈ 0.076

Do not reject H_0. There is not sufficient evidence to conclude that the two grade distributions are different.

(c) It can be a big change to adjust the entire curriculum, so making a Type I error and deciding that the grade distribution differs when in fact it does not can be very costly.

(d)

Grade	Observed	Expected
A	14	17.024
B	32	24.448
C	20	31.488
D	26	13.312
F	12	14.592
W	24	27.136

H_0: the grade distributions are the same
H_1: the grade distributions are not the same

$$\chi^2 = \frac{(14-17.024)^2}{17.024} + \frac{(32-24.448)^2}{24.448} +$$
$$\frac{(20-31.488)^2}{31.488} + \frac{(26-13.312)^2}{13.312} +$$
$$\frac{(12-14.592)^2}{14.592} + \frac{(24-27.136)^2}{27.136}$$
$$\approx 19.98$$

$df = 6-1 = 5$

$\chi^2_{0.01} = 15.09$

P-value ≈ 0.001

Reject H_0. There is sufficient evidence to conclude that the two grade distributions are different. Small sample sizes require overwhelming evidence against the null hypothesis in order to reject the statement in the null hypothesis.

25. (a) In a sample of 240 births, the expected number of low-birth-weight births for 35- to 39-year old mothers is 17.04 and the expected number of births that are not low-birth-weight is 222.96.

(b)

H_0: the birth rate of low-birth-rate babies is the same
H_1: the birth rate of low-birth-rate babies is higher

$$\chi^2 = \frac{(22-17.04)^2}{17.04} + \frac{(218-222.96)^2}{222.96}$$
$$\approx 1.55$$

$df = 2-1 = 1$

$\chi^2_{0.05} = 3.84$

P-value ≈ 0.213

Do not reject H_0. There is not sufficient evidence to conclude that the there are more low-birth-weight births for 35- to 39-year old mothers.

(c)

H_0: the birth rate of low-birth-rate babies is the same
H_1: the birth rate of low-birth-rate babies is higher

$$np_0(1-p_0) = 240(0.071)(0.929) \approx 15.83 > 10$$

$$z_0 = \frac{\hat{p}-p_0}{\sqrt{\dfrac{p_0(1-p_0)}{n}}}$$
$$\approx \frac{0.092-0.071}{\sqrt{\dfrac{0.071(1-0.071)}{240}}}$$
$$\approx 1.27 \text{ [Tech: 1.25]}$$

$z_{0.05} = 1.645$

P-value ≈ 0.102 [Tech: 0.106]

Do not reject H_0. There is not sufficient evidence to conclude that the there are more low-birth-weight births to 35- to 39-year old mothers.

27. (a)

x	Observed	P(x)
0	229	0.3976
1	211	0.3663
2	93	0.1615
3	35	0.0607
4	7	0.0122
5	0	0
6	0	0
7	1	0.0017
Total	576	1

$$\begin{aligned} \mu &= \sum xP(x) \\ &= 0(0.3976)+1(0.3663)+2(0.1615)+3(0.0607)+ \\ &\quad 4(0.0122)+5(0)+6(0)+7(0.0017) \\ &\approx 0.9321 \text{ [Tech: 0.9323]} \end{aligned}$$

(b) All expected counts need to be greater than or equal to 1 in order to conduct a goodness-of-fit test.

(c)

x	Observed	P(x)
0	229	0.3939
1	211	0.3670
2	93	0.1711
3	35	0.0532
4 or more	8	0.0148

The probability for $x = 4$ or more should be found by subtracting the sum of the other probabilities from 1 because "4 or more" is not a valid value of x for the Poisson formula.

(d)

x	P(x)	Expected number of regions
0	0.3939	226.8864
1	0.3670	211.392
2	0.1710	98.5536
3	0.0531	30.6432
4 or more	0.0151	8.5248

(e)

H_0: the rocket hits are modeled by a Poisson random variable
H_1: the rocket hits are not modeled by a Poisson random variable

$$\chi^2 = \frac{(229 - 226.8864)^2}{226.8864} + \frac{(211 - 211.392)^2}{211.392} +$$
$$\frac{(93 - 98.5536)^2}{98.5536} + \frac{(35 - 30.6432)^2}{30.6432} +$$
$$\frac{(8 - 8.5248)^2}{8.5248}$$
$$\approx 0.99$$

$df = 2 - 1 = 1$

$\chi^2_{0.05} = 9.49$

P-value ≈ 0.912

Do not reject H_0. There is not sufficient evidence to conclude that the rocket hits are not modeled by a Poisson random variable.

29. (a) The data is quantitative because it provides numerical measures that can be added or subtracted to provide meaningful results.

(b) $\bar{x} = \dfrac{\sum x}{n} = \$41,130.87$

$M = \$41,130.87$

(c) $s = \sqrt{\dfrac{\sum (x_i - \bar{x})}{n - 1}} = \$897.8$

$IQR = \$1592$

(d) The correlation between the raw data and normal scores is 0.991, which is greater that 0.939, so it is reasonable to conclude that the data come from a population that is normally distributed.

(e)

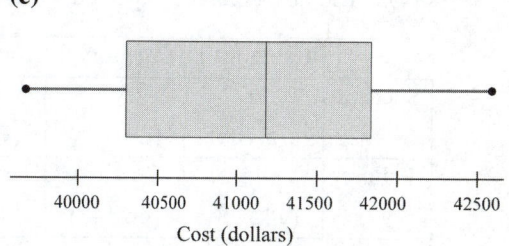

(g) A 90% confidence interval for all new 2015 domestic vehicles would be wider because there is more variability in the data. By removing the variability due to car type, the interval becomes more precise.

31. The χ^2 goodness-of-fit test tests are always right tailed because the numerator in the test statistic is squared, making every test statistic other than a perfect fit positive. Therefore, the test measures if $\chi^2_0 > \chi^2_\alpha$.

Section 12.2

1. True

3. (a) $\chi_0^2 = \sum \dfrac{(O_i - E_i)^2}{E_i} = \dfrac{(34 - 36.26)^2}{36.26} + \dfrac{(43 - 44.63)^2}{44.63} + \cdots + \dfrac{(17 - 20.89)^2}{20.89} = 1.701$ [Tech: 1.698]

   (b) <u>Classical Approach:</u>

There are 2 rows and 3 columns, so df $= (2-1)(3-1) = 2$ and the critical value is $\chi_{0.05}^2 = 5.991$. The test statistic, 1.701, is less than the critical value so we do not reject H_0.

<u>P-value Approach:</u>

There are 2 rows and 3 columns, so we find the P-value using $(2-1)(3-1) = 2$ degrees of freedom. The P-value is the area under the chi-square distribution with 2 degrees of freedom to the right of $\chi_0^2 = 1.701$. Using Table VIII, we find the row that corresponds to 2 degrees of freedom. The value of 1.701 is less than 4.605, which has an area under the chi-square distribution of 0.10 to the right. Therefore, we have P-value > 0.10 [Tech: P-value $= 0.427$]. Since P-value $> \alpha = 0.05$, we do not reject H_0.

<u>Conclusion:</u>

There is not sufficient evidence, at the $\alpha = 0.05$ level of significance, to conclude that the two variables are dependent. We conclude that X and Y are not related.

5. $H_0 : p_1 = p_2 = p_3$
$H_1 :$ At least one proportion differs from the others.

The expected counts are calculated as $\dfrac{\text{(row total)} \cdot \text{(column total)}}{\text{(table total)}} = \dfrac{229 \cdot 120}{363} \approx 75.702$ (for the first cell) and so on. The observed and expected counts are shown in the following table.

	Category 1	Category 2	Category 3	Total
Success	76	84	69	229
	(75.702)	(78.857)	(74.441)	
Failure	44	41	49	134
	(44.298)	(46.143)	(43.559)	
Total	120	125	118	363

Since none of the expected counts is less than 5, the requirements of the test are satisfied.

The test statistic is $\chi_0^2 = \sum \dfrac{(O_i - E_i)^2}{E_i} = \dfrac{(76 - 75.702)^2}{75.702} + \cdots + \dfrac{(49 - 43.559)^2}{43.559} = 1.989$.

<u>Classical Approach:</u>

df $= (2-1)(3-1) = 2$ so the critical value is $\chi_{0.01}^2 = 9.210$. The test statistic, 1.989, is less than the critical value so we do not reject H_0.

<u>P-value Approach:</u>

Using Table VIII, we find the row that corresponds to 2 degrees of freedom. The value of 1.989 is less than 4.605, which has an area under the chi-square distribution of 0.10 to the right. Therefore, we have P-value > 0.10 [Tech: P-value $= 0.370$]. Since P-value $> \alpha = 0.01$, we do not reject H_0.

<u>Conclusion:</u>

There is not sufficient evidence, at the $\alpha = 0.01$ level of significance, to conclude that at least one of the proportions is different from the others.

7. (a) The expected counts are calculated as $\dfrac{(\text{row total})\cdot(\text{column total})}{(\text{table total})} = \dfrac{199\cdot150}{380} \approx 78.553$ (for the first cell) and so on, giving the following table of observed and expected counts:

Sexual Activity	Family Structure				Total
	Both Parents	Single Parent	Parent and Stepparent	Nonparental Guardian	
Had intercourse	64	59	44	32	199
	(78.553)	(52.368)	(41.895)	(26.184)	
Did not have	86	41	36	18	181
	(71.447)	(47.632)	(38.105)	(23.816)	
Total	150	100	80	50	380

(b) All expected frequencies are greater than 5 so all requirements for a chi-square test are satisfied. That is, all expected frequencies are greater than or equal to 1, and no more than 20% of the expected frequencies are less than 5.

(c) $\chi_0^2 = \sum \dfrac{(O_i - E_i)^2}{E_i} = \dfrac{(64-78.553)^2}{78.553} + \cdots + \dfrac{(18-23.816)^2}{23.816} \approx 10.358$ [Tech: 10.357]

(d) H_0 : the row and column variables are independent.

H_1 : the row and column variables are dependent.

df $= (2-1)(4-1) = 3$ so the critical value is $\chi_{0.05}^2 = 7.815$.

Classical Approach:
The test statistic is 10.358 which is greater than the critical value so we reject H_0.

P-value Approach:
Using Table VIII, we find the row that corresponds to 2 degrees of freedom. The value of 10.358 is greater than 9.348, which has an area under the chi-square distribution of 0.025 to the right. Therefore, we have P-value < 0.025 [Tech: P-value = 0.016]. Since P-value $< \alpha = 0.05$, we reject H_0.

Conclusion:
There is sufficient evidence, at the $\alpha = 0.05$ level of significance, to conclude that family structure and sexual activity are dependent.

(e) The biggest difference between observed and expected occurs under the family structure in which both parents are present. Fewer females were sexually active than was expected when both parents were present. This means that having both parents present seems to have an impact on whether the child is sexually active.

(f) The conditional frequencies and bar chart show that sexual activity varies by family structure.

Sexual Intercourse	Both Parents	One Parent	Parent and Stepparent	Nonparent Guardian
Yes	$\dfrac{64}{150} \approx 0.427$	$\dfrac{59}{100} = 0.590$	$\dfrac{44}{80} = 0.550$	$\dfrac{32}{50} = 0.640$
No	$\dfrac{86}{150} \approx 0.573$	$\dfrac{41}{100} = 0.410$	$\dfrac{36}{80} = 0.450$	$\dfrac{18}{50} = 0.360$

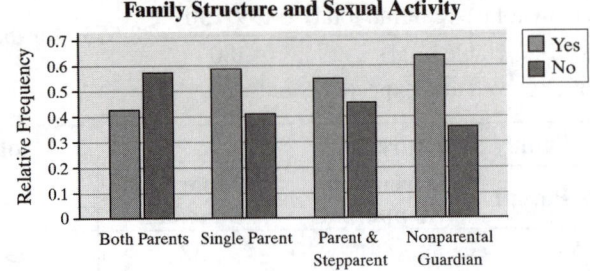

Family Structure and Sexual Activity

9. (a) The expected counts are calculated as $\dfrac{\text{(row total)} \cdot \text{(column total)}}{\text{(table total)}} = \dfrac{634 \cdot 551}{1996} \approx 175.017$ (for the first cell) and so on, giving the following table:

Happiness	Health				Total
	Excellent	Good	Fair	Poor	
Very Happy	271	261	82	20	634
	(175.017)	(295.718)	(128.642)	(34.622)	
Pretty Happy	247	567	231	53	1098
	(303.105)	(512.143)	(222.791)	(59.961)	
Not Too Happy	33	103	92	36	264
	(72.878)	(123.138)	(53.567)	(14.417)	
Total	551	931	405	109	1996

All expected frequencies are greater than 5 so the requirements for a chi-square test are satisfied.

$$\chi_0^2 = \sum \frac{(O_i - E_i)^2}{E_i} = \frac{(271 - 175.017)^2}{175.017} + \cdots + \frac{(36 - 14.417)^2}{14.417} \approx 182.173 \quad [\text{Tech: } 182.174]$$

$df = (3-1)(4-1) = 6$ so the critical value is $\chi_{0.05}^2 = 12.592$.

H_0 : health and happiness are independent.

H_1 : health and happiness are not independent.

Classical Approach:
The test statistic is 182.173 which is greater than the critical value so we reject H_0.

P-value Approach:
Using Table VIII, we find the row that corresponds to 6 degrees of freedom. The value of 182.173 is greater than 18.548, which has an area under the chi-square distribution of 0.005 to the right. Therefore, we have P-value < 0.005 [Tech: P-value < 0.001]. Since P-value < $\alpha = 0.05$, we reject H_0.

Conclusion:
There is sufficient evidence, at the $\alpha = 0.05$ level of significance, to conclude that health and happiness are related.

(b) The conditional frequencies and bar chart show that the level of happiness varies by health status.

Happiness	Health			
	Excellent	**Good**	**Fair**	**Poor**
Very Happy	$\frac{271}{551} \approx 0.492$	$\frac{261}{931} \approx 0.280$	$\frac{82}{405} \approx 0.202$	$\frac{20}{109} \approx 0.183$
Pretty Happy	$\frac{247}{551} \approx 0.448$	$\frac{567}{931} \approx 0.609$	$\frac{231}{405} \approx 0.570$	$\frac{53}{109} \approx 0.486$
Not Too Happy	$\frac{33}{551} \approx 0.60$	$\frac{103}{931} \approx 0.111$	$\frac{92}{405} \approx 0.227$	$\frac{36}{109} \approx 0.330$

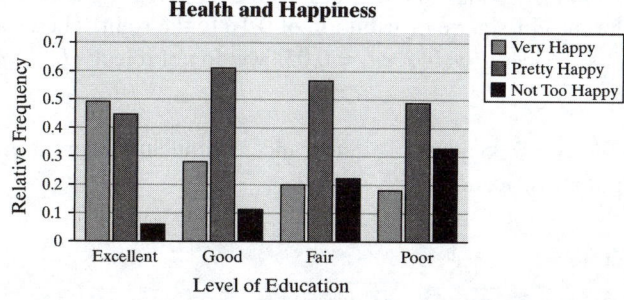

Health and Happiness

Relative Frequency vs Level of Education — Very Happy, Pretty Happy, Not Too Happy

(c) The proportion of individuals who are "very happy" is much higher for individuals in "excellent" health than any other health category. Further, the proportion of individuals who are "not too happy" is much lower for individuals in "excellent" health compared to the other health categories. The level of happiness seems to decline as health status declines.

11. (a) The data represent the measurement of two variables (weight classification and social well-being) on each individual in the study. Because two variables are measured on a single individual, use a chi-square test for independence.

(b) The expected counts are calculated as $\frac{(\text{row total}) \cdot (\text{column total})}{(\text{table total})} = \frac{554 \cdot 813}{2000} \approx 225.201$ (for the first cell) and so on, giving the following table:

Weight Classification	Social Well-Being			Total
	Thriving	**Struggling**	**Suffering**	
Obese	202	250	102	554
	(225.201)	(239.328)	(89.471)	
Overweight	294	302	110	706
	(286.989)	(304.992)	(114.019)	
Normal Weight	300	295	103	698
	(283.737)	(301.536)	(112.727)	
Underweight	17	17	8	42
	(17.073)	(18.144)	(6.783)	
Total	813	864	323	2000

All expected frequencies are greater than 5 so the requirements for a chi-square test are satisfied.

$$\chi_0^2 = \sum \frac{(O_i - E_i)^2}{E_i} = \frac{(202 - 225.201)^2}{225.201} + \cdots + \frac{(8 - 6.783)^2}{6.783} \approx 7.169$$

$df = (4-1)(3-1) = 6$ so the critical value is $\chi_{0.05}^2 = 12.592$.

H_0 : weight classification and social well-being are independent.

H_1 : weight classification and social well-being are not independent.

Classical Approach:
The test statistic is 7.169 which is less than the critical value so we do not reject H_0.

P-value Approach:
Using Table VIII, we find the row that corresponds to 6 degrees of freedom. The value of 7.169 is less than 10.645, which has an area under the chi-square distribution of 0.10 to the right. Therefore, we have P-value > 0.10 [Tech: P-value = 0.306]. Since P-value > $\alpha = 0.05$, we do not reject H_0.

Conclusion:
There is not sufficient evidence, at the $\alpha = 0.05$ level of significance, to conclude there is an association between social well-being and weight classification.

(c)

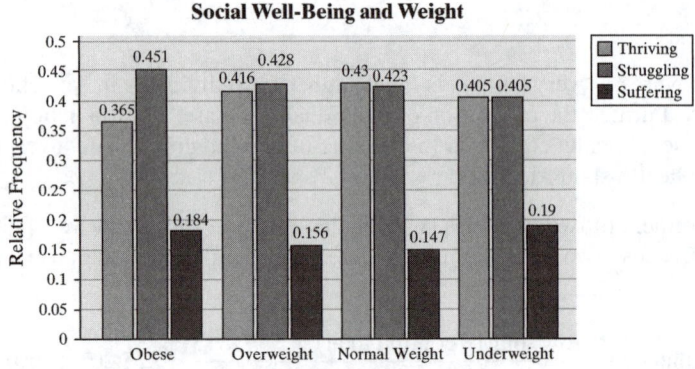

(d) There is not enough sample evidence to suggest that one's social well-being is associated with one's weight classification. However, it is worth noting some differences in the relative frequencies. For example, the lowest relative frequency for "suffering" is in the normal weight group

13. (a) This study used a completely randomized design with three levels of treatment.

(b) The response variable is whether the subject abstained from cigarette smoking, or not. It is qualitative with two possible outcomes, quitting or not quitting.

(c) The population is current smokers.

(d) $H_0 : p_1 = p_2 = p_3$
H_1 : At least one proportion differs from the others.

(e) The expected counts are calculated as $\dfrac{(\text{row total}) \cdot (\text{column total})}{(\text{table total})} = \dfrac{426 \cdot 499}{1466} \approx 145.003$ (for the first cell) and so on, giving the following table:

Qutting	Group			Total
	Group 1	**Group 2**	**Group 3**	
Did Not Quit	52	54	51	157
	(55.765)	(54.049)	(47.186)	
Quit	13	9	4	26
	(9.235)	(8.951)	(7.814)	
Total	65	63	55	183

All expected frequencies are greater than 5 so all requirements for a chi-square test are satisfied.

$$\chi_0^2 = \sum \frac{(O_i - E_i)^2}{E_i} = \frac{(52 - 55.765)^2}{55.765} + \cdots + \frac{(4 - 7.814)^2}{7.814} \approx 3.959 \text{ [Tech: 3.960]}$$

$df = (2 - 1)(3 - 1) = 2$ so the critical value is $\chi_{0.05}^2 = 5.991$.

<u>Classical Approach:</u>
The test statistic is 3.959 which is less than the critical value so we do not reject H_0.

<u>P-value Approach:</u>
Using Table VIII, we find the row that corresponds to 2 degrees of freedom. The value of 3.959 is less than 4.605, which has an area under the chi-square distribution of 0.10 to the right. Therefore, we have P-value > 0.10 [Tech: P-value = 0.138]. Since P-value > $\alpha = 0.05$, we do not reject H_0.

(f) The bar graph suggests that the proportion of individuals who abstain from cigarette smoking does not significantly differ by group. This supports the results from part (e).

(g) There is not sufficient evidence, at the $\alpha = 0.05$ level of significance, to suggest that the proportion of individuals who abstain from cigarette smoking in the three groups differs.

15. (a) Because there are three distinct populations that are being surveyed (Democrats, Republicans, and Independents) and the response variable is qualitative with two outcomes (positive/negative), we analyze the data using homogeneity of proportions.

(b)   The expected counts are calculated as $\dfrac{(\text{row total}) \cdot (\text{column total})}{(\text{table total})} = \dfrac{426 \cdot 499}{1466} \approx 145.003$ (for the first cell) and so on, giving the following table:

Reaction	Political Party			Total
	Democrat	**Independent**	**Republican**	
Positive	220	144	62	426
	(145.003)	(160.985)	(120.012)	
Negative	279	410	351	1040
	(353.997)	(393.015)	(292.988)	
Total	499	554	413	1466

All expected frequencies are greater than 5 so all requirements for a chi-square test are satisfied.

$$\chi_0^2 = \sum \frac{(O_i - E_i)^2}{E_i} = \frac{(220 - 145.003)^2}{145.003} + \cdots + \frac{(351 - 292.988)^2}{292.988} \approx 96.733 \text{ and df} = (2-1)(3-1) = 2$$

$H_0 : p_{\text{Democrats}} = p_{\text{Independents}} = p_{\text{Republicans}}$

$H_1 :$ at least one of the proportions differs.

Classical Approach:
With 2 degrees of freedom, the critical value is $\chi_{0.05}^2 = 5.991$. The test statistic is 96.733 which is greater than the critical value so we reject H_0.

P-value Approach:
Using Table VIII, we find the row that corresponds to 2 degrees of freedom. The value of 96.733 is greater than 10.597, which has an area under the chi-square distribution of 0.005 to the right. Therefore, we have P-value < 0.005 [Tech: P-value < 0.001]. Since P-value < $\alpha = 0.05$, we reject H_0.

Conclusion:
There is sufficient evidence, at the $\alpha = 0.05$ level of significance, to conclude that a different proportion of individuals within each political affiliation react positively to the word "socialism."

(c)   Dividing the observed counts within each party by the number of respondents identified with each party, we get:

Reaction	Political Party		
	Democrat	**Independent**	**Republican**
Positive	0.4409	0.2599	0.1501
Negative	0.5591	0.7401	0.8499
Total	1	1	1

(d)   Independents and Republicans are far more likely to react negatively to the word socialism than Democrats are. However, it is important to note that a majority of Democrats in the sample did have a negative reaction, so the word socialism has a negative connotation among all groups.

17. (a) Since there were no individuals who gave "career" as a reason for dropping, we have omitted that category from the analysis.

Gender	Drop Reason			Total
	Personal	**Work**	**Course**	
Female	5	3	13	21
	(4.62)	(6.72)	(9.66)	
Male	6	13	10	29
	(6.38)	(9.28)	(13.34)	
Total	11	16	23	50

(b) The expected counts are calculated as $\dfrac{(\text{row total}) \cdot (\text{column total})}{(\text{table total})} = \dfrac{21 \cdot 11}{50} = 4.62$ (for the first cell) and

so on (included in the table from part (a). All expected frequencies are greater than 1 and only one (out of six) expected frequency is less than 5. All requirements for a chi-square test are satisfied.

$$\chi_0^2 = \sum \frac{(O_i - E_i)^2}{E_i} = \frac{(5 - 4.62)^2}{4.62} + \cdots + \frac{(10 - 13.34)^2}{13.34} \approx 5.595$$

df $= (2-1)(3-1) = 2$ so the critical value is $\chi_{0.10}^2 = 4.605$.

Classical Approach:
The test statistic is 5.595 which is greater than the critical value so we reject H_0.

P-value Approach:
Using Table VIII, we find the row that corresponds to 2 degrees of freedom. The value of 5.595 is greater than 4.605, which has an area under the chi-square distribution of 0.10 to the right. Therefore, we have P-value < 0.10 [Tech: P-value = 0.061]. Since P-value < $\alpha = 0.10$, we reject H_0.

Conclusion:
There is sufficient evidence, at the $\alpha = 0.1$ level of significance, to conclude that gender and drop reason are dependent. Females are more likely to drop because of the course, while males are more likely to drop because of work.

(c) The conditional frequencies and bar chart show that the distribution of genders varies by reason for dropping. This supports the results from part (b).

	Personal	**Work**	**Course**
Female	$\dfrac{5}{21} \approx 0.238$	$\dfrac{3}{21} \approx 0.143$	$\dfrac{13}{21} \approx 0.619$
Male	$\dfrac{6}{29} \approx 0.207$	$\dfrac{13}{29} \approx 0.448$	$\dfrac{10}{29} \approx 0.345$

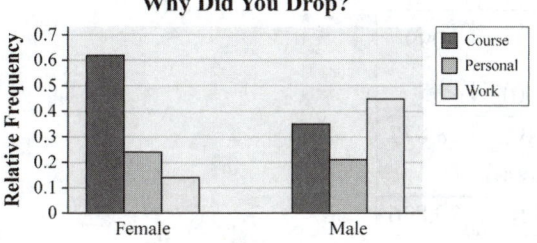

19. There are three different populations to study: (1) all capital letters, (2) all lower case letters, and (3) upper and lower case accurately. Obtain a random sample of applications from each population. Determine whether the loan corresponding to the application resulted in default. Conduct a hypothesis test for homogeneity of proportions to determine if the sample data suggest at least one population differs from the others.

21. (a) The population being studied is healthy adult women aged 45 years or older. The sample consists of the 39,876 women in the study.

(b) The response variable is whether or not the subject had a cardiovascular event (such as a heart attack or stroke). This is a qualitative variable because the values serve to simply classify the subjects.

(c) There are two treatments: 100 mg of aspirin and a placebo.

(d) Because the subjects are all randomly assigned to the treatments, this is a completely randomized design.

(e) Randomization controls for any other explanatory variables because individuals affected by these lurking variables should be equally dispersed between the two treatment groups.

(f) $H_0 : p_1 = p_2$

$H_1 : p_1 \neq p_2$

$\hat{p} = \dfrac{x_1 + x_2}{n_1 + n_2} = \dfrac{477 + 522}{19,934 + 19,942} = \dfrac{999}{39,876} \approx 0.025$, $\hat{p}_1 = \dfrac{477}{19,934} \approx 0.024$, $\hat{p}_2 = \dfrac{522}{19,942} \approx 0.026$

The test statistic is $z_0 = \dfrac{\hat{p}_1 - \hat{p}_2}{\sqrt{\hat{p}(1-\hat{p})}\sqrt{\dfrac{1}{n_1} + \dfrac{1}{n_2}}} = \dfrac{0.024 - 0.026}{\sqrt{0.025(0.975)}\sqrt{\dfrac{1}{19,934} + \dfrac{1}{19,942}}} \approx -1.28$ [Tech: -1.44]

Classical Approach:
The critical values for $\alpha = 0.05$ are $-z_{0.025} = -1.96$ and $z_{0.025} = 1.96$.
The test statistic is -1.28, which is not in the critical region so we do not reject H_0.

P-value Approach:
$P\text{-value} = 2 \cdot P(Z > 1.28)$
$= 2(0.1003)$
$= 0.2006$ [Tech: 0.1511]
Therefore, we have $P\text{-value} > \alpha = 0.05$ and we do not reject H_0.

Conclusion:
There is not sufficient evidence, at the $\alpha = 0.05$ level of significance, to conclude that a difference exists between the proportions of cardiovascular events in the aspirin group versus the placebo group.

(g) The expected counts are calculated as $\dfrac{(\text{row total}) \cdot (\text{column total})}{(\text{table total})} = \dfrac{999 \cdot 19,934}{39,876} \approx 499.400$ (for the first cell) and so on.

	aspirin	placebo	Total
cardiovascular event	477 (499.400)	522 (499.600)	999
no event	19,457 (19,434.600)	19,420 (19,442.400)	38,877
Total	19,934	19,942	39,876

All expected frequencies are greater than 5 so all requirements for a chi-square test are satisfied.

$$\chi_0^2 = \sum \frac{(O_i - E_i)^2}{E_i} = \frac{(477 - 499.400)^2}{499.400} + \cdots + \frac{(19,420 - 19,442.400)^2}{19,442.400} \approx 2.061$$

$df = (2-1)(2-1) = 1$ so the critical value is $\chi_{0.05}^2 = 3.841$.

Classical Approach:
The test statistic is 2.061 which is less than the critical value so we do not reject H_0.

P-value Approach:
Using Table VIII, we find the row that corresponds to 1 degree of freedom. The value of 2.061 is less than 2.706, which has an area under the chi-square distribution of 0.10 to the right. Therefore, we have $P\text{-value} > 0.10$ [Tech: $P\text{-value} = 0.1511$]. Since $P\text{-value} > \alpha = 0.05$, we do not reject H_0.

Conclusion:
There is not sufficient evidence, at the $\alpha = 0.05$ level of significance, to conclude that a difference exists between the proportions of cardiovascular events in the aspirin group versus the placebo group.

(h) Using technology in part (f), the test statistic to four decimal places is $z_0 = -1.4355$ which gives $(z_0)^2 \approx 2.061 = \chi_0^2$. Our conclusion is that for comparing two proportions, $z_0^2 = \chi_0^2$.

23. Answers may vary, but differences should include: chi-square test for independence compares two characteristics from a single population, whereas the chi-square test for homogeneity compares a single characteristic from two (or more) populations. Similarities should include: the procedures of the two tests and the assumptions of the two tests are the same.

Section 12.3

1. (a) $H_0 : p_A = p_B$
   $H_1 : p_A \neq p_B$

   (b) $f_{12} = 19$ and $f_{21} = 14$.

   $$\chi_0^2 = \frac{(19 - 14)^2}{19 + 14} \approx 0.758$$

   (c) $df = (2-1)(2-1) = 1$ so the critical value is $\chi_{0.05}^2 = 3.841$.

   (d) Using Table VIII, we find the row that corresponds to 1 degree of freedom. The value of 0.758 is between 0.016 and 2.706, which correspond to areas to the right of 0.90 and 0.10, respectively. Therefore, we have $0.10 < P\text{-value} < 0.90$ [Tech: $P\text{-value} = 0.3841$].

   Since the test statistic is less than the critical value, and since the P-value is greater than the $\alpha = 0.05$ level of significance, we do not reject H_0.

3. (a) This is a dependent sample because two variables (seat belt or no seat belt, and smoking or no smoking) are measured on the same individual.

   (b) $H_0 : p_{\text{smoke}} = p_{\text{no seat belt}}$
   $H_1 : p_{\text{smoke}} \neq p_{\text{no seat belt}}$

   $f_{12} = 448$ and $f_{21} = 327$.

   $$\chi_0^2 = \frac{(448 - 327)^2}{448 + 327} \approx 18.892$$

   $df = (2-1)(2-1) = 1$ so the critical value is $\chi_{0.05}^2 = 3.841$.

   Classical Approach:
   The test statistic is 18.892, which is greater than the critical value, so we reject H_0.

   P-value Approach:
   Using Table VIII, we find the row that corresponds to 1 degree of freedom. The value of 18.892 is greater than 7.879, which corresponds to areas to the right of 0.005. Therefore, we have $P\text{-value} < 0.005$ [Tech:

P-value < 0.001]. Since P-value < $\alpha = 0.05$, we reject H_0.

Conclusion:
There is sufficient evidence, at the $\alpha = 0.05$ level of significance, to conclude that there is a significant difference between the proportion of individuals who smoke and the proportion of individuals who do not wear a seat belt. The sample proportion of smokers is 0.17, while the sample proportion of those who do not wear a seat belt is 0.13. Smoking appears to be the more popular hazardous activity.

5. $H_0 : p_{NN} = p_{RN}$
    $H_1 : p_{NN} \neq p_{RN}$

    $f_{12} = 385$ and $f_{21} = 456$.

    $$\chi_0^2 = \frac{(385 - 456)^2}{385 + 456} \approx 5.994$$

    The critical value is $\chi_{0.05}^2 = 3.841$.

    Classical Approach:
    The test statistic is 5.994, which is greater than the critical value, so we reject H_0.

    P-value Approach:
    Using Table VIII, we find the row that corresponds to 1 degree of freedom. The value of 5.994 is between 5.024 and 6.635, which correspond to areas to the right of 0.025 and 0.01, respectively. Therefore, we have $0.01 < P$-value < 0.025 [Tech: P-value = 0.0144]. Since P-value < $\alpha = 0.05$, we reject H_0.

Conclusion:
There is sufficient evidence, at the $\alpha = 0.05$ level of significance, to conclude that there is a difference in the proportion of errors between the neural network and the remapped network.

7. $H_0 : p_{\text{Death Penalty}} = p_{\text{Gun Laws}}$
    $H_1 : p_{\text{Death Penalty}} \neq p_{\text{Gun Laws}}$

    $f_{12} = 69$ and $f_{21} = 58$.

    $$\chi_0^2 = \frac{(69 - 58)^2}{69 + 58} \approx 0.953$$

    The critical value is $\chi_{0.05}^2 = 3.841$.

    Classical Approach:
    The test statistic is 0.953, which is less than the critical value, so we do not reject H_0.

    P-value Approach:
    Using Table VIII, we find the row that corresponds to 1 degree of freedom. The value of 0.953 is between 0.016 and 2.706, which correspond to areas to the right of 0.90 and 0.10, respectively. Therefore, we have $0.10 < P$-value < 0.90 [Tech: P-value = 0.3291]. Since P-value > $\alpha = 0.05$, we do not reject H_0.

    Conclusion:
    There is not sufficient evidence, at the $\alpha = 0.05$ level of significance, to conclude that the proportion of adult Americans who favor a law which would require a person to obtain a permit prior to purchasing a gun is different from the proportion who favor the death penalty for persons convicted of murder.

Chapter 12 Review Exercises

1. H_0 : The wheel is balanced

    H_1 : The wheel is not balanced

    If the wheel is balanced, each slot is equally likely. Thus, we would expect the proportion of red to be $\frac{18}{38} = \frac{9}{19}$, the proportion of black to be $\frac{18}{38} = \frac{9}{19}$, and the proportion of green to be $\frac{2}{38} = \frac{1}{19}$. There are 500 spins. To determine the expected number of each color, multiply 500 by the given proportions for each color.

    For example, the expected count of red would be $500\left(\frac{9}{19}\right) \approx 236.842$. We summarize the observed and expected counts in the following table:

	O_i	E_i	$(O_i - E_i)^2$	$\dfrac{(O_i - E_i)^2}{E_i}$
Red	233	236.842	14.7610	0.0623
Black	237	236.842	0.0250	0.0001
Green	30	26.316	13.5719	0.5157
			$\chi_0^2 \approx$	0.578

Since all the expected cell counts are greater than or equal to 5, the requirements for the goodness-of-fit test are satisfied.

<u>Classical approach:</u> The critical value, with df $= 3 - 1 = 2$, is $\chi_{0.05}^2 = 5.991$. Since the test statistic is not in the critical region ($\chi_0^2 < \chi_{0.05}^2$), we do not reject the null hypothesis.

<u>P-value approach:</u> Using the chi-square table, we find the row that corresponds to 2 degrees of freedom. The value of 0.578 is less than 4.605, which has an area under the chi-square distribution of 0.10 to the right. Therefore, we have P-value > 0.10 [Tech: 0.749]. Since P-value $> \alpha = 0.05$, we do not reject the null hypothesis.

<u>Conclusion:</u> There is not enough evidence, at the $\alpha = 0.05$ level of significance, to conclude that the wheel is out of balance. That is, the evidence suggests that the wheel is in balance.

2. Using $\alpha = 0.05$, we want to test

H_0 : The teams are evenly matched.

H_1 : The teams are not evenly matched.

There are 84 series to consider. To determine the expected counts, multiply 84 by the given percentage for each number of games in the series. For example, the expected count for a 4 game series would be $84(0.125) = 10.5$. We summarize the observed and expected counts in the following table:

	O_i	E_i	$(O_i - E_i)^2$	$\dfrac{(O_i - E_i)^2}{E_i}$
4	16	10.5	30.25	2.88
5	17	21.0	16.0	0.76
6	19	26.25	52.5625	2.00
7	32	26.25	33.0625	1.26
			$\chi_0^2 =$	6.90

Since all the expected cell counts are greater than or equal to 5, the requirements for the goodness-of-fit test are satisfied.

<u>Classical approach:</u> The critical value, with df $= 4 - 1 = 3$, is $\chi_{0.05}^2 = 7.815$. Since the test statistic is not in the critical region ($\chi_0^2 < \chi_{0.05}^2$), we do not reject the null hypothesis.

<u>P-value approach:</u> Using the chi-square table, we find the row that corresponds to 3 degrees of freedom. The value of 6.90 is between 6.251 and 7.815, which correspond to areas to the right of 0.10 and 0.05, respectively. Therefore, we have $0.05 < P$-value < 0.10 [Tech: P-value $= 0.075$]. Since P-value $> \alpha = 0.05$, we do not reject H_0.

Conclusion: There is not enough evidence, at the $\alpha = 0.05$ level of significance, to conclude that the teams playing in the World Series have not been evenly matched. That is, the evidence suggests that the teams have been evenly matched. On the other hand, the P-value is suggestive of an issue. In particular, there are fewer six-game series than we would expect. Perhaps the team that is down "goes all out" in game 6, trying to force game 7.

3. **(a)** The expected counts are calculated as $\dfrac{(\text{row total}) \cdot (\text{column total})}{(\text{table total})} = \dfrac{499 \cdot 325}{1316} \approx 123.233$ (for the first cell) and so on, giving the following table:

	Class			Total
	First	**Second**	**Third**	
Survived	203	118	178	499
	(123.233)	(108.066)	(267.701)	
Did Not Survive	122	167	528	817
	(201.767)	(176.934)	(438.299)	
Total	325	285	706	1316

All expected frequencies are greater than 5 so all requirements for a chi-square test are satisfied. That is, all expected frequencies are greater than or equal to 1, and no more than 20% of the expected frequencies are less than 5.

$$\chi_0^2 = \sum \frac{(O_i - E_i)^2}{E_i} = \frac{(203 - 123.333)^2}{123.333} + \cdots + \frac{(528 - 438.299)^2}{438.299} \approx 133.053 \text{ [Tech: } 133.052]$$

H_0 : survival status and social class are independent.

H_1 : survival status and social class are not independent.

$df = (2-1)(3-1) = 2$ so the critical value is $\chi_{0.05}^2 = 5.991$.

Classical Approach:
The test statistic is 132.882 which is greater than the critical value so we reject H_0.

P-value Approach:
Using Table VIII, we find the row that corresponds to 2 degrees of freedom. The value of 132.882 is greater than 10.597, which has an area under the chi-square distribution of 0.005 to the right. Therefore, we have P-value < 0.005 [Tech: P-value < 0.001]. Since P-value < $\alpha = 0.05$, we reject H_0.

Conclusion:
There is sufficient evidence, at the $\alpha = 0.05$ level of significance, to conclude that survival status and social class are dependent.

(b) The conditional distribution and bar graph support the conclusion that a relationship exists between survival status and social class. Individuals with higher-class tickets survived in greater proportions than those with lower-class tickets.

	Class		
	First	**Second**	**Third**
Survived	$\dfrac{203}{325} \approx 0.625$	$\dfrac{118}{285} \approx 0.414$	$\dfrac{178}{706} \approx 0.252$
Did Not Survive	$\dfrac{122}{325} \approx 0.375$	$\dfrac{167}{285} \approx 0.586$	$\dfrac{528}{706} \approx 0.748$

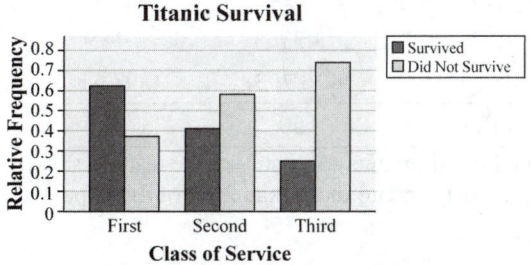

4. The expected counts are calculated as $\dfrac{\text{(row total)} \cdot \text{(column total)}}{\text{(table total)}} = \dfrac{1279 \cdot 40}{4953} \approx 10.329$ (for the first cell) and so on, giving the following table:

Less than	**Gestational Period (Weeks)**					**Total**
H.S. degree	**22–27**	**28–32**	**33–36**	**37–42**	**43+**	
Yes	14	34	140	1010	81	1279
	(10.329)	(25.565)	(124.724)	(1043.755)	(74.628)	
No	26	65	343	3032	208	3674
	(29.671)	(73.435)	(358.276)	(2998.245)	(214.372)	
Total	40	99	483	4042	289	4953

All expected frequencies are greater than 5 so all requirements for a chi-square test are satisfied. That is, all expected frequencies are greater than or equal to 1, and no more than 20% of the expected frequencies are less than 5.

$$\chi_0^2 = \sum \frac{(O_i - E_i)^2}{E_i} = \frac{(14 - 10.329)^2}{10.329} + \cdots + \frac{(208 - 214.372)^2}{214.372} \approx 10.238 \quad \text{[Tech: 10.239]}$$

H_0 : gestation period and completing high school are independent.

H_1 : gestation period and completing high school are not independent.

df $= (2-1)(5-1) = 4$ so the critical value is $\chi_{0.05}^2 = 9.488$.

Classical Approach:
The test statistic is 10.238 which is greater than the critical value so we reject H_0.

P-value Approach:
Using Table VIII, we find the row that corresponds to 4 degrees of freedom. The value of 10.238 is greater than 9.488, which has an area under the chi-square distribution of 0.05 to the right. Therefore, we have P-value < 0.05 [Tech: P-value < 0.037]. Since P-value < $\alpha = 0.05$, we reject H_0.

Conclusion:
There is sufficient evidence, at the $\alpha = 0.05$ level of significance, to conclude that gestation period and completing high school are dependent.

5. (a) The expected counts are calculated as $\dfrac{(\text{row total}) \cdot (\text{column total})}{(\text{table total})} = \dfrac{2310 \cdot 1243}{4194} \approx 684.628$ (for the first cell) and so on, giving the following table:

	Funding Level				Total
	High	**Medium**	**Low**	**No Funding**	
Roosevelt	745	641	513	411	2310
	(684.628)	(619.635)	(523.247)	(482.489)	
Landon	498	484	437	465	1884
	(558.372)	(505.365)	(426.753)	(393.511)	
Total	1243	1125	950	876	4194

All expected frequencies are greater than 5 so all requirements for a chi-square test are satisfied. That is, all expected frequencies are greater than or equal to 1, and no more than 20% of the expected frequencies are less than 5.

$$\chi_0^2 = \sum \frac{(O_i - E_i)^2}{E_i} = \frac{(745 - 684.628)^2}{684.628} + \cdots + \frac{(465 - 393.511)^2}{393.511} \approx 37.518 \text{, and df} = (2-1)(4-1) = 3$$

H_0 : $p_{\text{High}} = p_{\text{Medium}} = p_{\text{Low}} = p_{\text{No Funding}}$

H_1 : at least one proportion is different from the others.

Classical Approach:
With 3 degrees of freedom, the critical value is $\chi_{0.05}^2 = 7.815$. The test statistic is 37.518, which is greater than the critical value so we reject H_0.

P-value Approach:
Using Table VIII, we find the row that corresponds to 3 degrees of freedom. The value of 37.518 is greater than 12.838, which has an area under the chi-square distribution of 0.005 to the right. Therefore, we have P-value < 0.005 [Tech: P-value < 0.001]. Since P-value < α, we reject H_0.

Conclusion:
There is sufficient evidence, at the $\alpha = 0.05$ level of significance, to conclude that at least one proportion is different from the others. That is, the evidence suggests that the level of funding received by the counties is associated with the candidate.

(b) The conditional distribution and bar graph support the conclusion that the proportion of adults who feel morality is important when deciding how to vote is different for at least one political affiliation. It appears that a higher proportion of Republicans feel that morality is important when deciding how to vote than for Democrats or Independents.

	Funding Level			
	High	**Medium**	**Low**	**No Funding**
Roosevelt	0.5994	0.5698	0.5400	0.4692
Landon	0.4006	0.4302	0.4600	0.5308
Total	1	1	1	1

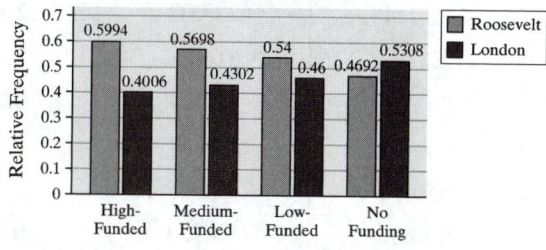

6. $H_0 : p_{\text{Jury}} = p_{\text{Voting}}$

$H_1 : p_{\text{Jury}} \neq p_{\text{Voting}}$

$f_{12} = 65$ and $f_{21} = 45$.

$$\chi_0^2 = \frac{(65-45)^2}{65+45} \approx 3.636$$

The critical value is $\chi_{0.05}^2 = 3.841$.

Classical Approach:
The test statistic is 3.636, which is less than the critical value, so we do not reject H_0.

P-value Approach:
Using Table VIII, we find the row that corresponds to 1 degree of freedom. The value of 0.953 is between 2.706 and 3.841, which correspond to areas to the right of 0.10 and 0.05, respectively. Therefore, we have $0.05 < P\text{-value} < 0.10$ [Tech: $P\text{-value} = 0.0565$]. Since $P\text{-value} > \alpha = 0.05$, we do not reject H_0.

Conclusion:
There is not sufficient evidence, at the $\alpha = 0.05$ level of significance, to suggest the proportion of individuals who believe voting is a civic duty differs from the proportion who believe jury duty is a civic duty.

Chapter 12 Test

1. H_0 : The dice are fair.

   H_1 : The dice are loaded.

   If the dice are fair, the sum of the two dice will follow the given distribution. There are 400 rolls. To determine the expected number of times for each sum, multiply 400 by the given relative frequency for each sum. For example, the expected count for the sum 2 would be $400\left(\dfrac{1}{36}\right) \approx 11.111$. We summarize the observed and expected counts in the following table:

Sum	Observed Count (O_i)	Expected Rel. Freq.	Expected Count (E_i)	$(O_i - E_i)^2$	$(O_i - E_i)^2 / E_i$
2	16	1/36	11.111	23.902	2.1512
3	23	2/36	22.222	0.605	0.0272
4	31	3/36	33.333	5.443	0.1633
5	41	4/36	44.444	11.861	0.2669
6	62	5/36	55.556	41.525	0.7474
7	59	6/36	66.667	58.783	0.8817
8	59	5/36	55.556	11.861	0.2135
9	45	4/36	44.444	0.309	0.0070
10	34	3/36	33.333	0.445	0.0133
11	19	2/36	22.222	10.381	0.4672
12	11	1/36	11.111	0.012	0.0011
				$\chi_0^2 \approx$	4.940

   Since all the expected cell counts are greater than or equal to 5, the requirements for the goodness-of-fit test are satisfied.

   <u>Classical approach:</u> The critical value, with df $= 11 - 1 = 10$, is $\chi_{0.01}^2 = 23.209$. Since the test statistic is not in the critical region ($\chi_0^2 < \chi_{0.01}^2$), we do not reject the null hypothesis.

   <u>P-value approach:</u> Using the chi-square table, we find the row that corresponds to 10 degrees of freedom. The value of 4.940 is less than 15.987, which has an area under the chi-square distribution of 0.10 to the right. Therefore, we have P-value > 0.10 [Tech: 0.895]. Since P-value $> \alpha$, we do not reject the null hypothesis.

   <u>Conclusion:</u> There is not enough evidence, at the $\alpha = 0.01$ level of significance, to conclude that the dice are loaded. The data indicate that the dice are fair.

2. H_0 : Educational attainment in the U.S. is the same as in 2000

  H_1 : Educational attainment is different today than in 2000.

  There are 500 Americans in the sample. To determine the expected number attaining each degree level, multiply 500 by the given relative frequency for each level. For example, the expected count for "not a high school graduate" would be $500(0.158) = 79$. We summarize the observed and expected counts in the following table:

Observed Count (O_i)	Expected Rel. Freq.	Expected Count (E_i)	$(O_i - E_i)^2$	$(O_i - E_i)^2 / E_i$
72	0.158	79	49.00	0.6203
159	0.331	165.5	42.25	0.2553
85	0.176	88	9.00	0.1023
44	0.078	39	25.00	0.6410
92	0.170	85	49.00	0.5765
48	0.087	43.5	20.25	0.4655
			$\chi_0^2 \approx$	2.661

  Since all the expected cell counts are greater than or equal to 5, the requirements for the goodness-of-fit test are satisfied.

  <u>Classical approach:</u> The critical value, with df $= 6 - 1 = 5$, is $\chi_{0.10}^2 = 9.236$. Since the test statistic is not in the critical region ($\chi_0^2 < \chi_{0.10}^2$), we do not reject the null hypothesis.

  <u>P-value approach:</u> Using the chi-square table, we find the row that corresponds to 5 degrees of freedom. The value of 2.661 is less than 9.236, which has an area under the chi-square distribution of 0.10 to the right. Therefore, we have P-value > 0.10 [Tech: 0.752]. Since P-value $> \alpha = 0.10$, we do not reject the null hypothesis.

  <u>Conclusion:</u> There is not enough evidence, at the $\alpha = 0.1$ level of significance, to conclude that the distribution of educational attainment has changed since 2000. That is, the data indicate that educational attainment has not changed.

3. (a) The expected counts are calculated as $\dfrac{\text{(row total)} \cdot \text{(column total)}}{\text{(table total)}} = \dfrac{344 \cdot 550}{1510} \approx 125.298$ (for the first cell)

  and so on, giving the following table:

Morality	Political Affiliation			Total
	Republican	Democrat	Independent	
Important	644	662	670	1976
	(592.631)	(691.685)	(691.685)	
Not important	56	155	147	358
	(107.369)	(125.315)	(125.315)	
Total	700	817	817	2334

  All expected frequencies are greater than 5 so all requirements for a chi-square test are satisfied. That is, all expected frequencies are greater than or equal to 1, and no more than 20% of the expected frequencies are less than 5.

$$\chi_0^2 = \sum \frac{(O_i - E_i)^2}{E_i} = \frac{(644 - 592.631)^2}{592.631} + \cdots + \frac{(147 - 125.315)^2}{125.315} \approx 41.767$$

$H_0: \; p_H = p_C = p_G$

$H_1:$ at least one proportion is different from the others

$df = (2-1)(3-1) = 2,$ so the critical value is $\chi_{0.05}^2 = 5.991.$

<u>Classical Approach:</u>
The test statistic is 41.767, which is greater than the critical value, so we reject H_0.

<u>P-value Approach:</u>
Using Table VIII, we find the row that corresponds to 2 degrees of freedom. The value of 5.605 is less than 10.597, which has an area under the chi-square distribution of 0.005 to the right. Therefore, we have P-value < 0.005 [Tech: P-value < 0.0001]. Since P-value $< \alpha = 0.05$, we reject H_0.

<u>Conclusion</u>
There is not sufficient evidence, at the $\alpha = 0.05$ level of significance, to conclude that at least one proportion is different from the others. That is, the evidence suggests that the proportion of adults who feel morality is important when deciding how to vote is different for at least one political affiliation.

(b) The conditional distribution and bar graph support the conclusion that a relationship exists between morality and party affiliation. A higher proportion of Republicans appears to feel that morality is important when deciding how to vote than do Democrats or Independents.

	Political Affiliation		
	Republican	**Democrat**	**Independent**
Important	$\frac{644}{700} = 0.92$	$\frac{662}{817} \approx 0.810$	$\frac{670}{817} \approx 0.820$
Not Important	$\frac{56}{700} = 0.08$	$\frac{155}{817} \approx 0.190$	$\frac{147}{817} \approx 0.180$

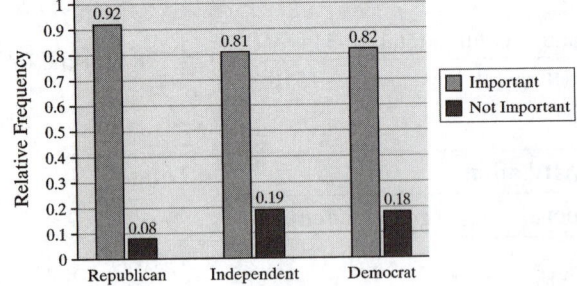

4. The expected counts are calculated as $\dfrac{(\text{row total})\cdot(\text{column total})}{(\text{table total})} = \dfrac{1997\cdot 547}{2712} \approx 402.787$ (for the first cell)

and so on, giving the following table:

Religion	Decades				Total
	1970s	**1980s**	**1990s**	**2000s**	
Affiliated	2,395	3,022	2,121	624	8,162
	(1950.23)	(2460.35)	(1809.08)	(1942.34)	
Unaffiliated	327	412	404	2,087	3,230
	(771.77)	(973.65)	(715.92)	(768.66)	
Total	2,722	3,434	2,525	2,711	11,392

All expected frequencies are greater than 5 so all requirements for a chi-square test are satisfied. That is, all expected frequencies are greater than or equal to 1, and no more than 20% of the expected frequencies are less than 5.

$$\chi_0^2 = \sum \frac{(O_i - E_i)^2}{E_i} = \frac{(2395 - 1950.23)^2}{1950.23} + \cdots + \frac{(2087 - 768.66)^2}{768.66} \approx 4155.584 \text{ and df} = (2-1)(4-1) = 3$$

$H_0 : p_{1970s} = p_{1980s} = p_{1990s} = p_{2000s}$

$H_1 :$ at least one of the proportions is not equal to the rest

Classical approach:

The critical value is $\chi_{0.05}^2 = 7.815$. The test statistic is 4155.584 which is greater than the critical value so we reject H_0.

P-value approach:

Using the chi-square table, we find the row that corresponds to 3 degrees of freedom. The value of 4155.584 is greater than 14.860, which has an area under the chi-square distribution of 0.005 to the right. Therefore, we have P-value < 0.005 [Tech: P-value < 0.0001]. Since P-value $< \alpha = 0.05$, we reject H_0.

Conclusion:

There is sufficient evidence, at the $\alpha = 0.05$ level of significance, to conclude that different proportions of 18-29 year-olds have been affiliated with religion in the past four decades.

5. The expected counts are calculated as $\dfrac{(\text{row total}) \cdot (\text{column total})}{(\text{table total})} = \dfrac{3179 \cdot 137}{8874} \approx 49.079$ (for the first cell) and

so on, giving the following table:

	Time Spent in Bars							Total
	Almost Daily	Several Times a Week	Several Times a Month	Once a Month	Several Times a Year	Once a Year	Never	
Smoker	80	409	294	362	433	336	1265	3179
	(49.079)	(271.902)	(241.094)	(298.412)	(360.387)	(323.847)	(1634.280)	
Nonsmoker	57	350	379	471	573	568	3297	5695
	(87.921)	(487.098)	(431.906)	(534.588)	(645.613)	(580.153)	(2927.720)	
Total	137	759	673	833	1006	904	4562	8874

All expected frequencies are greater than 5 so all requirements for a chi-square test are satisfied. That is, all expected frequencies are greater than or equal to 1, and no more than 20% of the expected frequencies are less than 5.

$$\chi_0^2 = \sum \frac{(O_i - E_i)^2}{E_i} = \frac{(80 - 49.079)^2}{49.079} + \cdots + \frac{(3297 - 2927.720)^2}{2927.720} \approx 330.803$$

H_0 : The distribution of bar visits is the same for smokers and nonsmokers

H_1 : The distribution of bar visits is different for smokers and nonsmokers

df $= (2-1)(7-1) = 6$ so the critical value is $\chi_{0.05}^2 = 12.592$,

Classical Approach:
The test statistic is 330.803 which is greater than the critical value so we reject H_0.

P-value Approach:
Using Table VIII, we find the row that corresponds to 6 degrees of freedom. The value of 330.803 is greater than 18.548, which has an area under the chi-square distribution of 0.005 to the right. Therefore, we have P-value < 0.005 [Tech: P-value < 0.001]. Since P-value < α, we reject H_0.

Conclusion:
There is sufficient evidence, at the $\alpha = 0.05$ level of significance, to conclude that the distributions of time spent in bars differ between smokers and nonsmokers.

To determine if smokers tend to spend more time in bars than nonsmokers, we can examine the conditional distribution of time spent in bars by smoker status.

	Time Spent in Bars						
	Almost Daily	Several Times a Week	Several Times a Month	Once a Month	Several Times a Year	Once a Year	Never
Smoker	$\frac{80}{137} \approx 0.584$	$\frac{409}{759} \approx 0.539$	$\frac{294}{673} \approx 0.437$	$\frac{362}{833} \approx 0.435$	$\frac{433}{1006} \approx 0.430$	$\frac{336}{904} \approx 0.372$	$\frac{1265}{4562} \approx 0.277$
Nonsmoker	$\frac{57}{137} \approx 0.416$	$\frac{350}{759} \approx 0.461$	$\frac{379}{673} \approx 0.563$	$\frac{471}{833} \approx 0.565$	$\frac{573}{1006} \approx 0.570$	$\frac{568}{904} \approx 0.628$	$\frac{3297}{4562} \approx 0.723$

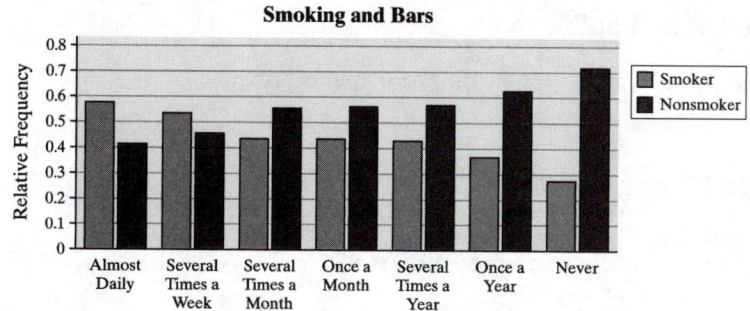

Based on the conditional distribution and the bar graph, it appears that smokers tend to spend more time in bars than nonsmokers.

6. $H_0 : p_{\text{Death Penalty}} = p_{\text{Abortion}}$

$H_1 : p_{\text{Death Penalty}} \neq p_{\text{Abortion}}$

$f_{12} = 2549$ and $f_{21} = 11,685$.

$$\chi_0^2 = \frac{(2549 - 11,685)^2}{2549 + 11,685} \approx 5863.882$$

The critical value is $\chi_{0.05}^2 = 3.841$.

<u>Classical Approach:</u>
The test statistic is 5863.882, which is greater than the critical value, so we reject H_0.

<u>P-value Approach:</u>
Using Table VIII, we find the row that corresponds to 1 degree of freedom. The value of 5863.882 is greater than 7.879, which corresponds to the area to the right of 0.005. Therefore, we have P-value < 0.005 [Tech: P-value < 0.0001]. Since P-value $< \alpha = 0.05$, we reject H_0.

<u>Conclusion:</u>
There is sufficient evidence, at the $\alpha = 0.05$ level of significance, to suggest the proportion of individuals who favor the death penalty for persons convicted of murder differs from the proportion of individuals who favor legal abortion.

Case Study:　Feeling Lucky? Well, Are You?

Computations for the case study were done using Minitab version 15. The output is provided.

Independence of Game and District

Chi-Square Test: Fantasy 5, Mega Money, Lotto

Expected counts are printed below observed counts
Chi-Square contributions are printed below expected counts

	Fantasy 5	Mega Money	Lotto	Total
1	87030	27221	270256	384507
	82777.17	28436.62	273293.21	
	218.497	51.966	33.754	
2	186780	56822	934451	1178053
	253612.79	87124.15	837316.06	
	17611.974	10539.218	11268.382	
3	267955	94123	1004651	1366729
	294231.21	101077.89	971419.91	
	2346.586	478.546	1136.795	
4	232451	61019	707934	1001404
	215583.56	74059.89	711760.55	
	1319.722	2296.314	20.572	
5	727390	262840	2746438	3736668
	804434.77	276349.22	2655884.01	
	7378.966	660.393	3087.494	
6	462874	135321	1563491	2161686
	465370.59	159869.77	1536445.64	
	13.394	3769.581	476.067	
7	353532	139421	1492549	1985502
	427441.47	146839.90	1411220.64	
	12779.783	374.830	4686.937	
8	377627	140285	1433077	1950989
	420011.46	144287.45	1386690.09	
	4277.128	111.026	1551.713	
9	470993	153591	1743370	2367954
	509776.23	175124.54	1683053.23	
	2950.587	2647.791	2161.615	
10	588154	185946	1852986	2627086
	565562.51	194288.92	1867234.58	
	902.421	358.251	108.729	
11	1283042	474067	2883453	4640562
	999026.25	343197.66	3298338.09	
	80743.569	49903.558	52186.778	
Total	5037828	1730656	16632656	23401140

Chi-Sq = 278452.937, DF = 20, P-Value = 0.000

Since the *P*-value is smaller than the significance level, $\alpha = 0.05$, we reject the null hypothesis. There is enough evidence at the $\alpha = 0.05$ level to conclude that lottery game and sales district are dependent. We conclude that there is some association between them.

Conditional Distribution Bar Graph (by Sales District)

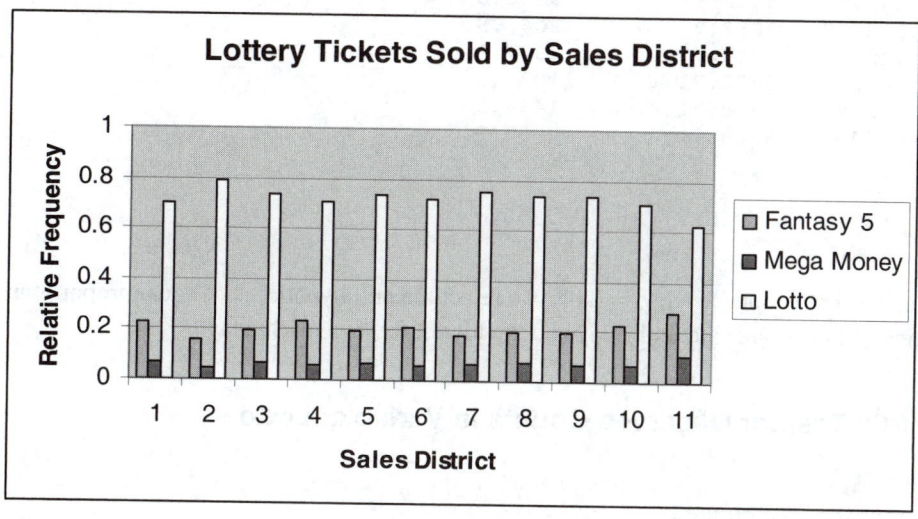

While the overall shape of the distribution is similar for each district, the bar graph tends to support the previous result. There are clear fluctuations in the lottery game as you change the sales district.

Daily Sales Structure

Chi-Square Goodness-of-Fit Test for Observed Counts in Variable: Fantasy 5

Category	Observed	Test Proportion	Expected	Contribution to Chi-Sq
1	12794	0.142857	12434.1	10.41
2	11564	0.142857	12434.1	60.89
3	12562	0.142857	12434.1	1.31
4	11408	0.142857	12434.1	84.68
5	15299	0.142857	12434.1	660.07
6	15684	0.142857	12434.1	849.40
7	7728	0.142857	12434.1	1781.21

N	DF	Chi-Sq	P-Value
87039	6	3447.98	0.000

Since the *P*-value is smaller than the significance level, $\alpha = 0.05$, we reject the null hypothesis of equal proportions. It is clear that a highest proportion of tickets are sold on Friday and Saturday, and the lowest on Sunday.

Chi-Square Goodness-of-Fit Test for Observed Counts in Variable: Mega Money

Category	Observed	Test Proportion	Expected	Contribution to Chi-Sq
1	2082	0.142857	3888.71	839.4
2	9983	0.142857	3888.71	9550.8
3	1040	0.142857	3888.71	2086.9
4	1677	0.142857	3888.71	1257.9
5	10439	0.142857	3888.71	11033.5
6	1502	0.142857	3888.71	1464.9
7	498	0.142857	3888.71	2956.5

N	DF	Chi-Sq	P-Value
27221	6	29189.8	0.000

Since the P-value is smaller than the significance level, $\alpha = 0.05$, we reject the null hypothesis of equal proportions. It is clear that a highest proportion of tickets are sold on Friday and the lowest (by far) on Sunday.

Chi-Square Goodness-of-Fit Test for Observed Counts in Variable: Lotto

Category	Observed	Test Proportion	Expected	Contribution to Chi-Sq
1	13093	0.142857	38608	16862
2	18200	0.142857	38608	10788
3	60746	0.142857	38608	12694
4	18054	0.142857	38608	10942
5	38513	0.142857	38608	0
6	115686	0.142857	38608	153880
7	5964	0.142857	38608	27601

N	DF	Chi-Sq	P-Value
270256	6	232768	0.000

Since the P-value is smaller than the significance level, $\alpha = 0.05$, we reject the null hypothesis of equal proportions. It is clear that a highest proportion of tickets are sold on Saturday and the lowest (by far) on Sunday.

Written reports will vary.

Chapter 13
Comparing Three or More Means

Section 13.1

1. Analysis of variance

3. True. One of the requirements to perform a one-way ANOVA is that the populations must have the same variance; that is, each treatment group has population variance σ^2.

5. Error; MSE

7.

Source of Variation	Sum of Squares	Degrees of Freedom	Mean Squares	F-Test Statistic
Treatment	387	2	193.5	0.650
Error	8042	27	297.852	
Total	8429	29		

9. $\bar{x} = \dfrac{10(40) + 10(42) + 10(44)}{10 + 10 + 10}$
$= 42$

$SST = 10(40 - 42)^2 + 10(42 - 42)^2 + 10(44 - 42)^2$
$= 80$

$MST = \dfrac{80}{3 - 1}$
$= 40$

$SSE = (10 - 1)(48) + (10 - 1)(31) + (10 - 1)(25)$
$= 936$

$MSE = \dfrac{936}{30 - 3}$
≈ 34.67

$F_0 = \dfrac{40}{34.67}$
≈ 1.15

11.

Population	Sample Size	Sample Mean	Sample Variance
1	4	27	8.67
2	4	21.75	11.58
3	4	24.5	20.33

$\bar{x} \approx 24.42$

$SST \approx 55.17$, $MST \approx 27.58$, $SSE = 121.75$,
$MSE \approx 13.53$

$F_0 = \dfrac{27.58}{13.53}$
≈ 2.04

13. (a) $H_0: \mu_{SP} = \mu_{SD} = \mu_{NT}$
H_1: At least one mean is different.

(b) There must be k random samples, one from each of k populations or a randomized experiment with k treatments; the k samples must be independent of each other; the populations must be normally distributed; and the populations must have the same variance.

(c) Since $0.007 < 0.05$, reject the null hypothesis. There is sufficient evidence to conclude that at least one mean is different from the others.

(d) Yes, because there is very little overlap between the Spring Disc boxplot and the other two boxplots.

(e) $F_0 = \dfrac{42.06}{5.92}$
≈ 7.10

(f) The correlation between the residuals and expected z-scores is 0.988. (Tech: 0.989) Since $0.988 > 0.946$ (the critical value from Table VI with $n = 18$), it is reasonable to conclude that the residuals are normally distributed.

15. (a) The response variable is the final exam score. It is quantitative because it provides a numerical measure of individuals.

(b) The factor in this study is the method of instruction. It has four levels (traditional, hybrid, online, emporium).

(c) $H_0: \mu_I = \mu_{II} = \mu_{III} = \mu_{IV}$
H_1: At least one mean is different.

(d) There must be k random samples, one from each of k populations or a randomized experiment with k treatments; the k samples must be independent of each other; the populations must be normally distributed; and the populations must have the same variance.

(e) Since $0.439 > 0.05$, do not reject the null hypothesis. There is not sufficient evidence to conclude that at least one mean is different from the others.

(f) Yes, because there is plenty of overlap among all the boxplots.

(g) If this experiment were repeated 100 times, about 44 of them would be expected to produce results at least as extreme as the results in this experiment.

(h) The correlation between the residuals and expected z-scores is 0.989. Since $0.989 > 0.960$ (the critical value from Table VI with $n = 30$), it is reasonable to conclude that the residuals are normally distributed.

17. (a) $H_0: \mu_F = \mu_E = \mu_U$
H_1: At least one mean is different.

(b) It is given that the samples are random. The samples are independent because they come from different sectors. It is given that the populations are normal.

$s_F \approx 5.12$
$s_E \approx 4.87$
$s_U \approx 4.53$

The largest sample standard deviation, 5.12, is less than twice the smallest sample standard deviation, 4.53, so it can be assumed that the populations have equal variances.

(c)

Source of Variation	Sum of Squares	Degrees of Freedom	Mean Squares	F-Test Statistic	P-value
Treatment	97.59	2	48.80	2.08	0.150
Error	493.26	21	23.49		
Total	590.85	23			

Since $0.150 > 0.05$, do not reject the null hypothesis. There is not sufficient evidence to conclude that at least one mean is different from the others.

(d)

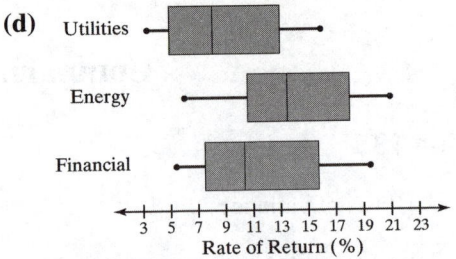

19. (a) The response variable is profit on initial value. It is quantitative because it provides a numerical measure of individuals.

(b) $H_0: \mu_{WM} = \mu_{BM} = \mu_{WF} = \mu_{BF}$
H_1: At least one mean is different.

(c) It is reasonable to assume that the samples are random. The samples are independent because they come from different races and genders. It is given that the populations are normal.

$s_{WM} \approx 226.72$
$s_{BM} \approx 241.52$
$s_{WF} \approx 157.15$
$s_{BF} \approx 180.56$

The largest sample standard deviation, 241.52, is less than twice the smallest sample standard deviation, 157.15, so it can be assumed that the populations have equal variances.

(d)

Source of Variation	Sum of Squares	Degrees of Freedom	Mean Squares	F-Test Statistic	P-value
Treatment	8,107,321	3	2,702,440	61.36	< 0.001
Error	1,189,048	27	44,038.8		
Total	9,296,369	30			

Since the P-value is very low, reject the null hypothesis. There is sufficient evidence to conclude that at least one mean is different from the others.

(e)

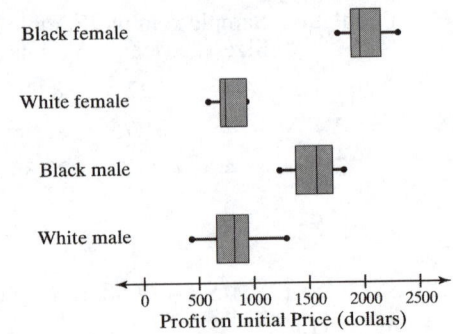

(f) The results suggest that car dealers make more of a profit off of black men and black women than white men and white women.

(g) The correlation between the residuals and expected z-scores is 0.994. (Tech: 0.993) Since $0.994 > 0.960$ (the critical value from Table VI with $n = 30$), it is reasonable to conclude that the residuals are normally distributed.

21. (a) All of the subjects were randomly assigned to a diet.

(b) The response variable is LDL cholesterol level. The explanatory variable is type of diet.

(c) Randomization avoids bias that might be generated by allowing the subjects to choose their own diets. Subjects may prefer one type of diet to another, or they might be more prone to cheating on their diet if it is a diet they chose than if it is a diet they were told to follow.

(d) $H_0: \mu_{SF} = \mu_M = \mu_{NCE}$
$H_1:$ At least one mean is different.

(e) It is given that the samples are random. The samples are independent because they come from different diets. It is given that the populations are normal.

$$s_{SF} \approx 51.94$$
$$s_M \approx 48.40$$
$$s_{NCE} \approx 36.33$$

The largest sample standard deviation, 51.94, is less than twice the smallest sample standard deviation, 36.33, so it can be assumed that the populations have equal variances.

(f)

Source of Variation	Sum of Squares	Degrees of Freedom	Mean Squares	F-Test Statistic	P-value
Treatment	34,045.2	2	17,022.6	8.03	0.001
Error	89,043.6	42	2120.09		
Total	123,088.8	44			

Since $0.001 < 0.05$, reject the null hypothesis. There is sufficient evidence to conclude that at least one mean is different from the others.

(g)

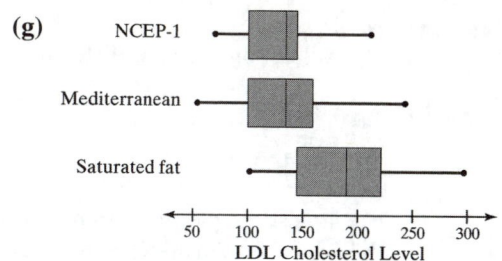

23. (a) Quantitative

(b) $H_0: \mu_{CB} = \mu_{RB} = \mu_S = \mu_{WR}$
$H_1:$ At least one mean is different.

(c) The samples are being treated as random. The samples are independent because they come from different positions. Normal probability plots suggest that each position's times are normally distributed.

$$s_{CB} \approx 0.086$$
$$s_{RB} \approx 0.102$$
$$s_S \approx 0.082$$
$$s_{WR} \approx 0.105$$

The largest sample standard deviation, 0.105, is less than twice the smallest sample standard deviation, 0.082, so it can be assumed that the populations have equal variances.

(d)

Source of Variation	Sum of Squares	Degrees of Freedom	Mean Squares	F-Test Statistic	P-value
Treatment	0.2341	3	0.0780	8.43	< 0.001
Error	0.9992	108	0.0093		
Total	1.2333	111			

Since the P-value is very low, reject the null hypothesis. There is sufficient evidence to conclude that at least one mean is different from the others.

(e)

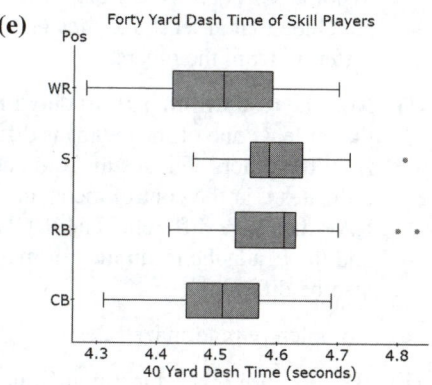

25. (a) $H_0: \mu_{301} = \mu_{400} = \mu_{353}$
H_1: At least one mean is different.

(b) $s_{301} \approx 107.83$
$s_{400} \approx 384.75$
$s_{353} \approx 195.75$

The largest sample standard deviation, 384.75, is more than twice the smallest sample standard deviation, 107.83, so it cannot be assumed that the populations have equal variances.

27. (a) $s_C \approx 4.6$
$s_R \approx 3.6$
$s_H \approx 2.9$

(b) Stratified sampling was used. This is because the researchers wanted to be sure that both children and adolescents were represented in the samples.

(c) Using technology and the unrounded values of the standard deviations results in a test statistic of $t_0 \approx 3.84$ and a P-value of about 0.0002. Since the P-value is very low, reject the null hypothesis. There is sufficient evidence to conclude that the means are different.

(d) No, because both samples have a sample size of 70, which is greater than 30.

(e) $\bar{x} \approx 11.03$

$SST \approx 451.269, \ MST \approx 225.6345,$
$SSE = 2967, \ MSE \approx 14.33$

$$F_0 = \frac{225.6345}{14.33} \approx 15.75$$

$F_{0.05,2,207} \approx 3.04$

Since $15.75 > 3.04$, reject the null hypothesis. There is sufficient evidence to conclude that at least one mean is different from the others.

(f) No. The result from part (e) only indicates that at least one of the means is different from the others. The result from part (c) indicates that the control mean and the RAP mean are different. The RAP mean and the headache mean may or may not also be different.

29. (a) Completely randomized design

(b) Whether or not a subject maintained smoking abstinence after six months

(c) The treatments are the e-cig with nicotine, the nicotine patch, and the e-cig with a placebo. There are three levels to this treatment.

(d) ANOVA is not appropriate because the response variable is qualitative. Since the response variable is qualitative with two possible outcomes, homogeneity of proportions should be used instead.

(e) $H_0: p_{\text{e-Cig/N}} = p_{\text{patch}} = p_{\text{e-Cig/placebo}}$
H_1: At least one proportion is different.

Using technology, the P-value for this test is about 0.548. Since $0.548 > 0.05$, do not reject the null hypothesis. There is not sufficient evidence to conclude that at least one of the proportions is different from the others.

31. The mean square due to treatment estimate of σ^2 is a weighted average of the squared deviations of each sample mean from the grand mean of all the samples. The mean square due to error estimate of σ^2 is the weighted average of the sample variances.

33. Rejecting the statement in the null hypothesis means that there is sufficient evidence to conclude that the mean of one of the populations is different than the other two or that all three population means are different.

Section 13.2

1. (a) $q_{0.05,10,3} = 3.877$

(b) $q_{0.05,24,5} = 4.166$

(c) $q_{0.05,32,8} \approx q_{0.05,30,8} = 4.602$

(d) $q_{0.05,65,5} \approx q_{0.05,60,5} = 3.977$

3. The critical value is $q_{0.05,12,3} = 3.773$.

$$q_{\mu_3,\mu_2} = \frac{18.1 - 9.1}{\sqrt{\dfrac{26.2}{2}\left(\dfrac{1}{5} + \dfrac{1}{5}\right)}} \approx 3.932 > 3.773$$

$$q_{\mu_3,\mu_1} = \frac{18.1 - 9.5}{\sqrt{\dfrac{26.2}{2}\left(\dfrac{1}{5} + \dfrac{1}{5}\right)}} \approx 3.757 < 3.773$$

$$q_{\mu_1,\mu_2} = \frac{9.5-9.1}{\sqrt{\frac{26.2}{2}\left(\frac{1}{5}+\frac{1}{5}\right)}} \approx 0.175 < 3.773$$

The results are ambiguous; $\underline{\mu_2\ \underline{\mu_1}\ \mu_3}$.

5. (a) H_1: At least one mean is different.

Source of Variation	Sum of Squares	Degrees of Freedom	Mean Squares	F-Test Statistic	P-value
Treatment	630.33	2	315.17	9.72	0.002
Error	486.50	15	32.43		
Total	1116.84	17			

Since $0.002 < 0.05$, reject the null hypothesis. There is sufficient evidence to conclude that at least one mean is different from the others.

(b) $q_{0.05,15,3} = 3.674$

$$q_{\mu_3,\mu_2} = \frac{47.72-34.05}{\sqrt{\frac{32.43}{2}\left(\frac{1}{6}+\frac{1}{6}\right)}} \approx 5.880 > 3.674$$

$$q_{\mu_3,\mu_1} = \frac{47.72-36.7}{\sqrt{\frac{32.43}{2}\left(\frac{1}{6}+\frac{1}{6}\right)}} \approx 4.740 > 3.674$$

$$q_{\mu_1,\mu_2} = \frac{36.7-34.05}{\sqrt{\frac{32.43}{2}\left(\frac{1}{6}+\frac{1}{6}\right)}} \approx 1.140 < 3.674$$

$\underline{\mu_1\ \mu_2}\ \underline{\mu_3}$

(c)

7. The confidence intervals for $\mu_{SD}-\mu_{SP}$ and $\mu_{SD}-\mu_{NT}$ did not contain zero, while the confidence interval for $\mu_{NT}-\mu_{SP}$ contains zero. These results indicate that

$\mu_{SP}=\mu_{NT}\neq\mu_{SD}$ $\left(\underline{\mu_{SP}\ \mu_{NT}}\ \underline{\mu_{SD}}\right)$. Since the spring disc method of planting has the highest mean, this is the recommended method of planting.

9. The critical value is $q_{0.05,27,4}\approx q_{0.05,24,4}=3.901$.

$$q_{\mu_{BF},\mu_{WF}} = \frac{2021.43-792.2}{\sqrt{\frac{44,038.8}{2}\left(\frac{1}{7}+\frac{1}{5}\right)}} \approx 14.15 > 3.901$$

$$q_{\mu_{BF},\mu_{WM}} = \frac{2021.43-829.8}{\sqrt{\frac{44,038.8}{2}\left(\frac{1}{7}+\frac{1}{15}\right)}} \approx 17.54 > 3.901$$

$$q_{\mu_{BF},\mu_{BM}} = \frac{2021.43-1554.5}{\sqrt{\frac{44,038.8}{2}\left(\frac{1}{7}+\frac{1}{4}\right)}} \approx 5.02 > 3.901$$

$$q_{\mu_{BM},\mu_{WF}} = \frac{1554.5-792.2}{\sqrt{\frac{44,038.8}{2}\left(\frac{1}{4}+\frac{1}{5}\right)}} \approx 7.66 > 3.901$$

$$q_{\mu_{BM},\mu_{WM}} = \frac{1554.5-829.8}{\sqrt{\frac{44,038.8}{2}\left(\frac{1}{4}+\frac{1}{15}\right)}} \approx 8.68 > 3.901$$

$$q_{\mu_{WM},\mu_{WF}} = \frac{829.8-792.2}{\sqrt{\frac{44,038.8}{2}\left(\frac{1}{15}+\frac{1}{5}\right)}} \approx 0.49 < 3.901$$

$\underline{\mu_{WF}\ \mu_{WM}\ \mu_{BM}}\ \underline{\mu_{BF}}$

11. (a) The individuals in the study were randomly assigned to one of five treatment groups. The factor drug was intentionally manipulated while other factors that may affect ADHD were either controlled by fixing them or dealt with through randomization.

(b) Score on ability to follow rules; the drug

(c) Randomization was used to "even out" those variables that may affect the symptoms associated with ADHD that were not controlled in the study.

(d) H_0: $\mu_P=\mu_{R10}=\mu_{R17.5}=\mu_{A7.5}=\mu_{A12.5}$
H_1: At least one mean is different.

(e)

Source of Variation	Sum of Squares	Degrees of Freedom	Mean Squares	F-Test Statistic	P-value
Treatment	7297.0	4	1824.2	5.75	0.002
Error	9518.6	30	317.3		
Total	16,815.5	34			

Since $0.002 < 0.05$, reject the null hypothesis. There is sufficient evidence to conclude that at least one mean is different from the others.

(f) $q_{0.05,30,5} = 4.102$

$$q_{\mu_{A12.5},\mu_P} = \frac{84.86 - 40.86}{\sqrt{\frac{317.3}{2}\left(\frac{1}{7}+\frac{1}{7}\right)}} \approx 6.535 > 4.102$$

$$q_{\mu_{A12.5},\mu_{A7.5}} = \frac{84.86 - 72.57}{\sqrt{\frac{317.3}{2}\left(\frac{1}{7}+\frac{1}{7}\right)}} \approx 1.825 < 4.102$$

$$q_{\mu_{A12.5},\mu_{R17.5}} = \frac{84.86 - 67.57}{\sqrt{\frac{317.3}{2}\left(\frac{1}{7}+\frac{1}{7}\right)}} \approx 2.271 < 4.102$$

$$q_{\mu_{A7.5},\mu_P} = \frac{72.57 - 40.86}{\sqrt{\frac{317.3}{2}\left(\frac{1}{7}+\frac{1}{7}\right)}} \approx 4.710 > 4.102$$

$$q_{\mu_{A7.5},\mu_{R10}} = \frac{72.57 - 68.71}{\sqrt{\frac{317.3}{2}\left(\frac{1}{7}+\frac{1}{7}\right)}} \approx 0.573 < 4.102$$

$$q_{\mu_{A7.5},\mu_{R17.5}} = \frac{72.57 - 67.57}{\sqrt{\frac{317.3}{2}\left(\frac{1}{7}+\frac{1}{7}\right)}} \approx 0.446 < 4.102$$

$$q_{\mu_{R10},\mu_{R17.5}} = \frac{68.71 - 67.57}{\sqrt{\frac{317.3}{2}\left(\frac{1}{7}+\frac{1}{7}\right)}} \approx 0.150 > 4.102$$

$$q_{\mu_{R17.5},\mu_P} = \frac{69.57 - 40.86}{\sqrt{\frac{317.3}{2}\left(\frac{1}{7}+\frac{1}{7}\right)}} \approx 4.264 > 4.102$$

$$q_{\mu_{R10},\mu_P} = \frac{68.71 - 40.86}{\sqrt{\frac{317.3}{2}\left(\frac{1}{7}+\frac{1}{7}\right)}} \approx 4.137 > 4.102$$

$$q_{\mu_{A12.5},\mu_{R10}} = \frac{84.86 - 68.71}{\sqrt{\frac{317.3}{2}\left(\frac{1}{7}+\frac{1}{7}\right)}} \approx 2.399 < 4.102$$

$\underline{\mu_P}\quad \underline{\mu_{R10}\ \ \mu_{R17.5}\ \ \mu_{A7.5}\ \ \mu_{A12.5}}$

(g)

13. (a) $H_0: \mu_{NS} = \mu_{LS} = \mu_{HS}$
H_1: At least one mean is different.

Source of Variation	Sum of Squares	Degrees of Freedom	Mean Squares	F-Test Statistic	P-value
Treatment	1494.86	2	747.43	19.62	< 0.001
Error	685.71	18	38.10		
Total	2180.57	20			

Since the *P*-value is very low, reject the null hypothesis. There is sufficient evidence to conclude that at least one mean is different from the others.

(b) $q_{0.05,18,3} = 3.609$

$$q_{\mu_{HS},\mu_{NS}} = \frac{78.57 - 58}{\sqrt{\frac{38.10}{2}\left(\frac{1}{7}+\frac{1}{7}\right)}} \approx 8.817 > 3.609$$

$$q_{\mu_{LS},\mu_{NS}} = \frac{70 - 58}{\sqrt{\frac{38.10}{2}\left(\frac{1}{7}+\frac{1}{7}\right)}} \approx 5.144 > 3.609$$

$$q_{\mu_{HS},\mu_{LS}} = \frac{78.57 - 70}{\sqrt{\frac{38.10}{2}\left(\frac{1}{7}+\frac{1}{7}\right)}} \approx 3.673 > 3.609$$

$\underline{\mu_{NS}}\quad \underline{\mu_{LS}}\quad \underline{\mu_{HS}}$

(c)

Heart Rate (per minute)

15. (a) $H_0: \mu_{SM} = \mu_{MM} = \mu_{WM}$
H_1: At least one mean is different.

Source of Variation	Sum of Squares	Degrees of Freedom	Mean Squares	F-Test Statistic	P-value
Treatment	117,596	2	58,798	70.51	< 0.001
Error	17,510.6	21	833.8		
Total	135,107	23			

Since the P-value is very low, reject the null hypothesis. There is sufficient evidence to conclude that at least one mean is different from the others.

(b) $q_{0.05,21,3} \approx q_{0.05,20,3} = 3.578$

$$q_{\mu_{MM},\mu_{WM}} = \frac{1018.9 - 867.1}{\sqrt{\dfrac{833.8}{2}\left(\dfrac{1}{8}+\dfrac{1}{8}\right)}} \approx 14.869 > 3.578$$

$$q_{\mu_{MM},\mu_{SM}} = \frac{1018.9 - 873.9}{\sqrt{\dfrac{833.8}{2}\left(\dfrac{1}{8}+\dfrac{1}{8}\right)}} \approx 14.203 > 3.578$$

$$q_{\mu_{SM},\mu_{WM}} = \frac{873.9 - 867.1}{\sqrt{\dfrac{833.8}{2}\left(\dfrac{1}{8}+\dfrac{1}{8}\right)}} \approx 0.666 < 3.578$$

$\underline{\mu_{WM}}\ \underline{\mu_{SM}}\ \underline{\mu_{MM}}$

(c)

Data

17. (a) This study is a cohort study because the children were selected and then observed over a long period of time. The response variable is time to complete a degree.

(b) $\overline{x}_{HS} = 4.05,\ \overline{x}_{S} = 4.2,\ \overline{x}_{NS} = 4.2,$
$\overline{x}_{OD} = 5.8,\ \overline{x}_{NR} = 4.65$

(c) $s_{HS} \approx 0.991,\ s_{S} \approx 0.554,\ s_{NS} \approx 1.099,$
$s_{OD} \approx 0.921,\ s_{NR} \approx 0.566;$ since the largest standard deviation, 1.099, is less than twice the smallest standard deviation, 0.554, it appears that the population variances are equal.

(d) Using technology, the confidence interval is $(3.341, 4.759)$.

(e) There are 10 possible pairwise comparisons.

(f) $(0.95)^{10} = 0.5987;\ 1 - 0.5987 = 0.4013$

The probability of a Type I error is 0.4013.

(g) H_1: At least one mean is different.

Source of Variation	Sum of Squares	Degrees of Freedom	Mean Squares	F-Test Statistic	P-value
Treatment	20.63	4	5.1575	7.03	< 0.001
Error	33.01	45	0.7336		
Total	53.64	49			

Since the P-value is very small, reject the null hypothesis. There is sufficient evidence to conclude that at least one mean is different from the others.

$q_{0.05,45,5} \approx q_{0.05,40,5} = 4.039$

$$q_{\mu_{OD},\mu_{HS}} = \frac{5.8 - 4.05}{\sqrt{\dfrac{0.7336}{2}\left(\dfrac{1}{10}+\dfrac{1}{10}\right)}} \approx 6.461 > 4.039$$

$$q_{\mu_{OD},\mu_{NS}} = \frac{5.8 - 4.2}{\sqrt{\dfrac{0.7336}{2}\left(\dfrac{1}{10}+\dfrac{1}{10}\right)}} \approx 5.907 > 4.039$$

$$q_{\mu_{OD},\mu_{NR}} = \frac{5.8 - 4.65}{\sqrt{\dfrac{0.7336}{2}\left(\dfrac{1}{10}+\dfrac{1}{10}\right)}} \approx 4.246 > 4.039$$

$$q_{\mu_{NR},\mu_{NS}} = \frac{4.65 - 4.2}{\sqrt{\dfrac{0.7336}{2}\left(\dfrac{1}{10}+\dfrac{1}{10}\right)}} \approx 1.661 < 4.039$$

$$q_{\mu_{HS},\mu_{NS}} = \frac{4.05 - 4.2}{\sqrt{\frac{0.7336}{2}\left(\frac{1}{10} + \frac{1}{10}\right)}} \approx 0.554 < 4.039$$

$$q_{\mu_{NR},\mu_{HS}} = \frac{4.65 - 4.05}{\sqrt{\frac{0.7336}{2}\left(\frac{1}{10} + \frac{1}{10}\right)}} \approx 2.215 < 4.039$$

Note that Selective and Non-Selective have the same mean, so some repetitive calculations have been omitted.

$\underline{\mu_{HS}\ \mu_{S}\ \mu_{NS}\ \mu_{NR}}\ \underline{\mu_{OD}}$

Section 13.3

1. **(a)** The treatment P-value=0.013 is sufficient evidence against the null hypothesis that the population means are equal. The researcher will conclude that at least one population mean is different from the others.

(b) The mean square error is $MSE = 17.517$.

(c) The P-values comparing treatment 3 with treatments 1 and 2 are small (less than 0.05) while the P-value comparing treatments 1 and 2 is large. Therefore, we conclude that means 1 and 2 are the same, but mean 3 is different. That is, we conclude $\mu_1 = \mu_2 \neq \mu_3$ or $\underline{\mu_1\ \mu_2}\ \ \underline{\mu_3}$.

3. **(a)** The treatment P-value=0.108 is not sufficient evidence against the null hypothesis that the population means are equal. In other words, the researcher does not reject the null hypothesis that population means are equal.

(b) The mean square error is $MSE = 0.944$.

(c) We would only use Tukey's test if the null hypothesis from part (a) had been rejected.

---

5. **(a)** Using Minitab, we obtain the following output:

Two-way ANOVA: Response versus Block, Treatment

Source	DF	SS	MS	F	P
Block	4	5.836	1.459	10.13	0.003
Treatment	2	4.468	2.234	15.51	0.002
Error	8	1.152	0.144		
Total	14	11.456			

Since the treatment P-value is 0.002, we reject the null hypothesis that the population means are equal. Thus, we conclude that at least one mean is different from the others.

(b) Using Minitab, we obtain the following output:

```
Tukey 95.0% Simultaneous Confidence Intervals
Response Variable Response
All Pairwise Comparisons among Levels of Treatment
Treatment = 1  subtracted from:
Treatment   Lower   Center    Upper   --------+---------+---------+---------
2          -1.606  -0.920  -0.2344            (---------*---------)
3          -1.986  -1.300  -0.6144   (--------*---------)
                                     --------+---------+---------+---------
                                        -1.40     -0.70     0.00


Treatment = 2  subtracted from:
Treatment   Lower   Center    Upper   --------+---------+---------+---------
3          -1.066  -0.3800  0.3056                (---------*--------)
                                     --------+---------+---------+---------
                                        -1.40     -0.70     0.00

Tukey Simultaneous Tests
Response Variable Response
All Pairwise Comparisons among Levels of Treatment
Treatment = 1  subtracted from:
            Difference    SE of             Adjusted
Treatment   of Means   Difference  T-Value  P-Value
2            -0.920      0.2400    -3.833    0.0123
3            -1.300      0.2400    -5.417    0.0016
```

```
Treatment = 2  subtracted from:
            Difference      SE of             Adjusted
Treatment   of Means    Difference  T-Value   P-Value
3            -0.3800       0.2400    -1.583    0.3066
```

The interval for Treatment 2 minus Treatment 1 does not include 0 and the P-value is 0.0123, for this comparison, which is small, so reject the null hypothesis that $\mu_1 = \mu_2$. Likewise, the interval for Treatment 3 minus Treatment 1 does not include 0 and the P-value is 0.0016, which is small, so we reject the null hypothesis that $\mu_1 = \mu_3$. Finally, the interval for Treatment 3 minus Treatment 2 does include 0 and the P-value is 0.3066, which is not small, so we do not reject the null hypothesis that $\mu_2 = \mu_3$. Thus, the Tukey's tests indicate that the means of Treatments 2 and 3 are equal, but the mean of Treatment 1 is different. That is, we conclude: $\mu_1 \neq \mu_2 = \mu_3$ or $\underline{\mu_1} \quad \underline{\mu_2 \quad \mu_3}$.

(c) The boxplots shown to the right support the results found in parts (a) and (b).

7. (a) The sample standard deviations for the provided data are $s_{87\ \text{Octane}} \approx 1.240$, $s_{89\ \text{Octane}} \approx 1.439$, and $s_{92\ \text{Octane}} \approx 1.225$. Since the largest standard deviation is less than twice the smallest, $1.439 < 2(1.225)$, the requirement for equal population variances is satisfied.

(b) Blocking is used to remove the effect car type (and driver) has on gas mileage.

(c) Using Minitab, we obtain the following output:

```
Two-way ANOVA: Miles per Gallon versus Car and Driver, Octane level
Source           DF      SS        MS        F        P
Car and Driver    5   25.0161   5.00322   92.84    0.000
Octane level      2    1.1411   0.57056   10.59    0.003
Error            10    0.5389   0.05389
Total            17   26.6961
```

Since the treatment (octane level) P-value is 0.003, there is sufficient evidence to conclude that the mean number of miles per gallon is different among the three octane levels at the $\alpha = 0.05$ level of significance.

(d) Using Minitab, we obtain the following output:

```
Tukey 95.0% Simultaneous Confidence Intervals
Response Variable Miles
All Pairwise Comparisons among Levels of Octane level
Octane level = 1  subtracted from:
Octane
level    Lower    Center    Upper    +---------+---------+---------+------
2       -0.3344   0.03333   0.4010   (---------*---------)
3        0.1823   0.55000   0.9177                  (---------*---------)
                                     +---------+---------+---------+------
                                   -0.35      0.00      0.35      0.70
Octane level = 2  subtracted from:
Octane
level    Lower    Center    Upper    +---------+---------+---------+------
3        0.1490   0.5167    0.8844                  (---------*---------)
                                     +---------+---------+---------+------
                                   -0.35      0.00      0.35      0.70
```

```
Tukey Simultaneous Tests
Response Variable Miles per Gallon
All Pairwise Comparisons among Levels of Octane level
Octane level = 1  subtracted from:
Octane          Difference      SE of                    Adjusted
level           of Means        Difference   T-Value     P-Value
2                  0.03333         0.1340      0.2487      0.9666
3                  0.55000         0.1340      4.1037      0.0055
Octane level = 2  subtracted from:
Octane          Difference      SE of                    Adjusted
level           of Means        Difference   T-Value     P-Value
3                  0.5167          0.13         3.855       0.0081
```

The interval for 89 octane minus 87 octane includes 0 and the P-value is 0.9666 for this comparison, which is not small, so we do not reject the null hypothesis that $\mu_{87} = \mu_{89}$. The interval for 92 octane minus 87 octane does not include 0 and the P-value is 0.0055, which is small, so we reject the null hypothesis that $\mu_{87} = \mu_{92}$. Finally, the interval for 92 octane minus 89 octane does not include 0 and the P-value is 0.0081, which is small, so we reject the null hypothesis that $\mu_{89} = \mu_{92}$. We conclude

$$\mu_{87} = \mu_{89} \neq \mu_{92} \text{ or } \underline{\mu_{87} \quad \mu_{89}} \quad \mu_{92}.$$

(e) The Tukey's tests indicate that the mean gasoline mileages for 87 octane and 89 octane are equal, but the mean gasoline mileage for 92 octane is different. In fact, since the confidence intervals for 92 octane minus 87 octane and 92 octane minus 87 octane contain all positive values, we can conclude that the mean gasoline mileage for 92 octane is higher than for 87 octane and 89 octane.

9. (a) The sample standard deviations for the provided data are $s_{\text{Jeep}} \approx 448$, $s_{\text{Saturn}} \approx 453$, $s_{\text{Toyota}} \approx 830$, and $s_{\text{Hyundai}} \approx 614$. Since the largest standard deviation is less than twice the smallest, $830 < 2(448)$, the requirement for equal population variances is satisfied.

(b) Using Minitab, we obtain the following output:

```
Two-way ANOVA: Cost versus Location, Brand
Source      DF        SS          MS          F          P
Location     3     3404468     1134823      10.09      0.003
Brand        3      926488      308829       2.75      0.105
Error        9     1012007      112445
Total       15     5342964
```

Since the treatment (brand) P-value is 0.105, there is not sufficient evidence to conclude that the mean cost of repair is different among the car brands at the $\alpha = 0.05$ level of significance.

(c) Since the null hypothesis was not rejected, there is no need to do a post hoc analysis.

11. (a) The response variable is water consumption (in ml/day). The treatment is the space flight, which has three levels: Lift-Off Minus 1, Return Plus 1, and Return Plus 1 Month.

(b) The sample standard deviations for the provided data are $s_{\text{lift-off}-1} \approx 5.65$, $s_{\text{return}+1} \approx 9.97$, and $s_{+1 \text{ month}} \approx 9.00$. Since the largest standard deviation is less than twice the smallest, $9.97 < 2(5.65)$, the requirement for equal population variances is satisfied.

(c) Using Minitab, we obtain the following output:

```
Two-way ANOVA: Water Consumption versus Rat, Day
Source      DF        SS          MS          F          P
Rat          5      861.78     172.356       8.62      0.002
Day          2      635.44     317.722      15.89      0.001
Error       10      199.89      19.989
Total       17     1697.11
```

Since the treatment (day) P-value is 0.001, there is sufficient evidence to conclude that the mean water consumption of rats is different among the experiment levels at the $\alpha = 0.05$ level of significance.

(d) Using Minitab with Treatment 1 representing Lift-off Minus 1, Treatment 2 representing Return Plus 1, and Treatment 3 representing Return Plus 1 Month, we obtain the following output:

```
Tukey 95.0% Simultaneous Confidence Intervals
Response Variable Water Consumption
All Pairwise Comparisons among Levels of Day
Day = 1   subtracted from:
Day    Lower   Center   Upper    -+---------+---------+---------+-----
2    -0.9152   6.167   13.25     (---------*---------)
3     7.4181  14.500   21.58                (---------*---------)
                                 -+---------+---------+---------+-----
                                0.0       7.0      14.0      21.0

Day = 2   subtracted from:
Day  Lower   Center   Upper     -+---------+---------+---------+-----
3    1.251   8.333   15.42           (---------*---------)
                                 -+---------+---------+---------+-----
                                0.0       7.0      14.0      21.0
Tukey Simultaneous Tests
Response Variable Water Consumption
All Pairwise Comparisons among Levels of Day
Day = 1   subtracted from:
       Difference      SE of              Adjusted
Day    of Means     Difference   T-Value   P-Value
2         6.167        2.581      2.389     0.0885
3        14.500        2.581      5.617     0.0006
Day = 2   subtracted from:
       Difference      SE of              Adjusted
Day    of Means     Difference   T-Value   P-Value
3         8.333        2.581      3.228     0.0225
```

The interval for Return Plus 1 minus Lift-off minus 1 includes 0 and the *P*-value is 0.0885 for this comparison, which is not small, so we do not reject the null hypothesis that $\mu_{\text{lift-off}-1} = \mu_{\text{return}+1}$. The interval for Return Plus 1 Month minus Lift-off Minus 1 does not include 0 and the *P*-value is 0.0006, which is small, so we reject the null hypothesis that $\mu_{\text{lift-off}-1} = \mu_{\text{return}+1 \text{ month}}$. Finally, the interval for Return Plus 1 Month minus Return Plus 1 does not include 0 and the *P*-value is 0.0225, which is small, so we reject the null hypothesis that $\mu_{\text{return}+1} = \mu_{\text{return}+1 \text{ month}}$. Thus, we conclude that

$$\mu_{\text{lift-off}-1} = \mu_{\text{return}+1} \neq \mu_{\text{return}+1 \text{ month}} \quad \text{or} \quad \underline{\mu_{\text{lift-off}-1} \quad \mu_{\text{return}+1}} \quad \underline{\mu_{\text{return}+1 \text{ month}}}.$$

13. (a) Using Minitab, we obtain the following output:

```
Two-way ANOVA: Wait Time versus Day, Procedure
Source        DF       SS          MS           F          P
Day            6    6538.75     1089.79      249.50      0.000
Procedure      1       3.02        3.02        0.69      0.438
Error          6      26.21        4.37
Total         13    6567.97
```

Since the treatment (procedure) *P*-value is 0.438, there is not sufficient evidence to conclude that the mean wait time is different for the two loading procedures at the $\alpha = 0.05$ level of significance.

(b)

Observation	1	2	3	4	5	6	7
Wait time before new procedure	11.6	25.9	20.0	38.2	57.3	32.1	81.8
Wait time after new procedure	10.7	28.3	19.2	35.9	59.2	31.8	75.3
$d_i =$ after − before	−0.9	2.4	−0.8	−2.3	1.9	−0.3	−6.5

Using technology, $\bar{d} \approx -0.93$ minutes and $s_d \approx 2.96$ minutes, where we measure differences as wait time after new procedure minus wait time before new procedure. The hypotheses are $H_0 : \mu_d = 0$ versus $H_1 : \mu_d \neq 0$. The level of significance is $\alpha = 0.05$. The test statistic is

$$t_0 = \frac{\bar{d}}{s_d / \sqrt{n}} = \frac{-0.93}{2.96 / \sqrt{7}} \approx -0.83.$$

Classical approach: Since this is a two-tailed test with 6 degrees of freedom, the critical values are $\pm t_{0.025} = \pm 2.447$. Since the test statistic $t_0 \approx -0.83$ falls between the critical values $\pm t_{0.025} = \pm 2.447$ (i.e., since the test statistic does not fall within the critical region), we do not reject H_0.

P-value approach: The P-value for this two-tailed test is the area under the t-distribution with 6 degrees of freedom to the left of the test statistic $t_0 = -0.83$ plus the area to the right of $t_0 = 0.83$. From the t-distribution table in the row corresponding to 6 degrees of freedom, 0.83 falls between 0.718 and 0.906 whose right-tail areas are 0.25 and 0.20, respectively. We must double these values in order to get the total area in both tails: 0.50 and 0.40. So, $0.40 < P\text{-value} < 0.50$. (Using technology, we find P-value is approximately equal to 0.438.) Because the P-value is greater than the level of significance, $\alpha = 0.05$, we do not reject H_0.

Conclusion: There is not sufficient evidence at the $\alpha = 0.05$ level of significance to conclude that the new loading procedure changed the wait time.

(c) The P-values for the two approaches are both 0.438. Yes, it appears that the method presented in this section is a generalization of the matched-pair t-test.

---

15. In a completely randomized design, the researcher manipulates a single factor fixed at $k \geq 2$ levels and randomly distributes the experimental units among the levels. The researcher may be able to control some factors, so they remain fixed. In a randomized complete block design, there is a factor whose level cannot be fixed. So the researcher partitions the experimental units according to this factor, forming blocks which may form homogeneous groups. Within each block, the experimental units are assigned to one of the treatments.

17. *Randomized* means that the order in which the treatment is applied within each block is random. *Complete* refers to the fact that each block gets every treatment.

19. A researcher can deal with explanatory variables by:

 (1) Controlling their levels so that they remain fixed throughout the experiment or setting them at fixed levels, and

 (2) Randomizing so that any effects that cannot be identified or controlled are minimized.

Section 13.4

1. 2; 3

3. Interaction effect

5. The plots are relatively parallel, so there is no significant interaction among the factors.

7. The plots cross, so there is significant interaction among the factors.

9. (a) Main effect of factor
$$A = \frac{20+14+1+3}{4} - \frac{12+8+6+7}{4} = 1.25;$$
Main effect of factor
$$B = \frac{6+7+1+3}{4} - \frac{12+8+20+14}{4} = -9.25.$$
Interaction effects: If factor B is fixed at low and factor A changes from low to high, the response variable increases by $\frac{20+14}{2} - \frac{12+8}{2} = 7$ units. On the other hand, if factor B is fixed at high and factor A changes from low to high, the response variable increases by $\frac{1+3}{2} - \frac{6+7}{2} = -4.5$ (decreases by 4.5) units.

So, the change in the response variable for factor A depends on the level of factor B, meaning there is an interaction effect between factors A and B. If factor A is fixed at low and factor B changes from low to high, the response variable

increases by $\dfrac{6+7}{2} - \dfrac{12+8}{2} = -3.5$

(decreases by 3.5) units. On the other hand, if factor A is fixed at high and factor B changes from low to high, the response variable increases by

$\dfrac{3+1}{2} - \dfrac{20+14}{2} = -15$ (decreases by 15)

units. So the change in the response variable for factor B depends on the level of factor A, meaning there is an interaction effect between factors A and B.

(b)

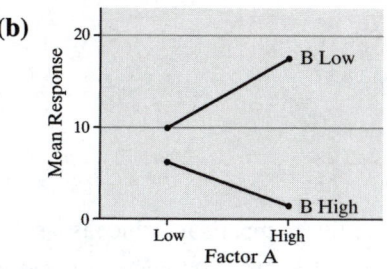

11. (a) Since the interaction P-value is 0.836, which is large, there is no evidence of an interaction effect.

(b) Since the P-value for factor A is 0.003, which is small, there is evidence of a difference in means from factor A. Since the P-value for factor B is less than 0.001, which is small, there is also evidence of a difference in means from factor B.

(c) MSE = 45.3

13. (a) Since the interaction P-value is less than 0.001, which is small, there is evidence of an interaction effect.

(b) In part (a), we found evidence of an interaction effect. Therefore, we do not consider main effects because they could be misleading.

(c) MSE = 31.9

---

15. Using Minitab with the provided data, we obtain the following output:

Two-way ANOVA: Response versus Factor A, Factor B

Source	DF	SS	MS	F	P
Factor A	1	30.15	30.150	0.32	0.579
Factor B	2	912.90	456.452	4.84	0.021
Interaction	2	6.80	3.402	0.04	0.965
Error	18	1699.27	94.404		
Total	23	2649.13			

(a) Since the interaction P-value is 0.965, which is large, there is no evidence of an interaction effect.

(b) Since the P-value for factor A is 0.579, which is large, there is no evidence of a difference in the means from factor A. Since the P-value for factor B is 0.021, which is small, there is evidence of a difference in the means from factor B.

(c) The lines in the interaction plot are close to parallel, supporting the conclusion that there is no interaction effect.

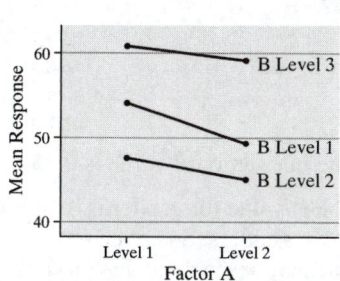

(d) The Tukey's test for factor A is unnecessary since there is no evidence for difference in means. For factor B, we use Minitab to perform Tukey's test. The results are:

```
Tukey 95.0% Simultaneous Confidence Intervals
Response Variable Response
All Pairwise Comparisons among Levels of Factor B
Factor B = 1  subtracted from:
Factor B  Lower  Center  Upper  --+---------+---------+---------+----
2        -18.55  -6.150  6.251  (-------*-------)
3         -3.53   8.875  21.276            (-------*-------)
                                 --+---------+---------+---------+----
                                  -15       0        15        30

Factor B = 2  subtracted from:
Factor B  Lower  Center  Upper  --+---------+---------+---------+----
3         2.624  15.03   27.43            (-------*-------)
                                 --+---------+---------+---------+----
                                  -15       0        15        30
```

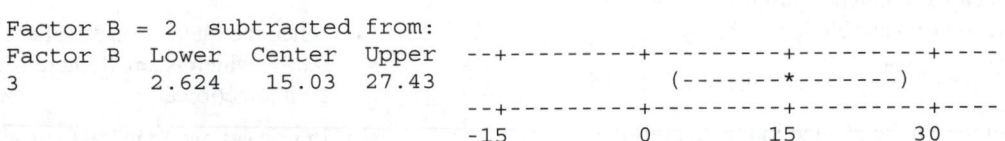

```
Tukey Simultaneous Tests
Response Variable Response
All Pairwise Comparisons among Levels of Factor B
Factor B = 1  subtracted from:
          Difference    SE of            Adjusted
Factor B  of Means   Difference  T-Value  P-Value
2          -6.150      4.858     -1.266    0.4317
3           8.875      4.858      1.827    0.1893
Factor B = 2  subtracted from:
          Difference    SE of            Adjusted
Factor B  of Means   Difference  T-Value  P-Value
3          15.03       4.858      3.093    0.0165
```

The interval for $B_2 - B_1$ includes 0 and the *P*-value is 0.4317 for this comparison, which is large, so we do not reject the null hypothesis that $B_1 = B_2$. Likewise, the interval for $B_3 - B_1$ includes 0 and the *P*-value is 0.1893, which is large, so we do not reject the null hypothesis that $B_1 = B_3$. Finally, the interval for $B_3 - B_2$ does not include 0 and the *P*-value is 0.0165, which is small, so we reject the null hypothesis that $B_2 = B_3$. Thus, the Tukey's test for factor B is ambiguous. It shows $B_1 = B_2$ and $B_1 = B_3$, but $B_2 \neq B_3$. That is, $\underline{B_2 \quad B_1} \quad B_3$.

17. (a) This is a 2×3 factorial design with 2 replications within each cell.

(b) The response variable is serum cholesterol (in mg/dL). The factors are gender and age (years).

(c) Using Minitab with the provided data, we obtain the following output:

```
Two-way ANOVA: Cholesterol versus Gender, Age
Source        DF       SS        MS       F       P
Gender         1    310.08    310.08    2.51   0.164
Age            2   2177.17   1088.58    8.81   0.016
Interaction    2    628.17    314.08    2.54   0.159
Error          6    741.50    123.58
Total         11   3856.92
```

Since the interaction *P*-value is 0.159, which is large, there is no evidence of an interaction effect.

(d) Since the *P*-value for gender is 0.164, which is large, there is no evidence of a difference in the means for the two genders. Since the *P*-value for age is 0.016, which is small, there is evidence of a difference in the means for the three age groups.

(e) The plots are not parallel, so there is significant interaction among the factors. This result contradicts the result found in part (b).

(f) Because a normal probability plot is roughly linear, it is reasonable to conclude the residuals are normally distributed.

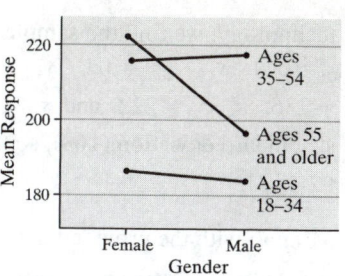

(g) The Tukey's test for gender is unnecessary since there is no evidence for difference in means for the two genders. For age, we use Minitab to perform Tukey's test:

```
Tukey 95.0% Simultaneous Confidence Intervals
Response Variable Cholesterol
All Pairwise Comparisons among Levels of Age
Age = 1  subtracted from:
Age   Lower    Center   Upper  ---+---------+---------+---------+---
2     7.3765   31.50    55.62                   (---------*--------)
3     0.1265   24.25    48.37        (---------*--------)
                             ---+---------+---------+---------+---
                             -25        0        25        50

Age = 2  subtracted from:
Age   Lower    Center   Upper  ---+---------+---------+---------+---
3     -31.37   -7.250   16.87  (---------*---------)
                             ---+---------+---------+---------+---
                             -25        0        25        50

Tukey Simultaneous Tests
Response Variable Cholesterol
All Pairwise Comparisons among Levels of Age
Age = 1  subtracted from:
      Difference   SE of              Adjusted
Age   of Means     Difference  T-Value  P-Value
2       31.50      7.861       4.007    0.0166
3       24.25      7.861       3.085    0.0490
Age = 2  subtracted from:
      Difference   SE of              Adjusted
Age   of Means     Difference  T-Value  P-Value
3       -7.250     7.861       -0.9223  0.6473
```

The interval for $\mu_{35-54} - \mu_{18-34}$ does not include 0 and the P-value is 0.0166 for this comparison, which is small, so we reject the null hypothesis that $\mu_{18-34} = \mu_{35-54}$. Likewise, the interval for $\mu_{55+} - \mu_{18-34}$ does not include 0 and the P-value is 0.0490, which is small, so we reject the null hypothesis that $\mu_{18-34} = \mu_{55+}$. Finally, the interval for $\mu_{55+} - \mu_{35-54}$ includes 0 and the P-value is 0.6473, which is large, so we do not reject the null hypothesis that $\mu_{35-54} = \mu_{55+}$. Thus, the Tukey's test for age shows $\mu_{18-34} \neq \mu_{35-54} = \mu_{55+}$. That is, $\underline{\mu_{18-34}} \quad \underline{\mu_{35-54} \quad \mu_{55+}}$.

19. **(a)** Using technology, we find the sample standard deviation of the data within each cell:

$s_{3.75/301} \approx 311$, $s_{3.75/400} \approx 320$, $s_{3.75/353} \approx 257$, $s_{4/301} \approx 298$, $s_{4/400} \approx 250$, $s_{4/353} \approx 168$,

$s_{5/301} \approx 236$, $s_{5/400} \approx 222$, and $s_{5/353} \approx 267$. The largest standard deviation (320) is less than twice the smallest standard deviation (168), so the requirement of equal population variances is satisfied.
$\left[320 < 2(168).\right]$

(b) Using Minitab with the provided data, we obtain the following output:

```
Two-way ANOVA: Strength versus Slump, Mixture
Source          DF        SS          MS         F        P
Slump            2    1234813      617406      8.95    0.002
Mixture          2    1824169      912084     13.23    0.000
Interaction      4     256415       64104      0.93    0.469
Error           18    1241100       68950
Total           26    4556496
```

Since the interaction P-value is 0.469, which is large, there is no evidence of an interaction effect.

(c) Since the P-value for slump is 0.002, which is small, there is sufficient evidence of difference in the means for the three types of slump. Since the P-value for mixture is less than 0.001, which is small, there is evidence of a difference in the means for the three mixtures.

(d)

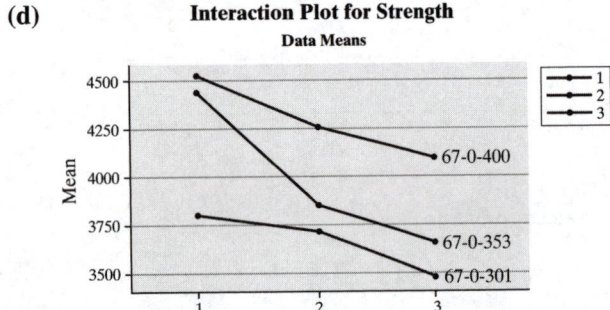

(e) Because a normal probability plot is roughly linear, it is reasonable to conclude the residuals are normally distributed.

(f) We use Minitab to perform Tukey's test. The results are:

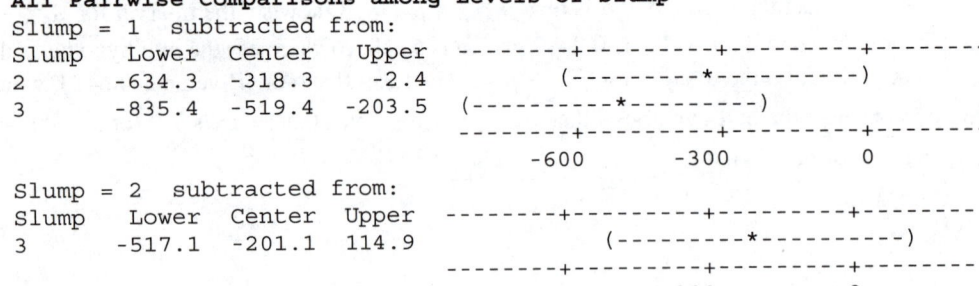

```
Tukey Simultaneous Tests
Response Variable Strength
All Pairwise Comparisons among Levels of Slump
Slump = 1  subtracted from:
        Difference       SE of                 Adjusted
Slump   of Means    Difference   T-Value       P-Value
2         -318.3        123.8     -2.572        0.0481
3         -519.4        123.8     -4.196        0.0015
Slump = 2  subtracted from:
        Difference       SE of                 Adjusted
Slump   of Means    Difference   T-Value       P-Value
3         -201.1        123.8     -1.625        0.2612
Tukey 95.0% Simultaneous Confidence Intervals
Response Variable Strength
All Pairwise Comparisons among Levels of Mixture
Mixture = 1  subtracted from:
Mixture    Lower   Center   Upper   ---+---------+---------+---------+---
2        320.691    636.7   952.6                   (------*-----)
3         -2.086    313.9   629.9             (-----*------)

                                           ---+---------+---------+---------+---
                                           -500        0       500      1000
Mixture = 2  subtracted from:
Mixture    Lower   Center   Upper   ---+---------+---------+---------+---
3         -638.8   -322.8  -6.803   (------*-----)
                                           ---+---------+---------+---------+---
                                           -500        0       500      1000

Tukey Simultaneous Tests
Response Variable Strength
All Pairwise Comparisons among Levels of Mixture
Mixture = 1  subtracted from:
          Difference       SE of               Adjusted
Mixture   of Means    Difference   T-Value     P-Value
2            636.7        123.8      5.143      0.0002
3            313.9        123.8      2.536      0.0517
Mixture = 2  subtracted from:
          Difference       SE of               Adjusted
Mixture   of Means    Difference   T-Value     P-Value
3           -322.8        123.8     -2.608      0.0448
```

Tukey's test for Slump: The interval for $\mu_4 - \mu_{3.75}$ does not include 0 and the *P*-value is 0.0481 for this comparison, which is small, so we reject the null hypothesis that $\mu_{3.75} = \mu_4$. Next, the interval for $\mu_5 - \mu_{3.75}$ does not include 0 and the *P*-value is 0.0015, which is small, so we reject the null hypothesis that $\mu_{3.75} = \mu_5$. Finally, the interval for $\mu_5 - \mu_4$ includes 0 and the *P*-value = 0.2612, which is large, so we do not reject the null hypothesis that $\mu_4 = \mu_5$. Thus, we have $\mu_{3.75} \neq \mu_4 = \mu_5$ or $\underline{\mu_{3.75}}$ $\underline{\mu_4}$ $\underline{\mu_5}$. That is, the Tukey's test indicates that the mean 28-day strength for slump 3.75 is significantly different from the mean 28-day strength for slump 4 and slump 5.

Tukey's test for Mixture: The interval for $\mu_{400} - \mu_{301}$ does not include 0 and the P-value is 0.0002, which is small, so we reject the null hypothesis that $\mu_{301} = \mu_{400}$. Next, the interval for $\mu_{353} - \mu_{301}$ includes 0 and the *P*-value is 0.0517, which is not small, so we do not reject the null hypothesis that $\mu_{301} = \mu_{353}$. Finally, the interval for $\mu_{353} - \mu_{400}$ does not include 0 and the *P*-value is 0.0448, which is small, so we reject the null hypothesis that $\mu_{400} = \mu_{353}$. Thus, we have $\mu_{301} = \mu_{353} \neq \mu_{400}$ or $\underline{\mu_{301}}$ $\underline{\mu_{353}}$ $\underline{\mu_{400}}$. That is, the Tukey's test indicates that the mean 28-day strength of mixture 67-0-400 is significantly different from the mean 28-day strength of the other two mixtures.

21. (a) Using technology, we find the sample standard deviation the data within each cell:
$s_{\text{specialty / Chicago}} \approx 4.00$, $s_{\text{general / Chicago}} \approx 2.50$, $s_{\text{specialty / Bolingbrook}} \approx 3.20$, $s_{\text{specialty / Peoria}} \approx 3.63$,

$s_{\text{general / Bolingbrook}} \approx 3.91$, and $s_{\text{general / Peoria}} \approx 2.52$. The largest standard deviation (4.00) is less than twice the smallest standard deviation (2.50), so the requirement of equal population variances is satisfied. $\left[4.00 < 2(2.50). \right]$

(b) Using Minitab with the provided data, we obtain the following output:

```
Two-way ANOVA: Cost versus Type, Location
Source          DF          SS          MS          F          P
Type            1           0.720       0.7200      0.06       0.804
Location        2           31.126      15.5630     1.39       0.287
Interaction     2           3.573       1.7867      0.16       0.854
Error           12          134.560     11.2134
Total           17          169.980
```

Since the interaction P-value is 0.854, which is large, there is no evidence of an interaction effect.

(c) Since the P-value for type is 0.804, which is large, there is no evidence of a difference in the means for the type of service center. Since the P-value for location is 0.287, which is large, there is no evidence of a difference in the means for the locations.

(d)

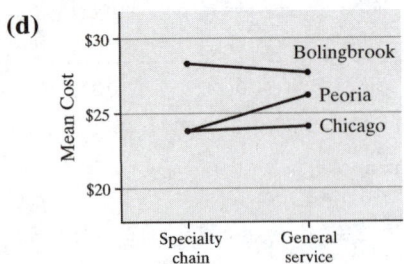

(e) Since there were no significant main effects, there is no need to do Tukey's tests.

---

23. In a completely randomized design, the researcher examines a single factor fixed at k levels and randomly distributes the experimental units among the levels. In a randomized complete block design, there is a factor whose level cannot be fixed. So the researcher partitions the experimental units according to this factor, forming blocks. The experimental units are then randomly assigned to the blocks, being sure each block has each treatment. In a factorial design, the researcher has two variables A and B, called factor A and factor B. The factors are fixed, so factor A has a levels and factor B has b levels. The experimental units are then uniformly distributed among the cells formed by the factors and the response variable is measured at each of the ab cells.

25. An interaction plot graphically displays the mean response for each treatment (combination of factor levels). The plots are useful for illustrating potential interaction between the factors.

Chapter 13 Review Exercises

1. (a) We look in the table of critical values for Tukey's Test with $\alpha = 0.05$ and where the row corresponding to $\nu = 16$ intersects with the column corresponding to $k = 7$ to find the critical value $q_{0.05,16,7} = 4.741$.

(b) We look in the table of critical values for Tukey's Test with $\alpha = 0.05$ and where the row corresponding to $\nu = 30$ intersects with the column corresponding to $k = 6$ to find the critical value $q_{0.05,30,6} = 4.302$.

(c) We look in the table of critical values for Tukey's Test with $\alpha = 0.01$ and where the row corresponding to $\nu = 40$ (since there is no row in the table for $\nu = 42$) intersects with the column corresponding to $k = 4$ to find the critical value $q_{0.01,42,4} \approx q_{0.01,40,4} = 4.696$.

(d) For this test, $k = 6$ and $\nu = n - k = 46 - 6 = 40$, so we look in the table with $\alpha = 0.05$ and where the row corresponding to $\nu = 40$ intersects with the column corresponding to $k = 6$ to find the critical value $q_{0.05,40,6} = 4.232$.

2. Arrange the sample means in ascending order: $\bar{x}_4 = 47$, $\bar{x}_2 = 64$, $\bar{x}_3 = 70$, $\bar{x}_1 = 78$. Now $n = 31 \cdot 4 = 124$ and $k = 4$, so $\nu = n - k = 120$ and the critical value is $q_{0.05,120,4} = 3.685$. We first test $H_0 : \mu_1 = \mu_4$. The test statistic is

$$q_0 = \frac{\bar{x}_1 - \bar{x}_4}{\sqrt{\frac{s^2}{2} \cdot \left(\frac{1}{n_1} + \frac{1}{n_4}\right)}} = \frac{78 - 47}{\sqrt{\frac{373}{2} \cdot \left(\frac{1}{31} + \frac{1}{31}\right)}} \approx 8.937.$$

Since this test statistic is greater than $q_{0.05,120,4} = 3.685$, we reject $H_0 : \mu_1 = \mu_4$. Second, we test $H_0 : \mu_1 = \mu_2$. The test statistic is $q_0 = \dfrac{\bar{x}_1 - \bar{x}_2}{\sqrt{\frac{s^2}{2} \cdot \left(\frac{1}{n_1} + \frac{1}{n_2}\right)}} = \dfrac{78 - 64}{\sqrt{\frac{373}{2} \cdot \left(\frac{1}{31} + \frac{1}{31}\right)}} \approx 4.036.$

Since this test statistic is greater than $q_{0.05,120,4} = 3.685$, we reject $H_0 : \mu_1 = \mu_2$. Third, we test $H_0 : \mu_1 = \mu_3$. The test statistic is

$$q_0 = \frac{\bar{x}_1 - \bar{x}_3}{\sqrt{\frac{s^2}{2} \cdot \left(\frac{1}{n_1} + \frac{1}{n_3}\right)}} = \frac{78 - 70}{\sqrt{\frac{373}{2} \cdot \left(\frac{1}{31} + \frac{1}{31}\right)}} \approx 2.306.$$

Since this test statistic is less than $q_{0.05,120,4} = 3.685$, we do not reject $H_0 : \mu_1 = \mu_3$.

Fourth, we test $H_0 : \mu_3 = \mu_4$. The test statistic is $q_0 = \dfrac{\bar{x}_3 - \bar{x}_4}{\sqrt{\frac{s^2}{2} \cdot \left(\frac{1}{n_3} + \frac{1}{n_4}\right)}} = \dfrac{70 - 47}{\sqrt{\frac{373}{2} \cdot \left(\frac{1}{31} + \frac{1}{31}\right)}} \approx 6.631.$

Since this test statistic is greater than $q_{0.05,120,4} = 3.685$, we reject $H_0 : \mu_3 = \mu_4$. Fifth, we test $H_0 : \mu_3 = \mu_2$. The test statistic is

$$q_0 = \frac{\bar{x}_3 - \bar{x}_2}{\sqrt{\frac{s^2}{2} \cdot \left(\frac{1}{n_3} + \frac{1}{n_2}\right)}} = \frac{70 - 64}{\sqrt{\frac{373}{2} \cdot \left(\frac{1}{31} + \frac{1}{31}\right)}} \approx 1.730.$$

Since this test statistic is less than $q_{0.05,120,4} = 3.685$, we do not reject $H_0 : \mu_3 = \mu_2$.

Finally, we test $H_0 : \mu_2 = \mu_4$. The test statistic is $q_0 = \dfrac{\bar{x}_2 - \bar{x}_4}{\sqrt{\frac{s^2}{2} \cdot \left(\frac{1}{n_2} + \frac{1}{n_4}\right)}} = \dfrac{64 - 47}{\sqrt{\frac{373}{2} \cdot \left(\frac{1}{31} + \frac{1}{31}\right)}} \approx 4.901.$

Since this test statistic is greater than $q_{0.05,120,4} = 3.685$, we reject $H_0 : \mu_2 = \mu_4$. The results of the test are ambiguous since we have that $\mu_1 \neq \mu_2$, but $\mu_1 = \mu_3 \neq \mu_4$ and $\mu_2 = \mu_3 \neq \mu_4$. Therefore, $\underline{\mu_1 \ \mu_3} \ \underline{\mu_2} \ \underline{\mu_4}$.

3. (a) Concentration of dissolved organic carbon (mg/L)

(b) The factor is the type of water, which has three levels: Organic, Mineral, and Surface.

(c) The hypotheses are:

$H_0 : \mu_{\text{organic}} = \mu_{\text{mineral}} = \mu_{\text{surface}}$ versus

H_1 : At least one of the means is different from the others.

(d) The largest standard deviation (6.246) is less than twice the smallest standard deviation (3.927), so the requirement of equal population variances is satisfied. $\left[6.246 < 2(3.927).\right]$

(e) Since P-value $= 0.0002$ is smaller than the $\alpha = 0.01$ level of significance, the researcher will reject the null hypothesis and conclude that the mean concentration of dissolved carbon is different in at least one of the collection areas.

(f) Results as extreme as these would only occur in about 2 out of 10,000 similar samples if the null hypothesis were true.

(g) Yes, the boxplots support the conclusion that at least one of the means is different. The plot for water collected from mineral soil appears to differ from the other two.

(h) From the Statcrunch output we have:

$-7.588 < \mu_{\text{mineral}} - \mu_{\text{organic}} < -2.092$;

$-4.606 < \mu_{\text{surface}} - \mu_{\text{organic}} < 0.964$; and

$0.527 < \mu_{\text{surface}} - \mu_{\text{mineral}} < 5.510$.

The first interval does not contain 0, so we reject $H_0 : \mu_{\text{organic}} = \mu_{\text{mineral}}$.

The second interval contains 0, so we do not reject $H_0 : \mu_{\text{organic}} = \mu_{\text{surface}}$.

The third interval does not contain 0, so we reject $H_0 : \mu_{\text{mineral}} = \mu_{\text{surface}}$. Thus, we have $\mu_{\text{organic}} = \mu_{\text{surface}} \neq \mu_{\text{mineral}}$ or

$\underline{\mu_{\text{organic}} \quad \mu_{\text{surface}}} \quad \underline{\mu_{\text{mineral}}}$. Thus, the

Tukey's test indicates that the mean dissolved carbon is equal in the surface water and in the ground water taken from organic soil, but it differs from the mean dissolved carbon concentration in ground water taken from mineral soil.

4. From the Minitab output we have:

$-6.914 < \mu_{\text{select structural}} - \mu_{\text{no. 1}} < 13.064$;

$1.936 < \mu_{\text{select structural}} - \mu_{\text{no. 2}} < 21.914$;

$-2.989 < \mu_{\text{select structural}} - \mu_{\text{below grade}} < 16.989$;

$-1.139 < \mu_{\text{no. 1}} - \mu_{\text{no. 2}} < 18.839$;

$-6.064 < \mu_{\text{no. 1}} - \mu_{\text{below grade}} < 13.914$; and

$-14.914 < \mu_{\text{no. 2}} - \mu_{\text{below grade}} < 5.064$. The first interval contains 0, so we do not reject

$H_0 : \mu_1 = \mu_2$. The second interval does not contain 0, so we reject $H_0 : \mu_1 = \mu_3$. The third interval contains 0, so we do not reject $H_0 : \mu_1 = \mu_4$. The fourth interval contains 0, so we do not reject $H_0 : \mu_2 = \mu_3$. The fifth interval contains 0, so we do not reject $H_0 : \mu_2 = \mu_4$. The sixth interval contains 0, so we do not reject $H_0 : \mu_3 = \mu_4$. Thus, Tukey's test is ambiguous. It indicates that $\mu_{\text{select structural}} \neq \mu_{\text{no. 2}}$, but that $\mu_{\text{select structural}} = \mu_{\text{no. 1}} = \mu_{\text{no. 2}} = \mu_{\text{below grade}}$ or

$\underline{\mu_{\text{select structural}} \quad \mu_{\text{no. 1}} \quad \mu_{\text{below grade}} \quad \mu_{\text{no. 2}}}$.

---

5. (a) There are three independent samples corresponding to Front/Middle/Back sear position. The response variable is quantitative (GPA). The equal variance condition is satisfied.

(b) The hypotheses are $H_0 : \mu_{\text{front}} = \mu_{\text{middle}} = \mu_{\text{back}}$ versus H_1 : at least one mean differs from the others. Using Minitab, we obtain the following output:

```
One-way ANOVA: Front, Middle, Back
Source      DF        SS         MS          F           P
Factor       2     2.350      1.175       7.50       0.003
Error       27     4.229      0.157
Total       29     6.580
```

The P-value is 0.003, so we reject the null hypothesis and conclude that at least one of the means is different.

(c) Using Minitab, we obtain the following output:

```
Tukey 95% Simultaneous Confidence Intervals
All Pairwise Comparisons
Individual confidence level = 98.04%
Front subtracted from:
            Lower    Center    Upper    --+---------+---------+---------+-------
Middle    -0.9779   -0.5386  -0.0993     (--------*--------)
Back      -1.0760   -0.6367  -0.1974   (--------*--------)
                                        --+---------+---------+---------+-------
                                       -1.00     -0.50      0.00      0.50

Middle subtracted from:
            Lower    Center    Upper    --+---------+---------+---------+-------
Back      -0.5374   -0.0981   0.3412              (--------*--------)
                                        --+---------+---------+---------+-------
                                       -1.00     -0.50      0.00      0.50
```

From the output we have: $-0.9779 < \mu_{\text{middle}} - \mu_{\text{front}} < -0.0993$; $-1.0760 < \mu_{\text{back}} - \mu_{\text{front}} < -0.1974$; and $-0.5374 < \mu_{\text{back}} - \mu_{\text{middle}} < 0.3412$.

The first interval does not contain 0, so we reject $H_0 : \mu_{\text{front}} = \mu_{\text{middle}}$. Likewise, the second interval does not contain 0, so we reject $H_0 : \mu_{\text{front}} = \mu_{\text{back}}$. Finally, the third interval contains 0, so we do not reject $H_0 : \mu_{\text{middle}} = \mu_{\text{back}}$. Thus, we have $\mu_{\text{front}} \neq \mu_{\text{middle}} = \mu_{\text{back}}$ or $\underline{\mu_{\text{front}}} \quad \underline{\mu_{\text{middle}} \quad \mu_{\text{back}}}$. That is, there is evidence that students who choose seats in the front of the mathematics classroom have a different mean GPA from that of students choosing seats in the middle or the back of the classroom.

(d) The boxplots shown to the right support the result of parts (a) and (b), indicating that GPAs of students who sit in front are generally higher than students who sit in the middle and back.

6. **(a)** Distance (in feet)

(b) There may be variability due to person firing. Blocking is used to remove any effect that the person firing the catapult may have on the distance that the water balloon travels.

(c) Using Minitab, we obtain the following output:

```
Two-way ANOVA: Distance versus Person, Catapult
Source      DF          SS          MS          F           P
Person       3      45.688     15.2292       16.01       0.001
Catapult     3     105.688     35.2292       37.03       0.000
Error        9       8.563      0.9514
Total       15     159.938
```

Since the treatment (catapult) *P*-value is less than 0.001, we reject the null hypothesis that the population means are equal. Thus, we conclude that at least one mean is different from the others.

(d) Using Minitab , we obtain the following output:

```
Tukey 95.0% Simultaneous Confidence Intervals
Response Variable Distance
All Pairwise Comparisons among Levels of Catapult
Catapult = 1  subtracted from:
Catapult   Lower   Center   Upper    ------+---------+---------+---------+
2         -1.656  0.5000   2.656    (-------*------)
3          1.844  4.0000   6.156              (------*-------)
4          4.094  6.2500   8.406                       (------*------)
                                    ------+---------+---------+---------+
                                        0.0       3.0       6.0       9.0

Catapult = 2  subtracted from:
Catapult   Lower   Center   Upper    ------+---------+---------+---------+
3          1.344   3.500   5.656              (-------*------)
4          3.594   5.750   7.906                       (------*------)
                                    ------+---------+---------+---------+
                                        0.0       3.0       6.0       9.0

Catapult = 3  subtracted from:
Catapult    Lower   Center   Upper   ------+---------+---------+---------+
4         0.09438   2.250   4.406        (------*-------)
                                    ------+---------+---------+---------+
                                        0.0       3.0       6.0       9.0

Tukey Simultaneous Tests
Response Variable Distance
All Pairwise Comparisons among Levels of Catapult
Catapult = 1  subtracted from:
           Difference     SE of               Adjusted
Catapult    of Means   Difference   T-Value    P-Value
2            0.5000      0.6897     0.7249     0.8847
3            4.0000      0.6897     5.7996     0.0012
4            6.2500      0.6897     9.0618     0.0000
Catapult = 2  subtracted from:
           Difference     SE of               Adjusted
Catapult    of Means   Difference   T-Value    P-Value
3            3.500       0.6897     5.075      0.0031
4            5.750       0.6897     8.337      0.0001
```

```
Catapult = 3  subtracted from:
          Difference      SE of              Adjusted
Catapult  of Means    Difference   T-Value   P-Value
4           2.250       0.6897      3.262     0.0405
```

The interval for $\mu_2 - \mu_1$ includes 0 and the *P*-value is 0.8847 for this comparison, which is large, so we do not reject the null hypothesis that $\mu_1 = \mu_2$. The interval for $\mu_3 - \mu_1$ does not include 0 and the *P*-value is 0.0012, which is small, so we reject the null hypothesis that $\mu_1 = \mu_3$. The interval for $\mu_4 - \mu_1$ does not include 0 and the *P*-value is less than 0.0001, which is small, so we reject the null hypothesis that $\mu_4 = \mu_1$. The interval for $\mu_3 - \mu_2$ does not include 0 and the *P*-value is 0.0031, which is small, so we reject the null hypothesis that $\mu_2 = \mu_3$. The interval for $\mu_4 - \mu_2$ does not include 0 and the *P*-value is 0.0001, which is small, so we reject the null hypothesis that $\mu_2 = \mu_4$. Finally, the interval for $\mu_4 - \mu_3$ does not include 0 and the *P*-value is 0.0405, which is small, so we reject the null hypothesis that $\mu_3 = \mu_4$. Thus, the Tukey's tests indicate that the means of Mark IV V2 and Balloon Ballista are equal, but the means of Hydrolaunch and Waterworks II are different. That is, we conclude: $\mu_1 = \mu_2 \neq \mu_3 \neq \mu_4$ or $\underline{\mu_1 \quad \mu_2} \quad \mu_3 \quad \mu_4$.

(e) The boxplots shown to the right support the results found in parts (a) and (b).

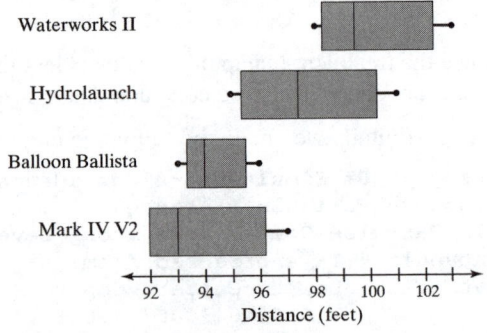

7. (a) Using technology, we find the sample standard deviation the data within each cell:
$s_{M/\text{one 8 hr}} \approx 3.42$, $s_{M/\text{two 4 hr}} \approx 2.92$, $s_{M/\text{two 2 hr}} \approx 2.39$, $s_{F/\text{one 8 hr}} \approx 4.72$, $s_{F/\text{two 4 hr}} \approx 2.70$, and $s_{F/\text{two 2 hr}} \approx 3.87$. The largest standard deviation (4.72) is less than twice the smallest standard deviation (2.39), so the requirement of equal population variances is satisfied. $\left[4.72 < 2(2.39).\right]$

(b) Using Minitab with the provided data, we obtain the following output:

Two-way ANOVA: Score versus Gender, Program

Source	DF	SS	MS	F	P
Gender	1	20.83	20.833	1.77	0.196
Program	2	890.60	445.300	37.90	0.000
Interaction	2	24.07	12.033	1.02	0.374
Error	24	282.00	11.750		
Total	29	1217.50			

Since the interaction *P*-value is 0.374, which is large, there is no evidence of an interaction effect.

(c) Since the *P*-value for gender is 0.196, which is large, there is no evidence of a significant difference in the means for gender. Since we have the program *P*-value program is less than 0.001, which is small, there is evidence of a significant difference in the means for type of driving program. The null hypothesis $H_0 : \mu_{8 \text{ hour}} = \mu_{4 \text{ hour}} = \mu_{2 \text{ hour}}$ is rejected.

(d) The lines in the interaction plot shown to the right are close to parallel, supporting the conclusion that there is no interaction effect.

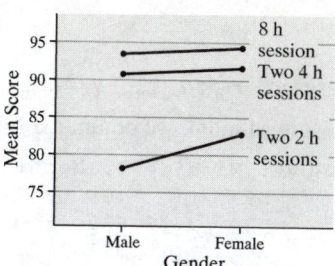

(e) The Tukey's test for gender is unnecessary since there is no evidence of a difference in means. For type of program, with Program 1 = One 8-hour session, Program 2 = Two 4-hour sessions, and Program 3 = Two 2-hour sessions, we use Minitab to perform Tukey's test. The results are:

```
Tukey 95.0% Simultaneous Confidence Intervals
Response Variable Score
All Pairwise Comparisons among Levels of Program
Program = 1  subtracted from:
Program   Lower   Center   Upper    ---+---------+---------+---------+---
2         -6.03   -2.20    1.626                           (-------*------)
3        -16.33  -12.50   -8.674    (-------*-------)
                                    ---+---------+---------+---------+---
                                    -15.0      -10.0      -5.0       0.0

Program = 2  subtracted from:
Program   Lower   Center   Upper    ---+---------+---------+---------+---
3        -14.13  -10.30   -6.474        (------*-------)
                                    ---+---------+---------+---------+---
                                    -15.0      -10.0      -5.0       0.0

Tukey Simultaneous Tests
Response Variable Score
All Pairwise Comparisons among Levels of Program
Program = 1  subtracted from:
          Difference    SE of              Adjusted
Program   of Means    Difference   T-Value  P-Value
2           -2.20       1.533      -1.435   0.3394
3          -12.50       1.533      -8.154   0.0000
Program = 2  subtracted from:
          Difference    SE of              Adjusted
Program   of Means    Difference   T-Value  P-Value
3          -10.30       1.533      -6.719   0.0000
```

The interval for $\mu_{\text{two 4 hr}} - \mu_{\text{one 8 hr}}$ includes 0 and the P-value is 0.3394 for this comparison, which is large, so we do not reject the null hypothesis that $\mu_{\text{one 8 hr}} = \mu_{\text{two 4 hr}}$. The interval for $\mu_{\text{two 2 hr}} - \mu_{\text{one 8 hr}}$ does not include 0 and the P-value is less than 0.0001, which is small, so we reject the null hypothesis that $\mu_{\text{one 8 hr}} = \mu_{\text{two 2 hr}}$. Finally, the interval for $\mu_{\text{two 2 hr}} - \mu_{\text{two 4 hr}}$ does not include 0 and the P-value is less than 0.0001, which is small, so we reject the null hypothesis that $\mu_{\text{two 4 hr}} = \mu_{\text{two 2 hr}}$. Thus, we have $\mu_{\text{one 8 hr}} = \mu_{\text{two 4 hr}} \neq \mu_{\text{two 2 hr}}$ or $\underline{\mu_{\text{one 8 hr}} \quad \mu_{\text{two 4 hr}}} \quad \mu_{\text{two 2 hr}}$. That is, the Tukey's test indicates that the mean score after taking one 8-hour defensive driving class is equal to the mean score after taking two 4-hour classes, but that the mean score after taking two 2-hour classes is different.

Chapter 13 Test

1. (a) The hypotheses are $H_0 : \mu_{\text{nonsmokers}} = \mu_{\text{light smokers}} = \mu_{\text{heavy smokers}}$ versus H_1 : at least one mean differs from the others. Using Minitab, we obtain the following output:

```
One-way ANOVA: Machine I, Machine II, Machine III, Machine IV
Source      DF       SS        MS       F        P
Factor       3     3.637     1.212    5.81     0.004
Error       24     5.011     0.209
Total       27     8.649
```

Since P-value $= 0.004$ is less than the $\alpha = 0.05$ level of significance, we reject H_0. There is sufficient evidence to support the hypothesis that at least one of the means is different from the others.

(b) Using Minitab, we obtain the following output:

```
Tukey 95% Simultaneous Confidence Intervals  All Pairwise Comparisons
Individual confidence level = 98.90%


Machine I subtracted from:
              Lower     Center     Upper
Machine II   -1.4307   -0.7571   -0.0836
Machine III  -0.4736    0.2000    0.8736
Machine IV   -0.7450   -0.0714    0.6022


              ------+---------+---------+---------+---
Machine II        (------*------)
Machine III          (-------*------)
Machine IV          (-----*------)
              ------+---------+---------+---------+---
                  -1.0       0.0       1.0       2.0


Machine II subtracted from:
              Lower     Center     Upper    ------+---------+---------+---------+---
Machine III  0.2836    0.9571    1.6307                        (------*-----)
Machine IV   0.0121    0.6857    1.3593                      (------*------)
                                           ------+---------+---------+---------+---
                                               -1.0       0.0       1.0       2.0


Machine III subtracted from:
              Lower     Center     Upper    ------+---------+---------+---------+--
Machine IV   -0.9450   -0.2714    0.4022              (-----*------)
                                           ------+---------+---------+---------+--
                                               -1.0       0.0       1.0       2.0
```

From the output, we have:

$-1.4307 < \mu_{\text{II}} - \mu_{\text{I}} < -0.0836$. This interval does not contain 0, so we reject $H_0 : \mu_{\text{I}} = \mu_{\text{II}}$.

$-0.4736 < \mu_{\text{III}} - \mu_{\text{I}} < 0.8736$. This interval contains 0, so we do not reject $H_0 : \mu_{\text{I}} = \mu_{\text{III}}$.

$-0.7450 < \mu_{\text{IV}} - \mu_{\text{I}} < 0.6022$. This interval contains 0, so we do not reject $H_0 : \mu_{\text{I}} = \mu_{\text{IV}}$.

$0.2836 < \mu_{\text{III}} - \mu_{\text{II}} < 1.6307$. This interval does not contain 0, so we reject $H_0 : \mu_{\text{II}} = \mu_{\text{III}}$.

$0.0121 < \mu_{\text{IV}} - \mu_{\text{II}} < 1.3593$. This interval does not contain 0, so we reject $H_0 : \mu_{\text{II}} = \mu_{\text{IV}}$.

$-0.9450 < \mu_{\text{IV}} - \mu_{\text{III}} < 0.4022$. This interval contains 0, so we do not reject $H_0 : \mu_{\text{III}} = \mu_{\text{IV}}$.

Thus, we have that $\mu_{\text{I}} = \mu_{\text{III}} = \mu_{\text{IV}} \neq \mu_{\text{II}}$. That is, we have: $\underline{\mu_{\text{I}} \quad \mu_{\text{III}} \quad \mu_{\text{IV}} \quad \mu_{\text{II}}}$. It appears that the mean amount dispensed by Machine II is less than the mean amount dispensed by the other machines.

(c) The boxplots shown to the right support the result of parts (a) and (b).

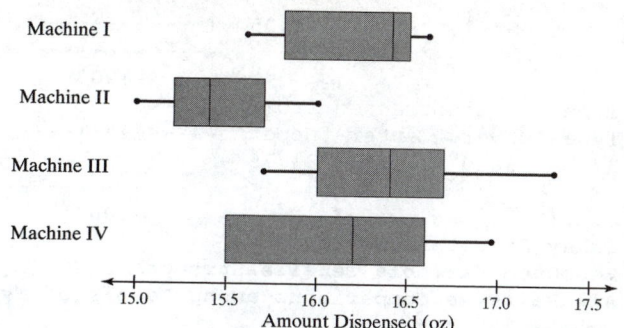

2. (a) Michael uses a completely randomized design for his experiment.

(b) The response variable is the height after 38 days; this is a quantitative variable.

(c) The explanatory variable is the color of the light bulb; this is a qualitative variable.

(d) Other factors identified that might impact plant growth included type of soil, amount of water, amount of light, and temperature. Michael kept these consistent for all three treatments.

(e) Michael randomly assigned the seeds to each treatment group.

(f) The hypotheses are $H_0 : \mu_B = \mu_R = \mu_G$ versus H_1 : at least one mean differs from the others.

(g) The largest standard deviation (0.950) is less than twice the smallest standard deviation (0.595) $\left[0.950 < 2(0.595) \right]$.

(h) We assume the level of significance is $\alpha = 0.05$. Since P-value $= 0.000$ (i.e., P-value < 0.0005), it is smaller than the $\alpha = 0.05$ level of significance, so we reject H_0. There is sufficient evidence to conclude that at least one of the mean heights is different for at least one group.

(i) Yes, it appears that red light promotes plant growth more than the others.

(j) From the output, we have:

$1.1460 < \mu_R - \mu_B < 2.0740$. This interval does not contain 0, so we reject $H_0 : \mu_R = \mu_B$.
$-0.9973 < \mu_G - \mu_B < -0.0694$. This interval does not contain 0, so we reject $H_0 : \mu_G = \mu_B$.
$-2.6073 < \mu_G - \mu_R < -1.6794$. This interval does not contain 0, so we reject $H_0 : \mu_G = \mu_R$.

Thus, we have that $\mu_G \neq \mu_B \neq \mu_R$. That is, we have: $\underline{\mu_G} \ \underline{\mu_B} \ \underline{\mu_R}$. Tukey's test indicates that all three means are different.

3. (a) Using Minitab, we obtain the following output:

Two-way ANOVA: Tensile Strength versus Thickness, Type

Source	DF	SS	MS	F	P
Thickness	3	97239906	32413302	28.27	0.001
Type	2	23628163	11814081	10.30	0.011
Error	6	6880388	1146731		
Total	11	127748456			

Since the treatment (type) P-value is 0.011, we reject the null hypothesis that the population means are equal. Thus, we conclude that at least one mean is different from the others.

(b) Using Minitab , we obtain the following output:

```
Tukey 95.0% Simultaneous Confidence Intervals
Response Variable Tensile Strength
All Pairwise Comparisons among Levels of Type
Type = 1  subtracted from:
Type  Lower  Center  Upper   ---------+---------+---------+-------
2     -4246   -1922   401               (----------*----------)
```

```
3       -5753    -3429   -1105   (------------*----------)
                                  ---------+---------+---------+-------
                                       -4000     -2000        0
Type = 2  subtracted from:
Type  Lower  Center  Upper    ---------+---------+---------+-------
3      -3830   -1506   817.5          (----------*-----------)
                                  ---------+---------+---------+-------
                                       -4000     -2000        0
Tukey Simultaneous Tests
Response Variable Tensile Strength
All Pairwise Comparisons among Levels of Type
Type = 1  subtracted from:
       Difference        SE of              Adjusted
Type    of Means    Difference   T-Value    P-Value
2         -1922         757.2     -2.539     0.0972
3         -3429         757.2     -4.528     0.0095
Type = 2  subtracted from:
       Difference        SE of              Adjusted
Type    of Means    Difference   T-Value    P-Value
3         -1506         757.2     -1.989     0.1956
```

The interval for $\mu_2 - \mu_1$ includes 0 and the P-value is 0.0972 for this comparison, which is large, so we do not reject the null hypothesis that $\mu_1 = \mu_2$. The interval for $\mu_3 - \mu_1$ does not include 0 and the P-value is 0.0095, which is small, so we reject the null hypothesis that $\mu_1 = \mu_3$. Finally, the interval for $\mu_3 - \mu_2$ includes 0 and the P-value is 0.1956, which is large, so we do not reject the null hypothesis that $\mu_2 = \mu_3$. Thus, the Tukey's test is ambiguous. It indicates that $\mu_3 \neq \mu_1 = \mu_2$, but $\mu_2 = \mu_3$. That is: $\underline{\mu_1 \ \mu_2} \ \underline{\ \mu_2 \ \mu_3}$.

(c)

4. **(a)** Using technology, we find the sample standard deviation the data within each cell:
$s_{M/20-34} \approx 0.46$, $s_{M/35-49} \approx 0.48$, $s_{M/50-64} \approx 0.53$, $s_{F/20-34} \approx 0.31$, $s_{F/35-49} \approx 0.28$, and $s_{F/50-64} \approx 0.32$.
The largest standard deviation (0.53) is less than twice the smallest standard deviation (0.28), so the requirement of equal population variances is satisfied. $\left[0.53 < 2(0.28) = 0.56. \right]$

(b) Using Minitab with the provided data, we obtain the following output:

```
Two-way ANOVA: Learning Time versus Gender, Age
Source         DF        SS          MS          F          P
Gender          1     0.05042    0.050417      0.30      0.589
Age             2     0.51083    0.255417      1.53      0.243
Interaction     2     0.27083    0.135417      0.81      0.459
Error          18     2.99750    0.166528
Total          23     3.82958
```

Since the interaction P-value is 0.459, which is larger than $\alpha = 0.05$, there is no evidence of an interaction effect.

(c) Since the P-value for gender is 0.589, which is large, there is no evidence of a difference in the means for gender. Since the P-value for age is 0.243, which is also large, there is no evidence of a difference in the means for the three age groups.

(d) The lines in the interaction plot shown to the right are not very parallel, contradicting the conclusion that there is no interaction effect.

(e) The Tukey's test is unnecessary since there is no evidence of a difference in means.

5. (a)

(b) The plot indicates a significant interaction effect since the graphs for the two levels of gender are not parallel.

(c) Since there is a significant interaction effect, any conclusion about main effects would be confounded.

Case Study: Hat Size and Intelligence

Does it appear that at least one of the head-circumference means is different from the others? Explain.

Yes. From the information provided, $k = 8$ and $n = 1672$. Using an $\alpha = 0.05$ significance level, the critical value is

$F_{\alpha, k-1, n} = F_{0.05, 7, 1672} \approx F_{0.05, 7, 1000} = 2.02$. From the results of the provided ANOVA table, we have $F_0 = 1998.6275$. Therefore, we reject the null hypothesis and conclude that at least one of the head-circumference means is different from the others.

Examine Epstein's table closely. Does it appear that the ANOVA procedure's requirement of equal variances is satisfied? Explain.

From Epstein's table, it does not appear that the ANOVA procedure's requirement of equal variance is satisfied. The largest standard deviation (2.5) is more than two times the smallest standard deviation (0.3). That is, $2.5 > 2(0.3) = 0.6$, so the requirement of equal variances is not satisfied.

Calculate the correct standard deviations by multiplying the tabled value by the square root of its sample size, n.

Vocational Status	n	Mean	Corrected S.D.
Professional	25	569.9	$1.9 \cdot \sqrt{25} = 9.5$
Semiprofessional	61	566.5	$1.5 \cdot \sqrt{61} \approx 11.7$
Clerical	107	566.2	$1.1 \cdot \sqrt{107} \approx 11.4$
Trades	194	565.7	$0.8 \cdot \sqrt{194} \approx 11.1$
Public service	25	564.1	$2.5 \cdot \sqrt{25} = 12.5$
Skill trades	351	562.9	$0.6 \cdot \sqrt{351} \approx 11.2$
Personal services	262	562.7	$0.7 \cdot \sqrt{262} \approx 11.3$
Laborers	647	560.7	$0.3 \cdot \sqrt{647} \approx 7.6$

If the ANOVA table was recalculated with the correct standard deviations, predict the effect on the F-test statistic. Would you be more or less likely to reject the null hypothesis? Explain.

Recalculating with the correct standard deviation will lower the F-test statistic, making it less likely to reject the null hypothesis. The corrected standard deviations are larger than those in the original table. These larger standard deviations will make the sum-of-squares error larger than it was originally, which will make the F-test statistic smaller.

Using the equation for MSE, calculate a revised sum-of-squares error, mean-squares error, and F-test statistic.

The revised sum of squares for error is

$$(25-1) \cdot 9.5^2 + (61-1) \cdot 11.7^2$$
$$+(107-1) \cdot 11.4^2 + (194-1) \cdot 11.1^2$$
$$+(25-1) \cdot 12.5^2 + (351-1) \cdot 11.2^2$$
$$+(262-1) \cdot 11.3^2 + (647-1) \cdot 7.6^2 = 166,228.74.$$

Therefore, the revised mean-squares error is

$$\text{MSE} = \frac{166,228.74}{1664} \approx 99.90, \text{ and the revised } F\text{-test}$$

statistic is $F_0 = \dfrac{1123.6284}{99.90} \approx 11.25.$

Retest the claim that all the head circumference population means are equal. Use an $\alpha = 0.05$ significance level.

Comparing $F \approx 11.25$ with the critical value $F_{0.05, 7, 1672} \approx F_{0.05, 7, 1000} = 2.02,$ we reject the null hypothesis at $\alpha = 0.05,$ and we conclude that at least one of the head-circumference means is different from the others.

From the corrected values for the standard deviation, does it appear that the ANOVA procedure's requirement of equal variances is satisfied?

Yes. The largest corrected standard deviation (12.5) is no more than two times the smallest corrected standard deviation (7.6). That is, $12.5 < 2(7.6) = 15.2,$ so the requirement of equal variances is satisfied.

Did any of your conclusions change in light of the new information? Explain.

The overall result concerning whether or not to reject the null hypothesis did not change. However, the strength of the rejection did change. Also, the new information allowed us to meet the requirement of equal variances so that we could perform the ANOVA.

If the null hypothesis of equal populations means was rejected, use Tukey's test to determine which of the pairwise means differ using a family wise error rate of $\alpha = 0.05$. Which of these categories appear to have significantly different means?

Arrange the sample means in ascending order are $\bar{x}_1 = 560.7,\ \bar{x}_2 = 562.7,\ \bar{x}_3 = 562.9,\ \bar{x}_4 = 564.1,$ $\bar{x}_5 = 565.7,\ \bar{x}_6 = 566.2,\ \bar{x}_7 = 566.5,$ and $\bar{x}_8 = 569.9.$ Now, $n = 1672$ and $k = 8,$ so $\nu = n - k = 1664$ and the critical value is

$$q_{\alpha, \nu, k} = q_{0.05, 1664, 8} \approx q_{0.05, 120, 8} = 4.363.$$

Test 1: $H_0 : \mu_1 = \mu_8.$ The test statistic is

$$q_0 = \frac{\bar{x}_8 - \bar{x}_1}{\sqrt{\frac{s^2}{2} \cdot \left(\frac{1}{n_1} + \frac{1}{n_8}\right)}} = \frac{569.9 - 560.7}{\sqrt{\frac{75.9}{2} \cdot \left(\frac{1}{647} + \frac{1}{25}\right)}} \approx 7.327. \text{ Since}$$

this test statistic is greater than the critical value $q_{\alpha, \nu, k} \approx 4.363,$ we reject $H_0 : \mu_1 = \mu_8.$

Test 2: $H_0 : \mu_1 = \mu_7.$ The test statistic is

$$q_0 = \frac{\bar{x}_7 - \bar{x}_1}{\sqrt{\frac{s^2}{2} \cdot \left(\frac{1}{n_1} + \frac{1}{n_7}\right)}} = \frac{566.5 - 560.7}{\sqrt{\frac{75.9}{2} \cdot \left(\frac{1}{647} + \frac{1}{61}\right)}} \approx 7.029. \text{ Since}$$

this test statistic is greater than the critical value $q_{\alpha, \nu, k} \approx 4.363,$ we reject $H_0 : \mu_1 = \mu_7.$

Test 3: $H_0 : \mu_1 = \mu_6.$ The test statistic is

$$q_0 = \frac{\bar{x}_6 - \bar{x}_1}{\sqrt{\frac{s^2}{2} \cdot \left(\frac{1}{n_1} + \frac{1}{n_6}\right)}} = \frac{566.2 - 560.7}{\sqrt{\frac{75.9}{2} \cdot \left(\frac{1}{647} + \frac{1}{107}\right)}} \approx 8.555. \text{ Since}$$

this test statistic is greater than the critical value $q_{\alpha, \nu, k} \approx 4.363,$ we reject $H_0 : \mu_1 = \mu_6.$

Test 4: $H_0 : \mu_1 = \mu_5.$ The test statistic is

$$q_0 = \frac{\bar{x}_5 - \bar{x}_1}{\sqrt{\frac{s^2}{2} \cdot \left(\frac{1}{n_1} + \frac{1}{n_5}\right)}} = \frac{565.7 - 560.7}{\sqrt{\frac{75.9}{2} \cdot \left(\frac{1}{647} + \frac{1}{194}\right)}} \approx 9.916. \text{ Since}$$

this test statistic is greater than the critical value $q_{\alpha, \nu, k} \approx 4.363,$ we reject $H_0 : \mu_1 = \mu_5.$

Test 5: $H_0 : \mu_1 = \mu_4.$ The test statistic is

$$q_0 = \frac{\bar{x}_4 - \bar{x}_1}{\sqrt{\frac{s^2}{2} \cdot \left(\frac{1}{n_1} + \frac{1}{n_4}\right)}} = \frac{564.1 - 560.7}{\sqrt{\frac{75.9}{2} \cdot \left(\frac{1}{647} + \frac{1}{25}\right)}} \approx 2.708. \text{ Since}$$

this test statistic is less than the critical value $q_{\alpha, \nu, k} \approx 4.363,$ we do not reject $H_0 : \mu_1 = \mu_4.$

Test 6: $H_0 : \mu_1 = \mu_3.$ The test statistic is

$$q_0 = \frac{\bar{x}_3 - \bar{x}_1}{\sqrt{\frac{s^2}{2} \cdot \left(\frac{1}{n_1} + \frac{1}{n_3}\right)}} = \frac{562.9 - 560.7}{\sqrt{\frac{75.9}{2} \cdot \left(\frac{1}{647} + \frac{1}{351}\right)}} \approx 5.387. \text{ Since}$$

this test statistic is greater than the critical value $q_{\alpha, \nu, k} \approx 4.363,$ we reject $H_0 : \mu_1 = \mu_3.$

Test 7: $H_0 : \mu_1 = \mu_2.$ The test statistic is

$$q_0 = \frac{\bar{x}_2 - \bar{x}_1}{\sqrt{\frac{s^2}{2} \cdot \left(\frac{1}{n_1} + \frac{1}{n_2}\right)}} = \frac{562.7 - 560.7}{\sqrt{\frac{75.9}{2} \cdot \left(\frac{1}{647} + \frac{1}{262}\right)}} \approx 4.433. \text{ Since}$$

this test statistic is greater than the critical value $q_{\alpha, \nu, k} \approx 4.363,$ we reject $H_0 : \mu_1 = \mu_2.$

Test 8: $H_0 : \mu_2 = \mu_8.$ The test statistic is

$$q_0 = \frac{\bar{x}_8 - \bar{x}_2}{\sqrt{\frac{s^2}{2} \cdot \left(\frac{1}{n_2} + \frac{1}{n_8}\right)}} = \frac{569.9 - 562.7}{\sqrt{\frac{75.9}{2} \cdot \left(\frac{1}{262} + \frac{1}{25}\right)}} \approx 5.583. \text{ Since}$$

this test statistic is greater than the critical value

$q_{\alpha,\nu,k} \approx 4.363$, we reject $H_0 : \mu_2 = \mu_8$.

Test 9: $H_0 : \mu_2 = \mu_7$. The test statistic is

$$q_0 = \frac{\overline{x}_7 - \overline{x}_2}{\sqrt{\frac{s^2}{2} \cdot \left(\frac{1}{n_2} + \frac{1}{n_7}\right)}} = \frac{566.5 - 562.7}{\sqrt{\frac{75.9}{2} \cdot \left(\frac{1}{262} + \frac{1}{61}\right)}} \approx 4.339. \text{ Since}$$

this test statistic is less than the critical value $q_{\alpha,\nu,k} \approx 4.363$, we do not reject $H_0 : \mu_2 = \mu_8$.

Test 10: $H_0 : \mu_2 = \mu_6$. The test statistic is

$$q_0 = \frac{\overline{x}_6 - \overline{x}_2}{\sqrt{\frac{s^2}{2} \cdot \left(\frac{1}{n_2} + \frac{1}{n_6}\right)}} = \frac{566.2 - 562.7}{\sqrt{\frac{75.9}{2} \cdot \left(\frac{1}{262} + \frac{1}{107}\right)}} \approx 4.952. \text{ Since}$$

this test statistic is greater than the critical value $q_{\alpha,\nu,k} \approx 4.363$, we reject $H_0 : \mu_2 = \mu_6$.

Test 11: $H_0 : \mu_2 = \mu_5$. The test statistic is

$$q_0 = \frac{\overline{x}_5 - \overline{x}_2}{\sqrt{\frac{s^2}{2} \cdot \left(\frac{1}{n_2} + \frac{1}{n_5}\right)}} = \frac{565.7 - 562.7}{\sqrt{\frac{75.9}{2} \cdot \left(\frac{1}{262} + \frac{1}{194}\right)}} \approx 5.141. \text{ Since}$$

this test statistic is greater than the critical value $q_{\alpha,\nu,k} \approx 4.363$, we reject $H_0 : \mu_2 = \mu_5$.

Test 12: $H_0 : \mu_2 = \mu_4$. The test statistic is

$$q_0 = \frac{\overline{x}_4 - \overline{x}_2}{\sqrt{\frac{s^2}{2} \cdot \left(\frac{1}{n_2} + \frac{1}{n_5}\right)}} = \frac{564.1 - 562.7}{\sqrt{\frac{75.9}{2} \cdot \left(\frac{1}{262} + \frac{1}{25}\right)}} \approx 1.086. \text{ Since}$$

this test statistic is less than the critical value $q_{\alpha,\nu,k} \approx 4.363$, we do not reject $H_0 : \mu_2 = \mu_4$.

Test 13: $H_0 : \mu_2 = \mu_3$. The test statistic is

$$q_0 = \frac{\overline{x}_3 - \overline{x}_2}{\sqrt{\frac{s^2}{2} \cdot \left(\frac{1}{n_2} + \frac{1}{n_5}\right)}} = \frac{562.9 - 562.7}{\sqrt{\frac{75.9}{2} \cdot \left(\frac{1}{262} + \frac{1}{351}\right)}} \approx 0.398. \text{ Since}$$

this test statistic is less than the critical value $q_{\alpha,\nu,k} \approx 4.363$, we do not reject $H_0 : \mu_2 = \mu_3$.

Test 14: $H_0 : \mu_3 = \mu_8$. The test statistic is

$$q_0 = \frac{\overline{x}_8 - \overline{x}_3}{\sqrt{\frac{s^2}{2} \cdot \left(\frac{1}{n_3} + \frac{1}{n_8}\right)}} = \frac{569.9 - 562.9}{\sqrt{\frac{75.9}{2} \cdot \left(\frac{1}{351} + \frac{1}{25}\right)}} \approx 5.489. \text{ Since}$$

this test statistic is greater than the critical value $q_{\alpha,\nu,k} \approx 4.363$, we reject $H_0 : \mu_3 = \mu_8$.

Test 15: $H_0 : \mu_3 = \mu_7$. The test statistic is

$$q_0 = \frac{\overline{x}_7 - \overline{x}_3}{\sqrt{\frac{s^2}{2} \cdot \left(\frac{1}{n_3} + \frac{1}{n_7}\right)}} = \frac{566.5 - 562.9}{\sqrt{\frac{75.9}{2} \cdot \left(\frac{1}{351} + \frac{1}{61}\right)}} \approx 4.213. \text{ Since}$$

this test statistic is less than the critical value $q_{\alpha,\nu,k} \approx 4.363$, we do not reject $H_0 : \mu_3 = \mu_7$.

Test 16: $H_0 : \mu_3 = \mu_6$. The test statistic is

$$q_0 = \frac{\overline{x}_6 - \overline{x}_3}{\sqrt{\frac{s^2}{2} \cdot \left(\frac{1}{n_3} + \frac{1}{n_6}\right)}} = \frac{566.2 - 562.9}{\sqrt{\frac{75.9}{2} \cdot \left(\frac{1}{351} + \frac{1}{107}\right)}} \approx 4.851. \text{ Since}$$

this test statistic is greater than the critical value $q_{\alpha,\nu,k} \approx 4.363$, we reject $H_0 : \mu_3 = \mu_6$.

Test 17: $H_0 : \mu_3 = \mu_5$. The test statistic is

$$q_0 = \frac{\overline{x}_5 - \overline{x}_3}{\sqrt{\frac{s^2}{2} \cdot \left(\frac{1}{n_3} + \frac{1}{n_5}\right)}} = \frac{565.7 - 562.9}{\sqrt{\frac{75.9}{2} \cdot \left(\frac{1}{351} + \frac{1}{194}\right)}} \approx 5.081. \text{ Since}$$

this test statistic is greater than the critical value $q_{\alpha,\nu,k} \approx 4.363$, we reject $H_0 : \mu_3 = \mu_5$.

Test 18: $H_0 : \mu_3 = \mu_4$. The test statistic is

$$q_0 = \frac{\overline{x}_4 - \overline{x}_3}{\sqrt{\frac{s^2}{2} \cdot \left(\frac{1}{n_3} + \frac{1}{n_4}\right)}} = \frac{564.1 - 562.9}{\sqrt{\frac{75.9}{2} \cdot \left(\frac{1}{351} + \frac{1}{25}\right)}} \approx 0.941. \text{ Since}$$

this test statistic is less than the critical value $q_{\alpha,\nu,k} \approx 4.363$, we do not reject $H_0 : \mu_3 = \mu_4$.

Test 19: $H_0 : \mu_4 = \mu_8$. The test statistic is

$$q_0 = \frac{\overline{x}_8 - \overline{x}_4}{\sqrt{\frac{s^2}{2} \cdot \left(\frac{1}{n_4} + \frac{1}{n_8}\right)}} = \frac{569.9 - 564.1}{\sqrt{\frac{75.9}{2} \cdot \left(\frac{1}{25} + \frac{1}{25}\right)}} \approx 3.329. \text{ Since}$$

this test statistic is less than the critical value $q_{\alpha,\nu,k} \approx 4.363$, we do not reject $H_0 : \mu_4 = \mu_8$.

Test 20: $H_0 : \mu_4 = \mu_7$. The test statistic is

$$q_0 = \frac{\overline{x}_7 - \overline{x}_4}{\sqrt{\frac{s^2}{2} \cdot \left(\frac{1}{n_4} + \frac{1}{n_7}\right)}} = \frac{566.5 - 564.1}{\sqrt{\frac{75.9}{2} \cdot \left(\frac{1}{25} + \frac{1}{61}\right)}} \approx 1.641. \text{ Since}$$

this test statistic is less than the critical value $q_{\alpha,\nu,k} \approx 4.363$, we do not reject $H_0 : \mu_4 = \mu_7$.

Test 21: $H_0 : \mu_4 = \mu_6$. The test statistic is

$$q_0 = \frac{\overline{x}_6 - \overline{x}_4}{\sqrt{\frac{s^2}{2} \cdot \left(\frac{1}{n_4} + \frac{1}{n_6}\right)}} = \frac{566.2 - 564.1}{\sqrt{\frac{75.9}{2} \cdot \left(\frac{1}{25} + \frac{1}{107}\right)}} \approx 1.535. \text{ Since}$$

this test statistic is less than the critical value $q_{\alpha,\nu,k} \approx 4.363$, we do not reject $H_0 : \mu_4 = \mu_6$.

Test 22: $H_0 : \mu_4 = \mu_5$. The test statistic is

$$q_0 = \frac{\overline{x}_5 - \overline{x}_4}{\sqrt{\frac{s^2}{2} \cdot \left(\frac{1}{n_4} + \frac{1}{n_5}\right)}} = \frac{565.7 - 564.1}{\sqrt{\frac{75.9}{2} \cdot \left(\frac{1}{25} + \frac{1}{194}\right)}} \approx 1.222. \text{ Since}$$

this test statistic is less than the critical value $q_{\alpha,\nu,k} \approx 4.363$, we do not reject $H_0 : \mu_4 = \mu_5$.

Test 23: $H_0 : \mu_5 = \mu_8$. The test statistic is

$$q_0 = \frac{\overline{x}_8 - \overline{x}_5}{\sqrt{\frac{s^2}{2} \cdot \left(\frac{1}{n_5} + \frac{1}{n_8}\right)}} = \frac{569.9 - 565.7}{\sqrt{\frac{75.9}{2} \cdot \left(\frac{1}{194} + \frac{1}{25}\right)}} \approx 3.208. \text{ Since}$$

this test statistic is less than the critical value $q_{\alpha,\nu,k} \approx 4.363$, we do not reject $H_0 : \mu_5 = \mu_8$.

Test 24: $H_0 : \mu_5 = \mu_7$. The test statistic is

$$q_0 = \frac{\overline{x}_7 - \overline{x}_5}{\sqrt{\frac{s^2}{2} \cdot \left(\frac{1}{n_5} + \frac{1}{n_7}\right)}} = \frac{566.5 - 565.7}{\sqrt{\frac{75.9}{2} \cdot \left(\frac{1}{194} + \frac{1}{61}\right)}} \approx 0.885. \text{ Since}$$

this test statistic is less than the critical value $q_{\alpha,\nu,k} \approx 4.363$, we do not reject $H_0 : \mu_5 = \mu_7$.

Test 25: $H_0 : \mu_5 = \mu_6$. The test statistic is

$$q_0 = \frac{\overline{x}_6 - \overline{x}_5}{\sqrt{\frac{s^2}{2} \cdot \left(\frac{1}{n_5} + \frac{1}{n_6}\right)}} = \frac{566.2 - 565.7}{\sqrt{\frac{75.9}{2} \cdot \left(\frac{1}{194} + \frac{1}{107}\right)}} \approx 0.674. \text{ Since}$$

this test statistic is less than the critical value $q_{\alpha,\nu,k} \approx 4.363$, we do not reject $H_0 : \mu_5 = \mu_6$.

Test 26: $H_0 : \mu_6 = \mu_8$. The test statistic is

$$q_0 = \frac{\overline{x}_8 - \overline{x}_6}{\sqrt{\frac{s^2}{2} \cdot \left(\frac{1}{n_6} + \frac{1}{n_8}\right)}} = \frac{569.9 - 566.2}{\sqrt{\frac{75.9}{2} \cdot \left(\frac{1}{107} + \frac{1}{25}\right)}} \approx 2.704. \text{ Since}$$

this test statistic is less than the critical value $q_{\alpha,\nu,k} \approx 4.363$, we do not reject $H_0 : \mu_6 = \mu_8$.

Test 27: $H_0 : \mu_6 = \mu_7$. The test statistic is

$$q_0 = \frac{\overline{x}_7 - \overline{x}_6}{\sqrt{\frac{s^2}{2} \cdot \left(\frac{1}{n_6} + \frac{1}{n_7}\right)}} = \frac{566.5 - 566.2}{\sqrt{\frac{75.9}{2} \cdot \left(\frac{1}{107} + \frac{1}{61}\right)}} \approx 0.304. \text{ Since}$$

this test statistic is less than the critical value $q_{\alpha,\nu,k} \approx 4.363$, we do not reject $H_0 : \mu_6 = \mu_7$.

Test 28: $H_0 : \mu_7 = \mu_8$. The test statistic is

$$q_0 = \frac{\overline{x}_8 - \overline{x}_7}{\sqrt{\frac{s^2}{2} \cdot \left(\frac{1}{n_7} + \frac{1}{n_8}\right)}} = \frac{569.9 - 566.5}{\sqrt{\frac{75.9}{2} \cdot \left(\frac{1}{61} + \frac{1}{25}\right)}} \approx 2.324. \text{ Since}$$

this test statistic is less than the critical value $q_{\alpha,\nu,k} \approx 4.363$, we do not reject $H_0 : \mu_7 = \mu_8$.

Thus, the Tukey's test results in the following pairwise differences: $\mu_{laborers} \neq \mu_{professional}$;

$\mu_{laborers} \neq \mu_{semiprofessional}$; $\mu_{laborers} \neq \mu_{clerical}$;

$\mu_{laborers} \neq \mu_{trades}$; $\mu_{laborers} \neq \mu_{skill\ trades}$;

$\mu_{laborers} \neq \mu_{personal\ services}$; $\mu_{personal\ services} \neq \mu_{professional}$;

$\mu_{personal\ services} \neq \mu_{clerical}$; $\mu_{personal\ services} \neq \mu_{trades}$;

$\mu_{skill\ trades} \neq \mu_{professional}$; $\mu_{skill\ trades} \neq \mu_{clerical}$; and

$\mu_{skill\ trades} \neq \mu_{trades}$.

The result of the Tukey's test is ambiguous since all other pairs were found to be equal.

Reports will vary.

Chapter 14
Inference on the Least-Squares Regression Model and Multiple Regression

Section 14.1

1. $\bar{y} = 4.302(20) - 3.293$
   $= 82.7$

3. 0; σ

5. (a) $\beta_0 \approx b_0 = -2.3256$
   $\beta_1 \approx b_1 = 2.0233$

   (b) The point estimate for σ is $s_e = 0.5134$.

   (c) $s_{b_1} = 0.1238$

   (d) The test statistic is
   $$t_0 = \frac{2.0233}{0.1238}$$
   $= 16.343$ [Tech: 16.344].

   Because the P-value $< 0.001 < \alpha = 0.05$, or
   $t_0 = 16.343$ [Tech: 16.344] $> t_{0.025} = 3.182$, we reject the null hypothesis and conclude that a linear relation exists between x and y.

7. (a) $\beta_0 \approx b_0 = 1.200$
   $\beta_1 \approx b_1 = 2.200$

   (b) The point estimate for σ is $s_e = 0.8944$.

   (c) $s_{b_1} = 0.2828$

   (d) The test statistic is
   $$t_0 = \frac{2.200}{0.2828}$$
   $= 7.779$ [Tech: 7.778]

   Because the
   P-value $= 0.0044 < \alpha = 0.05$, or
   $t_0 = 7.779 > t_{0.025} = 3.182$, we reject the null hypothesis and conclude that a linear relation exists between x and y.

9. (a) $\beta_0 \approx b_0 = 116.600$
   $\beta_1 \approx b_1 = -0.7200$

   (b) The point estimate for σ is $s_e = 3.2863$.

   (c) $s_{b_1} = 0.1039$

(d) The test statistic is
   $$t_0 = \frac{-0.7200}{0.1039}$$
   $= -6.930$ [Tech: -6.928].

   Because the P-value $= .0062 < \alpha = 0.05$, or
   $t_0 = -6.930$ [Tech: -6.928] $< -t_{0.025} = -3.182$, we reject the null hypothesis and conclude that a linear relation exists between x and y.

11. (a) $\beta_0 \approx b_0 = 69.0296$
   $\beta_1 \approx b_1 = -0.0479$

   (b) $s_e = 0.3680$

   (c) $s_{b_1} = 0.0043$

   (d) $H_0: \beta_1 = 0$
   $H_1: \beta_1 \neq 0$

   The test statistic is
   $$t_0 = \frac{-0.0479}{0.0043}$$
   $= -11.140$ [Tech: -11.157].

   Because the P-value $= 0.0001 < \alpha = 0.05$, or $t_0 = -11.140$ [Tech: -11.157] $< -t_{0.025} = -3.182$, we reject the null hypothesis and conclude that a linear relation exists between commute time and score on the well-being survey.

   (e) 95% confidence interval, lower bound:
   $$-0.0479 - 2.571 \cdot \frac{0.3680}{\sqrt{7344.858}} = -0.0589$$

   Upper bound:
   $$-0.0479 + 2.571 \cdot \frac{0.3680}{\sqrt{7344.858}} = -0.0369$$

13. (a) $\beta_0 \approx b_0 = 12.4932$
   $\beta_1 \approx b_1 = 0.1827$

   (b) $s_e = 0.0954$

(c) The correlation between residuals and expected z-scores is 0.986 [Tech: 0.987]. Because $0.986 > 0.923$ from Table VI, conclude the residuals are approximately normally distributed.

(d) $s_{b_1} = 0.0276$

(e) Since the residuals are normally distributed, the test can be conducted.

$H_0: \beta_1 = 0$
$H_1: \beta_1 \neq 0$

The test statistic is

$$t_0 = \frac{0.1827}{0.0276}$$
$$= 6.620 \text{ [Tech: 6.630]}.$$

Because the P-value $< 0.0001 < \alpha = 0.05$, or $t_0 = 6.620 > t_{0.005} = 3.250$, we reject the null hypothesis and conclude that a linear relation exists between height and head circumference.

(f) Since the residuals are normally distributed, the confidence interval can be calculated.

95% confidence interval, lower bound:

$$0.1827 - 2.262 \cdot \frac{0.0954}{\sqrt{11.9773}} = 0.1203$$

[Tech: 0.1204]

Upper bound:

$$0.1827 + 2.262 \cdot \frac{0.0954}{\sqrt{11.9773}} = 0.2451$$

(g) A good estimate of the child's head circumference would be $12.4932 + 0.1827(26.5) \approx 17.33$ inches.

This is a good estimate because the hypothesis test shows that a linear relation exists between the child's height and head circumference.

15. (a) $\beta_0 \approx b_0 = 2675.6$
$\beta_1 \approx b_1 = 0.6764$

(b) $s_e = 271.04$

(c) $s_{b_1} = 0.2055$

(d) $H_0: \beta_1 = 0$
$H_1: \beta_1 \neq 0$

The test statistic is

$$t_0 = \frac{0.6764}{0.2055}$$
$$= 3.292 \text{ [Tech: 3.291]}.$$

Because the P-value $= 0.011 < \alpha = 0.05$, or $t_0 = 3.292 > t_{0.025} = 2.306$, we reject the null hypothesis and conclude that a linear relation exists between 7-day strength and 28-day strength.

(e) 95% confidence interval, lower bound:

$$0.6764 - 2.306 \cdot \frac{271.04}{\sqrt{1,739,160}} = 0.2025$$

Upper bound:

$$0.6764 + 2.306 \cdot \frac{271.04}{\sqrt{1,739,160}} = 1.1504$$

(f) The mean 28-day strength of this concrete if the 7-day strength is 3000 psi is $2675.6 + 0.6764(3000) = 4704.8$ psi.

17. (a) $\beta_0 \approx b_0 = 6.4506$
$\beta_1 \approx b_1 = -0.0000108$

$H_0: \beta_1 = 0$
$H_1: \beta_1 < 0$

The test statistic is $t_0 = -7.38$. Since $t_0 = -7.38 < -t_{0.01} = -2.330$, and since P-value $< 0.0001 < \alpha = 0.01$, reject H_0 and conclude that a linear relation exists between 2013 cost and ROI.

(b) 90% confidence interval, lower bound: -0.000013; upper bound: -0.0000084

(c) $6.4506 - 0.0000108(180,000) = 4.51\%$

19. (a) $\beta_0 \approx b_0 = 51.1310$
$\beta_1 \approx b_1 = -0.1819$

(b) $s_e = 39.2647$; $s_{b_1} = 2.0688$

$H_0: \beta_1 = 0$
$H_1: \beta_1 \neq 0$

The test statistic is $t_0 = \dfrac{-0.1819}{2.0688}$

$= -0.088.$

Because the P-value $= 0.93 > \alpha = 0.05$, or $t_0 = -0.088 > -t_{0.025} = -2.228$, we do not reject the null hypothesis. There is insufficient evidence to conclude that a linear relation exists between compensation and stock return.

(c) 95% confidence interval, lower bound:

$$-0.1819 - 2.228 \cdot \frac{39.2647}{\sqrt{360.203}} = -4.7916$$

Upper bound:

$$-0.1819 + 2.228 \cdot \frac{39.2647}{\sqrt{360.203}} = 4.4278$$

(d) No, the results do not indicate that a linear relation exists, since the hypothesis test does not reject the null hypothesis and the 95% confidence interval for the slope contains 0. We could use $\bar{y} = 49.733\%$ as an estimate of the stock return.

21. (a)

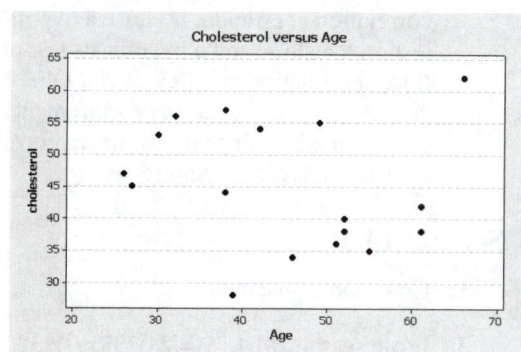

(b) $\hat{y} = 50.7841 - 0.1298x$

(c)

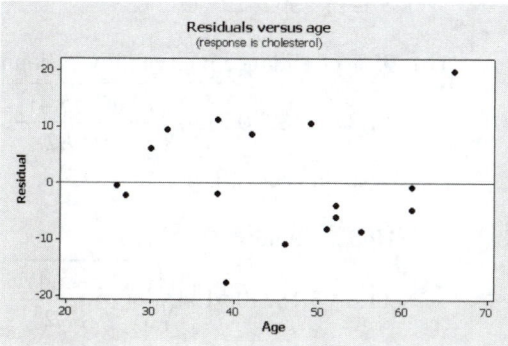

Because the residuals are evenly spread around the horizontal line drawn at 0, the requirement of constant error variance is satisfied, so the linear model seems appropriate.

(d) For age, the lower fence is 17 [Minitab: 7.25], and the upper fence is 73 [Minitab: 81.25], so there are no outliers.

For HDL cholesterol, the lower fence is 14 [Minitab: 10.75], and the upper fence is 78 [Minitab: 80.75], so there are no outliers.

The scatter diagram shows no influential observations.

(e) $s_e = 9.9534$; $s_{b_1} = 0.2021$

$$H_0: \beta_1 = 0$$
$$H_1: \beta_1 \neq 0$$

The test statistic is

$$t_0 = \frac{-0.1298}{0.2021}$$

$$= -0.642 \text{ [Tech:} -0.643].$$

Because the P-value $= 0.530 > \alpha = 0.05$, or $t_0 = -0.642 > -t_{0.005} = -2.947$, we do not reject the null hypothesis. Conclude that a linear relation does not exist between the age and HDL levels.

(f) 95% confidence interval, lower bound:
$$-0.1298 - 2.131 \cdot \frac{9.9534}{\sqrt{2426}} = -0.5604$$

[Tech: -0.5606]

Upper bound:

$$-0.1298 + 2.131 \cdot \frac{9.9534}{\sqrt{2426}} = 0.3008$$

[Tech: 0.3009]

(g) Do not recommend using the least-squares regression line to predict the HDL cholesterol levels, since we did not reject the null hypothesis. A good estimate for the HDL cholesterol level would be $\bar{y} = 44.9$.

23. (a) $\hat{y} = -12.4967 + 0.0657x$

(b) $H_0: \beta_1 = 0$
$H_1: \beta_1 \neq 0$

The test statistic is $t_0 = 17.61$. Because the P-value $< 0.0001 < \alpha = 0.05$, reject the null hypothesis and conclude that a linear relation exists between distance from the sun and sidereal year.

(c)

(d)

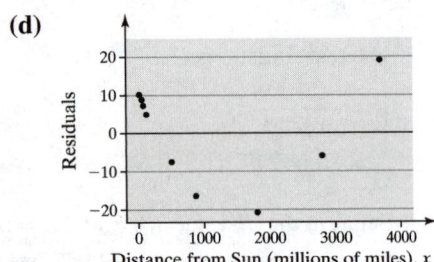

(e) A linear model would not be appropriate, since there is a distinct pattern in the residual plot.

(f) The moral is that the inferential procedures may lead us to believe that a linear relation between the two variables exists even though diagnostics tools such as residual plots indicate that a linear model is inappropriate.

25. (a)

Sale Price versus Zestimate

(b) $\hat{y} = 1.0228x - 0.759$

$H_0: \beta_1 = 0$
$H_1: \beta_1 \neq 0$

The test statistic is $t_0 = 105.45$. Because the P-value $< 0.0001 < \alpha = 0.05$, reject the null hypothesis and conclude that a linear relation exists between the Zestimate and the sale price.

(c) $\hat{y} = 0.5220x + 115.8094$

$H_0: \beta_1 = 0$
$H_1: \beta_1 \neq 0$

The test statistic is $t_0 = 1.441$. Because the P-value $= 0.193 > \alpha = 0.05$, do not reject the null hypothesis and conclude that a linear relation does not exist between the Zestimate and the sale price.

Since including this point changes the conclusion of the test for the significance of the slope, this observation is influential.

27. The y-coordinates of the least squares regression line represent the mean value of the response variable for any given value of the explanatory variable.

29. We do not conduct inference on the linear correlation coefficient because a hypothesis test on the slope and a hypothesis test on the linear correlation coefficient yield the same conclusion. Moreover, the requirements for conducting inference on the linear correlation coefficient are very hard to verify.

Section 14.2

1. Confidence; mean.

3. From section 14.1, $\hat{y} = 2.0233x - 2.3256$,
$s_e = 0.5134$, and $s_{b_1} = 0.1238$.

(a) $\bar{y} = 2.0233(7) - 2.3256$
$= 11.8$

(b) 95% confidence interval, lower bound:
$$11.8 - 3.182 \cdot 0.5134\sqrt{\frac{1}{5} + \frac{(7-5.4)^2}{17.2}}$$
$= 10.8 \text{ [Tech: 10.9]}$

Upper bound:
$$11.8 + 3.182 \cdot 0.5134\sqrt{\frac{1}{5} + \frac{(7-5.4)^2}{17.2}}$$
$= 12.8$

(c) $\hat{y} = 2.0233(7) - 2.3256$

$= 11.8$

(d) 95% prediction interval, lower bound:

$11.8 - 3.182 \cdot 0.5134\sqrt{1 + \dfrac{1}{5} + \dfrac{(7-5.4)^2}{17.2}}$

$= 9.9$

Upper bound:

$11.8 + 3.182 \cdot 0.5134\sqrt{1 + \dfrac{1}{5} + \dfrac{(7-5.4)^2}{17.2}}$

$= 13.7$

(e) The confidence interval is an interval estimate for the mean value of y at $x = 7$, whereas the prediction interval is an interval estimate for a single value of y at $x = 7$.

5. From section 14.1, $\hat{y} = 2.2x + 1.2$,

$s_e = 0.8944$, and $s_{b_1} = 0.2828$.

(a) $\bar{y} = 2.2(1.4) + 1.2$

$= 4.3$

(b) 95% confidence interval, lower bound:

$4.3 - 3.182 \cdot 0.8944\sqrt{\dfrac{1}{5} + \dfrac{(1.4-0)^2}{10}}$

$= 2.5$

Upper bound:

$4.3 + 3.182 \cdot 0.8944\sqrt{\dfrac{1}{5} + \dfrac{(1.4-0)^2}{10}}$

$= 6.1$

(c) $\hat{y} = 2.2(1.4) + 1.2$

$= 4.3$

(d) 95% prediction interval, lower bound:

$4.3 - 3.182 \cdot 0.8944\sqrt{1 + \dfrac{1}{5} + \dfrac{(1.4-0)^2}{10}}$

$= 0.9$

Upper bound:

$4.3 + 3.182 \cdot 0.8944\sqrt{1 + \dfrac{1}{5} + \dfrac{(1.4-0)^2}{10}}$

$= 7.7$

7. From section 14.1, $\hat{y} = 69.0296 - 0.0479x$ and $s_e = 0.3680$.

(a) $\bar{y} = 69.0296 - 0.0479(20)$

$= 68.07$

(b) 90% confidence interval, lower bound:

$68.07 - 2.015 \cdot 0.3680\sqrt{\dfrac{1}{7} + \dfrac{(20-43.857)^2}{7344.857}}$

$= 67.722$ [Tech: 67.723]

Upper bound:

$68.07 + 2.015 \cdot 0.3680\sqrt{\dfrac{1}{7} + \dfrac{(20-43.857)^2}{7344.857}}$

$= 68.418$ [Tech: 68.420]

(c) $\hat{y} = 69.0296 - 0.0479(20)$

$= 68.07$

(d) 90% prediction interval, lower bound:

$68.07 - 2.015 \cdot 0.3680\sqrt{1 + \dfrac{1}{7} + \dfrac{(20-43.857)^2}{7344.857}}$

$= 67.251$ [Tech: 67.252]

Upper bound:

$68.07 + 2.015 \cdot 0.3680\sqrt{1 + \dfrac{1}{7} + \dfrac{(20-43.857)^2}{7344.857}}$

$= 68.889$ [Tech: 68.891]

(e) The prediction made in part (a) is an estimate of the mean well-being index composite score for all individuals whose commute time is 20 minutes. The prediction made in part (c) is an estimate of the well-being index composite score of one individual, Jane, whose commute time is 20 minutes.

9. From section 14.1, $\hat{y} = 12.4932 + 0.1827x$ and $s_e = 0.0954$.

(a) $\bar{y} = 12.4932 + 0.1827(25.75)$

$= 17.20$ inches

(b) 95% confidence interval, lower bound:

$17.20 - 2.262 \cdot 0.0954\sqrt{\dfrac{1}{11} + \dfrac{(25.75-26.455)^2}{11.9773}}$

$= 17.12$ inches

Upper bound:

$17.20 + 2.262 \cdot 0.0954\sqrt{\dfrac{1}{11} + \dfrac{(25.75-26.455)^2}{11.9773}}$

$= 17.28$ inches

(c) $\hat{y} = 12.4932 + 0.1827(25.75)$

$\qquad = 17.20$ inches

(d) 95% prediction interval, lower bound:

$$17.20 - 2.262 \cdot 0.0954\sqrt{1 + \frac{1}{11} + \frac{(25.75 - 26.455)^2}{11.9773}}$$

$= 16.97$ inches

Upper bound:

$$17.20 + 2.262 \cdot 0.0954\sqrt{1 + \frac{1}{11} + \frac{(25.75 - 26.455)^2}{11.9773}}$$

$= 17.43$ inches

(e) The confidence interval is an interval estimate for the mean head circumference of all children who are 25.75 inches tall. The prediction interval is an interval estimate for the head circumference of a single child who is 25.75 inches tall.

11. From section 14.1, $\hat{y} = 2675.5619 + 0.6764x$ and $s_e = 271.04$.

(a) $\overline{y} = 2675.5619 + 0.6764(2550)$

$\qquad = 4400.4$ psi

(b) 95% confidence interval, lower bound:

$$4400.4 - 2.306 \cdot 271.04\sqrt{\frac{1}{10} + \frac{(2550 - 2882)^2}{1,739,160}}$$

$= 4147.8$ psi

Upper bound:

$$4400.4 + 2.306 \cdot 271.04\sqrt{\frac{1}{10} + \frac{(2550 - 2882)^2}{1,739,160}}$$

$= 4653.0$ psi [Tech: 4653.1 psi]

(c) $\hat{y} = 2675.5619 + 0.6764(2550)$

$\qquad = 4400.4$ psi

(d) 95% prediction interval, lower bound:

$$4400.4 - 2.306 \cdot 271.04\sqrt{1 + \frac{1}{10} + \frac{(2550 - 2882)^2}{1,739,160}}$$

$= 3726.3$ psi

Upper bound:

$$4400.4 + 2.306 \cdot 271.04\sqrt{1 + \frac{1}{10} + \frac{(2550 - 2882)^2}{1,739,160}}$$

$= 5074.5$ psi [Tech: 5074.6 psi]

(e) The confidence interval is an interval estimate for the mean 28-day strength of all concrete cylinders that have a 7-day strength of 2550 psi. The prediction interval is an interval estimate for the 28-day strength of a single cylinder whose 7-day strength is 2550 psi.

13. From Section 14.1, $\hat{y} = 6.4506 - 0.0000108x$ and $s_e = 2.3908$.

(a) $\overline{y} = 6.4506 - 0.0000108(180,000)$

$\qquad = 4.51\%$

(b)

95% confidence interval, lower bound:

$$4.51 - 1.962 \cdot 2.3908\sqrt{\frac{1}{1302} + \frac{(180,000 - 138,019.8)^2}{2.683 \times 10^{12}}}$$

$= 4.33\%$

Upper bound:

$$4.51 + 1.962 \cdot 2.3908\sqrt{\frac{1}{1302} + \frac{(180,000 - 138,019.8)^2}{2.683 \times 10^{12}}}$$

$= 4.69\%$

(c) $\hat{y} = 6.4506 - 0.0000108(180,000)$

$\qquad = 4.51\%$

(d)

95% prediction interval, lower bound:

$$4.51 - 1.962 \cdot 2.3908\sqrt{1 + \frac{1}{1302} + \frac{(180,000 - 138,019.8)^2}{2.683 \times 10^{12}}}$$

$= -0.18\%$

Upper bound:

$$4.51 + 1.962 \cdot 2.3908\sqrt{1 + \frac{1}{1302} + \frac{(180,000 - 138,019.8)^2}{2.683 \times 10^{12}}}$$

$= 9.20\%$

(e) Although the predicted annual ROIs in parts (a) and (c) are the same, the intervals are different because the distribution of the mean ROIs in part (a) has less variability than the distribution of the individual ROIs in part (c)

15. (a) It does not make sense to construct either a confidence interval or prediction interval based on the least-squares regression equation because the evidence indicated that there is no linear relation between CEO compensation and stock return.

(b) Since there is no linear relation between x and y, construct a confidence interval using only the y-data.

95% confidence interval, lower bound:

$$49.733 - 2.201 \cdot \frac{37.4519}{\sqrt{12}} = 25.938\%$$

Upper bound:

$$49.733 + 2.201 \cdot \frac{37.4519}{\sqrt{12}} = 73.529\%$$

17. (a)

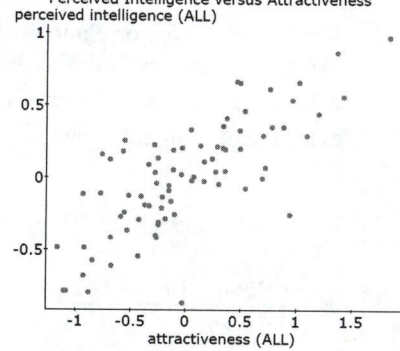

Perceived Intelligence versus Attractiveness

(b) $r = 0.762$; because $0.762 > 0.361$ (from Table II with $n = 30$), we conclude there is a linear relation between attractiveness and perceived intelligence.

(c) $\hat{y} = 0.4608x + 0.0000$

(d) A person of average attractiveness is perceived to be of average intelligence.

(e) $H_0: \beta_1 = 0$ vs. $H_1: \beta_1 > 0$

The test statistic is $t_0 = 10.40$. Since $P\text{-value} < 0.0001 < \alpha = 0.05$, reject the null hypothesis and conclude that perceived intelligence and attractiveness are associated.

(f)

Actual IQ versus Perceived Intelligence

$r = 0.086 < 0.361$. No linear relation between perceived intelligence and IQ.

(g) $\hat{y} = -1.8893x + 122.6038$

$H_0: \beta_1 = 0$
$H_1: \beta_1 > 0$

The test statistic is $t_0 = -0.34$. Since $P\text{-value} = 0.6239 > \alpha = 0.10$, do not reject the null hypothesis.

(h) $\hat{y} = 18.1100x + 128.8345$

$H_0: \beta_1 = 0$
$H_1: \beta_1 > 0$

The test statistic is $t_0 = 1.96$. Since $P\text{-value} = 0.0287 < \alpha = 0.10$, reject the null hypothesis.

(i) $\bar{y} = 18.1100(1.28) + 128.8345$
$\quad\;\; = 152.02$

95% confidence interval, lower bound: 126.1; upper bound: 177.9

(j) $\hat{y} = 18.1100(1.28) + 128.8345$
$\quad\;\; = 152.02$

95% prediction interval, lower bound: 106.3; upper bound: 197.7

Section 14.3

1. Correlation matrix

3. 4.39

5. (a) The slope coefficient of x_1 is 3. This indicates that $\hat{y}$ will increase by 3 units, on average, for every 1-unit increase in x_1, provided that x_2 remains constant.

The slope coefficient of x_2 is -4. This indicates that $\hat{y}$ will decrease by 4 units, on average, for every 1-unit increase in x_2, provided that x_1 remains constant.

(b) $\hat{y} = 5 + 3(10) - 4x_2$

$\quad = 35 - 4x_2$

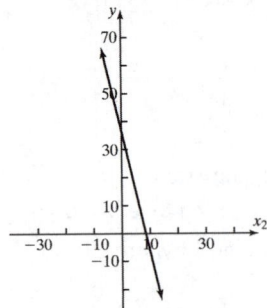

(c) $\hat{y} = 5 + 3(15) - 4x_2$

$\quad = 50 - 4x_2$

(d) $\hat{y} = 5 + 3(20) - 4x_2$

$\quad = 65 - 4x_2$

(e) Changing the value of x_1 has the effect of changing the y-intercept of the graph of the regression line.

7. (a) $R_{adj}^2 = 1 - \left(\dfrac{25-1}{25-3-1} \right)(1 - 0.653)$

$\quad = 0.603$

(b) $F = \dfrac{0.653}{1-0.653} \cdot \dfrac{25 - (3+1)}{3}$

$\quad = 13.173$

(c) $R_{adj}^2 = 1 - \left(\dfrac{25-1}{25-4-1} \right)(1 - 0.665)$

$\quad = 0.598$

Do not add the new variable. Its addition reduces the proportion of the variance explained by the model.

9. (a) Correlations: x_1, x_2, x_3, y

	x_1	x_2	x_3
x_2	-0.460		
x_3	0.183	-0.012	
y	-0.748	0.788	0.258

The largest linear correlation between any two explanatory variables is -0.460 between x_1 and x_2, so there is no evidence of multicollinearity.

(b)

$\hat{y} = 7.9647 - 0.1035x_1 + 0.9359x_2 + 0.1152x_3$

(c) $F = \dfrac{0.94}{1-0.94} \cdot \dfrac{10 - (3+1)}{3}$

$\quad = 31.33$

Since the P-value $< 0.001 < \alpha = 0.05$, we reject the null hypothesis and conclude that at least one of the explanatory variables is linearly related to the response variable.

(d) $t_{\beta_1} = \dfrac{-0.1035}{0.0208} = -4.98$

Reject $H_0: \beta_1 = 0$ since the P-value $= 0.003 < \alpha = 0.05$.

$t_{\beta_2} = \dfrac{0.9359}{0.2} = 4.68$

Reject $H_0: \beta_2 = 0$ since the P-value $= 0.003 < \alpha = 0.05$.

$t_{\beta_3} = \dfrac{0.1152}{0.0319} = 3.61$

Reject $H_0: \beta_3 = 0$ since the P-value $= 0.011 < \alpha = 0.05$.

11. (a)

Correlations: x_1, x_2, x_3, x_4, y

	x_1	x_2	x_3	x_4
x_2	−0.184			
x_3	0.182	−0.106		
x_4	−0.166	−0.588	0.291	
y	−0.046	0.932	−0.063	−0.750

There is no evidence that multicollinearity may be a problem.

(b) $\hat{y} = 279.02 - 0.068x_1 - 1.146x_2 + 0.3897x_3 - 2.9378x_4$

$$F = \frac{0.95897}{1 - 0.95897} \cdot \frac{12 - (4+1)}{4}$$
$$= 40.90$$

Since the P-value $< 0.001 < \alpha = 0.05$, we reject H_0 and conclude that at least one of the explanatory variables is linearly related to the response variable.

Do not reject $H_0: \beta_1 = 0$ since $t_{\beta 1} = 0.05$ and P-value $= 0.955 > \alpha = 0.05$.

Reject $H_0: \beta_2 = 0$ since $t_{\beta 2} = 7.41$ and P-value $< 0.001 < \alpha = 0.05$.

Do not reject $H_0: \beta_3 = 0$ since $t_{\beta 3} = 1.48$ and P-value $= 0.183 > \alpha = 0.05$.

Reject $H_0: \beta_4 = 0$ since $t_{\beta 4} = -3.57$ and P-value $= 0.009 < \alpha = 0.05$.

x_1 and x_3 have slope coefficients that are not significantly different from zero.

(c) Remove x_1 from the model.

$\hat{y} = 282.33 + 1.1426x_2 + 0.3941x_3 - 2.9570x_4$

Since $F = 62.29$ and P-value $< 0.001 < \alpha = 0.05$, the model is still significant.

Reject $H_0: \beta_2 = 0$ since $t_{\beta 2} = 8.49$ and P-value $< 0.001 < \alpha = 0.05$.

Do not reject $H_0: \beta_3 = 0$ since $t_{\beta 3} = 1.67$ and P-value $= 0.134 > \alpha = 0.05$.

Reject $H_0: \beta_4 = 0$ since $t_{\beta 4} = -4.18$ and P-value $= 0.003 < \alpha = 0.05$.

Remove x_3 from the model.

$\hat{y} = 281.62 + 1.1584x_2 - 2.6267x_4$

(d)

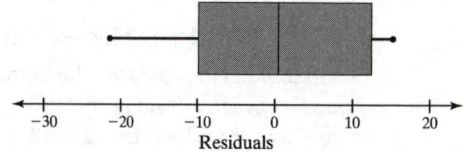

The residual plots and boxplot show no outliers.

(e) $\hat{y} = 281.62 + 1.1584(35.6) - 2.6267(29)$

$= 246.7$

(f)

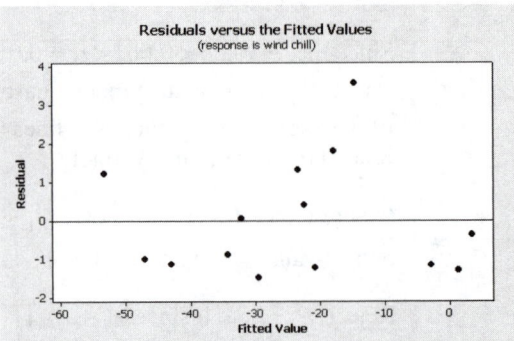

The residuals are normally distributed, so it is reasonable to construct confidence and prediction intervals.

(g) 95% confidence interval:
$(237.45, 255.91)$

95% prediction interval: $(214.93, 278.43)$

13. (a) Yes; conclude that at least one of the slope coefficients is linearly related to the response variable, computer anxiety.

(b) $b_1 = -0.87$; a 1-point increase in the computer confidence variable reduces the computer anxiety measurement by 0.87 point, on average, provided that all other explanatory variables remain constant.

$b_2 = -0.51$; a 1-point increase in the computer knowledge variable reduces the computer anxiety measurement by 0.51 point, on average, provided that all other explanatory variables remain constant.

$b_3 = -0.45$; a 1-point increase in the computer liking scale reduces the computer anxiety measurement by 0.45 point, on average, provided that all other explanatory variables remain constant.

$b_4 = 0.33$; a 1-point increase in the trait anxiety scale increases the computer anxiety measurement by 0.33 point, on average, provided that all other explanatory variables remain constant.

All of the slope coefficients are reasonable.

(c)

$\hat{y} = 84.04 - 0.87(25) - 0.51(19) - 0.45(20) + 0.33(43)$

$= 57.79$

(d) $R^2 = 0.69$ indicates that 69% of the variance in computer anxiety is explained by this model.

(e) The statement means that the requirements of a multiple regression model were checked and were met.

15. (a) $\hat{y} = -12.509 + 1.334x_1 - 0.414x_2$

(b)

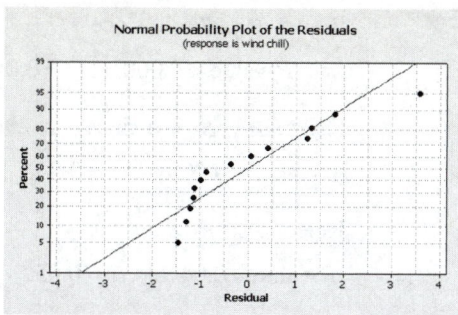

Because the residual plots show distinct patterns, conclude that the linear model is not appropriate.

17. (a) **Correlations: SLUMP, 7 DAY, 28 DAY**

```
           SLUMP      7 DAY
7 DAY     −0.460
28 DAY    −0.753     0.737
```

There is no reason to be concerned about multicollinearity based on the model.

(b) $\hat{y} = 3890.5 - 295.9x_1 + 0.552x_2$

(c)

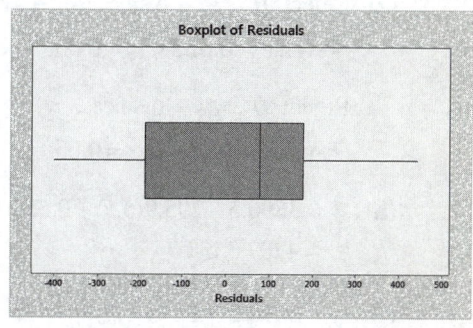

None of the residual plots show any pattern, and the boxplot does not show any outliers. Therefore, the linear model is appropriate.

(d) $b_1 = -295.9$: A 1-unit increase in slump of the concrete decreases the 28-day strength of the concrete by 295.9 pounds per square inch, on average, provided that the 7-day strength remains constant.

$b_2 = 0.552$: A 1-point per square increase in 7-day strength increases the 28-day strength of the concrete by 0.552 pound per square inch, on average, provided that slump remains constant.

(e) $R^2 = 76.0\%$ represents the variance in 28-day strength that is explained by the model.

$$R_{adj}^2 = 1 - \left(\frac{11-1}{11-2-1}\right)(1 - 0.7605)$$

$$= 70.1\%$$

$R_{adj}^2 = 70.1\%$ modifies the value of R^2 based on the sample size and the number of explanatory variables in the model.

(f) $F = \dfrac{0.7605}{1-0.7605} \cdot \dfrac{11-(2+1)}{2}$

$= 12.70$

The F-test has a
P-value $= 0.003 < \alpha = 0.005,$ so we reject the null hypothesis and conclude that at least one of the slope coefficients is linearly related to the 28-day strength of the concrete.

(g) Reject $H_0: \beta_1 = 0$ since $t_{\beta 1} = -2.69$ and
P-value $= 0.027 < \alpha = 0.05.$

Reject $H_0: \beta_2 = 0$ since $t_{\beta 2} = 2.54$ and
P-value $= 0.034 < \alpha = 0.05.$

(h) $\bar{y} = 3890.5 - 295.9(3.5) + 0.552(2450)$

$= 4207.3$ psi

(i) $\hat{y} = 3890.5 - 295.9(3.5) + 0.552(2450)$

$= 4207.3$ psi

(j) 95% confidence interval:
$(3988.6, 4427.2);$ for concrete for which slump is 3.5 inches and 7-day strength is 2450 psi, we are 95% confident that the true mean 28-day strength is between 3988.6 psi and 4427.2 psi.

95% prediction interval:
$(3533.6, 4882.2);$ for an individual concrete mixture for which slump is 3.5 inches and 7-day strength is 2450 psi, we are 95% confident that the 28-day strength will fall between 3533.6 psi and 4882.2 psi.

19. (a)
$\hat{y} = -1.1752x_1 + 2.3463x_2 - 2.2489x_3 + 539.4308$

(b) The hypotheses are $H_0: \beta_1 = \beta_2 = \beta_3 = 0$
versus $H_1:$ at least one of the coefficients differs from zero.

$F = \dfrac{0.78823}{1-0.78823} \cdot \dfrac{50-(3+1)}{3}$

$= 57.07$

The P-value $< 0.001,$ so reject the null hypothesis and conclude that at least one of the explanatory variables is linearly related to the response variable.

Reject $H_0: \beta_1 = 0$ since $t_{\beta 1} = -11.97$
and P-value $< 0.0001 < \alpha = 0.05.$

Reject $H_0: \beta_2 = 0$ since $t_{\beta 2} = 3.91$ and
P-value $= 0.0003 < \alpha = 0.05.$

Reject $H_0: \beta_3 = 0$ since $t_{\beta 3} = -2.10$ and
P-value $= 0.0416 < \alpha = 0.05.$

(c) $b_1 = -1.1752$: if the percent of students taking the SAT increases by 1, the average SAT score decreases by 1.1752, on average, holding average salary and starting salary constant.

$b_2 = 2.3463$: if average salary increases by 1 thousand dollars, average SAT scores increase by 2.3463, on average, holding percent of students taking the exam and starting salary constant.

$b_3 = -2.2489$: if starting salary increases by 1 thousand dollars, average SAT scores decrease by 2.2489, on average, holding percent of students taking the exam and average salary constant.

The coefficient for starting salary is the opposite sign we would expect, since one would think higher starting salaries would result in better teachers hired, which would translate into higher SAT scores. However, the correlation between average salary and starting salary is 0.811, which suggests the model suffers from multicollinearity.

(d) $\hat{y} = -1.2122x_1 + 1.4007x_2 + 511.674$

$H_0: \beta_1 = \beta_2 = 0$ vs. $H_1:$ at least one of the coefficients differs from zero.

$F = \dfrac{0.768}{1-0.768} \cdot \dfrac{50-(2+1)}{2}$

$= 77.79$

The P-value $< 0.001,$ so reject the null hypothesis and conclude that at least one of the explanatory variables is linearly related to the response variable.

Reject $H_0: \beta_1 = 0$ since $t_{\beta 1} = -12.12$
and P-value $< 0.0001 < \alpha = 0.05.$

Reject $H_0: \beta_2 = 0$ since $t_{\beta 2} = 3.41$ and
P-value $= 0.0013 < \alpha = 0.05$.

Each coefficient has a sign we would expect. The coefficient on percent taking the exam is negative. The higher the percentage of students who take the exam, the lower the average SAT score. The coefficient for average salary is positive. Higher salaries for teachers correspond to higher SAT scores.

21. When the explanatory variables have an additive effect, or do not interact, it means that the effect of any one explanatory variable on the response variable does not depend on the value of the other explanatory variables.

23. Multicollinearity exists when two explanatory variables are highly correlated. We use a correlation matrix to check if multicollinearity exists. A consequence of multicollinearity is that t-test statistics might be small and their P-values large, falsely indicating that neither explanatory variable is significant.

Section 14.4

1. Interaction

3. (a) $y_i = \beta_0 + \beta_1 x_1 + \beta_2 x_2 + \beta_3 x_3 + \varepsilon_i$

(b) $y_i = \beta_0 + \beta_1 x_1 + \beta_2 x_2 + \beta_3 x_1 x_2 + \varepsilon_i$

(c) $y_i = \beta_0 + \beta_1 x_1 + \beta_2 x_2 + \beta_3 x_3 + \beta_4 x_2 x_3 + \varepsilon_i$

5. (a)

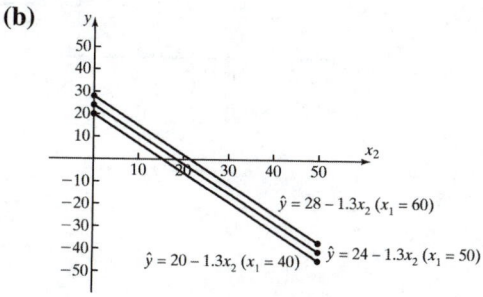

(b)

(c) The lines are parallel.

(d)

The lines are no longer parallel.

7. (a) $\hat{y} = 9.063 + 0.386(0) + 1.268(28) + 0.676(8)$
$\qquad - 0.913(18) - 0.175(8)(18)$
$\qquad = 8.3$ years

(b) $\hat{y} = 9.063 + 0.386(1) + 1.268(28) + 0.676(8)$
$\qquad - 0.913(18) - 0.175(8)(18)$
$\qquad = 8.7$ years

(c) The interaction term is $-0.175 x_3 x_4$. The number of teeth with root development, x_3, and the sum of the normalized heights of seven teeth on the left side of the mouth, x_4, interact.

(d) 86.3% of the variation in the response variable is explained by this least-squares regression model.

9. (a) $\hat{y} = 109.894 + 0.0000286 x_1$

(b) $H_0: \beta_1 = 0$
$H_1: \beta_1 \neq 0$

The test statistic is $t_0 = 2.78$. Since P-value $= 0.012 < 0.05$, reject the null hypothesis and conclude that there is a linear relation between MRI count and IQ.

(c)

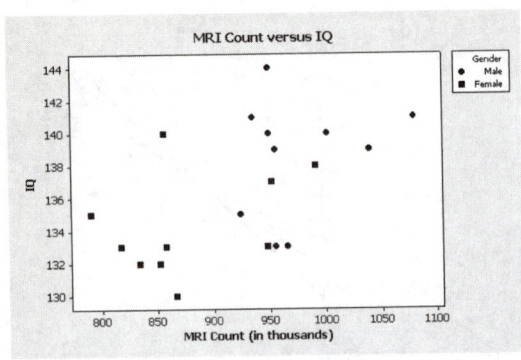

(d) $\hat{y} = 121.582 + 0.0000174x_1 - 2.486x_2$

(e) $H_0: \beta_1 = 0$
$H_1: \beta_1 \neq 0$

The test statistic is $t_0 = 1.27$. Since P-value $= 0.22 > 0.05$, do not reject the null hypothesis and conclude that $\beta_1 = 0$.

$H_0: \beta_2 = 0$
$H_1: \beta_2 \neq 0$

The test statistic is $t_0 = -1.23$. Since P-value $= 0.23 > 0.05$, do not reject the null hypothesis and conclude that $\beta_2 = 0$.

(f) The variable x_1, MRI count, originally had a slope coefficient that was significantly different from 0. Adding the variable x_2, gender, caused this relationship to change so that the slope coefficient of both variables are not significantly different from 0. This might lead us to conclude that gender is a lurking variable.

11. Use two indicator variables.

$$x_1 = \begin{cases} 1 & \text{if attached} \\ 0 & \text{otherwise} \end{cases}$$

$$x_2 = \begin{cases} 1 & \text{if detached} \\ 0 & \text{otherwise} \end{cases}$$

Section 14.5

1. (a)

There appears to be a quadratic relation between x and y.

(b) $\hat{y} = 36.9806 - 14.1746x + 1.7046x^2$

(c)

Boxplot of Residuals

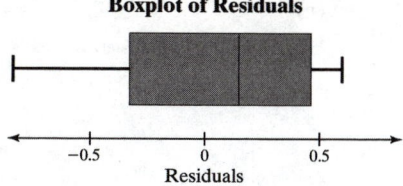

The quadratic model appears to be appropriate.

(d) $R^2 = 0.993$; 99.3% of the variation in y is explained by the least-squares regression model.

(e) $H_0: \beta_1 = \beta_2 = 0$

$H_1:$ at least one coefficient is different from 0.

Since the test statistic is $F_0 = 295.4$ and P-value < 0.0001, reject the null hypothesis and conclude that at least one coefficient is different from zero.

$H_0: \beta_1 = 0$
$H_1: \beta_1 \neq 0$

The test statistic is $t_0 = -21.70$, and P-value < 0.0001, so reject the null hypothesis.

$H_0: \beta_2 = 0$
$H_1: \beta_2 \neq 0$

The test statistic is $t_0 = 19.33$, and P-value < 0.0001, so reject the null hypothesis.

(f) Confidence interval: lower bound: 6.46, upper bound: 8.65

Prediction interval: lower bound: 5.54, upper bound: 9.57

3. (a)

(b) $\hat{y} = -25403.9 + 2756.98x - 29.7706x^2$

(c)

The quadratic model appears to be appropriate.

(d) $R^2 = 0.936$; 93.6% of the variability in income is explained by the least-squares regression model.

(e) $H_0: \beta_1 = \beta_2 = 0$

$H_1:$ at least one coefficient is different from 0.

Since the test statistic is $F_0 = 80.12$ and P-value < 0.0001, reject the null hypothesis and conclude that at least one coefficient is different from zero.

$H_0: \beta_1 = 0$
$H_1: \beta_1 \neq 0$

The test statistic is $t_0 = 11.88$, and P-value < 0.0001, so reject the null hypothesis.

$H_0: \beta_2 = 0$
$H_1: \beta_2 \neq 0$

The test statistic is $t_0 = -12.39$, and P-value < 0.0001, so reject the null hypothesis.

(f) Confidence interval: We are 95% confident the mean income of all 45-year-olds is between $36,934.24 and $39,814.83 [MINITAB: $36,934 and $39,815].

Prediction interval: We are 95% confident the income of a particular 45-year-old is between $34,199.07 and $42,550.01 [MINITAB: $34199 and $42550].

5. (a)

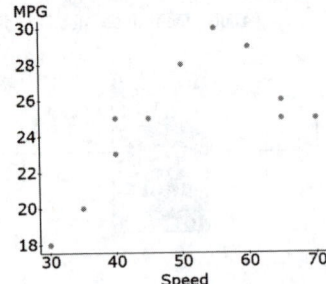

The relation appears to be quadratic.

(b) $\hat{y} = -25.3408 + 1.9346x - 0.01747x^2$

(c)

The quadratic model appears to be appropriate.

(d) $R^2 = 0.895$; 89.5% of the variability in miles per gallon is explained by the least-squares regression model.

(e) $H_0: \beta_1 = \beta_2 = 0$ vs. $H_1:$ at least one coefficient is different from 0.

Since the test statistic is $F_0 = 33.95$ and P-value < 0.0001, reject the null hypothesis and conclude that at least one coefficient is different from zero.

$H_0: \beta_1 = 0$ vs. $H_1: \beta_1 \neq 0$

The test statistic is $t_0 = 6.53$ and P-value $= 0.0002$, so reject the null hypothesis.

$H_0: \beta_2 = 0$ vs. $H_1: \beta_2 \neq 0$

The test statistic is $t_0 = -5.98$ and P-value $= 0.0003$, so reject the null hypothesis.

(f) 95% confidence interval: We are 95% confident that the mean miles per gallon of all Toyota Camrys is between 25.9 and 28.7.

95% prediction interval: We are 95% confident that the miles per gallon of a particular Toyota Camry is between 24.0 and 30.6.

7. (a) $\hat{y} = 0.0321(12) + 0.9848(1) + 0.5391(30)$
$$- 0.000145(75)^2 + 3.8483$$
$$= 20.6 \text{ years}$$

(b) 0.9848 year

(c) $\beta_1 = 0.0321$: If the percent of children aged 10 to 14 years involved in child labor increases by 1, the age gap at first marriage increases by 0.0321 year, on average, holding all other variables constant.

(d) 59.3% of the variability in the age gap at first marriage is explained by the least-squares regression model.

(e) Because the P-value is small, reject the null hypothesis. Percent of children aged 10 to 14 years involved in child labor is significant in explaining age gap at first marriage.

Section 14.6

1. $H_0: \beta_4 = \beta_5 = 0$

$H_1:$ at least one of β_4 and β_5 is different from zero.

For the full model, $SSE(2) = 0.266$ and $MSE(2) = 0.0665$. For the reduced model, $SSE(1) = 0.3605$.

$$F_0 = \frac{\dfrac{0.3605 - 0.266}{5 - 3}}{0.0665}$$
$$= 0.71$$

$F_0 = 0.71 < F_{0.05,2,4} = 6.94.$ Do not reject H_0.

There is not sufficient evidence to suggest that either x_4 or x_5 explain the variability in the response variable y.

3. (a)

$$\hat{y} = 25.628 + 0.07193x_1 - 0.8099x_2 + 55.3491x_3$$

(b) $H_0: \beta_2 = \beta_3 = 0$

$H_1:$ at least one of β_2 and β_3 is different from zero.

For the full model, $SSE(2) = 19247$ and $MSE(2) = 2138.5$. For the reduced model, $SSE(1) = 31160$.

$$F_0 = \frac{\dfrac{31160 - 19247}{3 - 1}}{2138.5}$$
$$= 2.79 \text{ [Minitab: 2.78]}$$

Since $F_{0.05,2,9} = 4.26$, do not reject H_0. There is not sufficient evidence to suggest number of bedrooms or number of baths explain the variability in asking price.

(c) $\hat{y} = 0.0718x_1 + 54.9163x_3 + 24.0920$

(d)

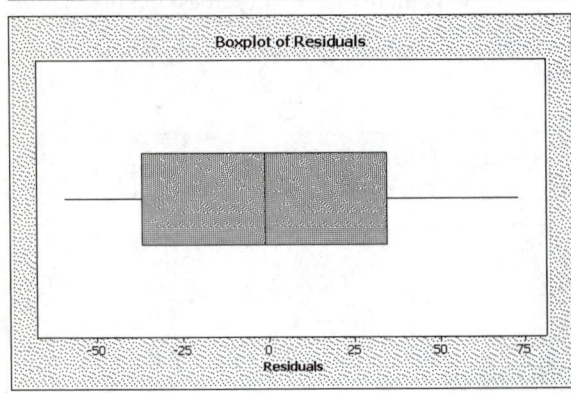

The residual plots do not show any obvious patterns nor a violation of the assumption of constant error variance. The boxplot of the residuals does not show any outliers. The normal probability plot indicates it is reasonable to conclude the residuals are normally distributed (because $0.991 > 0.932$ from Table VI).

(e) If the square footage increases by 1, the asking price will increase by 0.0718 thousand dollars, or $71.80, on average, assuming the number of baths is held constant.

If the number of baths increases by 1, the asking price will increase by 54.9163 thousand dollars, or $54,916.30, on average, assuming the square footage is held constant.

(f) Using the model in part (c), Confidence interval: lower bound: $346.8 thousand; upper bound: $404.3 thousand; prediction interval: lower bound: $273.6 thousand; upper bound: $477.4 thousand.

5. Answers may vary. Using stepwise regression with $\alpha = 0.05$:

$$y = 2.9176 \text{ WHIP} - 0.5504 \text{ G/F} - 3.0253 \text{ W\%} + 0.3577 \text{ RS} + 0.6176$$

7. Using stepwise regression and backward elimination:

$$y = -223.8081 \text{ Slalom} + 28.4755 \text{ Lap} + 403.0802$$

Using forward selection:

$$y = 558.2815 \text{ Cornering} - 325.2538 \text{ Slalom} + 64.2139 \text{ Lap} - 662.7484$$

Using forward selection adds the Cornering g-Force variable to the model, even though it has a high *P*-value.

Chapter 14 Review Exercises

1. The least-squares regression model is $y_i = \beta_0 + \beta_1 x_i + \varepsilon_i$. The requirements to perform inference on the least-squares regression line are (1) for any particular values of the explanatory variable x, the mean of the corresponding responses in the population depends linearly on x, and (2) the response variables, y_i, are normally distributed with mean $\mu_{y|x} = \beta_0 + \beta_1 x$ and standard deviation σ. We verify these requirements by checking to see that the residuals are normally distributed, with mean 0 and constant variance σ^2, and that the residuals are independent. We do this by constructing residual plots and a normal probability plot of the residuals.

2. (a) $\beta_0 \approx b_0 = 3.8589$

    $\beta_1 \approx b_1 = -0.1049$

    The mean GPA of students who choose a seat in the fifth row is $\hat{y} = 3.334$.

    (b) $s_e = 0.5102$

    (c)

    The residuals are normally distributed.

    (d) $s_{b_1} = 0.0366$

    (e) $H_0: \beta_1 = 0$

    $H_1: \beta_1 \neq 0$

    The test statistic is $t_0 = -2.866$ [Tech: -2.868].

Because the P-value $= 0.007 < \alpha = 0.05$, or $t_0 = -2.866 < -t_{0.025} = -2.028$, we reject the null hypothesis and conclude that a linear relation exists between the row chosen by students on the first day of class and their cumulative GPAs.

(f) 95% confidence interval: lower bound: -0.1791; upper bound: -0.0307.

(g) 95% confidence interval: lower bound: 3.159; upper bound: 3.509

(h) $\hat{y} = 3.334$

(i) 95% prediction interval: lower bound: 2.285; upper bound: 4.383 [Tech: 4.384]

(j) Although the predicted GPAs in parts (a) and (h) are the same, the intervals are different because the distribution of the mean GPAs, part (a), has less variability than the distribution of individual GPAs, part (h).

3. (a) $\beta_0 \approx b_0 = -399.25$

    $\beta_1 \approx b_1 = 2.5315$

    The mean rent of a 900-square-foot apartment in Queens is $\hat{y} = \$1897.10$.

    (b) $s_e = 229.547$

    (c)

    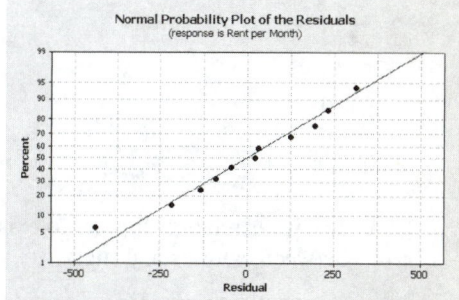

    The correlation between the residuals and expected z-scores is 0.985. Because $0.985 > 0.923$, the residuals are normally distributed.

    (d) $s_{b_1} = 0.2166$

    (e) $H_0: \beta_1 = 0$

    $H_1: \beta_1 \neq 0$

    The test statistic is $t_0 = 11.687$ [Tech: 11.690].

Because the P-value $< 0.0001 < \alpha = 0.05$, there is evidence that a linear relation exists between the square footage of an apartment in Queens, New York, and the monthly rent.

(f) 95% confidence interval about the slope: lower bound: 2.0416; upper bound: 3.0214.

(g) 90% confidence interval about the mean rent of 900-square-foot apartments: lower bound: \$1752.20 [Tech: \$1752.22]; upper bound: \$2006.00 [Tech: \$2005.96].

(h) When an apartment has 900 square feet, $\hat{y} = \$1897.10$.

(i) 95% prediction interval for the rent of a particular 900-square-foot apartment: lower bound: \$1439.60 [Tech: \$1439.59]; upper bound: \$2318.60 [Tech: \$2318.59].

(j) Although the predicted rents in parts (a) and (h) are the same, the intervals are different because the distribution of the means, part (a), has less variability than the distribution of the individuals, part (h).

4. (a)

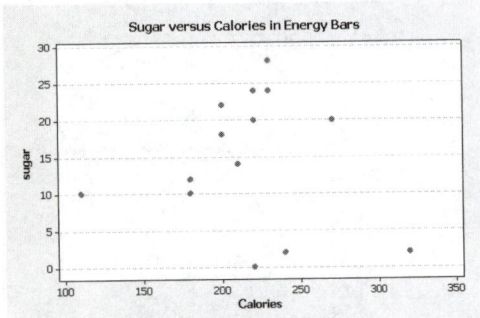

No linear relation appears to exist between calories and sugar.

(b) $\hat{y} = 17.8675 - 0.0146x$

(c) $s_e = 9.3749$

(d)

A normal probability plot shows that the residuals are approximately normally distributed. Therefore, the linear model is appropriate.

(e) $s_{b_1} = 0.0549$

(f) $H_0: \beta_1 = 0$
$H_1: \beta_1 \neq 0$

The test statistic is
$t_0 = -0.266$ [Tech: -0.265].

Because the P-value $= 0.795 > \alpha = 0.05$, do not reject the null hypothesis. Conclude that a linear relation does not exist between the number of calories per serving and the number of grams per serving in high-protein and moderate-protein energy bars.

(g) 95% confidence interval: lower bound: -0.1342 [Tech: -0.1343]; upper bound: 0.1050 [Tech: 0.1051].

(h) Do not recommend using the least-squares regression line to predict the sugar content of the energy bars, since we did not reject the null hypothesis. A good estimate for the sugar content is the mean sugar content, $\bar{y} = 14.7$ grams.

5. (a) $\hat{y} = 20,976.2 - 2054.6x$

(b) $s_e = 1247.18; s_{b_1} = 203.21.$

$H_0: \beta_1 = 0$
$H_1: \beta_1 \neq 0$

The test statistic is $t_0 = -10.111$.
Since P-value < 0.0001, reject the null hypothesis. There is sufficient evidence at the $\alpha = 0.05$ level of significance that a linear relation exists between the age of used Chevy Camaros and their selling price.

(c)

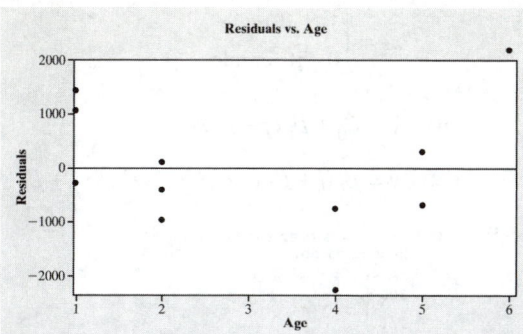

(d) The residual plot shows a pattern indicating that a linear model is not appropriate.

The moral is to perform graphical diagnostic tests along with the inferential procedures before drawing any conclusions regarding the model.

6. (a) Coefficient of x_1: If rainfall during the winter increases by 1 inch while temperature and rainfall during the harvest remain constant, then the wine quality rating increases by 0.00117.

Coefficient of x_2: If temperature increases by $1°$F while rainfall during the winter and rainfall during the harvest remain constant, then the wine quality rating increases by 0.0614.

Coefficient of x_3: If rainfall during the harvest increases by 1 inch while rainfall during the winter and temperature remain constant, then the wine quality rating decreases by 0.00386.

(b)

$$y = 12.15 + 0.00117(8) + 0.0614(58) - 0.00386(2.9)$$
$$= 15.7 \text{ inches}$$

7. (a) Correlations: Midterm, Project, Course Grade

	Midterm	Project
Project	−0.076	
Course Grade	0.704	0.526

There is no reason to be concerned with multicollinearity, since there is no correlation between the explanatory variables.

(b) $\hat{y} = -91.29 + 1.5211x_1 + 0.4407x_2$

(c)

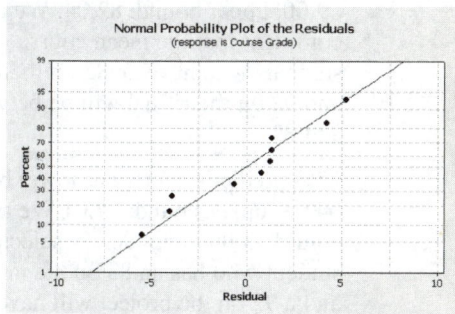

The residual plots show no patterns. The boxplot showns no outliers, and the normal probability plot shows that the residuals are approximately normally distributed.

(e) $R^2 = 0.833$

$$R_{adj}^2 = 1 - \left(\frac{10-1}{10-2-1}\right)(1-0.833)$$
$$= 0.785$$

83.3% of the variance in the course grade is explained by the linear regression model.

(f) $F_0 = 17.47$

Since P-value $= 0.002 < \alpha = 0.05,$ reject the null hypothesis and conclude that at least one of the coefficients differs from zero.

(g) $H_0: \beta_1 = 0$
$H_1: \beta_1 \neq 0$

The test statistic is $t_0 = 4.83,$ and the P-value $= 0.002.$ Reject the null hypothesis and conclude that the explanatory variable midterm is linearly related to course grade.

$H_0: \beta_2 = 0$
$H_1: \beta_2 \neq 0$

The test statistic is $t_0 = 3.77,$ and the P-value $= 0.007.$ Reject the null hypothesis and conclude that the explanatory variable project is linearly related to course grade.

(h) $\hat{y} = 71$

(i) 95% confidence interval: lower bound: 69.00, upper bound: 82.04. We are 95% confident that the mean course grade for students who have an 83 on their midterm and 92 on their final will be between 69 and 82.

95% prediction interval: lower bound: 64.03; upper bound: 87.01. We are 95% confident that a randomly selected student who has an 83 on the midterm and a 92 on the project will have a course grade between 64 and 87.

8. (a) $\hat{y} = \beta_1 x_1 + \beta_2 x_2 + \beta_3 x_3 + \beta_0,$ where

$$x_3 = \begin{cases} 1 & \text{if male} \\ 0 & \text{if female} \end{cases}$$

(b) $\hat{y} = \beta_0 + \beta_1 x_2 + \beta_2 x_2^2$

(c) $\hat{y} = \beta_1 x_1 + \beta_2 x_2 + \beta_3 x_3 + \beta_4 x_1 x_2 + \beta_0$

9. (a)

Apartments in New York

(b) $\hat{y} = 216.0137 + 1.8479 x_1 - 422.7573 x_2$

(c) The test statistic is $F_0 = 35.29.$ Since P-value $< 0.0001 < \alpha = 0.05,$ reject the null hypothesis. There is sufficient evidence to conclude that at least one coefficient is different from 0.

(d) $H_0: \beta_2 = 0$
$H_1: \beta_2 \neq 0$

The test statistic is $t_0 = -3.052,$ and the P-value $= 0.0066.$ Reject the null hypothesis and conclude that this coefficient is different from zero. The coefficient suggests that rents in Nassau County are $422.76 cheaper per month.

(e) Lower bound: $1765.1; upper bound: $2177.9. We are 95% confident that the mean monthly rent for 950-square-foot apartments in Queens is between $1765.10 and $2177.90.

10. (a)

$$\hat{y} = -493.2621 + 80.5515 x_1 + 0.7237 x_2 + 1.9681 x_3 + 49.1947 x_4 + 7.4312 x_5 + 7.8615 x_6$$

(b) From the full model, $SSE(2) = 42645.5$ and $MSE(2) = 292.0922.$ From the reduced model, $SSE(1) = 43034.7.$

$$F_0 = \frac{\dfrac{43034.7 - 42645.5}{6-4}}{292.0922}$$
$$= 0.666$$

Since $F_0 = 0.666 < F_{0.05,2,146} = 3.09$, do not reject the null hypothesis. There is not sufficient evidence to conclude that at least one of the coefficients is different from zero.

(c) Answers may vary. Using stepwise regression:

$$\hat{y} = -422.0052 + 83.3777x_1 + 2.0282x_3 + 52.5186x_4$$

Chapter 14 Test

1. (1) For any particular values of the explanatory variable x, the mean of the corresponding responses in the population depends linearly on x. (2) The response variable, y_i, is normally distributed with mean $\mu_{y|x_i} = \beta_0 + \beta_1 x$ and standard deviation σ.

2. (a) $\beta_0 \approx b_0 = -0.3091$
$\beta_1 \approx b_1 = 0.2119$

$$\hat{y} = -0.3091 + 0.2119(80.2)$$
$$= 16.685 \text{ [Tech: 16.687]}$$

The mean number of chirps per second when the temperature is $80.2\,°F$ is 16.69.

(b) $s_e = 0.9715$

(c)

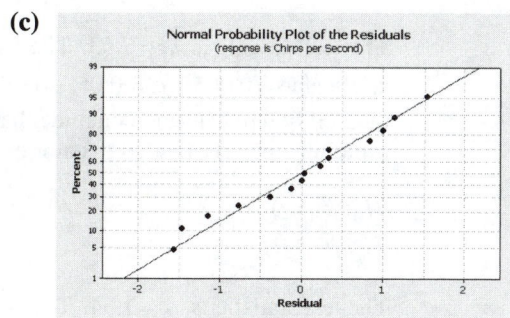

The residuals are normally distributed.

(d) $s_{b_1} = 0.0387$

(e) $H_0: \beta_1 = 0$
$H_1: \beta_1 \neq 0$

The test statistic is $t_0 = 5.475$. The P-value is $P\text{-value} = 0.0001 < \alpha = 0.05$, so reject the null hypothesis and conclude that a linear relation exists between temperature and the cricket's chirps.

(f) 95% confidence interval about the slope of the least-squares regression line: lower bound: 0.1283; upper bound: 0.2955.

(g) 90% confidence interval about the mean number of chirps at $80.2\,°F$: lower bound: 16.25 [Tech: 16.24]; upper bound: 17.13.

(h) When the temperature is $80.2\,°F$, $\hat{y} = 16.69$ chirps per second.

(i) 90% prediction interval for the number of chirps of a particular cricket at $80.2\,°F$: lower bound: 14.91; upper bound: 18.46.

(j) Although the predicted numbers of chirps in parts (a) and (h) are the same, the intervals are different because the distribution of the means, part (a), has less variability than the distribution of the individuals, part (h).

3. (a) $\beta_0 \approx b_0 = 29.705$
$\beta_1 \approx b_1 = 2.6351$

$$\hat{y} = 29.705 + 2.6351(7)$$
$$= 48.15 \text{ inches}$$

The mean height of a 7-year-old boy is 48.15 inches.

(b) $s_e = 2.45$

(c) $H_0: \beta_1 = 0$
$H_1: \beta_1 \neq 0$

The test statistic is $t_0 = 12.75$. The P-value is $P\text{-value} = 0.0001 < \alpha = 0.05$, so reject the null hypothesis and conclude that a linear relation exists between boys' ages and heights.

(d) 95% confidence interval about the slope of the true least-squares regression line: lower bound: 2.2009; upper bound: 3.0692.

(e) 90% confidence interval about the mean height of 7-year-old boys: lower bound: 47.11 inches; upper bound: 49.19 inches.

(f) The predicted height of a randomly chosen 7-year-old boy is $\hat{y} = 48.15$ inches.

(g) 90% prediction interval for the height of a particular 7-year-old boy: lower bound: 43.78 inches; upper bound: 52.52 inches.

(h) Although the predicted heights in parts (a) and (f) are the same, the intervals are different because the distribution of the means, part (a), has less variability than the distribution of the individuals, part (f).

4. (a) $\beta_0 \approx b_0 = 67.388$

   $\beta_1 \approx b_1 = -0.2632$

   (b) $H_0: \beta_1 = 0$
   $H_1: \beta_1 \neq 0$

   The test statistic is $t_0 = -1.567$. The P-value is P-value $= 0.138 > \alpha = 0.05$, so do not reject the null hypothesis and conclude that no linear relation exists between age and grip strength.

   (c) Because the null hypothesis was rejected in part (b), a good estimate of the grip strength of a 42-year-old female would be the mean strength of the population, $\bar{y} = 57$ psi.

5. (a) The student scored 1.52 standard deviations above the mean on the Math portion of the SAT.

   (b) Because $\beta_4 = 0.29$, the student's expected first-year college GPA will be 0.29 z-scores lower.

   (c) $\beta_4 = 0.29$: If the z-score of a student's high school GPA increases by 1, the z-score of their first-year college GPA will increase by 0.29, on average, holding all other variables constant.

   (d) 24% of the variability in first-year college GPA is explained by the least-squares regression model.

   (e) Since this indicates a relatively strong correlation between these two variables, this model may suffer from multicollinearity.

   (f)

   $\hat{y} = 0.06(-0.54) + 0.07(1.32) + 0.18(0.98) + 0.29(0.36)$
   $= 0.34$

   Since this is a positive z-score, we would expect a student with these credentials to have an above-average first-year college GPA.

6. (a)

   Correlations: Fat, Protein, Sugar, Carbs, Calories

	Fat	Protein	Sugar	Carbs
Protein	0.507			
Sugar	0.412	0.247		
Carbs	0.576	0.550	0.671	
Calories	0.950	0.661	0.542	0.784

   There is no reason to be concerned about multicollinearity.

   (b) $\hat{y} = -11.9 + 9.0765x_1 + 4.4676x_2$
   $+ 1.488x_3 + 3.5202x_4$

   (c) The test statistic is $F_0 = 1157.97$. The P-value is P-value $< 0.0001 < \alpha = 0.05$, so reject the null hypothesis and conclude that at least one of the explanatory variables is linearly associated with calories.

   (d) $H_0: \beta_1 = 0$
   $H_1: \beta_1 \neq 0$

   The test statistic is $t_0 = 37.49$. The P-value is P-value $< 0.0001 < \alpha = 0.05$, so reject the null hypothesis and conclude that fat is linearly related to calories.

   $H_0: \beta_2 = 0$
   $H_1: \beta_2 \neq 0$

   The test statistic is $t_0 = 7.53$. The P-value is P-value $= 0.005 < \alpha = 0.05$, so reject the null hypothesis and conclude that protein is linearly related to calories.

   $H_0: \beta_3 = 0$
   $H_1: \beta_3 \neq 0$

   The test statistic is $t_0 = 1.34$. The P-value is P-value $= 0.273 > \alpha = 0.05$, so do not reject the null hypothesis and conclude that sugar is not linearly related to calories.

   $H_0: \beta_4 = 0$
   $H_1: \beta_4 \neq 0$

   The test statistic is $t_0 = 11.60$. The P-value is P-value $= 0.001 < \alpha = 0.05$, so reject the null hypothesis and conclude that carbs is linearly related to calories.

Remove the explanatory variable sugar from the model.

(e)

$$\hat{y} = -4.42 + 9.1099x_1 + 4.2937x_2 + 3.7672x_4$$

$H_0: \beta_1 = 0$
$H_1: \beta_1 \neq 0$

The test statistic is $t_0 = 34.57$. The P-value is P-value $< 0.0001 < \alpha = 0.05$, so reject the null hypothesis and conclude that fat is still linearly related to calories.

$H_0: \beta_2 = 0$
$H_1: \beta_2 \neq 0$

The test statistic is $t_0 = 6.77$. The P-value is P-value $= 0.005 < \alpha = 0.05$, so reject the null hypothesis and conclude that protein is still linearly related to calories.

$H_0: \beta_4 = 0$
$H_1: \beta_4 \neq 0$

The test statistic is $t_0 = 14.28$. The P-value is P-value $= 0.0001 < \alpha = 0.05$, so reject the null hypothesis and conclude that carbs is still linearly related to calories.

All of the remaining slope coefficients are significantly different from zero.

(f)

None of the residual plots show any patterns. The boxplot of the residuals does not show any outliers, and the normal probability plot shows that the residuals are approximately normally distributed.

(g) Coefficient of x_1: If the fat content increases by 1 gram, the expected number of calories will increase by 9.1099 calories, holding protein and carbs constant.

Coefficient of x_2: If the protein content increases by 1 gram, the expected number of calories will increase by 4.2936 calories, holding fat and carbs constant.

Coefficient of x_4: If the carbohydrate content increases by 1 gram, the expected number of calories will increase by 3.7672 calories, holding fat and protein constant.

(h) $R^2 = 0.999$

$$R_{adj}^2 = 1 - \left(\frac{8-1}{8-3-1}\right)(1-0.999)$$
$$= 0.998$$

99.9% of the variation in calories is explained by the least-squares regression model.

(i) 95% confidence interval for the mean calories: lower bound: 656.16; upper bound: 668.17. We are 95% confident that the mean calories for a fast-food cheeseburger with 38 g of fat, 29 g of protein, 11 g of sugar, and 52 g of carbohydrates is between 656.16 and 668.17 calories.

95% prediction interval: lower bound: 647.53, upper bound: 676.80. We are 95% confident that a randomly selected fast-food cheeseburger with 38 g of fat, 29 g of protein, 11 g of sugar, and 52 g of carbohydrates will have between 647.53 and 676.80 calories.

Case Study: Housing Boom

(a) Correlations: acerage, beds, baths, footage, age, rooms, price

	acerage	beds	baths	footage	age	rooms
beds	0.482					
baths	0.543	0.686				
footage	0.692	0.823	0.834			
age	-0.222	-0.446	-0.500	-0.455		
rooms	0.606	0.762	0.668	0.881	-0.267	
price	0.800	0.638	0.736	0.908	-0.259	0.854

Several correlations among the explanatory variables are larger than 0.7. Therefore, multicollinearity may be a problem.

(b) $\hat{y} = -122.145 + 210.815x_1 - 60.4023x_2 + 10.1833x_3 + 0.1611x_4 + 0.7894x_5 + 23.5666x_6$

(c) $H_0: \beta_i = 0$

$H_1:$ at least one of the $\beta_i \neq 0$.

The test statistic is $F_0 = 45.60$. The P-value is P-value $< 0.0001 < \alpha = 0.05$, so reject the null hypothesis and conclude that at least

one of the explanatory variables is linearly related to price.

(d) $H_0: \beta_1 = 0$
$H_1: \beta_1 \neq 0$

The test statistic is $t_0 = 3.23$. The P-value is P-value $= 0.004 < \alpha = 0.05$, so reject the null hypothesis and conclude that acres is linearly related to the sale price.

$H_0: \beta_2 = 0$
$H_1: \beta_2 \neq 0$

The test statistic is $t_0 = -2.50$. The P-value is P-value $= 0.021 < \alpha = 0.05$, so reject the null hypothesis and conclude that the number of bedrooms is linearly related to the sale price.

$H_0: \beta_3 = 0$
$H_1: \beta_3 \neq 0$

The test statistic is $t_0 = 0.49$. The P-value is P-value $= 0.632 > \alpha = 0.05$, so do not reject the null hypothesis and conclude that the number of bathrooms is not linearly related to the sale price.

$H_0: \beta_4 = 0$
$H_1: \beta_4 \neq 0$

The test statistic is $t_0 = 3.53$. The P-value is P-value $= 0.002 < \alpha = 0.05$, so reject the null hypothesis and conclude that the square footage is linearly related to the sale price.

$H_0: \beta_5 = 0$
$H_1: \beta_5 \neq 0$

The test statistic is $t_0 = 1.48$. The P-value is P-value $= 0.155 > \alpha = 0.05$, so do not reject the null hypothesis and conclude that the age of the house is not linearly related to the sale price.

$H_0: \beta_6 = 0$
$H_1: \beta_6 \neq 0$

The test statistic is $t_0 = 1.80$. The P-value is P-value $= 0.086 > \alpha = 0.05$, so do not reject the null hypothesis and conclude that the number of rooms is not linearly related to the sale price.

(e) Remove the number of bathrooms, the age of the house, and the number of rooms from the model.

$$\hat{y} = 29.1459 + 222.375x_1 - 58.4898x_2 + 0.2038x_4$$

(f)

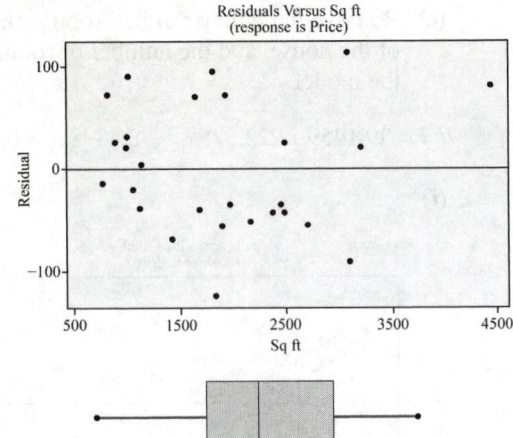

None of the residual plots show any patterns. The boxplot of the residuals does not show any outliers.

(g) $R^2 = 0.902$

$$R_{adj}^2 = 1 - \left(\frac{28-1}{28-3-1}\right)(1-0.902)$$
$$= 0.890$$

90.2% of the variation in selling price is explained by the model of best fit.

(h) $\hat{y} = 133.4$, or $133,400. This is more than the actual selling price of $99,900.

(i)

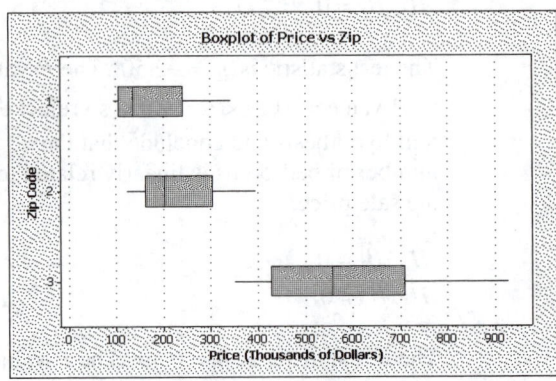

There is reason to believe that price varies by zip code. In particular, it appears that houses in the third zip code are more expensive than those in the first two zip codes.

(j) $\hat{y} = 91.4088 + 11.0067x_1 - 47.019x_2 + 17.0662x_3 + 0.1304x_4 + 0.1393x_5 + 23.1318x_6 - 180.178x_7 - 122.659x_8$

where $x_7 = \begin{cases} 1 & \text{1st zip code} \\ 0 & \text{otherwise} \end{cases}$ and $x_8 = \begin{cases} 1 & \text{2nd zip code} \\ 0 & \text{otherwise} \end{cases}$.

$H_0: \beta_i = 0$

$H_1:$ at least one of the $\beta_i \neq 0$.

The test statistic is $F_0 = 58.34$. The P-value is P-value $< 0.0001 < \alpha = 0.05$, so reject the null hypothesis and conclude that at least one of the explanatory variables is linearly related to price.

$H_0: \beta_1 = 0$
$H_1: \beta_1 \neq 0$

The test statistic is $t_0 = 0.136$. The P-value is P-value $= 0.893 > \alpha = 0.05$, so do not reject the null hypothesis and conclude that the acreage is not linearly related to the sale price.

$H_0: \beta_2 = 0$
$H_1: \beta_2 \neq 0$

The test statistic is $t_0 = -2.40$. The P-value is P-value $= 0.027 < \alpha = 0.05$, so reject the null hypothesis and conclude that the number of bedrooms is linearly related to the sale price.

$H_0: \beta_3 = 0$
$H_1: \beta_3 \neq 0$

The test statistic is $t_0 = 0.98$. The P-value is P-value $= 0.339 > \alpha = 0.05$, so do not reject the null hypothesis and conclude that the number of bathrooms is not linearly related to the sale price.

$H_0: \beta_4 = 0$
$H_1: \beta_4 \neq 0$

The test statistic is $t_0 = 3.397$. The P-value is P-value $= 0.003 < \alpha = 0.05$, so reject the null hypothesis and conclude that the square footage is linearly related to the sale price.

$H_0: \beta_5 = 0$
$H_1: \beta_5 \neq 0$

The test statistic is $t_0 = 0.268$. The P-value is P-value $= 0.791 > \alpha = 0.05$, so do not reject the null hypothesis and conclude that the age of the house is not linearly related to the sale price.

$H_0: \beta_6 = 0$
$H_1: \beta_6 \neq 0$

The test statistic is $t_0 = 2.27$. The P-value is P-value $= 0.035 < \alpha = 0.05$, so reject the null hypothesis and conclude that the number of rooms is linearly related to the sale price.

$H_0: \beta_7 = 0$
$H_1: \beta_7 \neq 0$

The test statistic is $t_0 = -3.82$. The P-value is P-value $= 0.0011 < \alpha = 0.05$, so reject the null hypothesis and conclude that whether the house is in the first zip code is linearly related to the sale price.

$H_0: \beta_8 = 0$
$H_1: \beta_8 \neq 0$

The test statistic is $t_0 = -2.53$. The P-value is P-value $= 0.021 < \alpha = 0.05$, so reject the null hypothesis and conclude that whether the house is in the second zip code is linearly related to the sale price.

Remove the acreage, the number of bathrooms, and the age of the house from the model.

$$\hat{y} = 123.503 - 48.8949x_2 + 0.1490x_4 + 22.1493x_6 - 183.772x_7 - 130.232x_8$$

None of the residual plots show any patterns. The boxplot of the residuals does not show any outliers.

$R^2 = 0.9587$

$$R_{adj}^2 = 1 - \left(\frac{28-1}{28-5-1} \right)(1 - 0.9587)$$
$$= 0.949$$

95.9% of the variation in sale price is explained by the least-squares regression model.

(k) $\hat{y} = 101.18$, or $101,180. This is a better prediction than the model that did not use indicator variables for location.

(l) Answers may vary.

(m) This model assumes that the house is in one of three zip codes. Further zip codes could be addressed by adding more indicator variables for each zip code.

(n) Answers may vary.

Chapter 15
Nonparametric Statistics

Section 15.1

1. Nonparametric statistics are inferential procedures that are not based on parameters. They do not require the population to follow a specific type of distribution. Parametric statistics test claims regarding parameters and often require that the distribution of the population follow a specific distribution.

3. The power of the test is the probability that the null hypothesis is rejected when the alternative hypothesis is true.

5. Disadvantages of nonparametric procedures include the following:

   1. They are often misused. That is, they are often used when more powerful parametric methods are appropriate.

   2. Nonparametric methods are less powerful than parametric procedures.

   3. Nonparametric methods are less efficient than parametric procedures.

Section 15.2

1. Run

3. (a) $n = 14$; $n_M = 6$; $n_F = 8$; $r = 6$

   (b) Since $n_M \leq 20$, $n_F \leq 20$, and $\alpha = 0.05$, Table XI should be used to find the critical values. The table gives a lower critical value of 3 and an upper critical value of 12.

   (c) Because the test statistic, $r = 6$, is not less than or equal to the lower critical value, 3, and is not greater than or equal to the upper critical value, 12, do not reject H_0. The sequence of data is random.

5. (a) $n = 19$; $n_Y = 9$; $n_N = 10$; $r = 5$

   (b) Since $n_Y \leq 20$, $n_N \leq 20$, and $\alpha = 0.05$, Table XI should be used to find the critical values. The table gives a lower critical value of 5 and an upper critical value of 16.

   (c) Because the test statistic, $r = 5$, is less than or equal to the lower critical value, 5, reject H_0. The sequence of data is not random.

7. $n = 20$, $n_F = 9$, $n_C = 11$, and $r = 12$. Since $n_F \leq 20$, $n_C \leq 20$, and $\alpha = 0.05$, Table XI should be used to find the critical values. The table gives a lower critical value of 6 and an upper critical value of 16. Since $6 < r < 16$, do not reject H_0. There is not enough evidence at the $\alpha = 0.05$ level of significance to indicate that Bumgarner's pitches are not random.

9. $n = 14$, $n_L = 6$, $n_O = 8$, and $r = 7$. Since $n_L \leq 20$, $n_O \leq 20$, and $\alpha = 0.05$, Table XI should be used to find the critical values. The table gives a lower critical value of 3 and an upper critical value of 12. Since $3 < r < 12$, do not reject H_0. There is not enough evidence at the $\alpha = 0.05$ level of significance to indicate that the arrival status is not random.

11. $n = 20$, $n_A = 14$, $n_R = 6$, and $r = 4$. Since $n_A \leq 20$, $n_R \leq 20$, and $\alpha = 0.05$, Table XI should be used to find the critical values. The table gives a lower critical value of 5 and an upper critical value of 13. Since $r \leq 5$, reject H_0. There is enough evidence at the $\alpha = 0.05$ level of significance to indicate that the way the machine overfills and underfills does not occur randomly.

13. $n = 45$, $n_P = 19$, $n_N = 26$, and $r = 19$. Since $n_N > 20$, the standard normal distribution should be used.

$$\mu_r = \frac{2(19)(26)}{45} + 1 = 22.956$$

$$\sigma_r = \sqrt{\frac{2(19)(26)[2(19)(26) - 45]}{45^2(45 - 1)}} = 3.234$$

$$z_0 = \frac{19 - 22.956}{3.234} = -1.22$$

The critical values at $\alpha = 0.05$ are $-z_{0.025} = -1.96$ and $z_{0.025} = 1.96$. Since the test statistic, $z_0 = -1.22$, does not lie within the critical region, do not reject H_0. There is not enough evidence at the $\alpha = 0.05$ level of significance to indicate that the stock price fluctuations do not behave in a random manner. (Minitab: P-value $= 0.221 > 0.05$.)

15. (a) AAABBABBBBBAA

   (b) $n = 12$, $n_A = 6$, $n_B = 6$, and $r = 5$.
Since $n_A \leq 20$, $n_B \leq 20$, and $\alpha = 0.05$, Table XI should be used to find the critical values. The table gives a lower critical value of 3 and an upper critical value of 11. Since $3 < r < 11$, do not reject H_0. There is not enough evidence at the $\alpha = 0.05$ level of significance to indicate that the residuals are not random.

17. Denote strengths above 75 with an A and those below 75 with a B to form a sequence.

   AABBBAAAAAABBBAAAABB

   $n = 20$, $n_A = 12$, $n_B = 8$, and $r = 6$. Since $n_A \leq 20$, $n_B \leq 20$, and $\alpha = 0.05$, Table XI should be used to find the critical values. The table gives a lower critical value of 6 and an upper critical value of 16. Since $r \leq 6$, reject H_0. There is enough evidence at the $\alpha = 0.05$ level of significance to indicate that the compression strengths are not random.

19. Random means that the sequence of outcomes follows no particular pattern. A run is a sequence of similar events, items, or symbols that is followed by an event, item, or symbol that is mutually exclusive from the first event, item, or symbol.

Section 15.3

1. $H_0 : M = 8$
$H_1 : M < 8$

Since the test is left-tailed, the test statistic is the number of plus signs, which is $k = 8$. Since $n \leq 25$, Table XII should be used to find the critical value. With $\alpha = 0.05$ and $n = 21$, the table gives a one-tailed critical value of 6. Since $k > 6$, do not reject H_0.

There is not sufficient evidence to support the hypothesis that the median is less than 8.

3. $H_0 : M = 100$
$H_1 : M \neq 100$

Since $n > 25$, Table V, the standard normal table, should be used to find the critical value. The table gives a two-tailed critical value of $-z_{0.025} = -1.96$. Since the test is two-tailed, k is the smaller of the number of minus signs and plus signs. So, use $k = 21$ and $n = 49$ to calculate the test statistic.

$$z_0 = \frac{(21 + 0.5) - \dfrac{49}{2}}{\dfrac{\sqrt{49}}{2}} = -0.86$$

Since $z_0 \geq -1.96$, do not reject H_0. There is not enough evidence at the $\alpha = 0.05$ level of significance to indicate that the median is different from 100.

5. $H_0 : M = 12$
$H_1 : M > 12$

First, convert the values into plus and minus signs. There are three values less than 12, one value equal to 12, and eleven values greater than 12. This results in 3 minus signs and 11 plus signs. Since the test is right-tailed, the test statistic is the number of minus signs, which is $k = 3$. Since $n \leq 25$, Table XII should be used to find the critical value. With $\alpha = 0.05$ and $n = 15$, the table gives a one-tailed critical value of 3. Since $k \leq 3$, reject H_0. There is sufficient evidence to support the hypothesis that the median is greater than 12.

7. $H_0 : M = 5.25$
$H_1 : M > 5.25$

First, convert the values into plus and minus signs. There are five values less than 5.25 and ten values greater than 5.25. This results in 5 minus signs and 10 plus signs. Since the test is right-tailed, the test statistic is the number of minus signs, which is $k = 5$. Since $n \leq 25$, Table XII should be used to find the critical value. With $\alpha = 0.05$ and $n = 15$, the table gives a one-tailed critical value of 3. Since $k > 3$, do not reject H_0. There is not sufficient evidence to support the hypothesis that the median is less than 5.25.

9. $H_0 : M = 57{,}283$
   $H_1 : M < 57{,}283$

   First, convert the values into plus and minus signs. There are eight values less than 57,283 and four values greater than 57,283. This results in 8 minus signs and 4 plus signs. Since the test is left-tailed, the test statistic is the number of plus signs, which is $k = 4$. Since $n \leq 25$, Table XII should be used to find the critical value. With $\alpha = 0.05$ and $n = 12$, the table gives a one-tailed critical value of 2. Since $k > 2$, do not reject H_0. There is not sufficient evidence to support the hypothesis that the median is less than 57,283.

11. $H_0 : M = 1000$
    $H_1 : M > 1000$

    First, convert the values into plus and minus signs. There are four values less than 1000 and ten values greater than 1000. This results in 4 minus signs and 10 plus signs. Since the test is right-tailed, the test statistic is the number of minus signs, which is $k = 4$. Since $n \leq 25$, Table XII should be used to find the critical value. With $\alpha = 0.05$ and $n = 14$, the table gives a one-tailed critical value of 3. Since $k > 3$, do not reject H_0. There is not sufficient evidence to support the hypothesis that the median is greater than 1000.

13. $H_0 : M = 5.25$
    $H_1 : M > 5.25$

    There were 5 minus signs and 10 plus signs, so use the binomial probability formula with $k = 5$, $n = 15$, and $p = 0.5$ to calculate the P-value.

    $$P\text{-value} = P(X \leq k) = P(X \leq 5)$$

    Using technology, we get P-value $= 0.1509$.

15. $H_0 : M = 57{,}283$
    $H_1 : M < 57{,}283$

    There were 8 minus signs and 4 plus signs, so use the binomial probability formula with $k = 4$, $n = 12$, and $p = 0.5$ to calculate the P-value.

$$P\text{-value} = P(X \leq k) = P(X \leq 4)$$

Using technology, we get P-value $= 0.1938$.

17. $H_0 : p = 0.5$
    $H_1 : p < 0.5$

    It was found that 38 of the 100 were women, so we would have 38 plus signs and 62 minus signs. So, use $k = 38$ and $n = 100$ to calculate the test statistic.

    $$z_0 = \frac{(38 + 0.5) - \dfrac{100}{2}}{\dfrac{\sqrt{100}}{2}} = -2.30$$

    Since $n > 25$, Table V, the standard normal table, should be used to find the critical value. The table gives a one-tailed critical value of $-z_{0.01} = -2.33$. Since $z_0 \geq -2.33$, do not reject H_0. There is not enough evidence at the $\alpha = 0.01$ level of significance to indicate that the proportion is less than 50%.

19. Since the participants had to agree to be part of the survey, the sample may not be truly random, which is required to use the sign test. Note that answers may vary.

21. If the true median, M, is less than M_0, then we would expect the majority of observations to be less than M_0 (giving minus signs) and few observations to be greater than M_0 (giving plus signs). In this case, k, the number of plus signs, will be much smaller than $\frac{1}{2}n$, resulting in a large negative value of the test statistic. If the true median, M, is greater than M_0, then we would expect the majority of observations to be greater than M_0 (giving plus signs) and few observations to be less than M_0 (giving minus signs). In this case, k is the number of minus signs, and again it will be much smaller than $\frac{1}{2}n$, resulting in a large negative value of the test statistic. So, in either case, a large negative value of the test statistic is evidence in favor of the alternative hypothesis and against the null hypothesis.

23. If the true median, M, is not equal to M_0, then either $M > M_0$ or $M < M_0$. If $M < M_0$, then we would expect the majority of observations to be less than M_0 (giving minus signs) and few observations to be greater than M_0 (giving plus signs). In this case, k, the smaller of the number of plus signs and minus signs, will be the number of plus signs and will be relatively small. Similarly, if $M > M_0$, then we would expect the majority of observations to be greater than M_0 (giving plus signs) and few observations to be less than M_0 (giving minus signs). In this case, k, the smaller of the number of plus signs and minus signs, will be the number of minus signs and will be relatively small. In either case, a small value of k is evidence that $M \neq M_0$.

Section 15.4

1. If we compute the differences as Population 2 – Population 1, then the researcher believes that $M_D > 0$, where M_D is the median of the differences between the observations in the matched pairs data.

3. Since $H_1 : M_D > 0$, this is a right-tailed test and the test statistic is $T = |T_-| = 16$. The critical value for $n = 12$ and $\alpha = 0.05$ is 17 (from the table of critical values). Since $T < 17$, reject H_0. There is enough evidence to conclude that $M_D > 0$.

5. Since $H_1 : M_D < 0$, this is a left-tailed test and the test statistic is $T = T_+ = 33$. The critical value for $n = 15$ and $\alpha = 0.05$ is 30 (from the table of critical values). Since $T \geq 30$, do not reject H_0. There is not enough evidence to conclude that $M_D < 0$.

7. Since $H_1 : M_D \neq 0$, this is a two-tailed test and the test statistic is
$T = \min\left(T_+, |T_-|\right) = \min(50, 121) = 50$. The critical value for $n = 18$ and $\alpha/2 = 0.025$ is 40 (from the table of critical values). Since $T \geq 40$, do not reject H_0. There is not enough evidence to conclude that $M_D \neq 0$.

9. Since $H_1 : M_D > 0$, this is a right-tailed test and $T = |T_-| = 300$. The critical value for $n = 40$ and $\alpha = 0.05$ is $-z_\alpha = -z_{0.05} = -1.645$ (since $n > 30$). The test statistic is
$$z_0 = \frac{T - \dfrac{n(n+1)}{4}}{\sqrt{\dfrac{n(n+1)(2n+1)}{24}}} = \frac{300 - \dfrac{40(40+1)}{4}}{\sqrt{\dfrac{40(40+1)(2 \cdot 40+1)}{24}}}$$
$= -1.48 > -1.645$
so we do not reject H_0. There is not enough evidence to conclude that $M_D > 0$.

11. From the computations in the table below we get $T_+ = 12$, $|T_-| = |-33| = 33$, and $n = 9$.

Before	After	After – Before	Absolute Value	Signed Rank
23.5	19.75	–3.75	3.75	–9
18.5	19.25	0.75	0.75	4.5
21.5	21.75	0.25	0.25	1.5
24	22.5	–1.5	1.5	–7.5
25	25	0	0	0
19.75	19.5	–0.25	0.25	–1.5
35	34.25	–0.75	0.75	–4.5
36.5	35	–1.5	1.5	–7.5
52	51.5	–0.5	0.5	–3
30	31	1	1	6

Since $H_1 : M_D < 0$, this is a left-tailed test and the test statistic is $T = T_+ = 12$. The critical value for $n = 9$ and $\alpha = 0.05$ is 8 (from the table of critical values). Since $T \geq 8$, do not reject H_0. There is not sufficient evidence to indicate that the median waistline has decreased. The exercise regimen does not appear to be effective in reducing waistline.

13. From the computations in the table below we get $T_+ = 5$, $|T_-| = |-16| = 16$, and $n = 6$.

Blue	Red	Red – Blue	Absolute Value	Signed Rank
0.582	0.408	–0.174	0.174	–4
0.481	0.407	–0.074	0.074	–1
0.841	0.542	–0.299	0.299	–6
0.267	0.402	0.135	0.135	3
0.685	0.456	–0.229	0.229	–5
0.450	0.533	0.083	0.083	2

Since $H_1 : M_D \neq 0$, this is a two-tailed test and the test statistic is $T = \min(T_+, |T_-|) = \min(5, 16) = 5$. The critical value for $n = 6$ and $\alpha/2 = 0.025$ is 0 (from the table of critical values). Since $T \geq 0$, do not reject H_0. There is not sufficient evidence to indicate that the median reaction times are different for blue versus red.

15. From the computations in the table below we get $T_+ = 31$, $|T_-| = |-5| = 5$, and $n = 8$.

Initial	5 Yrs	5 Yrs – Initial	Absolute Value	Signed Rank
38	52	14	14	8
58	60	2	2	2.5
65	72	7	7	5
74	72	–2	2	–2.5
56	54	–2	2	–2.5
36	48	12	12	7
56	58	2	2	2.5
52	60	8	8	6

Since $H_1 : M_D > 0$, this is a right-tailed test and the test statistic is $T = |T_-| = 5$. The critical value for $n = 8$ and $\alpha = 0.05$ is 5 (from the table of critical values). Since $T \geq 5$, do not reject H_0. There is not enough evidence to indicate that the median clarity of the lake has increased.

17. From the computations in the table below we get $T_+ = 20$, $|T_-| = |-35| = 35$, and $n = 10$.

Avis	Hertz	Hertz – Avis	Absolute Value	Signed Rank
67.54	59.14	–8.40	8.40	–10
97.16	97.15	–0.01	0.01	–1
103.43	103.14	–0.29	0.29	–5
38.79	36.63	–2.16	2.16	–9
87.90	87.80	–0.10	0.10	–2
109.86	110.09	0.23	0.23	4
109.35	109.35	0	0	0
82.86	83.03	0.17	0.17	3
112.65	113.32	0.67	0.67	6
116.46	114.53	–1.93	1.93	–8
105.49	107.04	1.55	1.55	7

Since $H_1 : M_D > 0$, this is a right-tailed test and the test statistic is $T = |T_-| = 35$. The critical value for $n = 10$ and $\alpha = 0.05$ is 10 (from the table of critical values). Since $T \geq 10$, do not reject H_0. There is not enough evidence to indicate that the median car rental cost for Avis is less than that for Hertz.

19. **(a)** $H_0 : M_D = 0$ and $H_1 : M_D \neq 0$.

(b) Since P-value $< 0.001 < 0.05 = \alpha$, we should reject H_0 and conclude that space flight affects the red blood cell count of rats.

21. Start by computing the difference between the length of stay for completers referred by employers and 107 days (the median for referrals from the criminal justice system). From the computations in the table below we get $T_+ = 8.5$, $|T_-| = |-57.5| = 57.5$, and $n = 11$.

Employer – Justice	Absolute Value	Signed Rank
–27	27	–11
1	1	1
–12	12	–7
0	0	0
–18	18	–9
–7	7	–4
–22	22	–10
–5	5	–3
8	8	5.5
2	2	2
–8	8	–5.5
–13	13	–8

Since $H_1 : M_D \neq 0$, this is a two-tailed test and the test statistic is
$T = \min(T_+, |T_-|) = \min(8.5, 57.5) = 8.5$. The critical value for $n = 11$ and $\alpha/2 = 0.025$ is 10 (from the table of critical values). Since $T < 10$, reject H_0. There is enough evidence at the $\alpha = 0.05$ level of significance to conclude that the median length of stay is different for employer referrals than those referred by the criminal justice system.

23. Her results are suspect because the data were not randomly selected.

Section 15.5

1. Since $H_1 : M_x \neq M_y$, this is a two-tailed test. As $n_1 = 12$, $n_2 = 15$, and $\alpha = 0.05$, the critical values are $w_{0.025} = 50$ (from the table of critical values) and
$w_{0.975} = n_1 n_2 - w_{0.025} = 12 \cdot 15 - 50 = 130$. The test statistic is
$T = S - \dfrac{n_1(n_1 + 1)}{2} = 170 - \dfrac{12 \cdot 13}{2} = 92$, which is not in the critical region, so we do not reject

H_0. There is not enough evidence to indicate that $M_x \neq M_y$.

3. Since $H_1 : M_x < M_y$, this is a left-tailed test. As $n_1 = 18$, $n_2 = 16$, and $\alpha = 0.05$, the critical value is $w_{0.05} = 96$ (from the table of critical values). The test statistic is
$T = S - \dfrac{n_1(n_1 + 1)}{2} = 210 - \dfrac{18 \cdot 19}{2} = 39$, which is less than $w_{0.05} = 96$, so we reject H_0. There is enough evidence to conclude that $M_x < M_y$.

5. Since $H_1 : M_x > M_y$, this is a right-tailed test. As $n_1 = 15$, $n_2 = 15$, and $\alpha = 0.05$, we first look up $w_{0.05} = 73$ (from the table of critical values). Then, the critical value is
$w_{0.95} = n_1 n_2 - w_{0.05} = 15 \cdot 15 - 73 = 152$. The test statistic is
$T = S - \dfrac{n_1(n_1 + 1)}{2} = 250 - \dfrac{15 \cdot 16}{2} = 130$, which is not greater than $w_{0.95} = 152$, so we do not reject H_0. There is not enough evidence to indicate that $M_x > M_y$.

7. Since $H_1 : M_x \neq M_y$, this is a two-tailed test. As $n_1 = 22 > 20$, $n_2 = 25 > 20$, and $\alpha = 0.05$, the critical values are $\pm z_{\alpha/2} = \pm z_{0.025} = \pm 1.96$. We first calculate
$T = S - \dfrac{n_1(n_1 + 1)}{2} = 590 - \dfrac{22 \cdot 23}{2} = 337$. Then, the test statistic is given by
$$z_0 = \frac{T - \dfrac{n_1 n_2}{2}}{\sqrt{\dfrac{n_1 n_2 (n_1 + n_2 + 1)}{12}}} = \frac{337 - \dfrac{22 \cdot 25}{2}}{\sqrt{\dfrac{22 \cdot 25 \cdot (22 + 25 + 1)}{12}}}$$
$= 1.32$

which is not in the critical region, so we do not reject H_0. There is not enough evidence to indicate that $M_x \neq M_y$.

9. From the computations in the table below we have $n_1 = 15$, $n_2 = 13$, and $S = 193.5$.

Women	Men	Women Ranks	Men Ranks
7	12	3.5	11.5
6	20	2	28
10	13	8.5	14.5
17	14	24.5	17.5
9	9	6.5	6.5
13	13	14.5	14.5
14	10	17.5	8.5
15	18	20.5	26.5
18	15	26.5	20.5
15	8	20.5	5
11	13	10	14.5
12	17	11.5	24.5
16	15	23	20.5
7		3.5	
3		1	
	Sum =	193.5	

Since $H_0 : M_W \neq M_M$ and $H_1 : M_W \neq M_M$, this is a two-tailed test. The critical values are $w_{0.025} = 55$ (from the table of critical values) and $w_{0.975} = n_1 n_2 - w_{0.025} = 15 \cdot 13 - 55 = 140$.

The test statistic is
$$T = S - \frac{n_1(n_1 + 1)}{2} = 193.5 - \frac{15 \cdot 16}{2} = 73.5,$$
which is not in the critical region, so we do not reject H_0. There is not enough evidence to indicate that the median sociability score is different for women and men.

11. From the computations in the table below we have $n_1 = 8$, $n_2 = 8$, and $S = 74.5$.

Carpeted	Uncarpeted	Carpeted Ranks	Uncarpeted Ranks
11.8	12.1	10	12
8.2	8.3	4	5
7.1	3.8	2	1
13.0	7.2	13	3
10.8	12.0	8	11
10.1	11.1	6.5	9
14.6	10.1	16	6.5
14.0	13.7	15	14
	Sum =	74.5	

Since $H_1 : M_x > M_y$, this is a right-tailed test. From the table of critical values, $w_{0.05} = 16$, and so the critical value is $w_{0.95} = n_1 n_2 - w_{0.05} = 8 \cdot 8 - 16 = 48$. The test statistic is
$$T = S - \frac{n_1(n_1 + 1)}{2} = 74.5 - \frac{8 \cdot 9}{2} = 38.5, \text{ which}$$
is not greater than $w_{0.95} = 48$, so we do not reject H_0. There is not enough evidence to conclude that the median amount of bacteria in carpeted rooms is greater than that in uncarpeted rooms.

13. From the computations in the table below we have $n_1 = 22$, $n_2 = 20$, and $S = 609$.

Lincoln	Clarendon	Lincoln Ranks	Clarendon Ranks
0.11	0.06	12	2
0.09	0.09	7.5	7.5
0.21	0.14	26.5	18
0.09	0.16	7.5	20.5
0.07	0.03	3	1
0.30	0.12	31.5	14
0.41	0.29	36.5	30
0.33	0.10	33.5	10.5
0.20	0.41	24.5	36.5
0.22	0.08	28	4.5
0.34	0.14	35	18
0.46	0.14	39	18
0.19	0.12	22.5	14
0.67	0.08	41	4.5
0.75	0.09	42	7.5
0.19	0.10	22.5	10.5
0.30	0.21	31.5	26.5
0.33	0.16	33.5	20.5
0.20	0.13	24.5	16
0.42	0.12	38	14
0.25		29	
0.47		40	
	Sum =	609	

Since $H_1 : M_x \neq M_y$, this is a two-tailed test. As $n_1 = 22 > 20$ and $\alpha = 0.05$, the critical values are $\pm z_{\alpha/2} = \pm z_{0.025} = \pm 1.96$. We first find $T = S - \dfrac{n_1(n_1+1)}{2} = 609 - \dfrac{22 \cdot 23}{2} = 356$. Then, the test statistic is given by

$$z_0 = \frac{T - \dfrac{n_1 n_2}{2}}{\sqrt{\dfrac{n_1 n_2 (n_1 + n_2 + 1)}{12}}}$$

$$= \frac{356 - \dfrac{22 \cdot 20}{2}}{\sqrt{\dfrac{22 \cdot 20 \cdot (22 + 20 + 1)}{12}}} = 3.43$$

which is in the critical region, so we reject H_0. There is enough evidence to conclude that the median calcium level in rainwater in Lincoln County, Nebraska, differs from the median level in Clarendon County, South Carolina.

15. (a) From the computations in the table below we have $n_1 = 15$, $n_2 = 15$, and $S = 262$.

Prof. A	Prof. B	Prof. A's Ranks	Prof. B's Ranks
A	A	3	3
A	A	3	3
B	A	10	3
B	B	10	10
B	B	10	10
C	B	19.5	10
C	B	19.5	10
C	B	19.5	10
C	B	19.5	10
C	C	19.5	19.5
C	C	19.5	19.5
D	C	26.5	19.5
D	C	26.5	19.5
D	D	26.5	26.5
F	F	29.5	29.5
	Sum =	262	

(b) This is a two-tailed test so $H_1 : M_x \neq M_y$, Assuming that $\alpha = 0.05$, the critical values are $w_{0.025} = 65$ (from the table of critical values) and $w_{0.975} = n_1 n_2 - w_{0.025} = 15 \cdot 15 - 65 = 160$. The test statistic is

$$T = S - \frac{n_1(n_1 + 1)}{2} = 262 - \frac{15 \cdot 16}{2} = 142,$$

which is not in the critical region, so we do not reject H_0. There is not enough evidence to support the claim that the median grade differs in the two classes.

17. If the two medians are the same, then the sum of the ranks of the observations from population X should be close to the sum of the ranks of the observations from population Y. The test statistic essentially measures the size of the sum of ranks for the sample from population X. If this sum is unusually small, then that is evidence that $M_x < M_y$ and if this

sum is unusually large then that is evidence that $M_x > M_y$.

19. The Wilcoxon signed-ranks test is used for dependent samples (matched pairs) while the Mann-Whitney test is used for independent samples.

Section 15.6

1. (a) The scatter diagram appears to show a positive linear association.

(b) The table shows the data and their ranks:

X	Y	Rank of X	Rank of Y	d_i	d_i^2
2	1.4	1	1	0	0
4	1.8	2	2	0	0
8	2.1	3.5	3	0.5	0.25
8	2.3	3.5	4	−0.5	0.25
9	2.6	5	5	0	0

$$\sum d_i^2 = 0 + 0 + 0.25 + 0.25 + 0 = 0.5$$

$$r_s = 1 - \frac{6 \sum d_i^2}{n(n^2 - 1)} = 1 - \frac{6(0.5)}{5(5^2 - 1)} = 1 - \frac{3}{120} = 0.975$$

(c) With $n = 5$ and $\alpha = 0.05$ (two tails), the critical values are ± 1.000. Since the test statistic, $r_s = 0.975$, is not greater than 1.000, we do not reject H_0. Note that when $n = 5$, this test is too insensitive to detect an association at the $\alpha = 0.05$ level. However at the $\alpha = 0.10$ level of significance, the critical values are ± 0.900 and since $r_s = 0.975 > 0.900$ we would reject H_0 and conclude that there is an association between X and Y.

3. (a) The scatter diagram appears to show a negative linear association.

(b) The table shows the data and their ranks:

X	Y	Rank of X	Rank of Y	d_i	d_i^2
0	5.8	1	6	−5	25
2	5.7	2	5	−3	9
3	5.2	3	4	−1	1
5	2.8	4	3	1	1
6	1.9	5.5	1	4.5	20.25
6	2.2	5.5	2	3.5	12.25

$$\sum d_i^2 = 25 + 9 + 1 + 1 + 20.25 + 12.25 = 68.5$$

$$r_s = 1 - \frac{6\sum d_i^2}{n(n^2 - 1)} = 1 - \frac{6(68.5)}{6(6^2 - 1)} = 1 - \frac{411}{210} \approx -0.957 \quad [\text{Tech: } r_s = -0.986]$$

Note: the text formula and technology values are not identical when there are ties among the ranks and the formula is only approximate.

(c) With $n = 6$ and $\alpha = 0.05$ (two tails), the critical values are ± 0.886. Since the test statistic is less than -0.886, we reject H_0 and conclude that there is an association between X and Y.

5. (a) The table shows the data and their ranks:

% with Bachelor's	Per Capita Income ($)	Rank of X	Rank of Y	d_i	d_i^2
18.8	31,946	1	1	0	0
29.6	42,325	9	9	0	0
24.3	36,751	4	7	−3	9
20.3	35,507	2	3	−1	1
25.4	36,745	6	6	0	0
27.1	38,081	7	8	−1	1
24.1	35,381	3	2	−1	1
28.1	35,667	8	4	4	16
25.3	36,484	5	5	0	0

$$\sum d_i^2 = 0 + 0 + 9 + 1 + 0 + 1 + 1 + 16 + 0 = 28$$

$$r_s = 1 - \frac{6\sum d_i^2}{n(n^2 - 1)} = 1 - \frac{6(28)}{9(9^2 - 1)} \approx 0.767$$

As we are testing for a positive association, this is a right-tailed test and with $n = 9$ and $\alpha = 0.05$ (one tail). The critical value is 0.600. Since the test statistic is greater than 0.600, we reject H_0. There is enough evidence to conclude that per capita personal income and percentage of the population with at least a bachelor's degree are positively associated.

(b) The scatter diagram appears to show a positive linear association, which supports our conclusion.

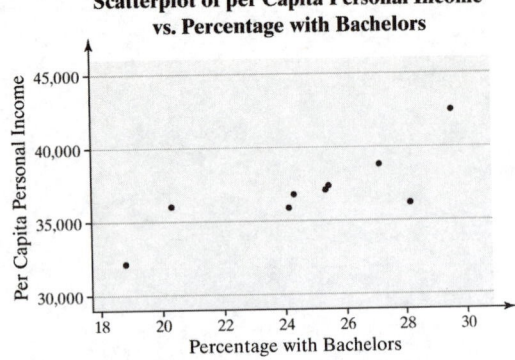

Scatterplot of per Capita Personal Income vs. Percentage with Bachelors

7. (a) The table shows the data and their ranks:

Per Capita Income ($)	Birth Rate	Rank of X	Rank of Y	d_i	d_i^2
39,817	14.1	6	4.5	1.5	2.25
48,285	13.9	8	3	5	25
66,000	15.1	9	8	1	1
43,874	14.1	7	4.5	2.5	6.25
32,219	12.1	2	1	1	1
34,453	14.5	4	7	−3	9
31,799	14.3	1	6	−5	25
33,786	15.8	3	9	−6	36
37,780	13.1	5	2	3	9

$$\sum d_i^2 = 2.25 + 25 + 1 + 6.25 + 1 + 9 + 25 + 36 + 9 = 114.5$$

$$r_s = 1 - \frac{6\sum d_i^2}{n(n^2 - 1)} = 1 - \frac{6(114.5)}{9(9^2 - 1)} = 1 - \frac{519}{720} \approx 0.046 \text{ [Tech: 0.042]}$$

As we are testing for a negative association, this is a left-tailed test and with $n = 9$ and $\alpha = 0.05$ (one tail), the critical value is −0.600. Our test statistic is positive so we do not reject H_0. There is not evidence to conclude that per capita personal income and birth rate are negatively associated.

(b) The scatter diagram appears show very little association, which supports our conclusion.

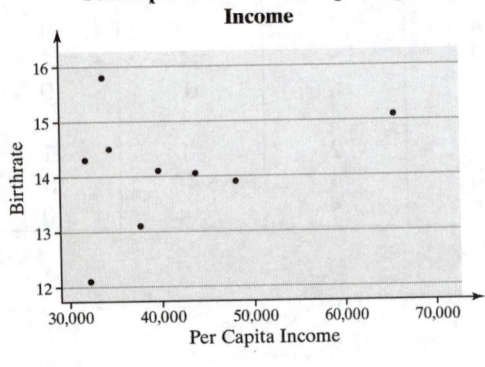

Scatterplot of Birthrate vs. per Capita Income

9. (a) The table shows the data and their ranks:

Family Income ($1000)	Commute Time (min)	Rank of X	Rank of Y	d_i	d_i^2
83	27.1	2.5	5	−2.5	6.25
82	17.0	1	1	0	0
86	26.9	4	4	0	0
90	26.0	5	3	2	4
101	32.1	6	6	0	0
83	22.6	2.5	2	0.5	0.25

$$\sum d_i^2 = 6.25 + 0 + 0 + 4 + 0 + 0.25 = 10.5$$

$$r_s = 1 - \frac{6 \sum d_i^2}{n(n^2 - 1)} = 1 - \frac{6(10.5)}{6(6^2 - 1)} = 1 - \frac{63}{210} = 0.7 \ \text{[Tech: 0.696]}$$

As we are testing for a positive association, this is a right-tailed test and with $n = 6$ and $\alpha = 0.10$ (one tail), the critical value is 0.657. Since the test statistic is greater than 0.657, we reject H_0. There is enough evidence to conclude that family income cost and commute time are positively associated.

(b) The scatter diagram perhaps shows a positive association, which supports our conclusion.

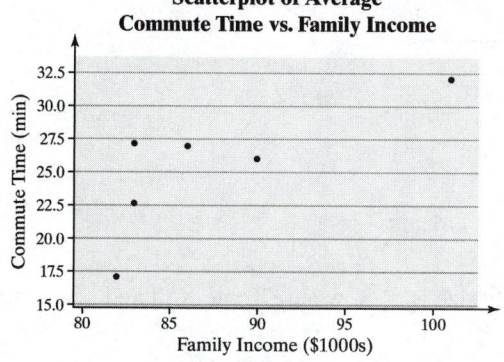

11. The table shows the data and their ranks:

X	Y	Rank of X	Rank of Y	$d = Y - X$	d^2
2	5	1	1	0	0
3	8	2	2	0	0
4	11	3	3	0	0
5	14	4	4	0	0
6	17	5	5	0	0
				$\sum d^2 =$	0

$r_s = 1.000$, the correlation coefficient of the ranks. X and Y are perfectly positively associated as we can see from the scatter diagram:

13. The Spearman rank-correlation test can be performed on qualitative data if the requirements are met and if the data can be ranked. For example, if a restaurant price can be ranked as expensive, moderate, or inexpensive, and the quality of the food can be ranked as good, fair, or poor, then the Spearman rank-correlation test is appropriate. Since Spearman's rank correlation test depends only on the ranks of the data values, as long as the data is ordinal, and can be ranked, we can perform this test to determine if there is an association between the variables.

Section 15.7

1. **(a)** The tables shows the ranks and the computations based on those ranks:

	X Ranks	Y Ranks	Z Ranks
	6.5	10	4.5
	1.5	12	8
	11	3	1.5
	4.5	6.5	9
R_i	23.5	31.5	23
R_i^2	552.25	992.25	529
n_i	4	4	4

The test statistic is calculated as:

$$H = \frac{12}{N(N+1)}\left[\frac{R_1^2}{n_1}+\cdots+\frac{R_k^2}{n_k}\right]-3(N+1)$$

$$= \frac{12}{12(12+1)}\left[\frac{552.25}{4}+\frac{992.25}{4}+\frac{529}{4}\right]-3(12+1)=0.875$$

(b) Since there are three populations and each sample size is ≤ 5, we use Table XVI and get a critical value of 5.6923 (corresponding to $\alpha = 0.049$).

(c) The test statistic is not greater than the critical value so we do not reject H_0. There is not enough evidence to indicate that the distributions are different.

3. **(a)** H_0: the distributions of the populations are the same.

H_1: the distributions of the populations are not the same.

(b) The calculations are summarized in the table:

	Monday	Tuesday	Wednesday	Thursday	Friday
n_i	8	8	8	8	8
R_i	48	226	144	194.5	207.5
R_i^2	2,304	51,076	20,736	37,830.25	43,056.25

$$H = \frac{12}{N(N+1)}\left[\frac{R_1^2}{n_1}+\cdots+\frac{R_k^2}{n_k}\right]-3(N+1)$$

$$= \frac{12}{40(40+1)}\left[\frac{2,304}{8}+\frac{51,076}{8}+\frac{20,736}{8}+\frac{37,830.25}{8}+\frac{43,056.25}{8}\right]-3(40+1)$$

$$= 18.771 \text{ [Tech: 18.772]}$$

(c) Since we are comparing more than 3 populations, we get the critical value from the χ^2 tables. With df $= k - 1 = 5 - 1 = 4$ we find $\chi^2_{0.05} = 9.488$.

(d) Since the test statistic is greater than the critical value, we reject H_0. There is enough evidence to conclude that the distribution of births varies by day of the week.

(e) Monday births clearly differ from births on other days of the week.

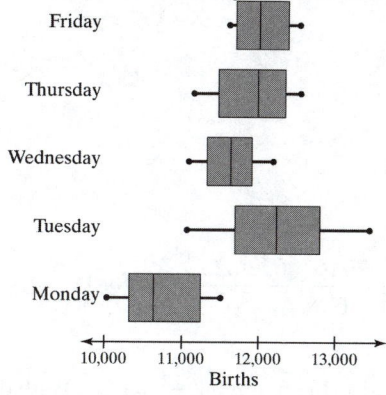

5. The table shows the ranks for the data and the calculations based on those ranks:

	Sludge plot	Spring disk	No till	Spring chisel	Great Lakes bt
	1.5	23.5	18	18	10
	6.5	18	3.5	23.5	23.5
	26.5	26.5	13.5	3.5	6.5
	18	30	23.5	10	18
	10	28.5	1.5	21	13.5
	6.5	28.5	13.5	13.5	6.5
R_i	69	155	73.5	89.5	78
R_i^2	4761	24025	5402.25	8010.25	6084
n_i	6	6	6	6	6

$$H = \frac{12}{N(N+1)}\left[\frac{R_1^2}{n_1} + \cdots + \frac{R_k^2}{n_k}\right] - 3(N+1)$$

$$= \frac{12}{30(30+1)}\left[\frac{4761}{6} + \frac{24{,}025}{6} + \frac{5402.25}{6} + \frac{8010.25}{6} + \frac{6084}{6}\right] - 3(30+1)$$

$$= 10.833 \text{ [Tech, adjusted for ties: } 10.975]$$

Since we are comparing more than 3 populations, we get the critical value from the χ^2 tables. With df $= k - 1 = 5 - 1 = 4$ we find $\chi^2_{0.05} = 9.488$. Since the test statistic is greater than the critical value, we reject H_0. There is enough evidence to support the conclusion that the distribution of corn plant yield varies by plot type.

7. The table shows the ranks for the data and the calculations based on those ranks:

	Simple	Go/No-Go	Choice
	7	16	15
	11.5	3	11.5
	9	6	13
	4	17	14
	5	10	8
	1	2	18
R_i	37.5	54	79.5
R_i^2	1406.25	2916	6320.25
n_i	6	6	6

$$H = \frac{12}{N(N+1)}\left[\frac{R_1^2}{n_1}+\cdots+\frac{R_k^2}{n_k}\right] - 3(N+1) = \frac{12}{18(18+1)}\left[\frac{1406.25}{6}+\frac{2916}{6}+\frac{6320.25}{6}\right] - 3(18+1)$$

$$= 5.237 \text{ [Tech, adjusted for ties: 5.242]}$$

Since at least one sample size is greater than 5, we get the critical value from the χ^2 tables. With df = $k-1$ = $3-1=2$ we find $\chi_{0.01}^2 = 9.210$. Since the test statistic is not greater than the critical value, we do not reject H_0. There is not enough evidence to indicate that the distribution of reaction time varies by stimulus.

9. The table shows the ranks for the data and the calculations based on those ranks:

	Large Family	Passenger Vans	Midsize Utility
	12.5	6	6
	20	3.5	18.5
	3.5	2	17
	8.5	8.5	11
	15	1	15
	15	12.5	21
	10	18.5	6
R_i	84.5	52	94.5
R_i^2	7140.25	2704	8930.25
n_i	7	7	7

$$H = \frac{12}{N(N+1)}\left[\frac{R_1^2}{n_1}+\cdots+\frac{R_k^2}{n_k}\right] - 3(N+1) = \frac{12}{21(21+1)}\left[\frac{7140.25}{7}+\frac{2704}{7}+\frac{8930.25}{7}\right] - 3(21+1)$$

$$= 3.664 \text{ [Tech, adjusted for ties: 3.693]}$$

Since at least one sample size is greater than 5, we get the critical value from the χ^2 tables. With df = $k-1=3-1=2$ we find $\chi_{0.01}^2 = 9.210$. Since the test statistic is not greater than the critical value, we do not reject H_0. There is not enough evidence to support the conclusion that the distribution of chest compression varies by vehicle category.

Chapter 15 Review Exercises

1. For the given data, $n = 25$, $n_1 = 11$ (number of E's), $n_2 = 14$ (number of W's), and $r = 11$. Since $n_1 \le 20$ and $n_2 \le 20$, the critical values from the table of critical values are 8 and 19. The test statistic is $r = 11$, which is not in the critical region, so we do not reject H_0. There is not enough evidence to conclude that the sequence is not random. It appears that the winning division occurs randomly.

2. We first convert the data to plus and minus signs (and zeros) by comparing the data values to $M_0 = 15$. This gives: $+ + + + + - + + + + + - + - + - + - - -$. There are 13 plus signs and 7 minus signs. Since $H_1 : M > 15$, this is a right-tailed test and $k =$ the number of minus signs $= 7$. We have $n = 20 \le 25$ and so the critical value, from the table of critical values, is 5. Since $k > 5$, we do not reject H_0. There is not enough evidence to indicate that the median number of hours students talk on the phone per week is greater than 15.

3. From the computations in the table below we get $T_+ = 32$, $|T_-| = |-23| = 23$, and $n = 10$.

Height (inches)	Arm Span (inches)	Arm Span – Height	Absolute Value	Signed Rank
59.5	62	2.5	2.5	6
69	65.5	–3.5	3.5	–9.5
77	76	–1	1	–3
59.5	63	3.5	3.5	9.5
74.5	74	–0.5	0.5	–1.5
63	66	3	3	7.5
61.5	61	–0.5	0.5	–1.5
67.5	69	1.5	1.5	4
73	70	–3	3	–7.5
69	71	2	2	5

$T_+ = 32$ and $T_- = -23$. Since $H_1 : M_D \ne 0$, this is a two-tailed test with test statistic $T = \min\left(T_+, |T_-|\right) = \min\left(32, 23\right) = 23$. The critical value for $n = 10$ and $\alpha / 2 = 0.025$ is 8 (from the table of critical values). Since $T > 8$, we do not reject H_0. There is not enough evidence to conclude that an individual's height and arm span are different.

4. From the computations in the table below we have $n_1 = 16$, $n_2 = 12$, and $S = 216$.

Female	Male	Female Ranks	Male Ranks
0.474	0.541	6	10
0.743	0.659	21	17
0.538	1.050	9	28
0.436	0.880	4	25
0.480	0.577	7	13
0.531	0.752	8	22
0.398	0.849	3	24
0.561	0.544	12	11
0.633	0.464	16	5
0.596	0.675	14	18
0.831	0.626	23	15
0.725	0.393	20	2
0.887		26	
0.905		27	
0.711		19	
0.338		1	
	Sum =	216	

Since $H_1 : M_x \neq M_y$, this is a two-tailed test. The critical values are $w_{0.005} = 42$ (from the table of critical values) and $w_{0.995} = n_1 n_2 - w_{0.005} = 16 \cdot 12 - 42 = 150$. The test statistic is $T = S - \dfrac{n_1(n_1+1)}{2} = 216 - \dfrac{16 \cdot 17}{2} = 80$ which is not in the critical region so we do not reject H_0. There is not enough evidence to indicate that the median reaction time of males differs from that of females.

5. The table shows the data and their ranks:

Engine Displacement	City mpg	Rank of X	Rank of Y	d_i	d_i^2
3.5	16	9	2.5	6.5	42.25
3.8	17	11.5	4.5	7	49
4.6	15	14.5	1	13.5	182.25
2.2	24	2.5	14	−11.5	132.25
3.5	18	9	7.5	1.5	2.25
2.4	22	5	11.5	−6.5	42.25
2.4	21	5	10	−5	25
2.7	18	7	7.5	−0.5	0.25
3.5	18	9	7.5	1.5	2.25
2.0	24	1	14	−13	169
4.0	17	13	4.5	8.5	72.25
4.6	16	14.5	2.5	12	144
3.8	18	11.5	7.5	4	16
2.2	24	2.5	14	−11.5	132.25
2.4	22	5	11.5	−6.5	42.25

$$\sum d_i^2 = 42.25 + 49 + 182.25 + \ldots + 132.25 + 42.25 = 1053.5$$

$$r_s = 1 - \frac{6 \sum d_i^2}{n(n^2 - 1)} = 1 - \frac{6(1053.5)}{15(15^2 - 1)} = 1 - \frac{6321}{3360} \approx -0.881 \quad [\text{Tech: } -0.930]$$

As we are testing for a negative association, this is a left-tailed test and with $n = 15$ and $\alpha = 0.05$ (one tail). The critical value is -0.446. Since the test statistic is less than -0.446, we reject H_0. There is enough evidence to conclude that engine displacement and fuel economy are negatively associated.

6. (a) H_0 : the distributions of the populations are the same.

   H_1 : the distributions of the populations are not the same.

   (b) The calculations are summarized in the table:

	Organic	Mineral	Surface
n_i	31	47	44
R_i	2,355.5	2,119.5	3,028.0
R_i^2	5,548,380	4,492,280	9,168,784

$$H = \frac{12}{N(N+1)} \left[\frac{R_1^2}{n_1} + \cdots + \frac{R_k^2}{n_k} \right] - 3(N+1)$$

$$= \frac{12}{122(122+1)} \left[\frac{5,548,380}{31} + \frac{4,492,280}{47} + \frac{9,168,784}{44} \right] - 3(122+1)$$

$$= 17.199 \ [\text{Tech, adjusted for ties: } 17.200]$$

   (c) Since at least one sample size is greater than 5, we get the critical value from the χ^2 tables. With df $= k - 1 = 3 - 1 = 2$ we find $\chi_{0.05}^2 = 5.991$.

   (d) Since the test statistic is greater than the critical value, we reject H_0. There is enough evidence to support the conclusion that the distribution of concentration of dissolved organic carbon varies by collection area.

   (e) The distributions do appear to differ.

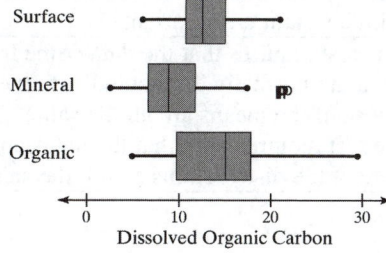

7. In general, parametric procedures are based on some assumptions about the distribution of the population that is being examined, and typically involve the estimation of a population parameter, such as the mean or variance of the population. Nonparametric methods, perhaps better called distribution-free methods, make no assumptions about the underlying population distributions and usually do not involve estimating any population parameters, although they may be used to perform inference about population parameters such as the population median. Nonparametric tests generally have fewer assumptions, some have simple computations and some can be applied to qualitative data such as rankings. However, they are typically also less powerful and less efficient than parametric tests. The tests are tabulated below with some of the differences.

Nonparametric Test	Parametric Test	Differences
Runs test for randomness	None	
One-sample sign test	z-Test or t-Test	The nonparametric test is a test of the median and the parametric test is a test of the mean. The parametric test requires that data come from a normal distribution or the sample size be large.
Wilcoxon Matched-Pairs Signed-Ranks Test	Paired t-Test for Dependent Samples	The nonparametric test is a test of the medians and the parametric test is a test of the means. The parametric test requires that the differences be normally distributed or the sample size be large. The nonparametric test requires that the differences have a symmetric distribution.
Mann-Whitney Test	Welch's t-Test for Independent Samples	The nonparametric test is a test of the medians and the parametric test is a test of the means. The parametric test requires that the differences be normally distributed or the sample sizes be large. The nonparametric test requires that the distributions of the two populations be the same, except possibly for the medians.
Spearman's Rank-Correlation Test	Hypothesis Test of the Significance of β_1	The nonparametric test is only for an association and so can also be used on ordinal data whereas the parametric test is only applicable to quantitative data. The parametric test requires that the residuals be normally distributed with constant error variance.
Kruskal-Wallis Test	F-test for One-Way ANOVA	The parametric test requires that the data come from populations that are normally distributed with the same variance and tests if the means are all the same. The nonparametric test requires only that the data can be ranked and tests if the distributions are all the same.

Chapter 15 Test

1. For the given data, $n = 25$, $n_1 = 11$ (number of E's), $n_2 = 14$ (number of W's), and $r = 12$. Since $n_1 \le 20$ and $n_2 \le 20$, the critical values are 8 and 19, from the table of critical values. The test statistic is $r = 12$, which is not in the critical region and so we do not reject H_0. There is not enough evidence to indicate that the sequence is not random. It appears that the winning division occurs randomly.

2. We first convert the data to plus and minus signs (and zeros) by comparing the data values to $M_0 = 15$. This gives: $- + - - - - - - - - + - + + - - - - - 0$. There are 4 plus signs and 15 minus signs (and one zero). Since $H_1 : M < 15$, this is a left-tailed test and $k =$ the number of plus signs = 4. We have $n = 4 + 15 = 19 \le 25$ and so the critical value, from the table of critical values, is 5. Since $k \le 5$, we reject H_0. There is enough evidence to conclude that the median number of hours students study per week is less than 15.

3. From the computations in the table below we get $T_+ = 11.5$, $|T_-| = |-33.5| = 33.5$, and $n = 9$.

Before	After	After – Before	Absolute Value	Signed Rank
136	128	−8	8	−5
120	111	−9	9	−6.5
129	129	0	0	0
143	148	5	5	3.5
115	110	−5	5	−3.5
113	112	−1	1	−1.5
89	98	9	9	6.5
122	103	−19	19	−8.5
102	103	1	1	1.5
122	103	−19	19	−8.5

Since $H_1 : M_D < 0$, this is a left-tailed test with test statistic $T = T_+ = 11.5$. The critical value for $n = 9$ and $\alpha = 0.05$ is 8 (from the table of critical values). Since $T \ge 8$, we do not reject H_0. There is not enough evidence to indicate that the median pulse rate has decreased after the exercise program.

4. From the computations in the table below we have $n_1 = 12$, $n_2 = 14$ and $S = 205$.

Houston	Chicago	Houston Ranks	Chicago Ranks
4.69	4.4	17	6.5
5.22	4.4	25.5	6.5
5.1	4.69	23	17
4.76	4.75	21	20
5.22	4.22	25.5	2
4.25	4.63	3	13
4.46	4.64	9	14
5.14	4.45	24	8
4.93	4.54	22	12
4.11	4.49	1	10
4.65	4.35	15	4
4.71	4.36	19	5
	4.69		17
	4.52		11
Sum =		205	146

Since $H_1 : M_H > M_C$, this is a right-tailed test. We first look up $w_{0.05} = 52$ (from the table of critical values) and then the critical value is $w_{0.95} = n_1 n_2 - w_{0.05} = 12 \cdot 14 - 52 = 116$. The test statistic is

$$T = S - \frac{n_1(n_1 + 1)}{2} = 205 - \frac{12 \cdot 13}{2} = 127 \text{ which}$$

is greater than $w_{0.95} = 116$ and so we reject H_0. There is enough evidence to indicate that the median pH of rain near Houston is greater than that of rain near Chicago.

5. The table shows the data and their ranks:

Temp.	Chirps per sec.	Rank of X	Rank of Y	d_i	d_i^2
88.6	20.0	14	15	−1	1
93.3	19.8	15	14	1	1
80.6	17.1	7.5	10.5	−3	9
69.7	14.7	2	2	0	0
69.4	15.4	1	4	−3	9
79.6	15.0	6	3	3	9
80.6	16.0	7.5	6.5	1	1
76.3	14.4	5	1	4	16
71.6	16.0	3	6.5	−3.5	12.25
84.3	18.4	13	13	0	0
75.2	15.5	4	5	−1	1
82.0	17.1	9	10.5	−1.5	2.25
83.3	16.2	11	8	3	9
82.6	17.2	10	12	−2	4
83.5	17.0	12	9	3	9

$$\sum d_i^2 = 1+1+9+...+4+9 = 83.5$$

$$r_s = 1 - \frac{6\sum d_i^2}{n(n^2-1)} = 1 - \frac{6(83.5)}{15(15^2-1)}$$

$$= 1 - \frac{501}{3360} \approx 0.851 \text{ [Tech: 0.850]}$$

As we are testing for an association, this is a two-tailed test and with $n = 15$ and $\alpha = 0.05$ (two tails). The critical values are ±0.521. Since the test statistic is greater than 0.521, we reject H_0. There is enough evidence to conclude that temperature and number of chirps per second are associated.

6. The table shows the ranks for the data and the calculations based on those ranks:

	Large Family	Passenger Vans	Midsize Utility
	18	16.5	4
	6.5	15	1.5
	12	8	13
	3	19	5
	14	21	1.5
	10	11	9
	16.5	20	6.5
R_i	80	110.5	40.5
R_i^2	6400	12,210.25	1640.25
n_i	7	7	7

$$H = \frac{12}{N(N+1)}\left[\frac{R_1^2}{n_1}+\cdots+\frac{R_k^2}{n_k}\right] - 3(N+1)$$

$$= \frac{12}{21(21+1)}\left[\frac{6,400}{7}+\frac{12,210.25}{7}+\frac{1640.25}{7}\right] - 3(21+1)$$

$$= 9.141 \text{ [Tech, adjusted for ties: 9.159]}$$

Since at least one sample size is greater than 5, we get the critical value from the χ^2 tables.

With df $= k-1 = 3-1 = 2$ we find $\chi_{0.05}^2 = 5.991$. Since the test statistic is greater than the critical value, we reject H_0. There is enough evidence to conclude that the distribution of femur force varies by vehicle category.

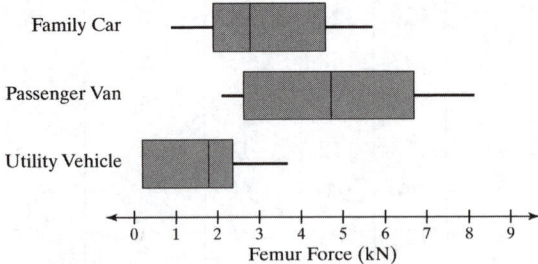

7. An efficiency of 0.713 means the nonparametric test would need a sample size of 1000 to achieve the same results as the corresponding parametric test with a sample size of 713.

Case Study: Evaluating Alabama's 1891 House Bill 504

1. Length of School Year:

1887 B	1887 W	D = W – B	|D|	Signed Ranks
89	61	–28	28	–16.0
85	71	–14	14	–7.0
106	70	–36	36	–18.0
82	65	–17	17	–10.0
131	120	–11	11	–5.0
65	65	0	0	0.0
82	72	–10	10	–3.5
75	60	–15	15	–8.0
106	74	–32	32	–17.0
80	64	–16	16	–9.0
95	71	–24	24	–13.5
83	60	–23	23	–12.0
70	60	–10	10	–3.5
95	70	–25	25	–15.0
59	63	4	4	2.0
100	97	–3	3	–1.0
96	72	–24	24	–13.5
60	60	0	0	0.0
83	71	–12	12	–6.0
79	79	0	0	0.0
78	60	–18	18	–11.0

$T_- = -169$ and $T_+ = 2$. Since $H_1 : M_D \neq 0$, this is a two–tailed test with test statistic $T = T_+ = 2$. The critical value for $n = 18$ and $\alpha = 0.05$ is 40 (from the table of critical values). Since $T < 40$, we reject H_0. There is enough evidence to conclude that the median school–year length was different for black and white children in 1887.

1915 B	1915 W	D = W – B	|D|	Signed Ranks
93	140	47	47	7.0
91	148	57	57	11.5
86	163	77	77	21.0
84	122	38	38	4.5
91	156	65	65	16.0
56	120	64	64	14.0
72	110	38	38	4.5
108	172	64	64	14.0
94	158	64	64	14.0
102	115	13	13	1.0
89	156	67	67	18.0
91	142	51	51	8.0
101	158	57	57	11.5
93	126	33	33	3.0
65	120	55	55	10.0
121	174	53	53	9.0
109	152	43	43	6.0
80	108	28	28	2.0
80	152	72	72	20.0
86	152	66	66	17.0
81	151	70	70	19.0

$T_- = 0$ and $T_+ = 231$. Since $H_1 : M_D \neq 0$, this is a two–tailed test with test statistic $T = |T_-| = 0$. The critical value for $n = 21$ and $\alpha = 0.05$ is 58 (from the table of critical values). Since $T < 58$, we reject H_0. There is enough evidence to conclude that the median school–year length was different for black and white children in 1915.

2. Teacher Monthly Salary (dollars):

1887 B	1887 W	D = W – B	|D|	Signed Ranks
22.08	26.50	4.42	4.42	9.0
29.30	23.33	–5.97	5.97	–11.0
27.50	20.00	–7.50	7.50	–13.0
26.04	25.56	–0.48	0.48	–1.0
31.57	38.00	6.43	6.43	12.0
25.93	22.95	–2.98	2.98	–7.0
22.33	24.75	2.42	2.42	5.0
34.33	17.00	–17.33	17.33	–19.5
29.22	19.30	–9.92	9.92	–17.0
36.86	19.53	–17.33	17.33	–19.5
20.75	18.00	–2.75	2.75	–6.0
29.47	26.44	–3.03	3.03	–8.0
20.00	20.00	0.00	0.00	0
34.50	25.18	–9.32	9.32	–16.0
19.93	21.77	1.84	1.84	3.0
22.00	30.00	8.00	8.00	14.0
32.39	20.57	–11.82	11.82	–18.0
14.00	16.00	2.00	2.00	4.0
31.00	36.95	5.95	5.95	10.0
27.75	27.07	–0.68	0.68	–2.0
25.60	16.30	–9.30	9.30	–15.0

$T_- = -153$ and $T_+ = 57$. Since $H_1 : M_D \neq 0$, this is a two–tailed test with test statistic $T = T_+ = 57$. The critical value for $n = 20$ and $\alpha = 0.05$ is 52 (from the table of critical values). Since $T > 52$, we do not reject H_0. There is not enough evidence to conclude that the average monthly pay was different for black and white teachers in 1887.

1915 B	1915 W	D = W – B	|D|	Signed Ranks
24.78	47.93	23.15	23.15	3
27.54	55.72	28.18	28.18	6
25.57	60.48	34.91	34.91	13
25.54	55.30	29.76	29.76	9
30.87	49.76	18.89	18.89	2
22.12	52.58	30.46	30.46	10
30.65	64.23	33.58	33.58	12
22.93	71.73	48.80	48.80	21
24.29	55.89	31.60	31.60	11
25.68	63.55	37.87	37.87	14
32.93	62.28	29.35	29.35	7
27.84	74.58	46.74	46.74	19
28.87	56.65	27.78	27.78	5
22.48	68.66	46.18	46.18	18
30.52	49.27	18.75	18.75	1
31.52	78.96	47.44	47.44	20
26.08	53.65	27.57	27.57	4
20.75	50.22	29.47	29.47	8
29.26	70.71	41.45	41.45	16
26.46	66.68	40.22	40.22	15
19.88	62.63	42.75	42.75	17

$T_- = 0$ and $T_+ = 231$. Since $H_1 : M_D \neq 0$, this is a two–tailed test with test statistic $T = |T_-| = 0$. The critical value for $n = 21$ and $\alpha = 0.05$ is 58 (from the table of critical values). Since $T < 58$, we reject H_0. There is enough evidence to conclude that the average monthly pay was different for black and white teachers in 1915.

Reports will vary.

Consumer Report®: Chapter 1

Emotional "Aspirin"

(a) Good science requires that all tests be performed in an objective and unbiased manner. The use of random codes ("blind testing") reduces the possibility of unconscious bias.

(b) In most experiments it is impossible to perfectly control all possible environmental influences that could conceivably affect test results. Randomizing the test order of replicate samples reduces the chance that any one treatment will be substantially more affected by external influences than any other treatment.

(c) Brand A: 231.2, 232.9, 261.1; Brand B: 224.6, 238.9, 263.7; Brand C: 207.8, 213.4, 227.1; Brand D: 192.8, 219.2, 251.1. This question requires generalization from the sample tested for a given brand (three lots) to the entire output of that brand. How one defines "meeting their label claim" is important. Certainly the brand average needs to be above the claimed value. In addition, it would seem necessary that few, if any of the lots fall below the claim. The following graph displays the observed amount of SAM-E for the various brands.

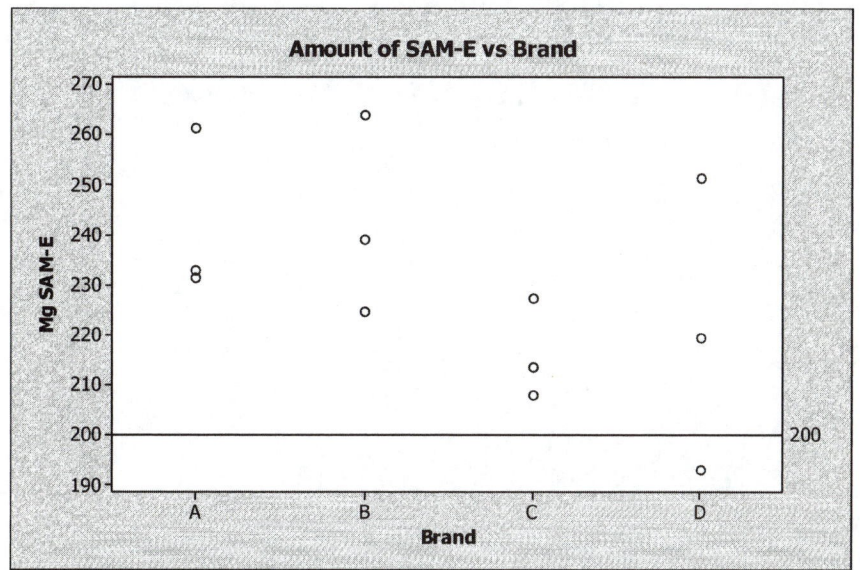

Brands A and B clearly appear to meet their label claim as all samples are well above the target of 200 mg. The brand C average is the closest of the four to the target, but the three observations are the most consistent of any brand, suggesting that it too meets its label claim. Brand D is the most variable of the brands and has one observation below 200 mg. Even though the sample mean is above 200, the lack of consistency suggests some chance that a larger sample could average below 200 or that a high proportion of lots might be below 200.

(d) **Step 1:** *Identify the problem to be solved. Consumer Reports®* would like to determine if the amount of SAM-E contained in several representative brands meets the label claims. Therefore, the response will be the amount of SAM-E (mg) in the pills.

Step 2: *Determine the explanatory variables that affect the response variable.* Some explanatory variables that affect the amount of SAM-E measured: Brand, lot from which pills are obtained, and testing mechanism.

Step 3: *Determine the number of experimental units.* In this experiment, we will measure the amount of SAM-E in 12 orders (4 brands) of mood-changing pills.

Step 4: *Determine the level of the explanatory variables:* We list the explanatory variables and their levels.

- **Brand** – We wish to determine the difference (if any) in the amount SAM-E as follows:
  Brand A: 3 orders
  Brand B: 3 orders
  Brand C: 3 orders
  Brand D: 3 orders

- **Lot from which pills are obtained** – Variability in the amount of SAM-E could possibly occur depending on the lot from which the pills are drawn. To account for this, we randomly select the samples from different lots.

- **Testing mechanism** – The measurement of SAM-E obtained could vary depending on the accuracy of the testing mechanism. We can control this somewhat by using the same testing approach for each order of pills. We also randomize the order in which the pills are tested. We also assign random codes to the various brands so that the laboratory will not know the manufacturer it is testing, which helps to avoid bias.

Step 5: *Conduct the experiment.* (a) We determine the order in which the pills will be tested. (b) We determine the amount of SAM-E for each order of pills.

Step 6: *Test the claim.* We analyze the data to determine whether the manufacturers are meeting their label claims.

Consumer Report®: Chapter 3

Basement Waterproofing Coatings

(a) $\bar{x}_A = \dfrac{\sum x_i}{n} = \dfrac{546.2}{6} \approx 91.03$ g

$M_A = \dfrac{90.9 + 91.2}{2} = \dfrac{182.1}{2} = 91.05$ g;

There are 2 modes: 90.8 g and 91.2 g (each value occurs twice).

(b) $\sum x_i = 546.2; \sum x_i^2 = 49{,}722.66; n = 6;$

$s_A = \sqrt{\dfrac{\sum x_i^2 - \dfrac{\left(\sum x_i\right)^2}{n}}{n-1}}$

$\quad = \sqrt{\dfrac{49{,}722.66 - \dfrac{(546.2)^2}{6}}{6-1}}$

$\quad \approx 0.23$ g

(c) $\bar{x}_B = \dfrac{\sum x_i}{n} = \dfrac{522.3}{6} = 87.05$ g

$M_B = \dfrac{87.0 + 87.1}{2} = \dfrac{174.1}{2} = 87.05$ g

There are 2 modes: 87.0 g and 87.2 g (each value occurs twice).

(d) $\sum x_i = 522.2; \sum x_i^2 = 45{,}466.33; n = 6;$

$s_B = \sqrt{\dfrac{\sum x_i^2 - \dfrac{\left(\sum x_i\right)^2}{n}}{n-1}}$

$\quad = \sqrt{\dfrac{45{,}466.33 - \dfrac{(522.3)^2}{6}}{6-1}}$

$\quad \approx 0.15$ g

(e)

```
      A              B
             |86 | 8
             |87 | 0  0  1  2  2
             |88 |
             |89 |
    9  8  8  |90 |
    3  2  2  |91 |
```

Yes, there appears to be a difference in these two products' abilities to mitigate water seepage. All 6 of the measurements for product B are less than the measurements for product A. Although it is not clear whether there is any practical difference in these two products' abilities to mitigate water seepage, product B appears to do a better job.

Consumer Report®: Chapter 4

Fit to Drink

(a)

(b) No, the relation appears to be nonlinear.

(c) $r = 0.868$. This measure is not very useful since the relation between No. Gallons Processed and % Lead Removed is nonlinear. $R^2 = 75.3\%$. This means that 75.3% of the variation in the percentage of lead removed can be explained by the least-squares regression model.

(d) Using Minitab, the linear regression model is: $\hat{y} = 0.04501x + 84.124$, where $x =$ No. Gallons Processed and $y =$ % Lead Removed.

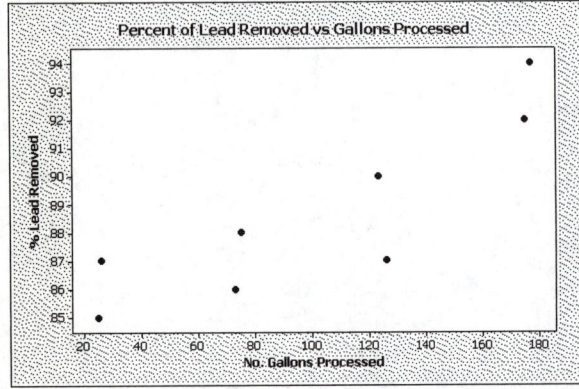

Notice the bowed pattern in the residual plot. This suggests that the linear model does not adequately describe the relation between No. Gallons Processed and % Lead Removed.

(e) Using Minitab, the quadratic regression model is: $\hat{y} = 86.46 - 0.02220x + 0.000333x^2$, where $x = $ No. Gallons Processed and $y = $ % Lead Removed.

$R^2 = 83.5\%$. Based on the regression plots, the residual plots, and the values of R^2, the quadratic model clearly fits the data better. However, one has to be careful using this model. For example, it isn't possible to remove more than 100% of the lead, regardless of how many gallons are processed.

(f) The shape of the true relationship is most likely s-shaped since the response; that is, the percent of lead removed, must be between 0% and 100%. Since you cannot remove more than 100% of the lead, this relation must eventually "level off," creating an s-shape.

Consumer Report®: Chapter 5

His N' Hers Razor?

(a) $P(\text{Excellent}) = \dfrac{21 + 19 + 13}{90} = \dfrac{53}{90} \approx 0.589$

(b) $P(\text{Poor}) = \dfrac{1 + 0 + 6}{90} = \dfrac{7}{90} \approx 0.078$

(c) $P(\text{Razor B}\,|\,\text{Fair}) = \dfrac{N(\text{Razor B and Fair})}{N(\text{Fair})}$

$\qquad\qquad\qquad = \dfrac{11}{30} \approx 0.367$

(d) $P(\text{Excellent}\,|\,\text{Razor C})$

$\qquad = \dfrac{N(\text{Excellent and Razor C})}{N(\text{Razor C})}$

$\qquad = \dfrac{13}{30} \approx 0.433$

(e) $P(\text{Poor}\,|\,\text{Razor A})$

$\qquad = \dfrac{N(\text{Poor and Razor A})}{N(\text{Razor A})}$

$\qquad = \dfrac{1}{30} \approx 0.033$

$P(\text{Poor}) = \dfrac{7}{90} \approx 0.078$

Since $P(\text{Poor}\,|\,\text{Razor A}) \neq P(\text{Poor})$, the razor type and rating are not independent.

(f) Razor A has the highest percentage of Very Good to Excellent ratings. However, the difference between razors A and B is small and could be due to the particular sample of panelists used for the study.

Consumer Report®: Chapter 6

Quality Assurance in Customer Relations

(a) Since we are interested in the number of draft responses that contains any number of errors (i.e., multiple errors are counted as one), X is a Binomial random variable. Note that there are only two possible outcomes for the individual drafts (i.e., the draft contains an error or it doesn't), the probability of finding a draft that contains errors is constant (i.e., 0.10 for each draft), and the individual drafts are independent from one another. If we were counting the number of errors in a single draft, the random variable would be Poisson.

(b) In practice, it may not be possible to find an n such that this equation is exactly true. Using Minitab with $p = 0.1$, we find the following:

With $n = 28$: $P(X \geq 1) = 1 - P(X = 0) = 1 - 0.052335 = 0.947665$

With $n = 29$: $P(X \geq 1) = 1 - P(X = 0) = 1 - 0.047101 = 0.952899$.

To be conservative, the manager needs to sample 29 documents to have (at least) a 95% probability of finding at least one document containing an error.

(c) Using Minitab with $p = 0.2$, we find the following:

With $n = 13$: $P(X \geq 1) = 1 - P(X = 0) = 1 - 0.054976 = 0.945024$

With $n = 14$: $P(X \geq 1) = 1 - P(X = 0) = 1 - 0.043980 = 0.956020$

If the error rate is really 20%, the sample size requirement reduces to 14.

(d) Y is a Poisson random variable with $\lambda = 0.3$ errors per document and $t = 1$ (document). Therefore, we have $\mu_Y = \lambda t = (0.3)(1) = 0.3$.

The possible outcomes for Y are $y = 0, 1, 2, \ldots, \infty$. However, since the expected error rate is low, we would expect the actual number of observed errors to be low.

(e) $$P(\text{no errors}) = P(0) = \frac{(0.3)^0}{0!} \cdot e^{-0.3} \approx 0.7408$$

$$P(\text{one error}) = P(1) = \frac{(0.3)^1}{1!} \cdot e^{-0.3} \approx 0.2222$$

$$P(\text{at least 2 errors}) = P(Y \geq 2) = 1 - P(Y \leq 1) = 1 - [P(0) + P(1)]$$
$$= 1 - [0.7408 + 0.2222] = 1 - 0.9630 = 0.0370$$

Consumer Report®: Chapter 7

Sunscreens

(a) There tends to be a great deal of variation in the way people respond to treatments such as sunscreen. Hence, we can either test sunscreens on a very large sample of people or block on people, that is, measuring each person before treatment and then again after treatment. By taking before and after measurements for each person we can determine if there is a treatment effect without testing a very large number of people.

(b) The random assignment of people and application sites to each treatment (the sunscreen) is important to avoid any potential sources of bias.

(c) Product A:
$$P(X < 15) = P\left(Z < \frac{15 - 15.5}{1.5}\right)$$
$$\approx P(Z < -0.33)$$
$$= 0.3707 \ [\text{Tech: } 0.3694]$$

Product B:
$$P(X < 15) = P\left(Z < \frac{15 - 14.7}{1.2}\right)$$
$$= P(Z < 0.25)$$
$$= 0.5987$$

(d) Product A:
$$P(X > 17.5) = P\left(Z > \frac{17.5 - 15.5}{1.5}\right)$$
$$= P(Z > 1.33)$$
$$= 0.0918 \ [\text{Tech: } 0.0912]$$

Product B:
$$P(X > 17.5) = P\left(Z > \frac{17.5 - 14.7}{1.2}\right)$$
$$= P(Z > 2.33)$$
$$= 0.0099 \ [\text{Tech: } 0.0098]$$

(e) Product A: $P(14.5 < X < 15.5)$
$$= P\left(\frac{14.5 - 15.5}{1.5} < Z < \frac{15.5 - 15.5}{1.5}\right)$$
$$= P(-0.67 < Z < 0)$$
$$= 0.5 - 0.2514 = 0.2486 \ [\text{Tech: } 0.2475]$$
Product B: $P(14.5 < X < 15.5)$
$$= P\left(\frac{14.5 - 14.7}{1.2} < Z < \frac{15.5 - 14.7}{1.2}\right)$$
$$= P(-0.17 < Z < 0.67)$$
$$= 0.7486 - 0.4325 = 0.3161 \ [\text{Tech: } 0.3137]$$

(f) Although Product B has a smaller standard deviation, it also has a smaller mean. Since it is important for a product to meet its advertised claim, it appears as if Product A is slightly superior to Product B. Note that approximately 37% of Product A fails to meet its claim while approximately 60% (or more than half) of Product B fails to meet its claim.

Consumer Report®: Chapter 8

Tanning Salons

(a) Selecting a random sample from a large number of facilities increases the likelihood of obtaining an estimate of the population parameter that is close to the true value. Since the reference frame is all American tanning salons, sampling from multiple metropolitan areas ensures representation from a broad spectrum of American tanning salons.

(b) Yes, the sample size, $n = 296$, is much less than 5% of the population (150,000).

(c) The sampling distribution of $\hat{p}$ is approximately normal with a mean of $\mu_{\hat{p}} = 0.25$ and a standard deviation of

$$\sigma_{\hat{p}} = \sqrt{\frac{(0.25)(1-0.25)}{296}} \approx 0.025.$$

$$
\begin{aligned}
P(\hat{p} < .226) &= P\left(Z < \frac{0.226 - 0.25}{\sqrt{0.25(0.75)/296}}\right)\\
&= P(Z < -0.95)\\
&= 0.1711 \ [\text{Tech: } 0.0.1701]
\end{aligned}
$$

About 17 out of 100 random samples of size $n = 296$ tanning facilities will result in less than 22.6% (i.e., less than 67) that would state that tanning in a salon is the same as tanning in the sun with respect to causing skin cancer.

(d) The sampling distribution of $\hat{p}$ is approximately normal with a mean of $\mu_{\hat{p}} = 0.18$ and a standard deviation of

$$\sigma_{\hat{p}} = \sqrt{\frac{(0.18)(1-0.18)}{296}} \approx 0.022.$$

$$
\begin{aligned}
P(\hat{p} \geq 0.142) &= P\left(Z \geq \frac{0.142 - 0.18}{\sqrt{0.18(0.82)/296}}\right)\\
&= P(Z \geq -1.70)\\
&= 1 - P(Z < -1.70)\\
&= 1 - 0.0446\\
&= 0.9554 \ [\text{Tech: } 0.9556]
\end{aligned}
$$

About 96 out of 100 random samples of size $n = 296$ tanning facilities will result in at least 14.2% (i.e., at least 42) that would state that tanning does not cause wrinkled skin.

$$\hat{p} = \frac{50}{296} \approx 0.169$$

$$
\begin{aligned}
P(X \leq 50) &= P(\hat{p} \leq 0.169)\\
&= P\left(Z \leq \frac{0.169 - 0.18}{\sqrt{0.18(0.82)/296}}\right)\\
&= P(Z \leq -0.49)\\
&= 0.3121 \ [\text{Tech: } 0.3111]
\end{aligned}
$$

About 31 out of 100 random samples of size $n = 296$ tanning facilities will result in 50 or fewer facilities (i.e., 16.9% or fewer) that would state that tanning does not cause wrinkled skin. Therefore, it is not unusual for 50 or less facilities to state that tanning in a salon does not cause wrinkled skin.

Consumer Report®: Chapter 9

Consumer Reports Tests Tires

(a) All the data values lie within the bounds on the normal probability plot, indicating that the data is roughly normally distributed.

(b) The five-number summary for the data is 123.2, 128.8, 131.8, 136.9, 140.3.

The lower fence $= Q_1 - 1.5 \cdot \text{IQR}$

$$= 128.8 - 1.5(136.9 - 128.8)$$

$$= 116.65 \text{ feet}$$

The upper fence $= Q_3 + 1.5 \cdot \text{IQR}$

$$= 136.9 + 1.5(136.9 - 128.8)$$

$$= 149.05 \text{ feet}$$

No data values are below the lower fence or above the upper fence, so the data set does not contain any outliers.

(c) Using technology, $\bar{x} \approx 132.12$ feet and $s \approx 5.47$ feet.

For 95% confidence, $\alpha/2 = 0.025$. Since $n = 9$, then df $= 8$ and $t_{0.025} = 2.306$.

Lower bound $= \bar{x} - t_{0.025} \cdot \dfrac{s}{\sqrt{n}}$

$$= 132.12 - 2.306 \cdot \frac{5.47}{\sqrt{9}}$$

$$\approx 132.12 - 4.20 = 127.92 \text{ feet}$$

Upper bound $= \bar{x} + t_{0.025} \cdot \dfrac{s}{\sqrt{n}}$

$$= 132.12 + 2.306 \cdot \frac{5.47}{\sqrt{9}}$$

$$\approx 132.12 + 4.20 = 136.32 \text{ feet}$$

Based on this test, *Consumer Reports* is 95% confident that the mean dry braking distance for this brand of tires is between 127.92 and 136.32 feet.

Consumer Report®: Chapter 10

Consumer Reports Tests Eyeglass Lenses

(a) The hypotheses are $H_0: \mu = 0.6$ versus $H_1: \mu < 0.6$. The test statistic is $t_0 = \dfrac{\bar{x} - \mu}{s/\sqrt{n}} = \dfrac{0.473 - 0.6}{0.166/\sqrt{6}} = -1.864$.

Classical approach: This is a left-tailed test with $n - 1 = 6 - 1 = 5$ degrees of freedom and $\alpha = 0.05$. The critical value is $-t_{0.05} = -2.015$. Since $t_0 = -1.864 > -t_{0.05} = -2.015$, the test statistic does not fall in the critical region, so we do not reject the null hypothesis.

P-value approach: This is a right-tailed test with 5 degrees of freedom. The P-value is the area under the t-distribution to the left of the test statistic $t_0 = -1.864$, which is equivalent to the area to the right of 1.864. From the t-distribution table (Table VI) in the row corresponding to 5 degrees of freedom, 1.864 falls between 1.476 and 2.015, whose right-tail areas are 0.1 and 0.05, respectively. Thus, $0.05 < P$-value < 0.1 [Tech: 0.0607]. Since P-value $> \alpha = 0.05$, we do not reject the null hypothesis.

Conclusion: There is insufficient evidence to conclude, from a statistical viewpoint, that the mean haze difference is less than 0.6 for the manufacturer's lenses. The data are not statistically significant. The data do not suggest that the manufacture's lenses are more scratch resistant than the competitor's.

(b) Answers will vary Here is a sample paragraph.

The manufacturer claims that their lenses are "the most scratch-resistant plastic lenses ever made." The data do not support this claim. Even though the mean haze difference for this manufacturer is less than the competitor's value of 0.6, it is not statistically significantly different. The data do not support the manufacturer's claim.

Consumer Report®: Chapter 11

The High Cost of Convenience

(a) The hypotheses are $H_0: \mu_{box} = \mu_{roll}$ versus $H_1: \mu_{box} < \mu_{roll}$.

(b) The normal probability plots are roughly linear, and all the data values lie within the bounds. We can conclude that the sample data come from a population that is normally distributed. Therefore it is reasonable to conduct a two-sample hypothesis test.

(c) The boxplot does not show any values less than the lower fence ($= Q_1 - 1.5 \cdot IQR$) or greater than the upper fence ($= Q_3 - 1.5 \cdot IQR$). Thus, we can conclude that the data set does not contain any outliers. Based on the boxplots, it would seem there does not appear to be much difference in the absorption times of the two paper towel products, although the boxed version may have a slightly lower absorption rate.

(d) The test statistic is $t = -1.09$ and the P-value of the test is 0.150.

Our tests revealed that the absorption rate for the boxed version of paper towels was not significantly lower than the absorption rate for the traditional roll. Therefore, unless the convenience of using boxed towels is worth the extra cost, we recommend using the traditional towels.

Consumer Report®: Chapter 12

Dirty Birds?

(a) Calculate the proportion of incidence for each of the brands shown in the table.

Brand	Present	Absent	Total	Proportion
A	8	192	200	0.040
B	17	183	200	0.085
C	27	173	200	0.135
D	14	186	200	0.070
E	20	180	200	0.100
Total	86	914	1000	1.000

(b) Using Minitab, a 95% confidence interval for the incidence of Salmonella for Brand C is (0.090887, 0.190309). Using a TI-84 Plus, a 95% confidence interval for the incidence of Salmonella for Brand C is (0.08764, 0.18236).

(c) Using Minitab or a TI-84 Plus, the test statistic is $\chi_0^2 = 12.646$ with 4 df. The P-value is 0.013, so there is evidence of different rates of incidence among the five brands.

(d) The sample proportion for Brand A is 0.04 and that for Brand D is 0.07. Our hypotheses would be $H_0: p_A = p_B$ vs. $H_1: p_A < p_B$. Using a two-sample test for proportions (not adjusted for multiple comparisons) we obtain a test statistic of $z_0 = -1.32$ and a P-value of 0.094. If this were adjusted for multiple comparisons, the significance level would be even lower, so there is no evidence that Brand A is cleaner than Brand D based on this sample.

(e) Answers will vary. Our tests revealed that there is evidence of different rates of incidence of Salmonella in the 5 brands of chicken studied. Incidence levels ranged from a high of 13.5% in Brand C to a low of 4% in Brand A.

Consumer Report®: Chapter 13

Tea Antioxidants: An Unexpected Side Benefit
to a Cheap and Tasty Beverage

(a) H_0: μ_i are all equal versus H_1: at least one of the μ_i is different from the others.

(b) The smallest value in Brand 4 and the largest value in Brand 6 are quite different from the other two observations in their respective brands and might be outliers. However, there are too few observations per group to use a boxplot to determine if these observations are outliers.

To perform a one-way ANOVA, the $k = 6$ samples must be independent of each other, the populations must be normally distributed, and the populations must have the same variance. (More sophisticated procedures exist that allow for unequal variances, but they are beyond the scope of this text.)

(c) There is a statistical difference among brands. The P-value is 0.001 which would be considered highly significant.

(d) Based on the overlapping confidence intervals, Brands 2, 3, 5, and 6 appear to have similar amounts of ORAC. Brand 4 may have slightly lower ORAC levels, but it's confidence interval overlaps with the confidence intervals of these other brands.

(e) Using Minitab, we obtain the following output:

```
Tukey 95% Simultaneous Confidence Intervals All Pairwise Comparisons among Levels
of Brand
Individual confidence level = 99.43%

Brand = 1 subtracted from:
Brand   Lower   Center   Upper   -----+---------+---------+---------+----
2       74.44   233.13   391.82                       (-------*-------)
3      104.44   263.13   421.82                        (-------*-------)
4      -28.79   129.90   288.59               (-------*-------)
5       40.61   199.30   357.99                    (-------*-------)
6       85.24   243.93   402.62                       (-------*-------)
                                        -----+---------+---------+---------+----
                                          -200        0       200      400


Brand = 2 subtracted from:
Brand    Lower    Center   Upper   -----+---------+---------+---------+----
3      -128.69    30.00   188.69                 (--------*------)
4      -261.92  -103.23    55.46        (-------*-------)
5      -192.52   -33.83   124.86             (--------*-------)
6      -147.89    10.80   169.49               (-------*------)
                                        -----+---------+---------+---------+----
                                          -200        0       200      400


Brand = 3 subtracted from:
Brand    Lower    Center   Upper   -----+---------+---------+---------+----
4      -291.92  -133.23    25.46     (-------*-------)
5      -222.52   -63.83    94.86          (-------*-------)
6      -177.89   -19.20   139.49            (-------*-------)
                                        -----+---------+---------+---------+----
                                          -200        0       200      400


Brand = 4 subtracted from:
Brand   Lower   Center   Upper   -----+---------+---------+---------+----
5      -89.29   69.40   228.09             (------*-------)
6      -44.66  114.03   272.72               (-------*-------)
                                      -----+---------+---------+---------+----
                                        -200        0       200      400


Brand = 5 subtracted from:
Brand    Lower   Center   Upper   -----+---------+---------+---------+----
6      -114.06   44.63   203.32             (-------*-------)
                                       -----+---------+---------+---------+----
                                         -200        0       200      400
```

Based on the results of our tests, there is sufficient evidence to believe that the mean ORAC levels for the six tested brands are not equal. In particular, Brand 1, with a mean of 445, has the lowest ORAC level. Brands 2, 3, 5, and 6 have similar mean levels of ORAC, ranging from a low of 645 to a high of 709. The mean level of ORAC for Brand 4, with a mean of 576, is somewhat lower, but there is insufficient evidence to conclude that the mean ORAC level of Brand 4 is different from Brands 2, 3, 5, and 6. There was no significant difference between Brands 1 and 4.

(f) Answers will vary.

Putting Kids to the Test

Using the Minitab output, we get the following:

(a) H_0: the distributions of preference scores for the eight scooters are the same.
H_1: the distributions of preference scores for the eight scooters are not the same.

(b) H = 68.00, or 69.07 (adjusted for ties), but both result in the same conclusion.

(c) P-value < 0.001.

(d) Based on the results of this test, there is sufficient evidence to believe that distribution of preference scores for the eight scooters is different. In particular, Scooter 6 (with a median preference score of 1.0) and Scooter 2 (with a median preference score of 2) were the most preferred scooters.